Insect
Neurochemistry
and Neurophysiology

Insect Neurochemistry and Neurophysiology

Edited by
Alexej B. Bořkovec
and
Thomas J. Kelly

United States Department of Agriculture
Beltsville, Maryland

Plenum Press • New York and London

Library of Congress Cataloging in Publication Data

International Conference on Insect Neurochemistry and Neurophysiology (1983: University
of Maryland)
   Insect neurochemistry and neurophysiology.

   "Proceedings of an International Conference on Insect Neurochemistry and Neurophysiology, held at the University of Maryland, College Park, Maryland"—T.p. verso.
   Includes bibliographical references and index.
   1. Insects—Physiology—Congresses. 2. Neurochemistry—Congresses. 3. Neurophysiology—Congresses.
I. Bořkovec, A. B. (Alexej B.), 1925–   II. Kelly, Thomas J. III. Title.
QL495.I535 1983   595.7'0188   83-24464

ISBN-13: 978-1-4684-4645-6   e-ISBN-13: 978-1-4684-4643-2
DOI: 10.1007/978-1-4684-4643-2

Proceedings of an international conference on Insect Neurochemistry
and Neurophysiology, held at the University of Maryland, College Park,
Maryland

©1984 Plenum Press, New York
Softcover reprint of the hardcover 1st edition 1984

A Division of Plenum Publishing Corporation
233 Spring Street, New York, N.Y. 10013

PREFACE

The function of the central nervous system as a
coordinator and regulator of cellular processes in
multicellular organisms is unequivocal. Until recently,
however, the chemical evidence necessary for validating
speculations on neurophysiological function in invertebrates has been lacking. In insects, because of their
small size, heroic efforts were needed to collect the
millions of tissues or organs necessary for isolation
and identification of neurochemicals. With the spectacular advances in physical and analytical technology
within the last decade and with significant advances in
radiochemical, radioimmunological and neurophysiological
assays, researchers are, for the first time, able to
handle microgram and nanogram quantities of complex
biological substances. More recent developments in
immunology promise to lower these levels further. It is
not surprising that these new opportunities accelerated
progress in insect neuroscience and that the time was
right for a rapid and personal exchange of ideas and
information on techniques. These considerations were
the primary impetus for convening the International
Conference on Insect Neurochemistry and Neurophysiology
(ICINN) at the University of Maryland, College Park, MD,
on August 1-3, 1983.

The Conference itself was to fulfill five basic needs:
(1) provide conceptual framework for understanding the
direction and purpose of recent and future research; (2)
present and describe, with examples of actual applications, the new and developing laboratory techniques for
separation and characterization of insect neurochemicals; (3) illustrate by examples of new research
results the progress being made in insect neuroscience;
(4) appraise the potential impact of research in this
field on insect control; and (5) bring together
academic, government, and industrial scientists from
various disciplines for personal interaction and
exchange of ideas. Written accounts of how the first

v

three purposes of the Conference were accomplished
constitute this book.

The first section presents contributions of nine
distinguished invited speakers who reviewed various
aspects of neuroscience as applied to insects.
Analytical, chemical, immunological, neurophysiological, and recombinant DNA techniques pertinent to
studies of insect neuropeptides were presented in two
workshops at the Conference by invited speakers and
are included in the following section. The final
section contains summaries of recent research results
that were presented by the Conference participants.
The ICINN Program Chairman, Larry L. Keeley, and the
workshop organizer, Timothy K. Hayes, deserve our
gratitude for providing the speakers and organizing
their presentations. The Conference, its proceedings
and indeed the publication of this volume would not
have been possible without dedicated efforts by the
ICINN Organizing Committee and without financial
support from the U.S. Government and industry. The
USDA's Agricultural Research Service and the National
Science Foundation on one hand and Monsanto Company,
American Cyanamid Company, Dow Chemical Company,
Shell Chemical Company, Stauffer Chemical Company,
E.I. Du Pont de Nemours and Company, FMC Corporation,
and Zoecon Corporation on the other, have our sincere
thanks and appreciation.

Alexej B. Bořkovec
Thomas J. Kelly

Beltsville, Maryland
August 1983

CONTENTS

NEUROSECRETORY ACTIVITY AND ITS REGULATION

L. H. Finlayson

Department of Zoology and Comparative Physiology
University of Birmingham
Birmingham B15 2TT, U.K.

INTRODUCTION

In the study of neurosecretion there have always been difficulties in the interpretation, in terms of function, of the activity
of the neurosecretory (NS) cell as revealed by its appearance in
light and electron microscopy. As the great variety of staining
reactions and ultrastructural phenomena and possible interpretations
of them have been well reviewed in recent years (e.g. BERLIND, 1977;
MADDRELL and NORDMANN, 1979; RAABE, 1982), only a few examples will
be discussed to highlight some of the problems. Much of the difficulty lies in determining whether a cell is in a state of quiescence
or is actively releasing its product into the blood. Only in studies
that involve neurohormonal or biological assay can the release
activity of the cell be monitored with confidence.

Relatively little is known about the mechanisms that regulate
the activities of NS neurons and the part played by the action potentials generated by these cells. There is evidence of both synaptic
and humoral regulation of neurosecretion but much remains to be done
before details of these mechanisms are fully elucidated. This paper
will deal briefly with the electrical activity of NS neurons in
relation to their secretory activity but principally with the regulation of these neurons by blood-borne factors.

Synthesis, Storage, Lysis and Release

The sequence of events that take place during synthesis of the
characteristic 'elementary' neurosecretory granules is well documented
and there is general agreement that sequestration of the material
that goes to form a granule occurs at the membranes of the rough

1

endoplasmic reticulum (RER). It is then transferred into a cisterna
of the RER and from there to a Golgi body where it is enclosed in a
membrane and then released into the lumen of the cell body. Further
changes may take place in the granule which are manifested by a
change in electron-density; usually it becomes denser but sometimes
paler as it progresses through the cell body and along one of the
processes (BERLIND, 1977). Differences in the electron-density of
the elementary granules may be indicative merely of differences in
treatment during preparation or of a true physiological change.
Problems of interpretation arise when NS cells show other types of
inclusion. A common phenomenon is the production of vesicles of a
variety of sizes and electron-opacity. In the stick insect sections
of neurohaemal tissue on the transverse branch of the median ('sympathetic') nerve show 3 types of process each containing a structurally
distinct type of NS granule. The bulk of the NS material (NSM) in
these axons is of 2 types, one in which the granules are ellipsoid
and contain tubular inclusions, and the other in which the granules
are spherical and contain very electron-dense finely granular
material. The ellipsoid ('tubulous') granules originate in the CNS
and the spherical granules in the link nerve neurosecretory neurons
(LNNs) (FIFIELD and FINLAYSON, 1978). In the normal well-fed animals
both types of granule are electron-dense but in an animal that has
been deprived of food and water there is a change in both types of
granule but a particularly striking change in the tubulous granules
(FINLAYSON and OSBORNE, 1975). The latter become pale and, as starvation progresses, they become fragmented, apparently spilling their
contents into the cytoplasm of the axon. The LNN granules also
become paler and some fragment. In the electron-micrographs of the
original description of the ultrastructure of this tissue by BRADY
and MADDRELL (1967) these neurohaemal processes were noticeably pale
in appearance, indicating that there had been a difference in the
treatment of the tissues or of the animals so that the sections were
paler than those from our 'normal' animals. Although such an example
highlights the difficulties of interpretation of cytological appear-
ances it also shows that changes induced in the appearance of NSM may
be of value in locating the neurons involved in water balance and
other metabolic processes. This example also indicates that there
may be other mechanisms of release of NSM besides the exocytosis
which undoubtedly takes place in *Carausius* as in many other insects.
On the other hand, the phenomena seen in starving stick insects may
be degenerative. This insect does not store large amounts of fat
body and so is sensitive to relatively short periods of starvation.

Large vesicles whose contents range from being extremely pale
to moderately dense have been described in several insects. In most
cases they appear to be dilations of RER (YIN and CHIPPENDALE, 1975;
KONO, 1975; PARK and SEONG, 1975) and have been interpreted as
direct secretion of NSM into cisternae of the RER or as a stage in
the production of NS granules by an invagination of the vacuole wall
which then receives NSM from the vacuole and is finally budded off

to form a NS granule (YIN and CHIPPENDALE, 1975). Similar configur-
ations have been seen in the neuron of the ecdysial gland (NEG) of
the larva of the moth *Agrotis* (GRIFFITHS and FINLAYSON, 1983) but
we have interpreted them as a stage in the autolysis of NS granules
because the granule is inside the vacuole. They are seen in the soma
of the NEG of the half-grown final instar larva and accompany the
early signs of degeneration of the NEG which, like the ecdysial gland
it innervates, disappears in the early pupa of *Agrotis*. In the final
instar larva, multilamellate bodies and deeply staining bodies are
numerous in the cell soma but not in the processes. The vacuoles
coalesce to form massive 'holes' in the soma. In the prepupa vacuol-
ation continues and massive lytic bodies, sometimes with enclosed
NS granules, are present. NS granules appear to fuse together to
form irregular dense bodies. The cell processes are by now empty
of NS granules and many have degenerated completely so that their
previous existence is revealed only by spaces in the sections. During
this time processes that originate in the CNS and which contain
smaller NS granules are unaffected and can be seen all round the
degenerating NEG and its processes. In the degenerating NEG the
following structures have, therefore, been identified in addition to
the vacuoles: i) irregular, deeply staining bodies, sometimes with
NS granules included (dense bodies); ii) multivesicular bodies;
iii) multilamellate bodies. Structures such as those described for
Agrotis have been seen in other insects, in other neurons. In
detailed studies of the posterior NS cells of the brain of *Rhodnius*
MORRIS and STEEL (1975, 1977) found swelling of RER cisternae as one
of the early signs of activation of the cells after feeding, as well
as an increase in the number of multivesicular bodies. Neurosecretory
granules are numerous in the vicinity of multivesicular bodies and
some small granules appear to fuse with these bodies. Dense bodies
are numerous in the resting cell and in the activated cell but fall
in number immediately after feeding. MORRIS and STEEL (1975) postulated that these dense bodies were lytic in nature and were indicative of degradation of NSM in the unfed insect so that lysis balanced
synthesis. · In their 1977 paper they added the further suggestion
that the dense bodies may be accumulations of NSM and may be concerned
with the maturation of NSM. As the dense bodies decrease the multivesicular bodies increase in number. At the peak of synthetic activity granules are most commonly associated with multivesicular bodies.
MORRIS and STEEL (1977) suggested that the sequence may be (a) formation of 'elementary' granule in Golgi body; (b) changes in size,
density and shape of granules in soma; (c) fusion of granules with
multivesicular bodies to form dense bodies; (d) fragmentation of
dense bodies to release smaller granules of NSM. The NSM would, in
their scheme, have passed through the lysosomal system in process of
maturation.

At the base of the antenna of *Rhodnius* there are neurosecretory
axons with terminations in the wall of the valvular pulsatile organ
which directs the flow of blood into the antenna (BEATTIE, 1976).

Autolysis takes place in some of these terminals with the formation
of many multilamellate bodies, some containing NS granules and a few
dense bodies. BEATTIE (1976) saw no signs of regeneration of these
axon terminals but it cannot be ruled out that there may be a cycle
of degeneration and regeneration.

   As the dense bodies in *Agrotis* become progressively larger until
they occupy a large part of the soma and as the cell is nearing the
end of its life there seems no doubt that these particular 'fusion
bodies' are lytic in nature. Similarly the multivesicular bodies
in *Agrotis* would also appear to be lytic rather than storage or processing bodies. The multilamellate bodies ('myelin' bodies) consisting of whorls of membranes are generally accepted as part of the
paraphernalia of cell autolysis.

   These examples emphasise the difficulties of interpreting cytological phenomena in the absence of other evidence.

Correlation of Electrical Activity, Cytological Appearance and Release of Neurohormones

   The first direct correlation to be established between electrical activity and neurohormonal secretory activity of NS cells
in an insect was shown in studies of the brain/corpus cardiacum
system of *Rhodnius* (ORCHARD and STEEL, 1980). The moulting cycle
is initiated by the intake of a blood meal. Within a few minutes
changes take place in the appearance of the medial neurosecretory
cells (MNCs) of the brain (STEEL, 1982). There are 17 of these cells,
subdivisible by their appearance in light and in electron microscopy
into 6 distinct types. All 6 types respond promptly to a meal by an
increase in synthetic activity, proliferation of microtubules amongst
the granules in the soma, appearance of granules in the axons, and a
corresponding increase in the intensity of staining by paraldehyde
fuchsin in the axons and a decrease in staining in the soma. All
these changes point to an increase in production of NSM and its transport along the axons. At the same time as this increase in neurosecretory activity as revealed by microscopy is proceeding, there are
concomitant changes in the electrical activity of the MNCs. These
were recorded by applying a suction electrode to the surface of the
corpus cardiacum (CC). Activity is low before feeding but within
minutes of the insect taking a blood meal there is a dramatic increase
in discharge of impulses. There is a continuous background and superimposed a series of bursts from other units. The bursting activity
declines after 2 h and remains at a low level for 5 days (at 28°C)
and then there is a second period of bursting activity for a few
hours. The continuous component of the discharge pattern which is
initiated at feeding remains at a high level throughout the 5 days.
Both periods of intense bursting activity precede an increase in ecdysterone in the blood (STEEL et al., 1982) and probably coincide with
the release of ecdysiotropic hormone from the CC into the blood.

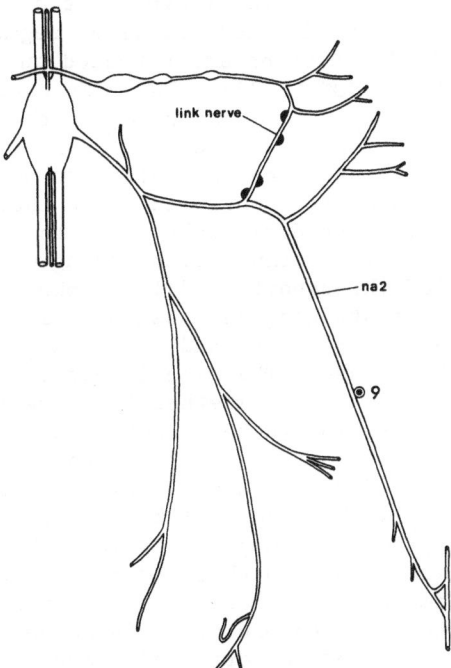

Fig. 1. Diagram of a ganglion and the principal nerves on the right
side of an abdominal segment of *Carausius* to show the position of the link nerve neurosecretory neurons and mechanoreceptive neuron 9. The LNNs conduct action potentials
centrifugally along their processes, some of which travel
along and terminate on nerve na_2 and some of its side
branches. Neuron 9 conducts action potentials along its
single axon in nerve na_2 towards the ganglion.

The burst of impulses rather than the continuous activity appears to
be correlated with the release of ecdysiotropic hormone. Recordings
made from the CC nerve showed that the impulses recorded from the
surface of the CC mainly originate in the MNCs of the brain (ORCHARD
and STEEL, 1980).

    In *Carausius* there is circumstantial evidence of a correlation
between the secretory activity of the extraganglionic neurosecretory
neurons and their electrical activity. In this insect the LNNs

(Fig. 1), a group of usually 4 cells, lie on the nerve that connects
the transverse branch of the median ('sympathetic') nerve with the
main lateral segmental nerve in each abdominal segment (FIFIELD and
FINLAYSON, 1978). Similar neurons are situated on nerves in the
thorax (ORCHARD and FINLAYSON, 1977). The LNNs have a cycle of
secretory activity which can be demonstrated by the use of the vital
dye acridine orange (FINLAYSON and ORCHARD, 1978). Inclusions in
the cell soma fluoresce orange and the cells can be graded by eye on
a five-point scale, ranging from those in which there are few orange
inclusions and the cell is predominantly green to those in which the
cell is packed with brilliant orange-fluorescing material. When
these cells are sampled throughout a 12:12 LD photoperiod it is found
that there is a cycle of staining response with a peak just before
the midpoint of the dark (D-) phase and a trough just before the midpoint of the light (L-) phase. The mean lowest value was around 2.3
for L-cells and 3.4 for D-cells. Whether these reactions to acridine
orange reflect an increase in synthesis during the dark period or a
cessation of release or whether the orange-fluorescing material is
directly correlated with NSM is not known but clearly there is a
cycle of major activity within the cell body which is revealed by
this technique. The primary assumption is that there is an increase
in NSM in the dark which reacts with acridine orange to produce an
orange-fluorescing complex, in contrast to the green fluroescence of
nucleic acids and other cell components. Evidence from similar
studies on the NEG of *Agrotis* supports this conclusion (GRIFFITHS
and FINLAYSON, unpublished).

In reasonable correspondence with the cycle in staining response
the LNNs undergo a cycle of electrical activity (FINLAYSON and
ORCHARD, 1978). They discharge impulses throughout the 12:12 LD
photoperiod but are more active around the midpoint of the D-phase
and least active around the midpoint of the L-phase. As the stick
insect is nocturnal and is exceptionally inactive during the day it
is reasonable to suppose that the cycle of secretory and electrical
activity of the LNNs is in some way correlated with the cycle of
locomotory and other activities of the insect that take place during
a 24 h period.

Less direct but strongly circumstantial evidence that release
of NSM is regulated by the electrical activity of the neuron is
provided by several studies in which the brain or CC has been stimulated electrically. These studies have been reviewed by BERLIND
(1977). The major change in appearance under the light microscope
after electrical stimulation is a decrease in the intensity of
staining (HODGSON and GELDIAY, 1959; HIGHNAM, 1961, 1962; GOSBEE
et al., 1968; GIRARDI et al., 1974). Similar studies using the
electron microscope showed changes after electrical stimulation
(SCHARRER and KATER, 1969; NORMANN, 1974). These include increases
in number of synaptoid complexes and exocytotic profiles and accumulation of electron-dense material.

Regulation of Electrical and Secretory Activity by Blood-Borne Factors

It is reasonable to suppose that the release of its product by a neurosecretory cell will be regulated by feedback of that product on to the cell so that it inhibits or reduces further release, depending on its concentration in the blood. This would be a one-step negative feedback system. Such a system may exist in vertebrates but none has so far been found in insects. The few feedback systems that have been described involve another substance which is released further along in a sequence. An example of such a system is the moulting cycle of *Rhodnius* (STEEL, 1978). As described above there is an initial release of ecdysiotropic hormone immediately after feeding and then again after 5 days at 28°C. Parabiotic experiments were carried out in which insects one day after feeding were joined to others that had been decapitated 8 days after feeding. After 7 days the MNCs of the first insects were examined and some cells were found to contain a massive amount of NSM. Synthesis of NSM is stimulated by feeding but accumulation does not normally occur until after day 8. The massive amounts seen in some of the MNCs must have accumulated because it was not released during the 7 days of parabiosis to the physiologically older insects. STEEL (1982) suggested that the inhibition of release of NSM from these cells was caused by ecdysone because ecdysone climbs to its highest level in the blood during the period when the decapitated insects were joined in parabiosis i.e. from day 7 - day 14 (STEEL et al., 1982). Confirmation that ecdysone is the feedback inhibitor was obtained by injecting ecdysterone and thereby producing a similar accumulation of NSM in these neurons (STEEL, 1975).

In the adult female *Rhodnius* there is a positive feedback system involving ecdysterone and 10 MNCs of the brain. The ovaries are necessary for the production of ecdysterone that appears in the blood 5 days after feeding in animals kept at 28°C (RUEGG et al., 1981). Extracellular recordings were made with suction electrodes on the corpus cardiacum on the 6th day after feeding. In mated females a higher frequency of impulses was recorded than in virgin females. Ovariectomy reduced the impulse frequency and eliminated bursting activity in mated females and reduced the impulse frequency, but not significantly, in virgin females. Injection of ecdysterone into ovariectomized, mated females on the 6th day after feeding produced a doubling of impulse frequency within 1 h, a maximum at 2 h and a maintained high level for 12 h or more. Ecdysterone had a similar effect on the electrical activity of the CC in isolated head preparations. Recordings were made of the electrical activity of the MNCs and the CC in isolated head preparations incubated in ecdysterone and it was shown that the impulses originating in the MNCs could be recorded from the contralateral side of the CC (RUEGG et al., 1982). Ecdysterone does not act directly on the neurosecretory cells of the brain and CC but via aminergic interneurons. Its stimulating action

on the MNCs can be mimicked by dopamine (ORCHARD et al., 1983).
The direct evidence from these studies on *Rhodnius* is of a positive
feedback of ecdysone from the ovaries which stimulates the brain
MNCs via aminergic neurons to produce the myotropic ovulation factor.
Indirect evidence from studies on the moulting cycle suggests a
negative feedback of ecdysone from the blood on to the MNCs (perhaps
indirectly also) to inhibit release of NSM but also positively to
promote synthesis of NSM.

Hormonal interaction between the ecdysial gland and neurosecretory cells of the brain has also been shown in the Lepidoptera.
In *Mamestra* there are 2 groups of MNCs which can be distinguished on
morphological grounds by light microscopy (AGUI, 1976). One of these
groups (I) is subdivisible into Types 1 and 2 which have been shown
to respond to alpha ecdysone and to ecdysterone *in vitro*. When
brains from diapausing pupae were cultured in a medium containing
alpha ecdysone or ecdysterone or active ecdysial glands, the stainability with paraldehyde fuchsin (PAF) of Type 2 cells was increased
over the control brain cultured on its own (AGUI and HIRUMA, 1977b).
Similar results were obtained when ecdysterone was injected into
final instar larvae of various ages from which the ecdysial glands
had been removed by cautery (AGUI and HIRUMA, 1977a). The stainability of Type 2 cells decreased in larvae in which the ecdysial
glands had been destroyed but increased again after ecdysterone was
injected. Injection of ecdysterone into such larvae also caused an
increase in PAF-staining of NSM in the axon tracts from the NS cells.
The conclusions that may be drawn from these results are a) reduction
in the level of ecdysone in the blood causes Type 2 cells to lose
their PAF stained contents, probably because they have been released;
b) ecdysterone injection into larvae without ecdysial glands increases
the stainability of Type 2 cells and is, therefore, promoting synthesis of NSM; it also causes increased staining in NSC axons, indicating that it stimulates transport of NSM and possibly release. Extirpation of ecdysial glands did not affect Type 1 cells but injection
of ecdysterone into intact larvae caused an increase in numbers of
strongly-stained tear-shaped cells.

In *Rhodnius* and in *Mamestra* ecdysones have been shown to stimulate the neurosecretory activity of brain cells but the relationship
between synthesis and release is not entirely clear. An increase in
NSM in a cell can mean that synthesis has increased but also that
release has been inhibited. Transport of NSM into cell processes may
also be correlated with release but not necessarily so.

Another example of a control system that appears at first sight to
be possibly one in which there is stimulation of release of a neurohormone by an ecdysone is the control of vitellogenin synthesis in
the female mosquito. However, the evidence is that ecdysone is not
the factor responsible for activating the CC but is involved in the
chain of events leading to vitellogenesis.

In the female mosquito yolk-protein precursor (vitellogenin) is synthesised and released from the fat body in response to a blood meal (or spontaneously in autogenous species). Implantation of an ovary from a blood-fed donor into an intact recipient will stimulate vitellogenin synthesis but if the recipient is decapitated prior to implantation of the ovary no vitellogenin will be produced (BOROVSKY, 1982). The factor that induces vitellogenesis is stored in the CC and released in response to a blood meal (LEA, 1972). According to BOROVSKY (1982) the ovary produces a CC stimulating factor (CCSF) after the female has fed. CCSF stimulates the CC to release its stored 'egg development neurosecretory hormone' (EDNH). When the ovary receives EDNH from the blood it releases alpha ecdysone which is converted into ecdysterone in the fat body. It is not clear whether ecdysterone is the factor responsible for vitellogenin synthesis in the fat body or whether there is another link in the chain of control. CCSF is not an ecdysone because injection of ecdysterone does not release EDNH from the CC. How the blood meal induces the ovary to release CCSF is unknown but it also involves a blood-borne factor because the system works when the ovaries are denervated.

In *Acheta* ovariectomy leads to a reduction in PAF-stained NSM in certain brain cells. Injection of ecdysterone into such animals partially restores the NSCs to their normal deeply stained condition. Whether this is an example of positive feedback stimulating the NSCs to synthesise or whether it is negative feedback preventing the release of NSM is not known but it is probably another example of positive feedback because in the absence of the ovaries it is unlikely that NSM would be released. The studies summarised above indicate that it is likely that the ovaries produce a factor (CCSF or ecdysone) that stimulates the CC directly or indirectly to release an egg maturation factor (BRADLEY and SIMPSON, 1981).

There are other examples of feedback, both negative and positive, on to neurosecretory cells e.g. via juvenile hormone in *Manduca* (NIJHOUT and WILLIAMS, 1974), in *Locusta* (McCAFFERY and HIGHNAM, 1975 a,b) and in *Hyalophora* (TRUMAN, 1978).

Some of the previous examples involve an inhibitory effect or negative feedback on to neurosecretory neurons but in general inhibitory factors appear to be less common than excitatory ones. The 'cockroach paralysing agent' described by BEAMENT (1958) remains the most spectacular example of a severely inhibitory factor that is released into the blood of the insect when it is 'stressed' by being immobilised or by being over-stimulated mechanically or electrically. This factor can be transferred to an 'unstressed' cockroach by injection or parabiosis. HODGSON and GELDIAY (1959) found that such stresses induce the disappearance of NSM from the CC and that extracts made from the CC of stressed animals lost their inhibitory effect on the electrical activity of the central nervous system. CC extracts

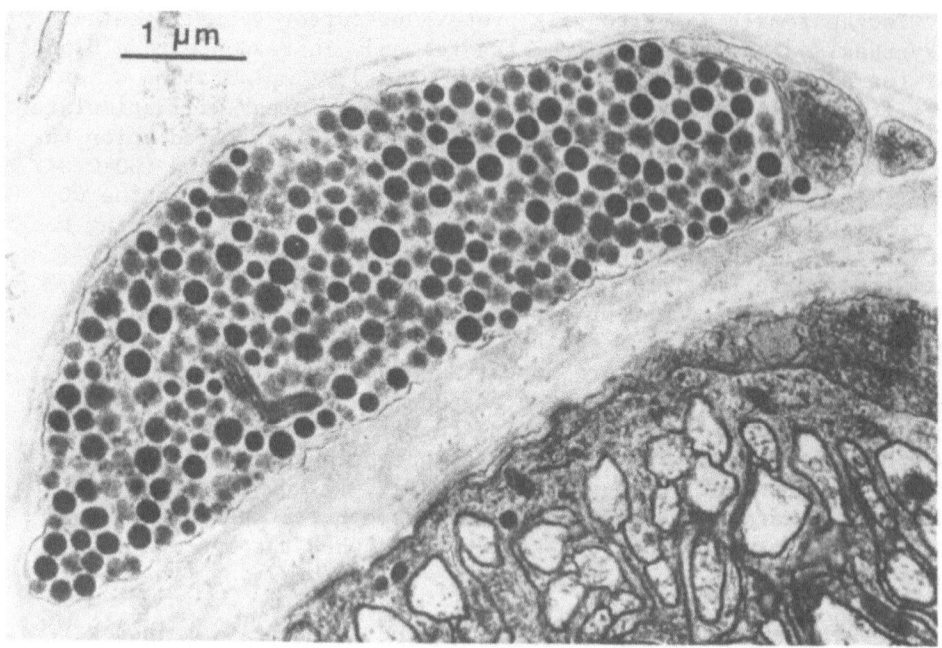

Fig. 2. Section through part of nerve na$_2$ of *Carausius* to show a
process from a link nerve neuron on the outside of the
nerve.

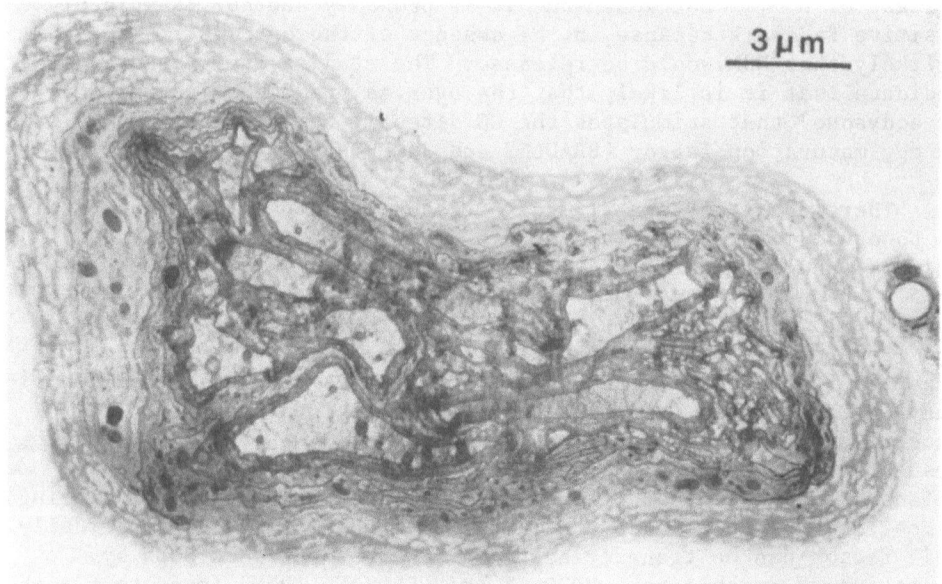

Fig. 3. Section through a principal branch of nerve na$_2$ of
Carausius near the muscles of the body wall to show that
there are no processes as far along as this.

also lose their neurodepressing activity when the CC have been dis-
connected from the brain, suggesting that the neurodepressing factor
originates in the brain and not in the CC (OZBAS and HODGSON, 1958).
It is not proven that the 'paralysing agent' of BEAMENT (1958) is the
same as the neurodepressing factor from the brain, but it is a reasonable inference that it is. A comparable inhibitory system is also
found in *Locusta* but no stress is involved. After feeding, locust
nymphs become quiescent (the 'post-prandial rest') within a few
minutes after feeding. The primary stimulus appears to be distension
of the crop because a reduction in locomotory activity can be induced
by filling the gut with agar via a cannula. Blood from a recently-
fed donor injected into a hungry recipient reduces the latter's locomotory activity showing that the inhibitor is a blood-borne factor.
Homogenates of the storage lobes of the CC proved effective in
reducing activity and it was found that CC from hungry locusts were
more effective than CC from recently-fed locusts. This result is to
be expected if the inhibitory factor accumulates in the hungry animal
and is then released immediately after feeding (BERNAYS, 1980). A
similar post-prandial inhibitory factor in *Phormia* was also found to
be blood-transmissible (GREEN, 1964) but was not traced to the CC.

In studies of the extraganglionic neurosecretory neurons of
Carausius, described above, the regulation of the diel cycle of
electrical and cytological activities was investigated to see if a
blood-borne factor is involved (FINLAYSON et al., 1983). Cultures
of stick insects were kept in a 12:12 LD regime. The times of the
cycles of illumination were so arranged that animals could be obtained
in the middle of the light phase and in the middle of the dark phase
at the same time. The link nerve on which the 4 peripheral neurosecretory neurons are located (FINLAYSON and OSBORNE, 1968) was dissected out, plus a convenient length of the long nerve (na_2) that
runs to the dorsal region of each abdominal segment (Fig. 1).
Previous studies (FIFIELD and FINLAYSON, 1978) have shown that processes from only the LNNs run along nerve na_2 (as well as along other
nerves) (Fig. 2) and conduct action potentials centrifugally from the
soma towards the periphery (ORCHARD and FINLAYSON, 1976). Their
processes terminate on na_2 and some of its side branches (Fig. 3).
The arrangement of the processes, which release their contents into
the blood by exocytosis (FIFIELD and FINLAYSON, 1978), means that
the LNNs have an extremely diffuse and widespread system for release.
This appears to be an arrangement suitable for rapid dissemination
of NSM from these neurons throughout the blood system. Non-neurosecretory neuron No. 9 (FIFIELD and FINLAYSON, 1978) lies on the
length of nerve na_2 which is used in these preparations (Fig. 1).
Normally its activity is not recorded because it conducts action
potentials along its axon only towards the CNS. Occasionally, however, it is taken into the end of the suction electrode and its
activity is then picked up. Fortunately the action potentials from
neuron 9, which is probably a stretch receptor (ORCHARD and FINLAYSON,
1976), are shorter in duration than those of the LNNs and can easily

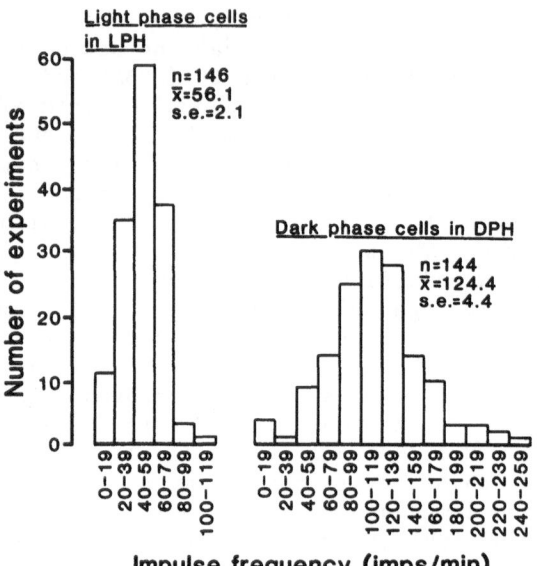

Fig. 4   Frequency of action potentials in link nerve neurons of
         Carausius from animals in the midpoints of the dark and
         light phases of a 12:12 LD photoperiod.   The mean frequency
         is greater in D-cells than in L-cells.   LPH - light phase
         haemolymph;   DPH - dark phase haemolymph.

be recognised.   Such a preparation is discarded or neuron 9 is cut
off.

     Preparations of LNNs taken from animals in the middle of the
dark phase of the photoperiod (D-cells) and of the light phase
(L-cells) (Fig. 4) were placed in samples of their own haemolymph
and of haemolymph taken from animals in the opposite photophase.
The results showed a fall in impulse frequency in D-cells in L-
haemolymph (Fig. 5) and an increase in L-cells in D-haemolymph.
These results could be explained in 3 different ways: a) L-haemolymph contains an inhibitory hormone;   b) D-haemolymph contains an
excitatory hormone or c) there is a diel fluctuation in haemolymph
potassium which affects the excitability of the LNNs.

     The possibility that the cyclical changes in impulse frequency
could be caused by variations in K^+ level was investigated.   Such
fluctuations have been described in the cockroach *Leucophaea* (LETTAU
et al., 1977), but no correlation could be found between K^+ levels
and locomotory activity.   The resting potential of these cells is
determined largely by the K^+ concentration of the haemolymph

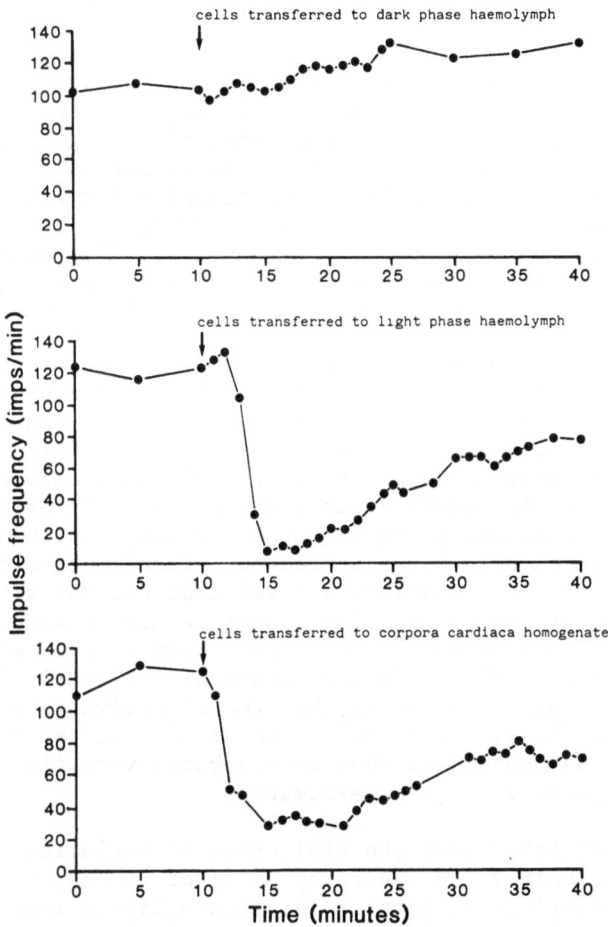

Fig. 5. Records of action potentials from 3 preparations of link
         nerve neurons of *Carausius* taken from animals at the mid-
         point of the dark phase to show typical responses to
         transfer into another sample of dark phase haemolymph,
         into light phase haemolymph, and into homogenate of
         corpus cardiacum. The rise in activity of the control
         preparation (top graph) is probably not significant.

(ORCHARD, 1976) and any fluctuations would render the cell more or less prone to depolarisation and to the generation of action potentials. Analysis of haemolymph by emission flame photometry showed no significant difference in mean values between the D- and L-haemolymph, although there were considerable differences between individuals. The mean value for total K^+ concentration in whole blood was around 21 mM/l for L-phase animals and around 20 mM/l for D-phase animals. The values for centrifuged blood were naturally lower (L = 18.5 mM/l and D = 19 mM/l). These values give no support to the hypothesis that K^+ fluctuations control firing frequency of the neurons. Total haemolymph K^+, however, is not an accurate measure of the K^+ ions available in the haemolymph because flame photometry measures bound as well as unbound K^+. We did not measure unbound K^+ (= K^+ activity) but LETTAU et al. (1977), using implanted ion-selective electrodes, found that the K^+ activity of the haemolymph of *Leucophaea* was about 50% of the total K^+ concentration.

The saline solution normally used for *Carausius* is based on the work of WOOD (1957) who also analysed whole blood by flame photometry and based his saline upon the results, using a concentration of 18 mM/L K^+ in the final formula. It is not surprising, therefore, that this saline causes an increase in the firing frequency of the LNNs (FINLAYSON et al., 1983) because the K^+ activity of the haemolymph is likely to be nearer to 50% of the concentration of K^+ in whole blood. Experiments were carried out, therefore, with lower concentrations of K^+ to establish a level that did not affect the firing frequency of the cells. That level should be around the K^+ activity level of the haemolymph *in vivo*. Such a level was found to lie between 12 and 15 mM/l. Salines containing 12 mM/l K^+ were definitely inhibitory, indicating that the K^+ activity level of *Carausius* may not be as low as 50% of total K^+. As the readings in both *Carausius* and *Leucophaea* show considerable variation, it is not possible to be precise in these estimates.

Having established that the diel cycle of variations in impulse frequency is unlikely to be caused by K^+ fluctuations, the next problem was to distinguish between the possibilities that there was an inhibitor in L-haemolymph or an excitator in D-haemolymph. This was done by diluting L-haemolymph with saline so that the final mixture would have a sufficiently high level of K^+ activity to be potentially stimulatory. A dilution of 20% haemolymph was made in saline with WOOD's original concentration of 18 mM/l K^+. Assuming that the K^+ activity of the haemolymph was 50% of the total K^+ in whole haemolymph (20.82) the K^+ activity of the mixture of haemolymph and saline was calculated to be 12.59 mM/l. The equivalent total K^+ in a saline would be 17.24 mM/l. Such a saline would definitely be stimulatory. In fact, LNNs from D-phase animals were inhibited by the 20% L-phase blood/saline mixture (Fig. 6), indicating that there was an inhibitory factor present in the haemolymph and that the inhibition was not a result of a low K^+ activity. In 50% L-haemolymph the depression of

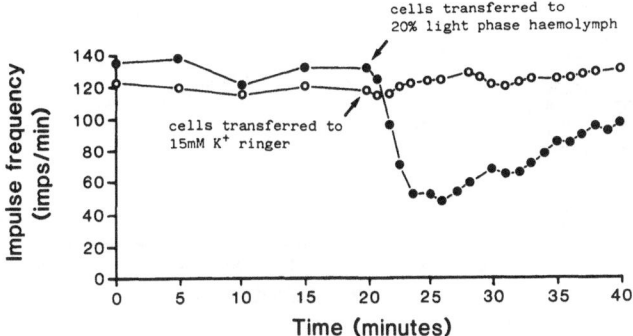

Fig. 6. Response of dark phase link nerve neurons of *Carausius* to 20% light phase (inhibitory) haemolymph (•) in saline and to a saline of comparable K^+ activity (o).

activity is even greater despite the fact that such a mixture is equivalent to a saline with a K^+ level of 16.14 mM/l, which should be stimulatory if K^+ concentration is the critical factor.

Daily cycling inhibitory factors have been described in other arthropods. In the scorpion *Heterometrus* haemolymph and extracts of the cephalothoracic nerve mass from L-phase animals inhibit spontaneous electrical activity of the isolated nerve cord. Conversely, haemolymph and extracts from D-phase animals stimulate electrical activity (RAO and GROPALAKRISHNAREDDY, 1967). The stimulatory response could be illusory and simply a result of the removal of the inhibitor if L-phase animals were the donors of the isolated nerve cord in which the assays were made. Another example is the much more thoroughly investigated neurodepressing factor of crayfish (ARÉCHIGA et al., 1977a), which reduces the impulse frequency of a wide range of motor neurons in the CNS. This factor is a peptide which is produced mainly in the sinus gland of the eyestalk, a neurohaemal organ. It is non-specific; the hormone extracted from eyestalks of the crayfish *Nephrops* and of the crab *Carcinus* has a neurodepressing effect on both (ARÉCHIGA et al., 1977b). The neurodepressor may control the diel cycle of locomotory activity in these crustaceans but in the stick insect its function may be different although it is obviously correlated with the diel cycle of locomotory activity. The factor is present during the day when the insects are motionless. However, L-haemolymph, when applied to neuron 9 of the peripheral system, a sensory mechanoreceptive neuron, has no effect on its rate of firing (Fig. 7). Similarly L-haemolymph has no effect on motor discharge from CNS ganglia of the abdominal nerve cord. It is noteworthy that neuron 9 is well sheathed by glial cell membranes, as are the motor neurons of the CNS; both are protected by a 'blood-

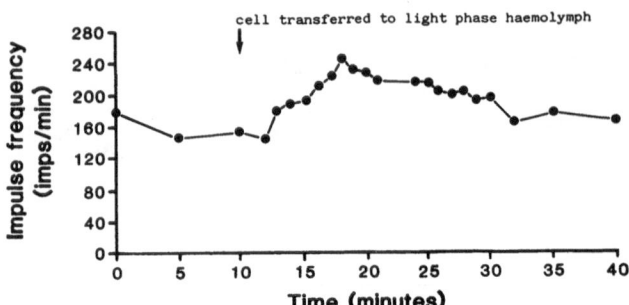

Fig. 7.   Records from peripheral neuron 9 of a dark phase *Carausius*
          in its own haemolymph and then in light phase haemolymph.
          Unlike the link nerve neurons, neuron 9 is not inhibited
          by light phase haemolymph (compare with Fig. 5). (FINLAYSON
          and DAY, unpublished).

brain' barrier. The extraganglionic neurosecretory neurons of the
link nerve, on the other hand, are covered only by a flimsy stroma
and are exposed to the immediate influence of factors in the haemolymph.

     The site of production of the neurodepressing factor of
Carausius is not known but homogenates of CC have a similar depres-
sing effect on the firing frequency of the LNNs (Fig. 5). It is not
clear that this is the influence of the same factor; a complication
is that CC homogenate also depresses the firing frequency of sensory
neuron 9. The CC is the source of a number of active substances,
however, and there is no reason to suppose that the same substance
is acting on LNNs and on neuron 9   (FINLAYSON and DAY, unpublished).

     There is a suggestion from these experiments on *Carausius* that
the LNNs can be 'activated' in some way so that they become more
sensitive to the effect of the inhibitor. It is noticeable that
preparations kept in saline before being transferred to inhibitory
(L) haemolymph respond with a greater fall in firing frequency than
preparations that are transferred directly from D-haemolymph to L-
haemolymph. It was also found that preparations of cells of low
activity from L-animals which had been transferred to D-haemolymph
and consequently had become more active were depressed below their
original level when returned to L-haemolymph. Similarly the firing
frequency of L-cells which had been immersed in 15 mM/l K^+ saline
fell below the original level when the cells were returned to L-
haemolymph.

SUMMARY

1.   Changes in the appearance of neurosecretory granules may be indicative of a cycle of synthesis, maturation, storage and then autolysis of surplus neurosecretory material.   There may be a mechanism of release that involves the incorporation of neuro-secretory granules into dense bodies by fusion or by being engulfed in membranous bodies.   The dense bodies may give up their contents by fragmentation but there is no evidence to show how the neuro-secretory material then leaves the cell.   Exocytosis as a release mechanism is well founded.   In the one documented example of an insect neurosecretory neuron that undergoes total autolysis (the ecdysial gland neuron of Lepidoptera) similar structures seem clearly to be lytic in nature.   Cisternae of endoplasmic reticulum are considered to be synthetic structures, but in the case of the ecdysial gland neuron they may be lytic in nature because granules are incorporated in them and they are formed at the beginning of the autolytic phase of the neuron.

2.   The electrical activity of neurosecretory neurons in insects can be correlated with their secretory activity (as shown by light and electron microscopy) and by the release of neurohormones.   An increase in the frequency of discharge of impulses precedes or accompanies the release of their secretions.   Bursts of impulses are commonly recorded from these neurons.

3.   The secretory and electrical activity of neurosecretory neurons in insects is influenced by a variety of blood-borne factors including ecdysterone, juvenile hormone, octopamine, corpus cardiacum stimu-lating factor and a neurodepressing factor.   The level of potassium in the blood can influence electrical activity and the release of neurohormones but there is no proof that fluctuations in potassium level play a part in the normal neurosecretory activity of the insect.

REFERENCES

AGUI N. (1976) Studies on insect endocrinology by means of organ culture. *Mem. Fac. Agr. Tokyo Univ. Educ.* 22, 173-235 (Japanese with English summary).

AGUI N. and HIRUMA K. (1977a) *In vitro* activation of neurosecretory brain cells in *Mamestra brassicae* by β-ecdysone. *Gen. comp. Endocrinol.* 33, 467-472

AGUI N. and HIRUMA K. (1977b) Ecdysone as a feedback regulator for the neurosecretory brain cells in *Mamestra brassicae*. *J. Insect Physiol.* 23, 1393-1396.

ARECHIGA H., CABRERA-PERALTA C. and HUBERMAN A. (1977a) Functional characterization of the neurodepressing hormone in the crayfish. *J. Neurobiol.* 10, 409-422.

ARÉCHIGA H., WILLIAMS J. A., PULLIN R. S. V. and NAYLOR E. (1977b) Cross-sensitivity to neuro-depressing hormone and its effect on locomotor rhythmicity in two different groups of crustaceans. *Gen. comp. Endocrinol.* 37, 350-357.

BEAMENT J. W. L. (1958) A paralysing agent in the blood of cockroaches. *J. Insect Physiol.* 2, 199-224.

BEATTIE T. M. (1976) Autolysis in axon terminals of a new neurohaemal organ in the cockroach *Periplaneta americana*. *Tissue and Cell* 8, 305-310.

BERLIND A. (1977) Cellular dynamics in invertebrate neurosecretory systems. *Int. Rev. Cytol.* 49, 171-251.

BERNAYS E. A. (1980) The post-prandial rest in *Locusta migratoria* nymphs and its hormonal regulation. *J. Insect Physiol.* 26, 119-123.

BOROVSKY D. (1982) Release of egg development neurosecretory hormone in *Aëdes aegypti* and *Aëdes taeniorhynchus* induced by an ovarian factor. *J. Insect Physiol.* 28, 311-316.

BRADLEY J. T. and SIMPSON T. A. (1981) Brain neurosecretion during ovarian development and after ovariectomy in adult *Acheta domestica* L. *Gen. comp. Endocrinol.* 44, 117-127.

BRADY J. and MADDRELL S. H. P. (1967) Neurohaemal organs in the median nervous system of insects. *Z. Zellforsch.* 76, 389-404.

FIFIELD S. M. and FINLAYSON L. H. (1978) Peripheral neurons and peripheral neurosecretion in the stick insect, *Carausius morosus*. *Proc. R. Soc. Lond. B.* 200, 63-85.

FINLAYSON L. H. and ORCHARD I. (1978) Neurosecretory and electrical activity of extra-ganglionic neurons in the stick insect. In *Comparative Endocrinology*. (Ed. by GAILLARD P. J. and BOER H. H.), pp. 323-326. Elsevier/North Holland Biomedical Press, Amsterdam.

FINLAYSON L. H. and OSBORNE M. P. (1968) Peripheral neurosecretory cells in the stick insect (*Carausius morosus*) and the blowfly larva (*Phormia terraenovae*). *J. Insect Physiol.* 14, 1793-1801.

FINLAYSON L. H. and OSBORNE M. P. (1975) Secretory activity of neurons and related electrical activity. *Adv. Comp. Physiol. Biochem.* 6, 165-258.

FINLAYSON L. H., ORCHARD I. and DAY N. (1983) A factor in the haemolymph of the stick insect that depresses the activity of extra-ganglionic neurosecretory cells. *J. Comp. Physiol.* (in press).

GIRARDIE A., MOULINS M. and GIRARDIE J. (1974) Rupture de la diapause ovarienne d'*Anacridium aegyptium* par stimulation électrique des cellules neurosécrétrices medianes de la pars intercerebralis. *J. Insect Physiol.* 20, 2261-2275.

GOSBEE J. L., MILLIGAN J. V. and SMALLMAN B. N. (1968) Neural properties of the protocerebral neurosecretory cells of the adult cockroach, *Periplaneta americana*. *J. Insect Physiol.* 14, 1785-1792.

GREEN G. W. (1964) The control of spontaneous locomotor activity in *Phormia regina* Meigen. II. Experiments to determine the mechanism involved. *J. Insect Physiol.* 10, 727-752.

GRIFFITHS A. C. and FINLAYSON L. H. (1983) Ultrastructural changes in, and involution of, the ecdysial gland neuron at metamorphosis in *Agrotis, Spodoptera,* and *Manduca* (Lepidoptera). *Int. J. Insect Morphol. & Embryol.* (in press).

HIGHNAM K. C. (1961) Induced changes in the amounts of material in the neurosecretory system of the desert locust. *Nature, Lond.* 191, 199-200.

HIGHNAM K. C. (1962) Neurosecretory control of ovarian development in *Schistocerca gregaria. Q. J. microsc. Sci.* 103, 57-72.

HODGSON E. S. and GELDIAY (1959) Experimentally induced release of neurosecretory materials from roach corpora cardiaca. *Biol. Bull.* 117, 275-283.

KONO Y. (1975) Daily changes of neurosecretory type-II cell structure of *Pieris* larvae entrained by short and long days. *J. Insect Physiol.* 21, 249-264.

LEA A. O. (1972) Regulation of egg maturation in the mosquito by the neurosecretory system: the role of the corpus cardiacum. *Gen. Comp. Endocrinol. Supp.* 3, 602-608.

LETTAU J, FOSTER W. A., HARKER J. E. and TREHERNE J. E. (1977) Diel changes in potassium activity in the haemolymph of the cockroach *Leucophaea maderae. J. Exp. Biol.* 71, 171-186.

McCAFFERY A. R. and HIGHNAM K. C. (1975a) Effects of corpora allata on the activity of the cerebral neurosecretory system of *Locusta migratoria* R. and F. *Gen. Comp. Endocrinol.* 25, 358-372.

McCAFFERY A. R. and HIGHNAM K. C. (1975b) Effects of corpus allatum hormone and its mimics on the cerebral neurosecretory system of *Locusta migratoria migratorioides* R. and F. *Gen. Comp. Endocrinol.* 25, 373-386.

MADDRELL S. H. P. and NORDMANN J. J. (1979) Neurosecretion. Blackie, Glasgow and London.

MORRIS G. P. and STEEL C. G. H. (1975) Ultrastructure of neurosecretory cells in the pars intercerebralis of *Rhodnius prolixus* (Hemiptera). *Tissue and Cell* 7, 73-90.

MORRIS G. P. and STEEL C. G. H. (1977) Sequence of ultrastructural changes induced by activation in the posterior neurosecretory cells in the brain of *Rhodnius prolixus* with special reference to the role of lysosomes. *Tissue and Cell* 9, 547-561.

NIJHOUT H. F. and WILLIAMS C. M. (1974) Control of moulting and metamorphosis in the tobacco hornworm *Manduca sexta* (L): cessation of juvenile hormone secretion as a trigger for pupation. *J. Exp. Biol.* 61, 493-501.

NORMANN T. C. (1974) Calcium dependence of neurosecretion by exocytosis. *J. Exp. Biol.* 61, 401-409.

ORCHARD I. (1976) Calcium dependent action potentials in a peripheral neurosecretory cell of the stick insect. *J. Comp. Physiol.* 112, 95-102

ORCHARD I. and FINLAYSON L. H. (1976) The electrical activity of mechanoreceptive and neurosecretory neurons in the stick insect *Carausius morosus. J. Comp. Physiol.* 107, 327-338.

ORCHARD I. and FINLAYSON L. H. (1977) Studies on peripheral neurons
and neurohaemal tissue in the thorax of the stick insect (*Carausius
morosus*). *Experientia* 33, 1440-1442.

ORCHARD I. and STEEL C. G. H. (1980) Electrical activity of neurosecretory axons from the brain of *Rhodnius prolixus:* relation of
changes in the pattern of activity to endocrine events during the
moulting cycle. *Brain Research* 191, 53-65.

ORCHARD I., RUEGG R. P. and DAVEY K. G. (1983) The role of central
aminergic neurons in the action of 20-hydroxyecdysone on neurosecretory cells of *Rhodnius prolixus*. *J. Insect Physiol*. 29, 387-
391.

OZBAS S. and HODGSON S. (1958) Action of insect neurosecretion upon
central nervous system in vitro and upon behaviour. *Proc. nat.
Acad. Sci. U.S.A.* 44, 825-830.

PARK K. E. and SEONG S. I. (1975) Fine structure of median neurosecretory cell in diapause and non-diapause brain of the silkworm,
Bombyx mori. *J. Insect Physiol*. 21, 1311-1317.   •

RAABE M. (1982) Insect Neurohormones. Plenum Press, New York and
London.

RAO K. P. and GROPALAKRISHNAREDDY T. (1967) Blood borne factors in
circadian rhythms of activity. *Nature, London.* 213, 1047-1048.

RUEGG R. P., KRIGER F. L., DAVEY K. G. and STEEL C. G. H. (1981)
Ovarian ecdysone elicits release of a myotropic ovulation hormone
in *Rhodnius* (Insecta: Hemiptera). *Int. J. Invert. Reprod.* 3, 357-
361.

RUEGG R. P., KRIGER F. L., DAVEY K. G. and STEEL C. G. H. (1982)
20-hydroxyecdysone as a modulator of electrical activity in neurosecretory cells of *Rhodnius prolixus*. *J. Insect Physiol*. 28, 243-
248.

SCHARRER B. and KATER S. B. (1969) Neurosecretion. XV. An electron
microscopic study of the corpora cardiaca of *Periplaneta americana*
after experimentally induced hormone release. *Z. Zellforsch*. 95,
177-186.

STEEL C. G. H. (1975) A neuroendocrine feedback mechanism in the
insect moulting cycle. *Nature* 253, 267-269.

STEEL C. G. H. (1978) Nervous and hormonal regulation of neuro-
secretory cells in the insect brain. In *Comparative Endocrinology*.
(Ed. by GAILLARD P. J. and BOER H. H.), pp. 327-330. Elsevier/North
Holland Biomedical Press, Amsterdam.

STEEL C. G. H. (1982) Parameters and timing of synthesis, transport,
and release of neurosecretion in the insect brain. In *Neurosecretion:
Molecules, Cells, Systems*. (Ed. by FARNER D. S. and LEDERIS K.),
pp. 221-231. Plenum Publ. Co.

STEEL C. G. H., BOLLENBACHER W. E., SMITH S. L. and GILBERT L. I.
(1982) Haemolymph ecdysteroid titres during larval-adult development
in *Rhodnius prolixus:* correlations with moulting hormone action and
brain neurosecretory cell activity. *J. Insect Physiol*. 28, 519-525.

TRUMAN J. W. (1978) Rhythmic control over endocrine activity in
insects. In *Comparative Endocrinology*.(Ed. by GAILLARD P. J. and
BOER H. H.), pp. 123-136. Elsevier/North Holland Biomedical Press,
Amsterdam.

WOOD D. W. (1957) The effects of ions upon neuromuscular transmission in a herbivorous insect. *J. Physiol (Lond)*. 128, 119-139.
YIN C.-M. and CHIPPENDALE G. M. (1975) Insect frontal ganglion: fine structure of its neurosecretory cells in diapause and non-diapause larvae of *Diatraea grandiosella*. *Can. J. Zool.* 53, 1093-1100.

IONIC BASIS OF ELECTRICAL ACTIVITY

IN INSECT NERVE CELLS AND SYNAPSES

Yves Pichon

Département de Biophysique
Laboratoire de Neurobiologie Cellulaire
du C.N.R.S.
F - 91190 - Gif sur Yvette (France)

INTRODUCTION

After the pioneering experiments of Hodgkin, Huxley and Katz (1952) on squid axons and those of Fatt and Katz (1951) on frog neuromuscular junctions, it is now well established that electrical activity in nerve and muscle is associated with changes in membrane ionic conductances. The following chapter summarizes the evidences which have been obtained so far on insect on the ionic basis of electrical activity in nerve cells and synapses. Not very long after Hodgkin et al. (1952), Boistel and Coraboeuf (1958) were able to show that action potentials recorded from giant axons in the desheathed nerve cord of the american cockroach, Periplaneta americana were sensitive to external sodium concentration. They were soon followed by Yamasaki and Narahashi (1959) who studied systematically the effects of sodium and potassium ions on resting and action potentials of these same axons. The ionic basis of electrical activity was fully established on isolated axons using the voltage-clamp technique (Pichon, 1967, 1968, 1974). These results will be summarized in the first two parts of this chapter together with relevant informations on nerve cell bodies.

Synaptic events have also been recorded in insect ganglia were both excitatory synaptic potentials (EPSP) and inhibitory synaptic potentials (IPSP) were found to occur. Unfortunately, for technical reasons, the ionic basis of these synaptic events have not been yet fully established at neuro-neuronal synapses. Neuromuscular junctions have revealed more suitable for electrophysiological investigation and the mechanisms involved at this level have been recently analysed. The main findings will be

23

examined in the third part of this chapter.

The fourth and last part of the chapter will deal with the ionic aspects of nerve function "in situ".

This chapter is in no way an attempt to cover all ionic aspects of nerve function. Further details will be found in other chapters of this book. Relevant informations will also be found in three volumes of "Comprehensive Insect Physiology, Biochemistry and Pharmacology" : vol.5 (Lane, 1984; Pichon and Ashcroft, 1984; Treherne, 1984; Callec, 1984), vol.10 (Pichon and Manaranche, 1984) and vol.11 (Pitman, 1984).

IONIC BASIS OF RESTING POTENTIAL PRODUCTION IN AXONS AND NERVE-CELL-BODIES.

Resting potential in axons

When the tip of a microelectrode is pushed through the membrane into the axoplasm of a giant axon of the cockroach, the potential moves abruptly to a new value which is about 70 mV more negative, demonstrating the existence of a resting potential difference between the two faces of the axonal membrane. This value is of -70.1 mV according to Yamasaki and Narahashi (1959) or - 67.4 mV according to Pichon and Boistel (1967b) in desheathed connectives of _Periplaneta americana_ bathed in a saline containing 3.1 mM K^+. It is lower in isolated axons (-58.4 mV, Pichon 1969). All these values are significantly smaller than E_K, the equilibrium potential for potassium ions, which lies around -90 mV according to Pichon, Poussart and Lees (1983).

Resting potentials have also been measured in giant axons in situ, in sheathed connectives of different insect species, but these values are not reliable because of the interposition between the recording electrode and the reference electrode of the nerve sheath which may be polarized. Furthermore, as pointed out by Treherne and Pichon (1972) and Treherne (1984), the ionic medium which surrounds nerve cells in situ is likely to be significantly different from the bathing medium so that it is not possible to interpret the data in terms of potassium concentrations in the bath.

The possibility of describing the resting potential of insect axons in terms of potassium, sodium and chloride permeabilities and gradients across the nerve membrane has been examined (Yamasaki and Narahashi, 1959; Pichon, 1969). The slope relating the resting potential to the logarithm of the external potassium concentration is about 42 mV for a ten fold change (fig.1). For potassium concentrations below 8 mM, the curve tends to level up. The slope differs by 17 mV from the 59 mV slope predicted by the

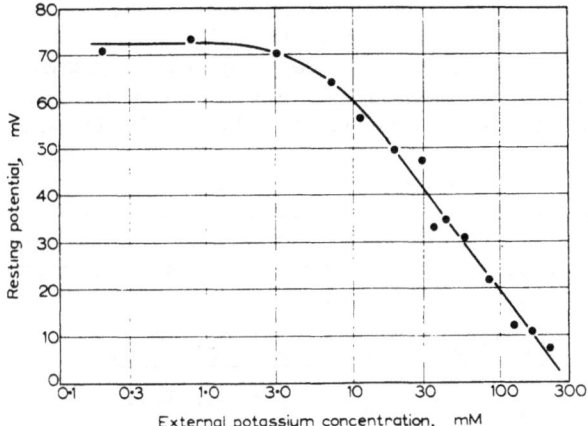

Fig.1.   Effects of external potassium concentration on the
          resting potential of cockroach giant axons in a
          desheathed connective (from Yamasaki and Narahashi,
          1959).

Nernst equation for potassium ions :

$$E_K = \frac{RT}{F} \ln \frac{IKI_o}{IKI_i} \qquad (1)$$

where R is the gas constant, T the absolute temperature, F the
Faraday and IKI_o and IKI_i the potassium concentrations
respectively outside and inside. This indicates that the axonal
membrane does not behave as a perfect potassium electrode and that
ions other than potassium participate to the generation of the
resting potential. A better description of the resting potential
(E_r) of the cockroach axon is derived, as for the squid axon
(Hodkin and Katz, 1949), from the constant field simplification of
Goldman (1943) of the Plank (1890) equation :

$$E_r = \frac{RT}{F} \ln \frac{P_K \, IKI_o + P_{Na} \, INaI_o + P_{Cl} \, IClI_i}{P_K \, IKI_i + P_{Na} \, INaI_i + P_{Cl} \, IClI_o} \qquad (2)$$

where $I I_o$ and $I I_i$ are the concentrations in the outside and
the inside of the axon respectively and P_K, P_{Na} and P_{Cl} are
the relative permeabilities of the membrane to potassium, sodium
and chloride ions. A fourfold reduction in the external sodium
concentrations results in a hyperpolarization of the axonal

membrane by 7.25 mV (Boistel and Coraboeuf, 1958), indicating that
the resting sodium permeability of the giant axon of the cockroach
is relatively important. A large resting sodium permeability has
also been observed in desheathed nerve cords of the hawk moth,
Manduca sexta (Pichon, Sattelle and Lane, 1972) where a 15 mV
change in resting potential corresponded to a tenfold change in
the external sodium concentration. As pointed out by Pichon
(1969), a more complete description of the resting potential
should also take into account the existence in the nerve membrane
of an active and possibly electrogenic pumping mechanism so that
equation (2) should be modified into :

$$E_r = \frac{RT}{F} \ln \frac{P_K IKI_o + P_{Na} INaI_o + P_{Cl} IClI_i}{P_K IKI_i + P_{Na} INaI_i + P_{Cl} IClI_o} + E_p \qquad (3)$$

where E_p is the contribution of the pump, if electrogenic.

Resting potential in nerve cell bodies

The resting potential recorded in nerve cell bodies is
slightly lower than that recorded in axons in similar conditions.
In Periplaneta (Jego, Callec, Pichon and Boistel, 1970) and
Schistocerca (Goodman and Heitler, 1979; Gwilliam and Burrows,
1980), it lies around -60 mV against -46 mV in Carausius (Treherne
and Maddrell, 1967; Orchard, 1976) and -50 mV in Bombyx
(Monticelli and Depretto (1979).
The relationship between E_r and the external potassium
concentrations is similar to that reported for the axon, the slope
of the linear portion of the curve being of about 42 mV in
excitable nerve cell bodies of the cockroach (Jego et al., 1970)
against 37 mV in unexcitable cells in desheathed ganglia of the
stick insect (Treherne and Maddrell, 1967) and 46 mV in
neurosecretory cells of this same insect (Orchard, 1976). The
significantly higher slope of 58.4 mV reported by Monticelli and
Depretto (1979) for unidentified cells from the ventral nerve cord
of last instar larvae of the silkworm, Bombyx mori, could result
from the absence of a significant driving force for sodium ions
(the external solution contained 1.7 mM Na^+ against 36 mM K^+).

IONIC BASIS OF ELECTRICAL ACTIVITY IN AXONS AND NERVE-CELL-BODIES

The fundamental property of nerve cells is their excitability,
i.e. their ability to respond to a depolarization with a graded
and/or an all-or-none action potential. In both cases the effect
of the stimulus is to modify the ionic permeability of the
membrane. As mentioned earlier, these modifications have been
studied in isolated axons and the major part of the present
subchapter will concern ionic currents and conductances in
cockroach axons.

Graded activity and action potentials in axons and nerve cell bodies

When the membrane of a cockroach axon is depolarized above a certain level, a hump appears on top of the initial phase of the catelectrotonic potential. This "local response" which is not actively propagated along the axon increases with stimulus intensity and, above the critical depolarization (which lies around -40 mV), give rise to an all-or-none depolarization : the action potential. The action potential, an example of which is shown in fig.2, is propagated along the axon. Its amplitude and time course are essentially similar to those recorded in other unmyelinated axons. In a series of experiments on isolated axons of the american cockroach, Periplaneta americana, the mean amplitude of the spike (the large and fast portion of the action potential which follows the stimulus) was found to approximate 95 mV, the membrane potential change its polarity (overshoot) by about +35 mV (Pichon, 1969). The maximum rate of rise and rate of fall of action potential were of respectively 1370 V/sec and 640 V/sec. The spike was followed by a slight posthyperpolarization (the positive after potential) followed in turn by an even smaller and slower postdepolarization (the negative afterpotential). Both after potentials are significantly smaller in isolated axons than in situ (Pichon, 1969). Such a diffrence most probably reflects modifications in the driving force for K$^+$ ions and in the

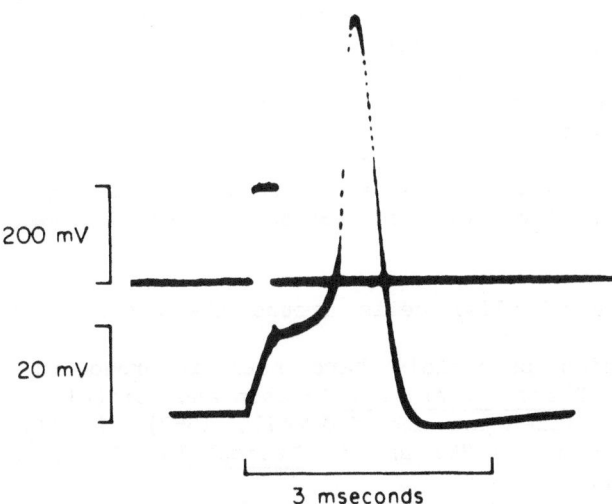

200 mV

20 mV

3 mseconds

Fig.2.   Membrane action potential recorded from an isolated giant axon of Periplaneta americana (lower tracing) induced by a short duration depolarizing pulse (upper tracing). (From Pichon and Boistel, 1967a).

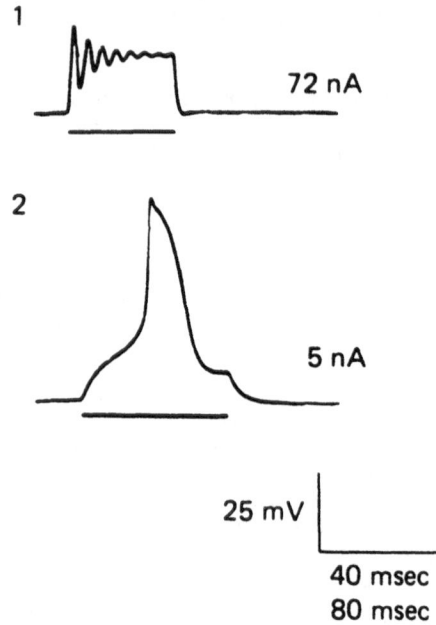

Fig.3.   Effects of external application of TEA$^+$ on the electri-
          cal response of a nerve cell body to a depolarizing
          current pulse. The neurone was impaled with two
          microelectrodes filled with 1 M potassium acetate, one
          for   stimulation,   the   other   for   recording.   Resting
          potential : -57 mV (1) : response in normal saline to a
          72 nA square pulse of current : only a series of oscilla-
          tions in seen (horizontal calibration : 40 msec); (2)
          response after bathing the preparation during 5 min in a
          saline containing 50 mM TEA$^+$: a large all-or-none
          spike is elicited by a 5 nA square pulse of current
          (horizontal calibration : 80 msec). (From Pitman, 1979).

geometry (absence of glial cells around the axons in isolated
axons).

        Similar action potentials have been recorded in another
cockroach species Blabera craniifer (Pichon and Boistel, 1966), in
the stick insect (Treherne and Maddrell, 1967), in the locust
(Gwilliam and Burrows, 1980) and in Drosophila (Tanouye, Ferrus
and Fujita, 1981).

        Active electrical responses have also been recorded from
nerve cell bodies. In some cases, these action potentials arise
spontaneously or following depolarization of the cell (Callec and
Boistel, 1966; Kerkut, Pitman and Walker, 1968; Jego et al., 1970;
Orchard, 1976; Orchard and Finlayson, 1976, 1977; Goodman and

Heitler, 1979). Most cell bodies in insect ganglia are however inexcitable under normal conditions, they can be purely passive or respond to direct electrical stimulation by graded oscillatory responses. These cells can give action potentials following various treatments such as axotomy, colchicine, injection of citrate or superfusion with tetraethylammonium ions (TEA$^+$), Pitman, 1975 a and b, 1979; Goodman and Heitler, 1979 (fig.3).

In spontaneously firing cell bodies, such as those of the dorsal surface of the sixth abdominal ganglion of the cockroach (Jego et al., 1970), the action potential is rather large (90 mV) and is followed by a large positive after potential. It resembles in many respects the action potentials recorded in nerve cell bodies of molluscs. Its duration at room temperature is significantly longer than that of the axonal spike (3 msec against 0.5 msec). Calcium spikes such as those observed in neurosecretory cells of Carausius (Orchard, 1976) are even slower (7-20 msec at 50 percent spike height).

Effects of ions on action potentials in axons and nerve-cell-bodies

From the pioneering work of Hodgkin and Katz (1949) it is known that the action potential in the squid axon is correlated with the existence of sodium ions in the external solution. As mentionned earlier, Boistel and Coraboeuf (1958) observed that the action potential recorded intracellularly from a cockroach axon falled from 74 mV in 154 mM Na Cl to 38 mV in 38.5 M NaCl. Almost simultaneously, Yamasaki and Narahashi (1959) found on this same preparation that the slope relating the overshoot of the action potential to the external sodium concentration did not differ significantly from the theoretical 59 mV slope for a ten-fold change predicted from the Nernst equation for sodium ions :

$$E_{Na} = \frac{RT}{F} \ln \frac{INaI_o}{INaI_i} \qquad (4)$$

A similar sensitivity to sodium ions was found in desheathed connectives of other insect species such as Blabera craniifer (Pichon and Boistel, 1966), Carausius morosus (Treherne and Maddrell, 1967), Manduca sexta (Pichon et al., 1972). Action potentials were also blocked in sodium free-tris solutions in desheathed or stretched crural nerves of Periplaneta and Locusta (Pichon and Treherne, 1973).

The action potential of nerve cell bodies was found to be sensitive to sodium in a few cases (Jego et al., 1970, Pitman, 1975a) whereas calcium seems to be responsible for action potential production in all other cases (Pitman, 1975b; Orchard, 1976; Orchard and Finlayson, 1976 , 1977; Pitman, 1979).

The ionic channels which carry the inward current are different in axons and cell bodies.

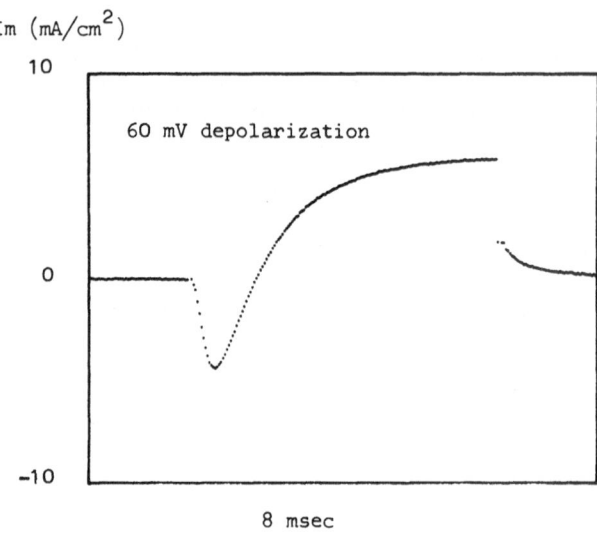

Im (mA/cm^2)

Fig.4   Ionic currents corresponding to a 5 msec depolarizing
        voltage pulse to 0 mV in a giant axon of Periplaneta
        americana under voltage-clamp conditions. Leak current
        substracted. HP = -60 mV. (From Pichon and Ashcroft,
        1984).

        Axonal sodium channels were found to be permeable to Na^+
and Li^+ ion (Pichon, unpublished) but not to Ca^{++}. Thus, in a
careful study of the effects of divalent cations on the maximum
rate of rise of the action potential of Periplaneta, Narahashi
(1966) found that Ca^{++}, Mg^{++}, Sr^{++} and Ba^{++} ions do not
contribute significantly to the action potential by shift the
sodium inactivation curve ($h\infty$) toward lower membrane potentials.
        Calcium channels from nerve cell bodies are blocked by Mg^{++}
ions (Pitman, 1979) and are generally much less selective than the
sodium channels. Another important difference between sodium and
calcium channels is the high sensitivity of the former to TTX
(Narahashi, 1965; Treherne and Maddrell, 1967; Pichon, 1969;
Pichon et al., 1972; Tanouye et al., 1981).
        It is clear that, when rapid conduction of informations is
necessary, as in axons, fast all-or-none action potentials are
needed. Graded responses are conversely an obvious advantage in
nerve cell bodies since it allows integration of several inputs.
Calcium ions which enter during the local response or the spike in
neurosecretory cells may also play a direct role in stimulus -
secretion coupling.

Ionic currents and conductances in axons

Direct analysis of the mechanisms which underlie electrical activity in axons is not possible using the conventional microelectrode technique. One simplification is already obtained by eliminating lateral spread of current between adjacent regions of the membrane (space-clamp).

Another simplification consists in maintaining the membrane potential V_m constant. Under these conditions (voltage-clamp), it is relatively simple to analyse the relationship between membrane conductances (which are thought to reflect the opening and closing of membrane ionic channels) and membrane potential and time. As will be illustrated at the end of this subchapter, it is possible to reconstruct normal membrane behaviour from voltage-clamp experiments.

Voltage-clamp experiments have been carried-out on isolated giant axons of Periplaneta americana using a double oil-gap technique. Typical membrane currents generated under those conditions in an isolated axon are illustrated in fig.4. When the membrane potential is held at its resting level (-60 mV), no current crosses the membrane. When the membrane potential is

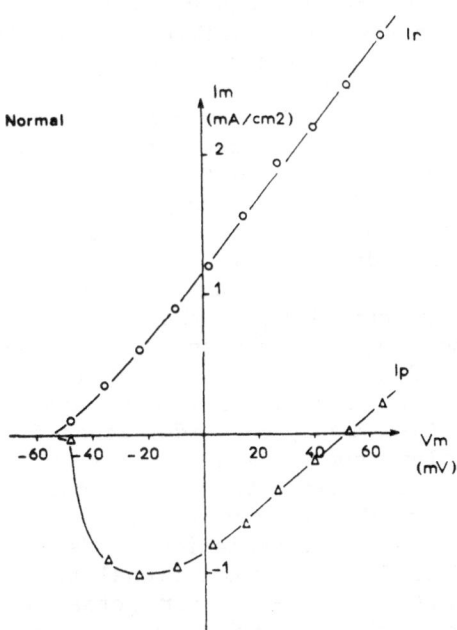

Fig.5   Current-voltage relations in the cockroach giant axon under voltage-clamp conditions after substraction of the leak component. (From Pichon, 1968).

stepped to 0 mV, a transient inward current turns on, reaches
a peak value of about -5 mA cm^{-2}, then decreases and is followed
by outward current which reaches a plateau of about 6 mA cm^{-2}.
When the membrane is stepped back to its original holding
potential value, the outward current returns to its resting value
following an exponential time course.

Such current records can be obtained for various potential
levels. The turning on and off of the inward current as well as
the turning on of the delayed outward current become faster and
faster as the membrane steps are made more positive. The inward
current increases first, reaches a maximum for V_m : -10 mV, then
decreases. The outward current increases continuously with step
size. In all cases, the outward tails of current which follow
return to holding potential level are exponential and porportional
to the pulse size.

Current-voltage relations are illustrated in fig.5 where I_p
is the peak (usually inward) current and I_r the delayed outward
current. It can be seen that the peak current reverses around +50
mV and is outwardly directed for larger pulses. The current
densities as well as the current-voltage characteristics are very
similar to those described by Hodgkin, Huxley and Katz (1952) for
the squid axon (Pichon, 1967; Pichon and Boistel, 1967a). As in
other axons, the peak current is carried by Na$^+$ (Pichon, 1968,
1969, 1974). Since the sodium current is selectively blocked by
TTX (Pichon, 1969), it has been possible to separate the two ionic
currents and study their kinetics (Pichon, 1969, 1974). These
experiments have shown that the model proposed by Hodgkin and
Huxley (1952) to describe their experimental findings on the squid
axon and reconstruct most of the reactions of the nerve membrane
can be used with only minor changes for the cockroach axon.

The insect nerve membrane can be represented phenomenologi-
cally as being made of a capacitance C_m in parallel with three
ionic channels : the sodium channel with a variable conductance
g_{Na} to sodium ions which are more concentrated outside than
inside (equilibrium potential E_{Na}), the potassium channel with a
variable conductance g_K to potassium ions which are more
concentrated inside than outside (equilibrium potential E_K) and
a leak channel exhibiting a constant and relatively low
conductance to leak (presumably chloride) ions (fig.6).

Under voltage-clamp conditions, the membrane potential being
held constant during and between the pulses, one is dealing with a
simplified system in which the driving force for each ion is kept
constant (when there is no accumulation or depletion). The ionic
currents are therefore directly proportional to the driving force
for each given ion (i.e. $E_m - E_K$ for potassium ions) and the
conductance of the membrane for this ion (g_K) and one can
write :

$$I_K = g_K (E_m - E_K) \tag{5}$$

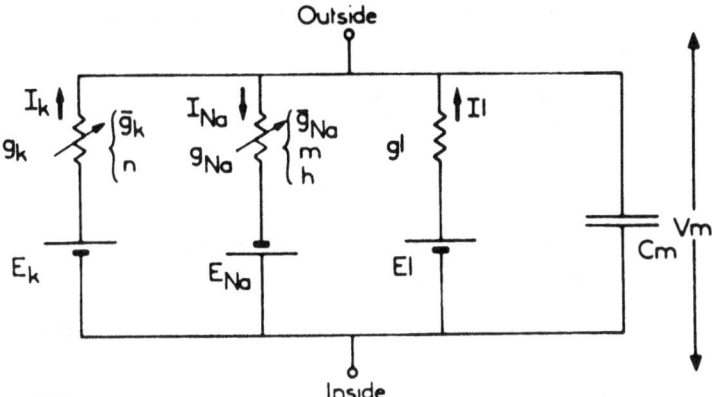

Fig.6.   Electrical equivalent circuit of the axonal membrane
          based on the Hodgkin-Huxley model. (From Pichon, 1974).

Similarly, for sodium :

$$I_{Na} = g_{Na} \ (E_m - E_{Na})$$ (6)

and for the leak current :

$$I_1 = g_1 \ (E_m - E_1)$$ (7)

According to the Hodgkin and Huxley formulation, the
potassium and sodium variable conductances are function of a
maximum conductance ($\bar{g}_K$ or $\bar{g}_{Na}$) and their time course and
voltage-dependence can be described by a set of three time and
voltage dependent parameters (n,m and h) which can vary between 0
and 1. These parameters which are called respectively potassium
activation (n), sodium activation (m) and sodium inactivation
(1-h) might correspond the probability of a charged particle or
site to be located in given region of the membrane. In cockroach
axons, one potassium channel would open when three "n" particles
move simultaneously to this region of the membrane. The potassium
conductance would then be given by :

$$g_K = \bar{g}_K \ n^3$$ (8)

The proportion of "n" particles which move in the membrane
following a step change in membrane potential is a function of
time and potential :

$$dn/dt = \alpha_n \ (1-n) - \beta_n n$$ (9)

where α_n and β_n are voltage dependent rate constants. The solution of equation 4.15 is :

$$n(t) = n_\infty - (n_\infty - n_o) \exp(-t/\tau_n) \qquad (10)$$

where n_o and n_∞ are the steady-state values of n respectively before and after the change in potential and τ_n the potassium activation time constant. The steady-state values of n as well as those of the time constants τ_n are related to the rate constants according to the following equations :

$$n_\infty = \alpha_n / (\alpha_n + \beta_n) \qquad (11)$$

$$\tau_n = 1/(\alpha_n + \beta_n) \qquad (12)$$

Fig.7 illustrates the fit between potassium current traces recorded in the presence of 10^{-7} M TTX and theoretical curves computed according to equation 8 and 10 after curve fitting of the linearized experimental curves with $E_K = -74.5$ mV and $\bar{g}_K = 35$ mmho cm^{-2} (from Pichon et al., 1983).

Membrane current fluctuations corresponding to the opening and closing of individual ionic channels have also been studied (Pichon et al., 1983) and it has been found that the TTX

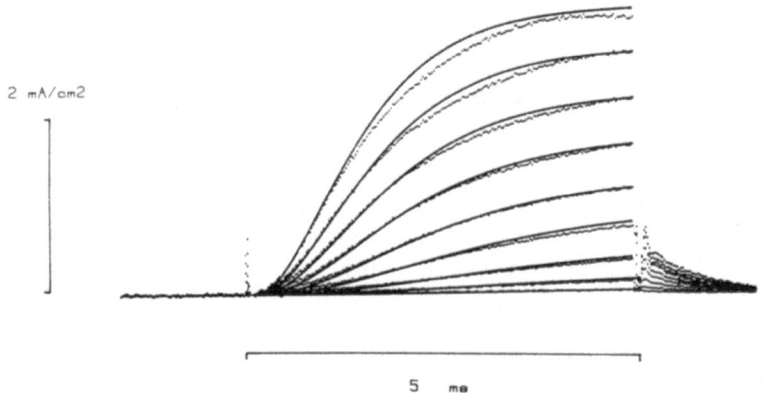

2 mA/cm2

5   ms

Fig.7   Superimposed tracings of the potassium currents corresponding to 5 msec step depolarizations of the axonal membrane of _Periplaneta americana_ under voltage-clamp conditions. The membrane was brought from its holding value of -60 mV to -50, -40, -30, -20, -10, 0, 10, 20 and 30 mV. Leak current substracted and Na$^+$ current blocked with 0.1 μM TTX. (From Pichon, Poussart and Lees, 1983).

insensitive component of the main "noise" could be fitted with the combination of a 1/f component, proportional to $(E_m - E_k)$, and a lorentzian component $1/(1 + (f/f_c)^2)$. This is illustrated in fig.8 for three different values of V_m. It can be seen that the relative contribution of the lorentzian component ot the overall noise decreases with membrane depolarization whereas the corner frequency (f_c) increases.

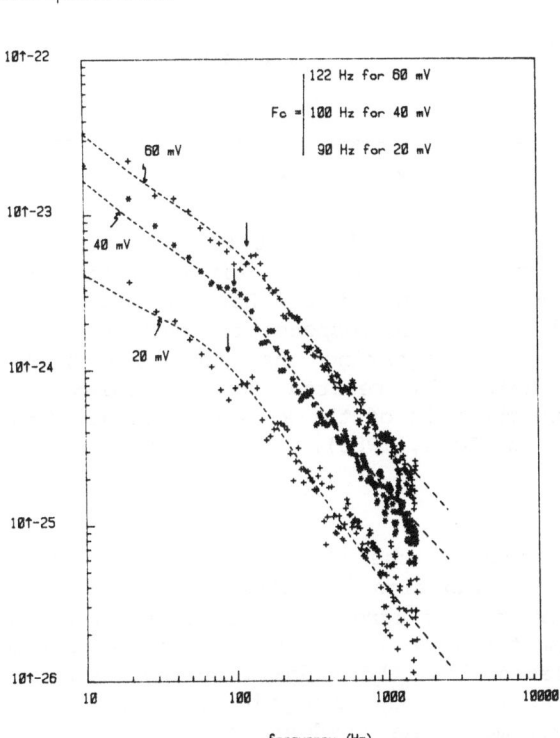

Fig.8   Family of fitted difference spectra of membrane noise
         from a patch of axonal membrane of Periplaneta americana
         under voltage-clamp conditions. Peaks corresponding to
         50 Hz (AC line) and its harmonics were eliminated and
         the corrected data smoothed once. Experimental points
         were then fitted with a combination of a 1/f and a
         lorentzian $(1/(1+(f/f_c)^2))$ components. The relative
         proportion of Lorentzian noise decreases with increased
         membrane depolarization whereas the corner frequency
         (f_c) increases. (From Pichon, Poussart and Lees,
         1983).

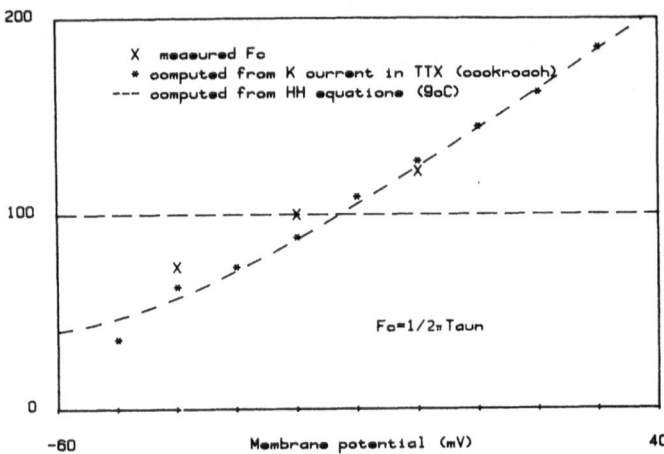

Fig.9   Measured (crosses) and computed (stars and interrupted
         line) corner frequencies against membrane potential.
         Data from fig.7 and 8 and Hodgkin and Huxley (1952).
         Note the good correspondence between measured and
         computed f_c, suggesting that the lorentzian component
         of the noise is related to the relaxation of the
         potassium channels of the axonal membrane. (From Pichon,
         Poussart and Lees, 1983).

A comparison between the measured values of f_c and those
predicted from the relaxation kinetics of the potassium system
($f_c = 1/2 \pi\tau_n$) shows a good agreement between the two sets of
data obtained from the same preparation. They also compare well
with those calculated from the Hodgkin Huxley equations for the
squid axon (fig.9). As for the synaptic channels (see below), it
is possible to estimate the conductance of a single potassium
channel from the power spectra. According to Pichon et al. (1983),
the single channel conductance of a potassium channel would
approximate 2.5 pS for small depolarizations. Division of the
maximum potassium conductance by this value yields a potassium
channel density of about 100 μm^{-2}.

In cockroach axons, g_{Na} may be described by the following
equation :

$$g_{Na} = \bar{g}_{Na} \, m^5 \, h \qquad\qquad\qquad (4.19)$$

Steady-state values of n,m and h for different potential
values have been calculated and are illustrated in fig.10. From
these data, it has been possible to reconstruct by calculation the
ionic currents corresponding to a step depolarization in a

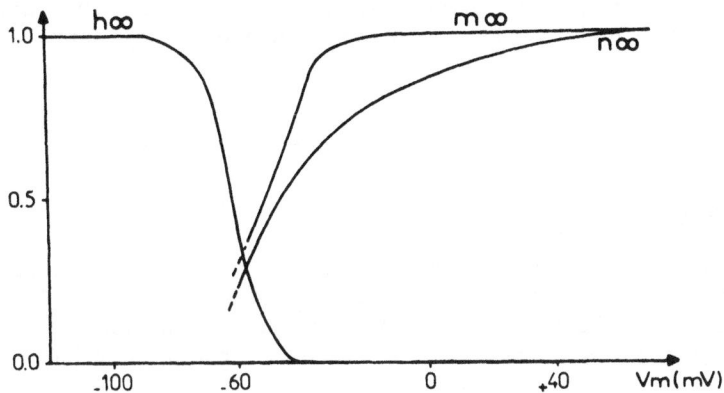

Fig.10   Steady-state values of m, n and h in a typical giant
            axon of the cockroach calculated from experimental
            measurements of sodium and potassium conductances and
            based on the assumption that $g_{Na} = \bar{g}_{Na} \, m^5 h$ and
            $g_K = \bar{g}_K n^3$. (From Pichon, 1974).

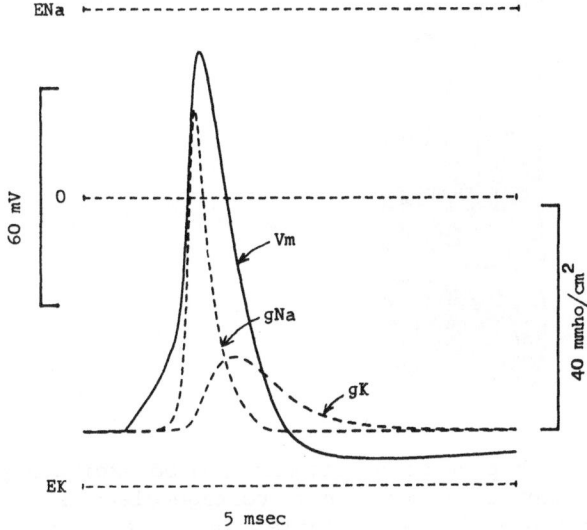

Fig.11   Computer reconstruction of a membrane action potential
            (continuous curve) and underlying conductance changes in
            a cockroach giant axon (interrupted curves). (From
            Pichon and Ashcroft, 1984).

voltage-clamped cockroach axon and to reconstruct the action
potential and the underlying conductance changes as illustrated in
fig.11.

IONIC BASIS OF SYNAPTIC ACTIVITY AT THE NEUROMUSCULAR JUNCTION

Excitatory synapses

There is good evidence that L-glutamate is the excitatory
transmitter at the insect neuromuscular junction (Pichon, 1974;
Usherwood and Cull-Candy, 1975, Pichon and Manaranche, 1984).
Voltage-clamp studies of the muscle fibres of the locust have

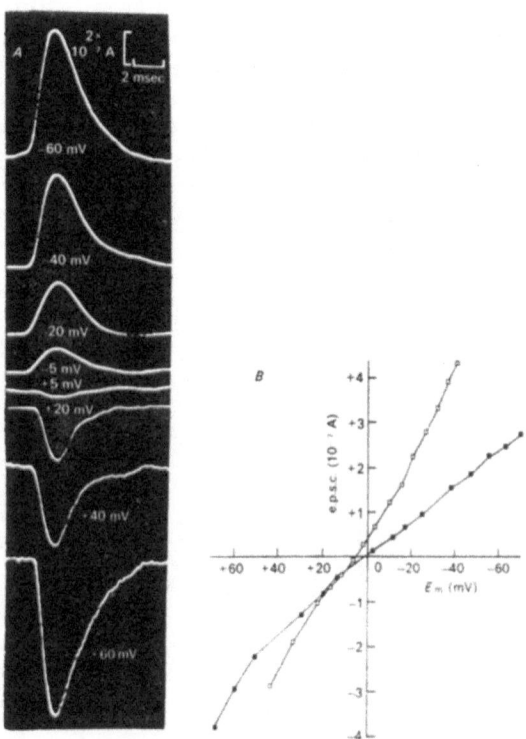

Fig.12   Relationship between neurally evoked excitatory
          postsynaptic current in a voltage-clamped locut muscle
          fibre and membrane potential level. A : time course of
          the epsc at the indicated potentials; B : current-
          voltage relationship for two epscs (□ and ■). Inward
          currents are shown as positive. (From Anwyl, 1977).

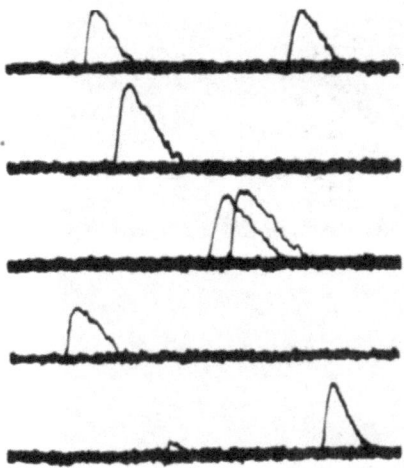

Fig.13   Miniature excitatory junctional currents (mejcs) from a
          locust   fibre   in   normal   saline   recorded   under
          voltage-clamp conditions. Membrane (= clamp) potential :
          -100 mV. Calibration bar represents 10 msec or 5 nA.
          Temperature : 22°C. (From Cull-Candy and Miledi, 1982).

shown that the neurally evoked excitatory postsynaptic current
(epsc) reverses sign arounbd 0 mV (Anwyl and Usherwood, 1974,
1975; Anwyl, 1977) (fig.12). This value (+ 3 mV) is very close
from that obtained for the glutamate current under similar
experimental conditions (+4 mV, Anwyl, 1977). Comparable results
have been obtained for the larval neuromuscular junction of
Drosophila (Jan and Jan, 1976). In both cases, the value of the
reversal potential suggests that both glutamate and the natural
transmitter increase the membrane permeability to all ions (Na^+,
K^+, Ca^{++}, Mg^{++} and Cl^-). The constant field equation (2)
was found to describe fairly well the ionic dependence of the
reversal potential. The P_{Na}/P_K value was found to be of 0.9 in
the locust (Anwyl, 1977) and of 1.3 in the fruit fly (Jan and Jan,
1976). In this last species, Mg^{++} ions were found to participate
to a large extent to the ejp, the permeability ratio P_{Mg}/P_K
being of 4.7. On the other hand, in both insect species, the
contribution of Cl^- ions to the ejp is thought to be
comparatively small.
    As at vertebrate neuromuscular junctions, small "spontaneous"
ejp have been observed in the absence of neural activity. The
corresponding currents, the miniature excitatory junctional
currents (mejcs) have been studied under voltage-clamp at the

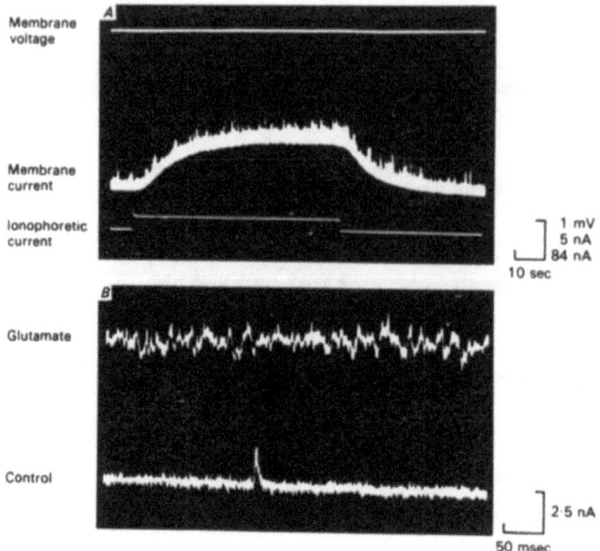

Fig.14   Membrane current noise at the locust neuromuscular
          junction in the absence (lower trace) and during
          ionophoretic application of glutamate (upper trace). The
          current noise record was taken during a mean membrane
          current of 32 nA. (From Anderson, Cull-Candy and Miledi,
          1978).

locust nerve-muscle junction (Cull-Candy and Miledi, 1982). At a
membrane potential of -80 mV and a temperature of 22°C, these
currents, some of which are illustrated in fig.13, were found to
have a mean amplitude of 2.34 nA and a decay time constant around
2.6 msec. This decay time which is only slightly longer than that
of the neurally evoked ejcs decreases exponentially with membrane
hyperpolarization. The equilibrium potential for transmitter
action at the locust neuromuscular junction is also close to 0 mV.
Each mejc is supposed to correspond to the release of a single
packet of transmitter and to open an average of 250 ionic channels
(Cull-Candy and Miledi, 1982).

    The properties of these synaptic ion channels have been
studied in the locust using noise analysis. As shown in fig.14,
the increase in conductance induced by glutamate is accompanied by
small current fluctuations. These fluctuations may be attributed
to the statistical variations in the number of ion channels in the
open state (Anderson, Cull-Candy and Miledi, 1978). This "noise"
has been analysed into its frequency components using the Fourier
analysis. The spectral density at different frequencies (S (f))
was found to fit well the theoretical (lorentzian) relationship.

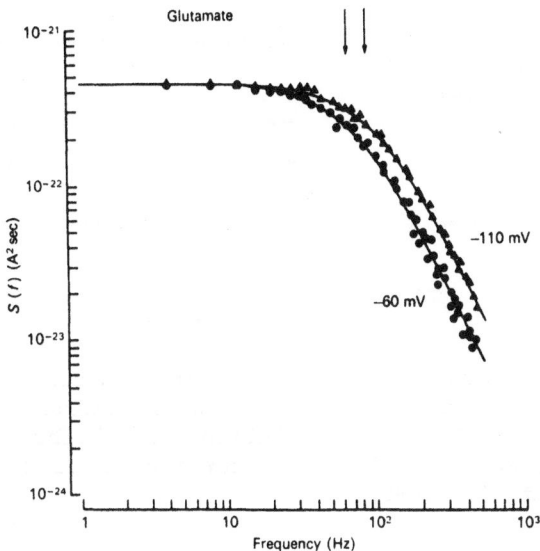

Fig.15   Power spectra of the noise induced by iontophoretic
          application of glutamate at the neuromusclar junction of
          the locust. The spectra were fitted with f_c = 64 Hz at
          -60 mV and f_c = 91 Hz at -110 mV. The -110 mV data and
          corresponding curve have been shifted along the vertical
          axis to allow comparison between curves. To get correct
          values for -110 mV, the ordinate values should be
          multiplied by 0.508. Temperature : 23°C. (From Anderson
          et al., 1978).

This fit is illustrated in fig. 15 for two membrane potential
values (-60 and -110 mV). The single channel conductance
calculated from the experimental curve is of about 125 mS. It is
independent of the membrane potential and of the amplitude of the
glutamate induced current. The channel life time which has a mean
value of 2.5 msec at -60 mV decreases exponentially with membrane
hyperpolarization (Anderson et al., 1978).
     The opening and closing of single glutamate activated ion
channels have been also studied directly using the so called
"patch-clamp" technique. In these experiments which were performed
on extrajunctional channels of the locust (Patlak, Gration and
Usherwood, 1979; Cull-Candy, Miledi and Parker, 1980), the muscle
fibre was voltage-clamped with two intracellular microelectrodes
and a patch electrode, filled with 100 µM L-glutamate in saline,
pressed against the surface of the muscle and used to measure the
current passing through the membrane under the patch. The muscle
was denervated to increase the number of extrajunctional D
receptors and treated with concanavalin A to reduce

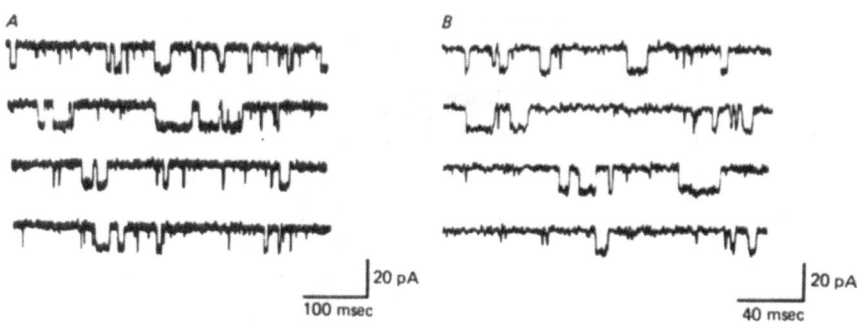

Fig.16   Patch-clamp recordings of glutamate induced single
           channel activity from two locust muscle fibres treated
           with ConA in Cl⁻ free solution. The fibres were
           voltage-clamped at -110 mV. Temperature : 22°C. (From
           Cull-Candy et al. (1980).

desensitization. All-or-nothing square current pulses were
obtained in response to glutamate, indicating that glutamate
activated ionic channels exist in either the open or the closed
state. This is illustrated in fig.16. The single channel
conductance of these extrajunctional channels approximated 150 pS,
i.e. a value similar to that extrapolated from noise at the
junctional area. It was not voltage-dependent. The lifetimes of
the channels were exponentially distributed with a mean value of
2.3 msec at 23°C and for V_m =-60 mV (Cull-Candy et al., 1980).
All these values are in good agreement with those obtained from
noise analysis on junctional channels. The single channel
conductance is the highest reported so far (the single channel
conductance at the frog neuromuscular junction is only about 25 pS
according to Neher and Sakmann (1976).

Inhibitory synapses

     Inhibition at the neuromuscular junction of insects have been
studied using conventional electrophysiological techniques by
Usherwood and Grundfest (1964, 1965), Grundfest and Usherwood
(1965) and Kerkut and Walker (1966, 1967). The pharmacological
properties of the extensor tibialis muscle of Romalea microptera
and Shistocerca gregaria were found to be remarkably similar to
those of crustacean muscle fibres studied by Boistel and Fatt
(1958) and Grundfest, Reuben and Rickles (1959). Stimulation of
the inhibitory nerve of Romalea gives rise to a hyperpolarizing
postsynaptic potential (ipsp). This potential is due to a
selective increase in the chloride permeability of the
postsynaptic membrane (it is inverted if proponiate is substituted
for chloride in the saline, Usherwood and Grundfest, 1965). Gamma

amino butyric acid (GABA) which is the putative inhibitory transmitter at the neuromuscular junction of insects slightly hyperpolarizes the membrane. This hyperpolarization is also due to an increased permeability of the membrane to chloride ions and its reversal potential approximates -67 mV (Usherwood and Grundfest, 1965).

Some extrajunctional receptors to glutamate (H receptors) also mediate a hyperpolarizing response which is due to an increase in chloride permeability (Cull-Candy and Usherwood, 1973; Cull-Candy, 1976).

ULTRASTRUCTURE AND IONIC REGULATION IN THE CENTRAL NERVOUS SYSTEM OF INSECTS.

This last subchapter considers the functioning of the nervous elements in their normal environment. This aspect is of specia importance in insects since (1) the axonal as well as the synaptic structures are isolated from the haemolymph by a complex of overlying tissues, (2) insect nerve cells function in a fluid environment that can differ substantially from that of the external medium, (3) the chemical composition of the brain microenvironment is actively controlled. THe main components of the nervous system together with the different categories of junctions which occur within the CNS will be described first. This description will be followed by a short development on various ionic problems related with the existence of the blood-brain barrier. More details will be found in Lane (1984) and Treherne (1984).

Ultrastructure of the brain

The various elements which constitute the CNS in insects are illustrated in fig.17 (from Lane and Treherne, 1980).

The CNS is separated from the outside medium by the (acellular) neural lamella (NL) which surrounds the specialized outer glial layer, the peripeurium (PN). Beneath these two layers, the axons (A) and nerve cell bodies (NCB) are surrounded by inner glial cells (G) which are spirally arranged around the axons. The mesaxon channels which spiral up to six or seven times around each giant axon (Treherne and Pichon, 1972) show occasional dilatations. Adjacent folds of inner glial cells and axons are associated by gap junctions (GJ) whereas the perineurial cells are closely apposed by tight junctions (TJ) and septate junctions (SJ). Hemidesmosomes (HD) occur at the periphery between peripeurial cells and the neural lamella. The glial lacunar system (S), which is reduced in size, contains tracheoles (T).

The respective roles of these various elements, which havve been observed using conventional electron microscopy as well as freeze fracture studies, are not yet completely elucidated.

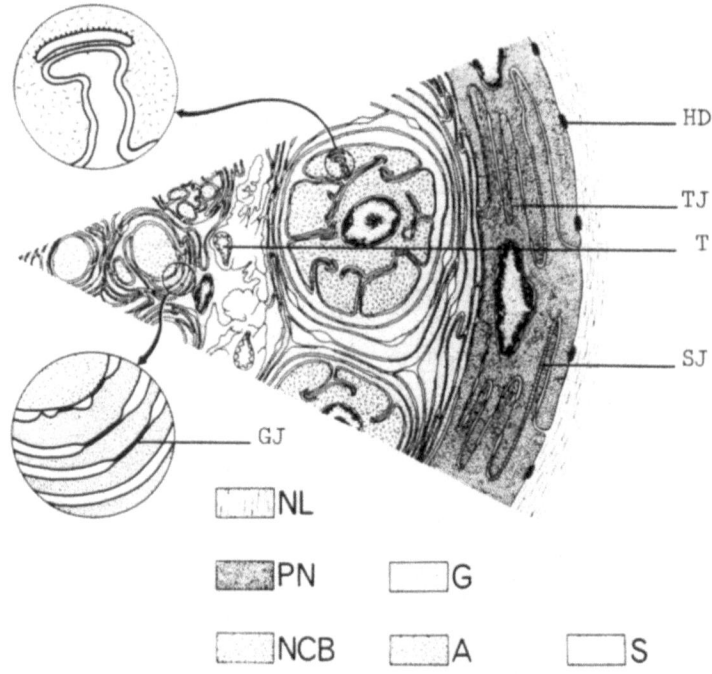

Fig.17   Schematic representation of the constitutive elements of
           the nervous system of insects (From Lane and Treherne,
           1980). For legend, see text.

According to Lane and Skaer (1980), tight junctions in the
perineurium would restrict movements from the haemolymph into the
extracellular fluid wherease gap junctions would ensure cell to
cell communication.

Blood brain barrier and ionic regulation in the CNS

     It is now well established that insects possess a well
developed blood brain barrier which protects the nerve elements
from abnormal concentrations of ions and toxic compounds in the
haemolymph (Treherne and Pichon, 1972; Abbott and Treherne, 1977;
Treherne, 1980, 1984; Treherne and Schofield, 1981).
     The presence of this peripheral barrier imposes some
physiological constraints on the underlying tissues : insect
neurones and axons are isolated in an extremely restricted
microenvironment. Short-term local homeostasis should be able to
cope with fluctuations in the ionic composition resulting from
neuronal activity such as increases in potassium concentrations or
sodium depletion. Furthermore, the distance between the nerve
membranes and the blood-brain interface which can exceed 200 µm

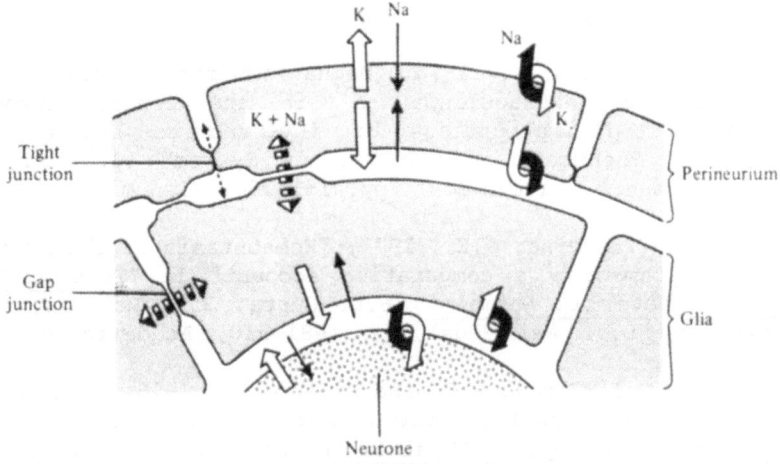

Fig.18   Dynamic model for ionic regulation in the CNS of the
           cockroach. Solid arrows indicate passive diffusion,
           dotted arrows intercellular diffusion through
           tight-junction, broken arrows intracellular diffusion
           across gap-junction and linked arrows active Na/K
           transport. A high Na^+ - low K^+ extracellular fluid
           is maintained by the pumps located on the inwardly
           facing perineurial membrane and the glial and neuronal
           membranes (From Treherne and Schofield, 1981).

represents physiological problems for the supply of nutrients.
    Ionic regulation has been studied almost exclusively on the
nerve cord of the cockroach. In this preparation, the combination
of several techniques have led to the construction of a dynamic
model illustrated fig.18 which takes into account most
observations on this preparation. Sodium transport inhibitors such
as ouabain, stophanthidine and ethacrynic acid were found to
modify ionic fluxes, providing evidence for the dynamic nature of
the blood-brain-barrier (see Treherne, 1984). Recent experiments
on the drone retina (Cole and Tsacopoulos, 1979, 1981) indicate
that regulation may occur via spacial buffering as suggested by
Kuffler, Nicholls and Orkand (1966) for amphibia and Abbott and
Pichon (1976) for crayfishes.

ACKNOWLEDGEMENTS

Thanks are due to Dr. J.E. Treherne and Dr. N.J. Lane for helpful
suggestions and to Mrs D. Chaslard for typing the manuscript. Part
of the work presented here was supported by the C.N.R.S. and the
D.G.R.S.T. (grant n°79-7-1068).

REFERENCES

Abbott, N.J. and Pichon, Y., 1976, "Mechanisms for the passive regulation of extracellular K^+ in the central nervous system : the implications of invertebrate studies" in "Transport Phenomena in the Nervous System" (G. Levi, L. Battistin and A. Lajtha eds), pp. 151-164, Plenum Press : New York.

Abbott, N.J. and Treherne, J.E., 1977, "Homeostasis of the brain microenvironment : a comparative account" in "Transport of Ions and Water in Animals" (B.L. Gupta, R.B. Moreton, J.L. Oshman and B.J. Wall eds), pp. 481-510, Academic Press : London.

Anderson, C.R., Cull-Candy, S.G. and Miledi, R., 1978, Glutamate current noise : postsynaptic channel kinetics investigated under voltage clamp. J. Physiol., London, 282 : 219-242.

Anwyl, R. and Usherwood, P.N.R., 1975, The ionic permeability changes caused by the excitatory transmitter at the insect neuromuscular junction. J. Physiol., London, 249 : 24-25P.

Boistel, J. and Coraboeuf, E., 1958, Rôle joué par les ions sodium dans la genèse de l'activité électrique du tissu nerveux d'insecte. C.R. Acad. Sci., Paris, 247 : 1781-1783.

Boistel, J. and Fatt, P., 1958, Membrane permeability change during inhibiotry transmitter action in crustacean muscle. J. Physiol., London, 144 : 176-191.

Callec, J.J., 1984, "Synaptic transmission" in "Comprehensive Insect Physiology, Biochemistry and Pharmacology" (G.A. Kerkut and L.I. Gilbert eds), vol.5, Pergamon Press : London (in press).

Callec, J.J. and Boistel, J., 1966, Etude de divers types d'activités électriques enregistrées par microélectrodes capillaires au niveau du dernier ganglion abdominal de Periplaneta americana L. C.R. Soc. Biol., 160 : 1943-1947.

Cole, J.A. and Tsacoupoulos, M., 1979, K^+ activity in photoreceptors, glial cells and extracellular space in the drone retina. J. Physiol., London, 290 : 525-549.

Cole, J.A. and Tsacoupoulos, M., 1981, Ionic and possible metabolic interactions between sensory neurones and glial cells in the retina of the honey-bee drone. J. Exp. Biol., 95 : 75-92.

Cull-Candy, S.G., 1976, Two types of extrajunctional L-glutamate receptors in locust muscle. J. Physiol., London, 255 : 449-464.

Cull-Candy, S.G. and Miledi, R., 1982, Properties of miniature excitatory junctional currents at the locust nerve-muscle junction. J. Physiol., London, 326 : 527-551.

Cull-Candy, S.G. and Usherwood, P.N.R., 1973, Two populations of glutamate receptors on locust muscle fibres. Nature New Biol., 246 : 62-64.

Cull-Candy, S.G., Miledi, R. and Parker, I., 1980, Single glutamate-activated channels recorded from locust muscle fibres with perfused patch-clamp electrodes. J. Physiol., London, 321 : 195-210.

Fatt, P. and Katz, B., 1951, An analysis of the end-plate potential recorded with an intracellular electrode. J. Physiol., London, 115 : 320-370.

Goldman, D.E., 1943, Potential, impedance and rectification in membranes. J. gen. Physiol., 27 : 37-60.

Goodman, C.S. and Heitler, W.J., 1979, Electrical properties of insect neurones with spiking and non spiking somata : normal, axotomized, and colchicine treated neurones, J. exp. Biol., 83 : 95-121.

Grundfest, H., Reuben, P. and Rickles, W.H., 1959, The electrophysiology and pharmacology of lobster neuromuscular synapse. J. gen. Physiol., 42 : 1301-1323.

Gwilliam, G.F. and Burrows, M., 1980, Electrical characteristics of the membrane of an identified insect motor neurone. J.exp.Biol., 86 : 49-61.

Hodgkin, A.L. and Katz, B., 1949, The effect of sodium ions on the electrical activity of the giant axon of the squid, J. Physiol., London, 108 : 37-77.

Hodgkin, A.L. and Huxley, A.F., 1952, Quantitative description of membrane current and its application to conduction and excitation in nerve. J. Physiol., London, 117 : 500-544.

Hodgkin, A.L., Huxley, A.F. and Katz, B., 1952, Measurement of current-voltage relations in the membrane of the giant axon of Loligo. J. Physiol., London, 116 : 424-448.

Hoyle, G., 1953, Potassium ions and insect nerve muscle. J. exp. Biol., 30 : 121-135.

Jan, L.Y. and Jan, Y.N., 1976, L-glutamate as an excitatory transmitter at the Drosophila larval neuromuscular junction. J. Physiol.,London, 262 : 215-236.

Jego, P., Callec, J.J., Pichon, Y. and Boistel, J., 1970, Etude électrophysiologique de corps cellulaires excitables du VIème ganglion abdominal de Periplaneta americana; Aspects électriques et ioniques. C.R. Soc. Biol., 164 : 893-904.

Kerkut, G.A., Pitman, R.M. and Walker, R.J., 1968, Electrical activity in insect nerve cell bodies. Life Science, 7 : 605-608.

Kerkut, G.A. and Walker, R.J., 1966, The effect of L-glutamate, acetylcholine and -aminopbutyric acid on the end plate potential and contractions of the coxal muscle of the cockroach Periplaneta americana. Comp. Biochem. Physiol., 17 : 435-454.

Kerkut, G.A. and Walker, R.J., 1967, The effect of iontophoretic injection of L-glutamic acid and aminobutyric acid on the miniature endplate potentials and contractions of the coxal muscles of the cockroach Periplaneta americana. Comp. biochem. Physiol., 20 : 999-1003.

Kuffler, S.W., Nicholls, J.G. and Orkand, R.K., 1966,
    Physiological properties of glial cells in the central
    nervous system of amphibia. J. Neurophysiol., 29 : 768-787.
Lane, N.J., 1984, "Structure of components of the nervous system"
    in "Comprehensive Insect Physiology, Biochemistry and
    Pharmacology (G.A. Kerkut and L.I. Gilbert eds), vol. 5,
    Pergamon Press : London (in press).
Lane, N.J. and Skaer, H. leB., 1980, Intercellular junctions in
    insect tissues. Adv. Insect Physiol., 15, 35-213.
Lane, N.J. and Treherne, J.E., 1980, "Functional Organization of
    Arthropod Neuroglia" in "Insect Biology in the Future VBW
    80".M.Locke and D.S. Smith eds), pp 765-795, Academic Press :
    London.
Monticelli, G. and Depretto G., 1979, Electrophysiological
    characteristics of Bombyx mori L. ventral nerve cord (effects
    of sodium and potassium on the membrane potential).
    Experientia, 35 : 62-64.
Narahashi, T., 1965, "The Physiology of Insect Axons" in The
    Physiology of the Insect Central Nervous System" (Treherne
    J.E. and Beament J.W.L. eds.), p. 1-22, Academic Press :
    London.
Narahashi, T., 1966, Dependence of excitability of cockroach giant
    axons on external divalent cations. Comp. Biochem. Physiol.,
    19 : 759.
Neher, E. and Sakmann, B., 1976, Single-channel currents recorded
    from membrane of denervated frog muscle fibres. Nature, 260 :
    799-802.
Orchard, I., 1976, Calcium dependent action potentials in a
    peripheral neurosecretory cell of the stick insect. J. comp.
    Physiol., 112 : 95-102.
Orchard, I. and Finlayson, L.H., 1976, The electrical activity of
    mechanoreceptive and neurosecretory neurons of the stick
    insect, Carausius morosus. J. comp. Physiol., 107 : 327-338.
Orchard, I. and Finlayson, L.H., 1977, Electrical properties of
    identified neurosecretory cells in the stick insect. Comp.
    Biochem. Physiol., 58 : 87-91.
Patlak, J.B., Gration, K.A.F. and Usherwood, P.Ṅ.R., 1979, Single
    glutamate activated channels in locust muscle. Nature, 278 :
    643-465.
Pichon, Y., 1967, Application de la technique du voltage imposé à
    l'étude de la fibre nerveuse isolée d'insecte. J. Physiol.,
    Paris, 9 : 282.
Pichon, Y., 1968, Nature des courants membranaires dans une fibre
    nerveuse d'insecte : l'axone géant de Periplaneta americana
    C.R. Soc. Biol., 162 :2233-2240.
Pichon, Y., 1969, Aspects électriques et ioniques du
    fonctionnement nerveux chez les insectes. Cas particulier de
    la chaîne nerveuse abdominale d'une blatte, Periplaneta
    americana L. Thèse de Doctorat-ès-Sciences, Rennes.

Pichon, Y., 1974, "Axonal Conduction in Insects", in "Insect Neurobiology" (J.E. Treherne ed.) pp 73-117; North Holland : Amsterdam.

Pichon, Y. and Ashcroft, F.M., 1984, "Nerve and Muscle : Electrical Activity" in "Comprehensive Insect Physiology, Biochemistry and Pharmacology, (G.A. Kerkut and L.I. Gilbert eds), vol.5, Pergamon Press : London (in press).

Pichon, Y. and Boistel, J., 1966, Application aux fibres géantes de blattes (Periplaneta americana L. et Blabera craniifer Bürm) d'une technique permettant l'introduction d'une microélectrode dans le tissu nerveux sans résection préalable de la gaine. J. Physiol., Paris, 58 : 592.

Pichon, Y. and Boistel, J., 1967a, Current-voltage relations in the isolated giant axons of the cockroach under voltage-clamp conditions. J. exp. Biol., 47 : 343-356.

Pichon, Y. and Boistel, J., 1967b, Microelectrode study of the resting and action potentials of the cockroach giant axons with special reference to the role played by the nerve sheath. J. exp. Biol., 47 : 357-373.

Pichon, Y. and Manaranche, R., 1984, "Biochemistry of the Nervous System" in "Comprehensive Insect Physiology, Biochemistry and Pharmacology" (G.A. Kerkut and L.I. Gilbert eds), vol.10, Pergamon Press London (in press).

Pichon, Y., Poussart, D. and Lees, G.V., 1983, "Membrane Ionic Currents, Current Noise and Admittance in Isolated Cockroach Axons" in "Structure and Function in Excitable cells" (Chang D.C., Tasaki I., Adelman W.J. and Leuchtag H.R. eds), Plenum Press : New York (in press).

Pichon, Y., Sattelle, D.B. and Lane, N.J., 1972, Conduction processes in the nerve cord of the moth, Manduca sexta in relation to its ultrastructure and haemolymph ionic composition. J. exp. Biol., 56 : 717-734.

Pichon, Y. and Treherne, J.E., 1973, An electrophysiological study of the sodium and potassium permeabilities of insect peripheral nerves. J.exp.Biol., 59 : 447-461.

Pitman, R.M., 1975a, The ionic dependence of action potentials induced by colchicine in an insect motorneurone cell body. J. Physiol., London, 247 : 511-520.

Pitman, R.M., 1975b, Calcium-dependent action potentials in the cell body of an insect motorneurone. J. Physiol., London, 247 : 62-63 P.

Pitman, R.M., 1979, Intracellular citrate and externally applied tetraethylammonium ions produce calcium-dependent action potential in an insect motoneurone cell body. J. Physiol., London 291 : 327-337.

Pitman, R.M., 1984, "Nervous System" in "Comprehensive Insect Physiology, Biochemistry and Pharmacology" (G.A. Kerkut and L.I. Gilbert eds-, vol.11, Pergamon Press : London (in press).

Plank, M., 1890, Über die Potentialdifferenz zwischen zwei
    verdünnten Lösungen binärer ELektrolyte. Ann. Physik. Chem,
    40 : 561-576.
Tanouye, M.A., Ferrus, A. and Fujita, S.C., 1981, Abnormal action
    potentials associated with the Shaker complex locus of
    Drosophila. Proc. Nat. Acad. Sci., U.S.A., 78 : 6548-6552.
Treherne, J.E., 1984, "Blood-Brain-Barrier" in "Comprehensive
    Insect Physiology, Biochemistry and Pharmacology" (G.A.
    Kerkut and L.I. Gilbert eds), vol.5, Pergamon Press : London
    (in press).
Treherne, J.E. and Maddrell, S.H.P., 1967, Membrane potentials in
    the central nervous system of the stick insect, Carausius
    morosus. J.exp. Biol., 46 : 413-421.
Treherne, J.E. and Pichon, Y., 1972, The insect
    blood-brain-barrier. Adv. Insect.Physiol., 9 : 257-313.
Treherne, J.E. and Schofield, P.K., 1981, Mechanisms of ionic
    homeostatis in the central nervous system of an insect. J.
    Exp. Biol, 95 : 61-73.
Usherwood, P.N.R. and Cull-Candy, S.G., 1975, "Pharmacology of
    somatic nerve-muscle synapses" in Insect Muscle" (P.N.R.
    Usherwood ed.) pp 207-281, Academic Press : London.
Usherwood, P.N.R. and Grundfest, H., 1964, Inhibitory postsynaptic
    potentials in grasshopper muscle. Science, 143 : 817-818.
Usherwood, P.N.R. and Grundfest, H., 1965, Peripheral inhibition
    in skeletal muscles of insects. J. Neurophysiol.,
    28 : 497-518.
Yamasaki, T. and Narahashi, T., 1959, The effects of potassium and
    sodium ions on the resting and action potentials of the
    cockroach giant axon. J. Insect Physiol., 3 : 146-158.

PHARMACOLOGY OF NEUROTRANSMITTER RECEPTORS AND

ION CHANNELS IN THE INSECT CNS

David B. Sattelle

Agricultural Research Council Unit of
Insect Neurophysiology and Pharmacology
Department of Zoology, Downing Street
Cambridge CB2 3EJ, U.K.

INTRODUCTION

The study of electrical and chemical excitation of insect neurones is of comparative and evolutionary interest (Pichon, 1974a,b; Callec, 1974) and is also an essential prerequisite to an understanding of the mechanism of action of insecticidally-active molecules, many of which act on insect neurones (Sattelle, 1978; Pelhate & Sattelle, 1982; Eldefrawi et al., 1982). The economic importance of this group of invertebrate animals is one stimulus to such studies, often almost overshadowing the unique advantages insects offer for fundamental investigations in neurobiology. These include the use of the growing numbers of identifiable neurones and pathways in relating putative neurotransmitter molecules and neurohormones to specific functions. Insects also provide experimental material well-suited to explore genetic (Dudai, 1982; Hall et al., 1982) and developmental (Goodman & Spitzer, 1980; Hildebrand, 1980) studies on neurones and the molecular components involved in nerve impulse conduction and communication between cells. Most striking have been the recent advances in insect neurochemical pharmacology (cf. Sattelle et al., 1980; Hildebrand, 1982).

A wide range of synthetic molecules and toxins of both plant and animal origins have now been shown to exert specific actions on insect excitable tissues. Amongst these chemically diverse molecules are many which act on the voltage-dependent sodium and potassium channels involved in action potential generation. Considerable progress in our understanding of the physiology and pharmacology of the nerve axon membrane has resulted from experiments on the giant axons of invertebrate animals (Narahashi,

51

CNS (PERIPLANETA AMERICANA)   GANGLIA

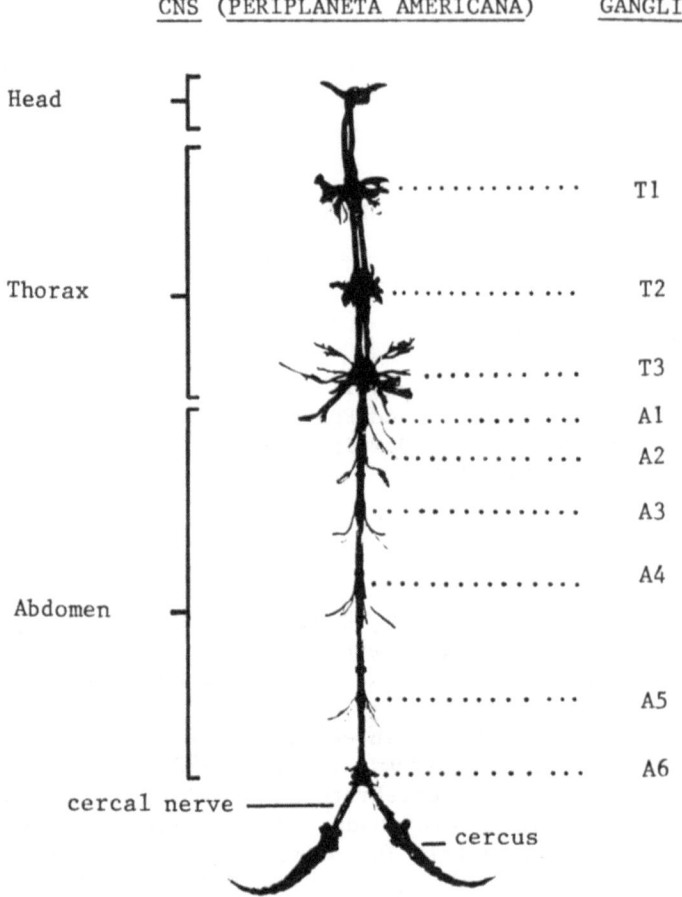

Fig. 1. The isolated central nervous system of the cockroach *Periplaneta americana* showing ganglia of the head thoracic (T1-T3) and abdominal (A1-A6) regions of the animal. The cercal nerves and cerci are also shown.

1974, 1975). In recent years, advances in methods for recording the electrical properties of insect giant axons have revealed their value in determining the molecular mechanisms of novel neuroactive agents including insecticides. The majority of investigations on insect axons have been performed using the giant interganglionic interneurones, the axons of which are located in the connectives linking the abdominal ganglia in the central nervous system of the cockroach *Periplaneta americana* (Fig. 1). By means of the oil-gap, single-fibre recording technique external recording electrodes are

used to voltage-clamp (space-clamp) a localized area of axon
membrane, the remainder of the axon being bathed in mineral oil
(Pichon & Boistel, 1966, 1967). Here we assess recent findings
using aminopyridines and related compounds, new synthetic
insecticides, several alkaloids of plant origin and finally venoms,
and toxins of animal origin.

    Chemical signalling is one mechanism by which neurones
communicate. By means of precisely regulated release of chemical
messenger molecules, nerve cells can interact with each other and
with muscle and gland cells (Hall et al., 1975). Information is
conveyed rapidly between cells at specific sites (chemical
synapses) by means of neurotransmitter molecules which are
discharged into the intercellular spaces separating the presynaptic
neurone and the postsynaptic (follower) cell. These primary
messenger molecules are particularly effective on specialized,
highly-sensitive, regions of the postsynaptic cell. They may
trigger in the postsynaptic cell direct changes in ion conductance,
leading to modifications of membrane potential or spike discharge.
Alternatively they may induce biochemical changes including enzyme
activation, leading to changes in the level of an intracellular
secondary messenger molecule which in turn may result in ion
permeability changes. Nevertheless, neurotransmitters represent
only a narrow band in the broad spectrum of intercellular chemical
communication mechanisms. Other primary chemical messengers are
also employed by nerve cells to mediate a range of longer-term
cellular interactions (Shain & Carpenter, 1981) and these types of
communication will be widely discussed in this symposium.
Nevertheless synaptic transmission, though a highly specialized,
often rapid, form of communication, is the best understood of all
the types of chemical signalling studied to date. For this reason,
and since an increasing number of the molecular constituents of
pre- and postsynaptic membranes, particularly receptors, have been
characterized (and in some cases isolated and purified), it merits
consideration in a volume largely devoted to neurohormones and
neuromodulator substances. The bulk of the work to date on insect
central synapses has focussed on the neurotransmitter candidates
acetylcholine and γ-aminobutyric acid (GABA) and this is
necessarily reflected in the present survey.

A. PHARMACOLOGY OF VOLTAGE-DEPENDENT AXONAL POTASSIUM AND SODIUM
CHANNELS

    Ion channels in cell membranes can be categorized on the basis
of whether or not they are voltage-dependent. Voltage-dependent
channels are primarily those involved in action potential
generation, mediating specific conductances to sodium, potassium
and calcium. The bulk of the data reviewed here was obtained on
isolated axons of giant interneurones (Fig. 2) of the cockroach
Periplaneta americana. The axons of these cells which extend from

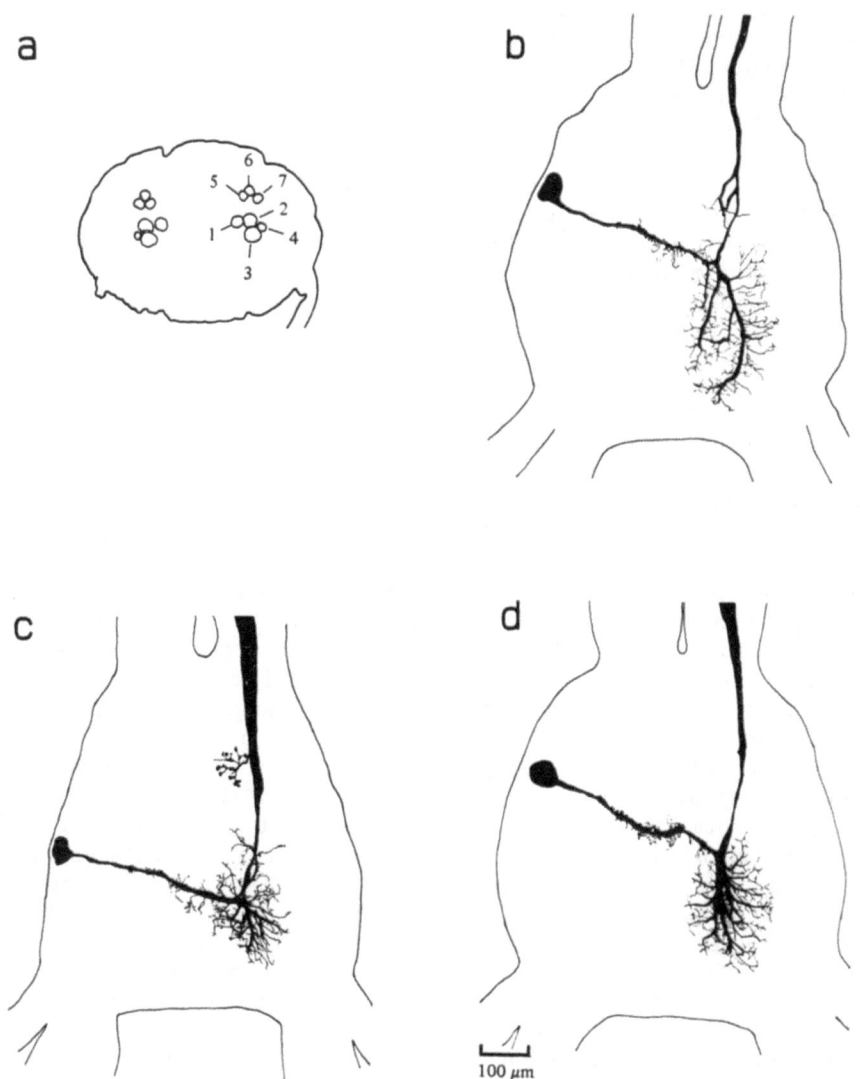

Fig.2.   Camera lucida drawings of cobalt-filled giant interneurones of the cockroach _Periplaneta americana_.   (a) Section through the fifth abdominal ganglion showing relative positions of axons of giant interneurones 1-7.   (b-d) Three giant interneurones in the sixth abdominal ganglion, each showing distinctive morphology: (b) GI 1; (c) GI 2; (d) GI 3.   Modified from Harrow, Hue, Pelhate & Sattelle (1980).

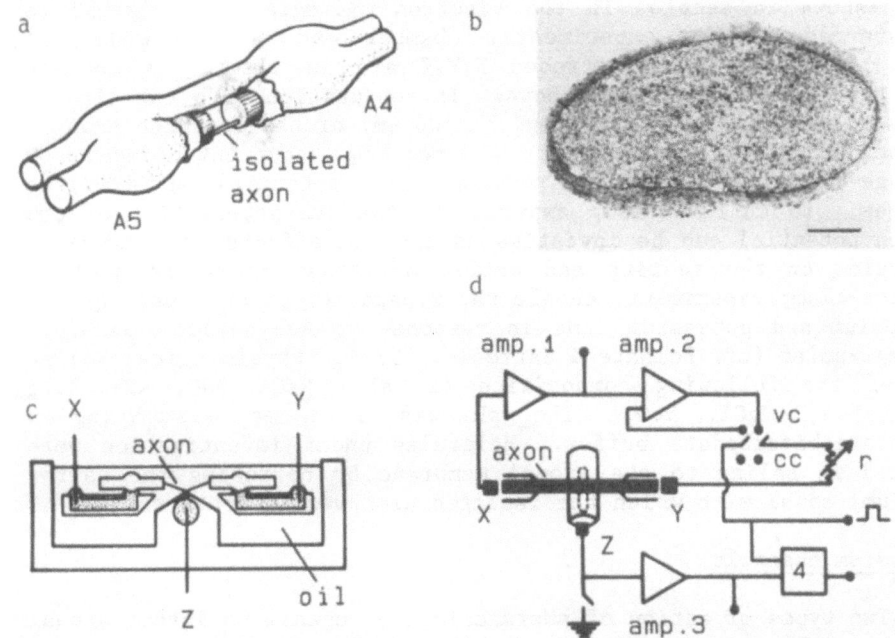

Fig. 3. The isolated cockroach giant axon preparation used in current-clamp and voltage-clamp experiments. (a) Schematic representation of the dissection of the axon — one of the ventrally located giant interneurone axons from a desheathed connective linking the fourth (A4) and fifth (A5) abdominal ganglia. (b) Low-power composite electron micrograph (scale bar 10 μm) prepared from transverse thin sections of an isolated giant interneurone axon. Only a thin layer of glial material remains adhering to the axonal membrane. (c) Sectional view of the experimental chamber used for oil-gap, single-fibre experiments and (d) the associated electronic apparatus. X, Y and Z are Ag–AgCl electrodes. Z is connected to the perfusion canal containing saline or test solution. Amplifier 1 is of high-input impedance; amplifier 2 is a high gain, differential amplifier; amplifier 3 is a current to voltage converter. The device labelled 4 is an analogue compensator for leakage and fast and slow capacity currents. Under current-clamp conditions (cc), electrode Z is grounded and the value of r is maximum (2.2 MΩ); under voltage-clamp, r is annulled and the switches are in position vc. Modified from Pelhate & Sattelle (1982).

the terminal abdominal ganglion to the thoracic and head ganglia reach diameters of 40–50 μm in the connectives linking the abdominal ganglia. A short length of a single giant axon (Fig. 3a) is isolated from connectives linking the fifth (A5) and fourth (A4) abdominal ganglia, using finely-sharpened needles. The isolated axon, shown in section in the electron micrograph in Fig. 3b is transferred to the experimental chamber and submerged in oil (Fig. 3c). External electrodes X,Y,Z are in electrical contact with the axon at positions shown. In conjunction with the circuit shown in Fig. 3d, the short length (100 μm) of the isolated axon in contact with the perfusion channel was current-clamped or voltage-clamped to record, respectively, potential and current changes. In current-clamp experiments the characteristics of the action potential can be investigated and the effects of a range of molecules on the resting and active membrane can be determined. Voltage-clamp experiments enable the separation of currents carried by sodium and potassium ions in response to the application of a voltage-pulse (cf. Pelhate & Sattelle, 1982). Physiological saline was of the following composition (in mM): NaCl, 200; KCl, 3.1; $CaCl_2$, 5.4; $MgCl_2$, 5.0. The pH was held at 7.2 using a phosphate-bicarbonate buffer. Molecules under investigation were applied in saline to the axonal membrane by perfusing the narrow (100 μm) canal with which the isolated axon was in contact.

Potassium channels

Two types of action of pharmacological agents on insect axonal potassium channels can readily be distinguished: potassium channel blocking agents (4-aminopyridine (4-AP) and related compounds), and potassium channel modifiers which induce inactivation (for example 9-aminoacridine (9-AA), atropine and strychnine). Whereas the aminopyridines appear to be specific to potassium channels, the actions of the potassium channel modifiers are not confined to a single site of action.

At micromolar concentrations, 4-AP prolonged the action potential by a slowing of the repolarizing phase without any change in resting potential, though sometimes a small increase in the amplitude of the action potential was detected (Pelhate et al., 1972, 1974a). The increase in excitability is reflected in the striking decrease in the intensity of the current needed to trigger the action potential (Fig. 4). These actions pointed to a decrease in potassium conductance induced by 4-AP. This view was supported by the results of voltage-clamp experiments (Pelhate et al., 1974b; Pelhate & Pichon, 1974). The specific block of potassium current resulting from the application of 4-AP is shown in Fig. 5a,b. Without any change in the peak sodium current, 4-AP blocked all the outward potassium current following a voltage-clamp pulse to −10mV. As shown in Fig. 5c, inwardly directed potassium currents were also blocked by 4-AP (cf. also Pelhate et al., 1975). Specific block of potassium currents was partly reversible and though inhibition was noted over a wide range of membrane potentials, the degree of suppression decreased as the membrane was clamped to more positive potentials (Fig. 5d,e,f). Thus 4-AP, the axonal actions of which were first described in insects, has emerged as a major

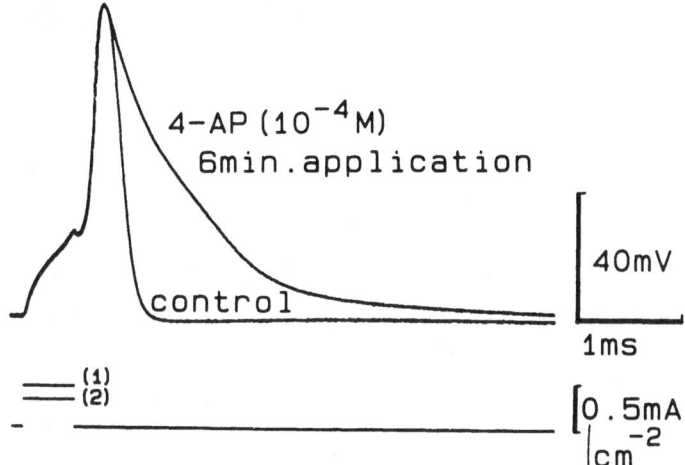

Fig. 4. Effects of 4-aminopyridine (4-AP) on the axonal action
potential of a cockroach giant interneurone recorded under
current-clamp conditions. After a 6 min application of 1.0 x
10^{-4}M 4-AP, the applied current required to trigger an action
potential was reduced (2 in the lower trace) by 37% compared to the
current needed in normal saline (1 in lower trace). The action
potential amplitude was unchanged (upper trace) but the
repolarizing (recovery) phase was drastically slowed, resulting in
a 3-fold increase in the duration of the action potential after
4-AP treatment. From Pelhate & Sattelle (1982).

pharmacological tool, though one limitation is the slowing of the
rate of rise of the potassium current observed in the presence of
low concentrations (1.0 x 10^{-6}M - 5.0 x 10^{-5}M) of 4-AP. As
shown for axons of squid (Meves & Pichon, 1974) and cockroach
(Pichon et al., 1982), the rise-time can be accelerated, however,
by the application of a burst of long duration depolarizing pulses.
Meves & Pichon (1977a,b) propose that the effects of repetitive
pulsing and the voltage-dependence of the 4-AP block can be
explained if 4-AP molecules are displaced from their blocking sites
during the pulse and slowly rebound afterwards. Recently Lees et
al. (1981) and Pichon et al. (1982) showed that 4-AP reduced the
membrane current noise recorded from the cockroach axon in a manner
indicating that the molecule can bind to closed potassium channels.

Dose-response curves have been constructed for the actions of
various aminopyridines and derivative molecules on the potassium

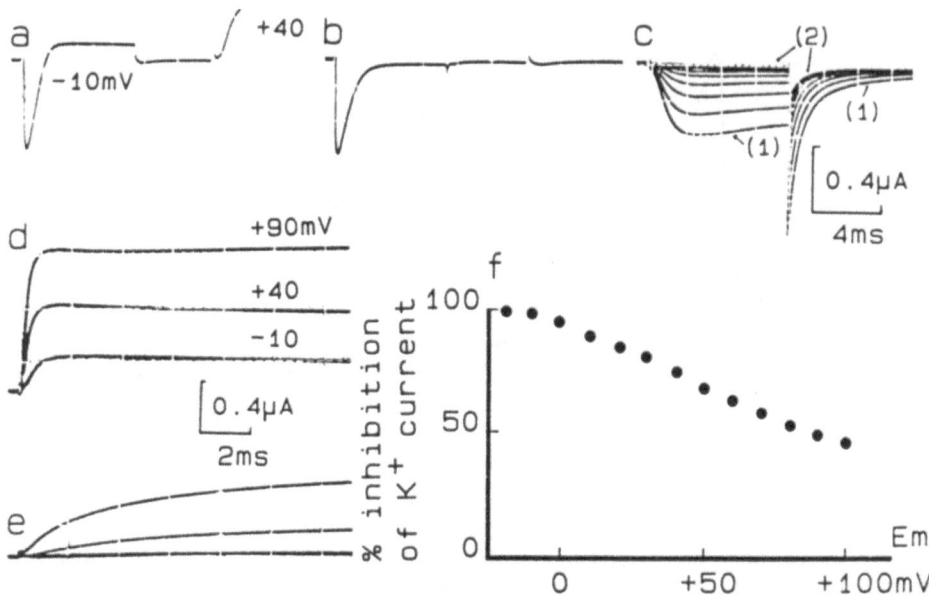

Fig. 5. Actions of 4-aminopyridine (4-AP) on ionic currents in voltage-clamped cockroach giant axons. (a) and (b) Ionic currents (after leak current corrections) corresponding to successive voltage-clamp pulses from E_h (holding potential) = -60 mV to E_m (membrane potential) = -10 mV (first pulse) and E_m = +40 mV (second pulse): (a) in normal saline; (b) after 6 min of application of 4-AP. The outward (potassium) currents are almost completely suppressed without any corresponding change in the peak inward (sodium) current. (c) In this experiment the axon is bathed in a high potassium (200 mM K^+) saline. A voltage-clamp pulse to E_m = -10 mV results in the development of an inward potassium current followed by an inward tail current. These potassium currents are rapidly reduced in the presence of 2.0×10^{-4} M 4-AP. (d) and (e) Potassium currents recorded in the presence of a 1.0×10^{-6} M concentration of the sodium channel blocking agent tetrodotoxin (TTX), for three different voltage-clamp pulses, before (d) and after (e) treatment with 1.0×10^{-3} M 4-AP. (f) Plot showing the membrane potential dependence of the inhibition of the potassium current by 4-AP (1.0×10^{-4} M). This shows a reduced inhibition for high (positive) values of membrane potential (voltage-dependent block). From Pelhate & Sattelle (1982).

conductance of cockroach axons. The results are summarized in Table 1. Whereas the pyridine nucleus was practically inactive, 4-AP and its isomers together with the derivative molecule 3-4-diaminopyridine (3-4-di-AP) depressed potassium currents at micromolar concentrations. As shown in Table 1, 4-aminoquinoline

Table 1. Inhibition of potassium conductance in cockroach axons by aminopyridines and their derivatives.

Molecule	K_d(M)	n
3-4-diaminopyridine (3-4-di-AP)	1.8×10^{-5}	19
4-aminopyridine (4-AP)	3.2×10^{-5}	40
3-aminopyridine (3-AP)	5.0×10^{-5}	12
2-aminopyridine (2-AP)	2.0×10^{-4}	12
9-aminoacridine (9-AA)	2.7×10^{-4}	31
4-aminoquinoline (4-AQ)	3.7×10^{-4}	14

Potassium currents were obtained in response to voltage-clamp pulses (12 ms in duration, applied every 10 s) to E_m = +40 mV. Currents were measured at a range of drug concentrations. The number of responses used to construct each dose-response curve is indicated (n). The apparent dissociation constant (K_d) was estimated as the concentration required to suppress by 50% the amplitude of the steady-state, outwardly - directed, potassium current.

(4-AQ) was also a potent potassium channel blocking agent but was less selective for potassium channels (cf. Pelhate et al., 1982; Pelhate & Sattelle, 1982). Another derivative 9-AA also reduced both sodium and potassium conductance (Pelhate et al., 1976). Its effects on the potassium conductance can be explained by an inactivation of the potassium channels following an apparently normal opening.

Thus at least three factors are involved in the blocking of potassium channels by the aminopyridines and derivative molecules: (a) lipid solubility; (b) the presence of a positively charged amino group ($-NH_3^+$); (c) molecular size. By analogy with the hypothesis of Armstrong (1971) to account for the actions of tetraethylammonium and related ions on the squid axon, the potassium channel blocking actions of aminopyridines and their derivatives could be attributed to the presence of the positively charged amino group. It is envisaged that this hydrophilic group is located in the wider (innermost) part of the channel where it would compete with potassium ions preventing them crossing the more restricted (outermost) part of the channel. The aminopyridine molecule, being small and highly lipid soluble, could be displaced slowly from the potassium channel as the membrane is depolarized resulting in reduction of potassium channel block. The smaller molecules of those tested such as 3-4-di-AP and aminopyridine isomers are selective for potassium channels whereas the larger molecules (4-AQ and 9-AA) also partially inhibit sodium channels. Thus these potassium channel probes, some of which were first investigated in insects, have added to our understanding of nerve membrane ion conductance mechanisms.

Sodium channels

At least three distinct categories of action are discernible amongst the wide range of pharmacological agents tested on insect axonal sodium channels. Thus, sodium channel blocking agents (tetrodotoxin, saxitoxin), sodium channel modifiers (DDT, pyrethroids, aconitine, veratridine), and sodium channel inactivation inhibitors (sea anemone toxins, together with mammal toxin II and insect toxin from the venom of the scorpion Androctonus australis) have been detected in experiments on cockroach axons. Whereas the axonal actions of tetrodotoxin (TTX), saxitoxin (STX) pyrethroids, purified Anemonia sulcata toxin (ATX$_{II}$), aconitine and ervatamine appear to be confined to the sodium channel, DDT, veratridine and crude sea anemone and scorpion toxins also exhibit other actions.

Tetrodotoxin (TTX) the puffer fish poison, and the dinoflagellate toxin saxitoxin (STX) and its synthetic form (sSTX) selectively inhibit sodium conductance in cockroach giant axons (Pelhate & Sattelle, 1978; Sattelle et al., 1979). The actions of synthetic saxitoxin on ion currents in the cockroach axon are illustrated in Fig. 6. Dose-response curves for the suppression of the inward sodium current by each of these toxins yielded apparent K_d values (concentrations at which the sodium current was blocked by 50%) of 3.0×10^{-9} M (sSTX); 7.0×10^{-9} M (STX); 2.0×10^{-8} M (TTX). From the concentration dependence of sodium current inhibition it was concluded (Sattelle et al., 1979) that individual sodium channels were blocked by single toxin molecules. The development of synthetic saxitoxin has provided the first chemically synthesized probe suitable for investigating the molecular pharmacology of sodium channels in cell membranes.

Working on the fruit fly Drosophila melanogaster, Gitschier et al. (1980) demonstrated a saturable specific component of [^{3}H]-saxitoxin binding to head extracts with the pharmacological properties expected for binding to the ion selectivity filter of the voltage-sensitive sodium channel. In the same study nerve conduction in Drosophila was blocked by submicromolar concentrations of TTX and STX. An equilibrium dissociation constant (K_d) of 1.9×10^{-9} M was calculated from binding studies on Drosophila.

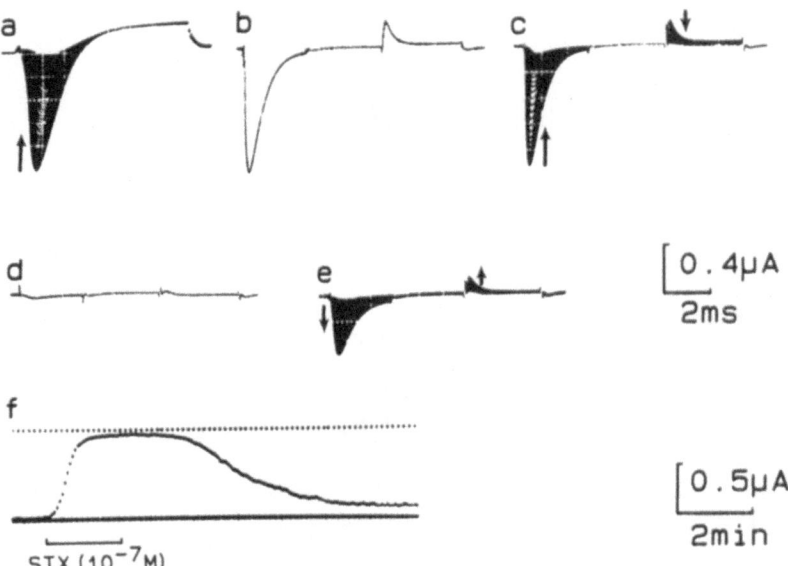

Fig. 6. Actions of synthetic saxitoxin (sSTX) on the ionic currents of the isolated cockroach axon. (a) Superimposed records of membrane currents for voltage-clamp pulses to E_m = -10 mV during a 5 min application of 1.0 x 10^{-7} M sSTX. The inward sodium current is completely suppressed without any reduction of the late outward (potassium) current. (b) Sodium currents (isolated pharmacologically by treatment of the axon with 5.0 x 10^{-4} M 4-AP) associated with paired voltage-clamp pulses first to E_m = -10 mV, and secondly to E_m = +60 mV. (c) Progressive actions of sSTX (1.0 x 10^{-7} M) on the sodium currents during a 2 min application of the toxin in the presence of 5.0 x 10^{-4} M 4-AP. (d) After 2 min in the presence of sSTX more than 90% of the sodium currents are suppressed. (e) Within 3 min of rebathing the axon in normal saline, the sodium currents have recovered to 50% of their normal amplitude. (f) Plot of the changes in amplitude of the peak inward sodium current during exposure to and recovery from a 2 min application of 1.0 x 10^{-7} M sSTX. From Pelhate & Sattelle (1982).

Recently several pyrethroid insecticides have been tested on the cockroach giant axon (Pelhate et al., 1980; Laufer et al., 1983) and differences have emerged in their actions on axonal sodium channels. Deltamethrin at micromolar concentrations induced a slow progressive depolarization of the axon membrane accompanied by a gradual reduction in action potential amplitude. The deltamethrin-induced depolarization was enhanced by an increase in stimulation frequency and was reduced in the presence of 1.0 x 10^{-7} M STX. Other synthetic pyrethroids (biopermethrin, biotetra-

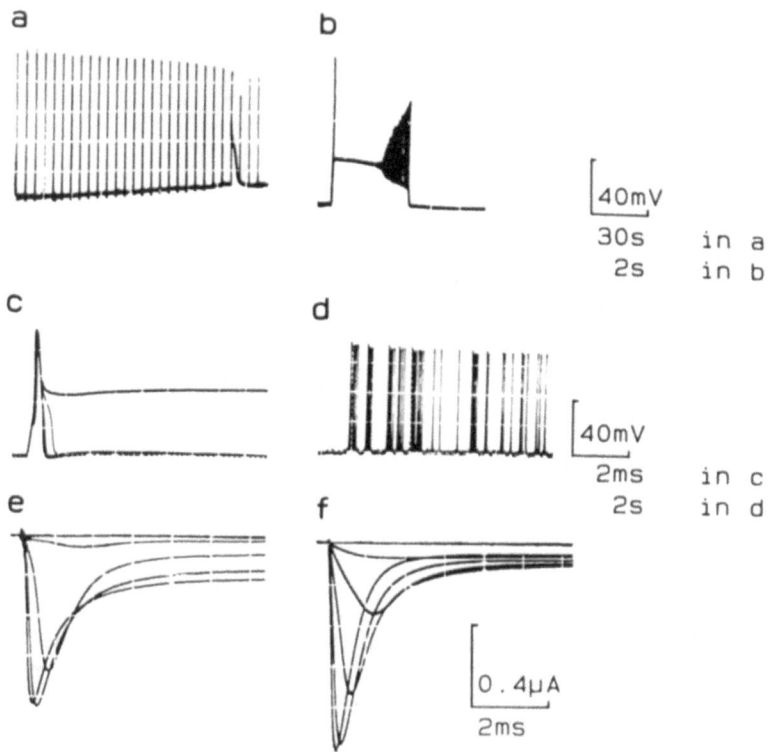

Fig. 7. Actions of venom and toxins extracted from the scorpion
Androctonus australis (Hector) on the isolated cockroach giant
axon. (a) Continuous recordings of action potentials induced by
short current pulses in the presence of 200 μg/ml of crude venom.
(b) Plateau potential during the falling phase of the action
potential with a superimposed burst of repetitive firing observed
following treatment with the crude venom. (c) Prolongation of the
action potential induced by the application of 3.5×10^{-6} M mammal
toxin II (MT$_{II}$). (d) Insect toxin (IT) at a concentration of 1.3
$\times 10^{-6}$ M induces a 10 mV depolarization of the axon and bursts of
repetitive activity. (e) and (f) Sodium currents recorded in the
presence of MT$_{II}$ (e) and IT (f) corresponding to voltage-clamp
pulses from E_h = -60 mV to E_m = -10 mV in 10 mV steps. From
Pelhate & Sattelle (1982).

methrin, s-bioallethrin, bioresmethrin, cismethrin and kadethrin)
induced prolonged depolarizing after-potentials, in contrast to the
findings with deltamethrin. Deltamethrin appears to affect a small
fraction of sodium channels which are held in a modified open-
state, whereas the pyrethroids which generate large depolarizing

after-potentials apparently induce a brief alteration of the open-state sodium channels with a larger number of channels affected. These findings for insect axonal sodium channels resemble those reported for sodium channels of other invertebrates and vertebrates in the case of deltamethrin (Vijverberg & Van den Bercken, 1979; Duclohier & Geogescauld, 1979) and other pyrethroids (Lund & Narahashi, 1981).

Polypeptide neurotoxins including the venom from Condylactis gigantea (CTX), the purified toxin II from Anemonia sulcata (ATX_{II}) and Anthopleurin A (AP_A) purified from Anthopleura xanthogrammica have all been tested on cockroach giant axons (Pelhate et al., 1979, 1980; Pelhate & Sattelle, 1982). Whereas CTX showed partial suppression (10-20%) of the amplitude of the potassium current this was not the case for the purified toxins. The primary axonal actions of all three was to slow the recovery of the sodium current which normally inactivates in 4 ms. Sauviat & Pichon (1976) showed that palytoxin (PTX) irreversibly increases the resting sodium permeability of the cockroach axonal membrane either by modifying pre-existing sodium channels or by forming new channels with distinct properties. The alkaloid aconitine modified a fraction of the sodium channels so they did not inactivate over the range of membrane potentials -50 to -30 mV (Pelhate & Sattelle, 1982), whereas another alkaloid ervatamine induced a frequency-dependent inhibition of peak sodium conductance. Both molecules were without effects on potassium channels. Various fractions from scorpion venoms have been tested on cockroach axons (Fig. 7). Of these the mammal toxin (MT_{II}) and insect toxin (IT) from the venom of Androctonus australis appear to be sodium inactivation inhibitors (Pelhate & Zlotkin, 1982). In addition, veratridine converted a proportion of the sodium channels into 'slow' channels and the insecticide DDT also considerably slowed the turn-off of sodium conductance, but in both cases effects on potassium channels were also noted (Pichon, 1969; Pichon & Boistel, 1969).

It is of interest to note that in some cases widely differing chemical structures can yield very similar changes in ion currents in the insect axon. Also, in cases where structure-activity data is available (e.g. pyrethroid-sodium channel interactions and aminopyridine-potassium channel interactions), a role can be ascribed to certain key features of the molecule in producing the observed effects on ion channel function.

B. PHARMACOLOGY OF NEUROTRANSMITTER REGULATED VOLTAGE-DEPENDENT CHANNELS IN NEURONAL CELL BODIES

A great deal of our knowledge of chemical synaptic transmission in insect CNS stems from the results obtained using the cercal afferent, giant interneurone pathway in the terminal abdominal ganglion of the cockroach <u>Periplaneta americana</u> (Callec, 1974). A modification of the oil-gap, single-fibre technique described in section A has enabled recording of excitatory (epsp) and inhibitory (ipsp) postsynaptic potentials from identifiable cockroach interneurones (Callec, 1974). Reversal potentials for epsps and ipsps correspond, respectively, to those for the depolarizing response to ionophoretically-applied acetylcholine and the hyperpolarizing response to ionophoretically-applied GABA (Callec, 1974). A particular advantage of this preparation is that precise control over presynaptic input via cercal nerves can be achieved. For example, unitary synaptic potentials corresponding to the activity of single cercal afferents can be detected.

Pharmacological studies on this ganglion using external hook electrode recordings (Shankland et al., 1971) and sucrose-gap recordings (Callec & Sattelle, 1973; Sattelle et al., 1976; Sattelle, 1978) indicated that nicotinic cholinergic ligands were particularly active. More recently the cercal afferent, giant interneurone 2 (GI 2) pathway has been studied in detail using the oil-gap technique. α-Bungarotoxin, the vertebrate nicotinic cholinergic receptor probe, is the most potent receptor ligand tested to date, blocking cercal afferent input to GI 2 at nanomolar concentrations (Sattelle et al., 1980; Harrow et al., 1982; Sattelle et al., 1983). Three radiolabelled putative cholinergic receptor ligands ^{125}I-α-bungarotoxin, [^{3}H]-quinuclidinyl benzilate and [^{3}H]-decamethonium have enabled characterization of three pharmacologically distinct binding sites in CNS particulate extracts (cf. Sattelle, 1980). In contrast to the findings for α-bungarotoxin, unlabelled quinuclidinyl benzilate and decamethonium were many orders of magnitude less effective in modifying cercal afferent, giant interneurone synaptic transmission (Harrow et al., 1982; Sattelle et al., 1983). Thus the specific, saturable component of ^{125}I-α-bungarotoxin binding characterized in the cockroach CNS (Gepner et al., 1978) appears to be a functional component of a CNS acetylcholine receptor involved in synaptic transmission at cercal afferent, giant interneurone 2 synapses. Although studied in less detail responses of giant interneurones to bath-applied and ionophoretically-applied GABA were blocked by micromolar picrotoxin, indicating a functional role for GABA receptors at inhibitory synapses on to cockroach giant interneurones (Callec, 1974).

However, although the demonstration of a synaptic role is essential in providing evidence for the functions of a putative neurotransmitter receptor detailed pharmacological experiments on insect CNS synaptic membranes are hampered by problems of access of pharmacological agents and the possibilities of indirect drug actions via other cells which may in turn modify the response of the neurone under investigation. A complicating factor common to pharmacological studies of all synapses is the possibility of drug actions on the presynaptic membrane. No chemical synapses on to cell bodies have been detected in the sixth abdominal ganglion of the cockroach (Smith & Treherne, 1965). Therefore extrasynaptic, cell body membranes which are peripherally located, facilitating drug accessibility, and are sensitive to locally applied putative neurotransmitters, provide suitable preparations for detailed investigations of neurotransmitter regulated voltage-dependent channels.

Earlier studies on the cell bodies of unidentified dorsal midline neurones in the cockroach sixth abdominal ganglion have shown that these cells are highly sensitive to ionophoretically-applied acetylcholine and GABA (Kerkut et al., 1968, 1969a,b, 1970; Pitman & Kerkut, 1970; Callec & Boistel, 1967, 1971a,b). Moto-neurones in the metathoracic ganglion were also shown to be sensitive to GABA. The present discussion is largely confined to recent studies on the identified cells giant interneurone 2 (GI 2) and the fast coxal depressor motoneurone (D_f).

An acetylcholine receptor/ion channel complex

In studies on the extrasynaptic cell body receptors of GI 2 in desheathed sixth abdominal ganglia, it was shown that the depolarizing response to ionophoretically-applied acetylcholine was blocked by submicromolar concentrations of α-bungarotoxin. By contrast micromolar concentrations of the muscarinic cholinergic antagonist quinuclidinyl benzilate were ineffective (Harrow & Sattelle, 1983). Thus of these two highly specific cholinergic receptor probes, the nicotinic antagonist was the more effective in blocking the depolarization induced by ionophoretic application of acetylcholine (Fig. 8).

The cell body membrane of the fast coxal depressor motoneurone (Fig. 9) provides an even more accessible cell body membrane. Situated on the ventral surface of the cockroach metathoracic ganglion, the cell body can be located by visual inspection. Recently David & Pitman (1982) reported the relatively high sensitivity of this cell body membrane to ionophoretically-applied acetylcholine and a reversal potential of -35 mV. Using the same cell it was shown that the acetylcholine-induced current was blocked by exposure to sodium-free saline (Harrow et al., 1982). A direct indication of the involvement of sodium ions was recently obtained by demonstrating with a sodium selective microelectrode a transient increase in intracellular sodium in response to ionophoretically-applied acetylcholine (David & Sattelle, unpublished observations).

When α-bungarotoxin at concentrations of 1.0×10^{-8} M and higher was applied to desheathed metathoracic ganglia, the sensitivity of the cell body of D_f to ionophoretically-applied

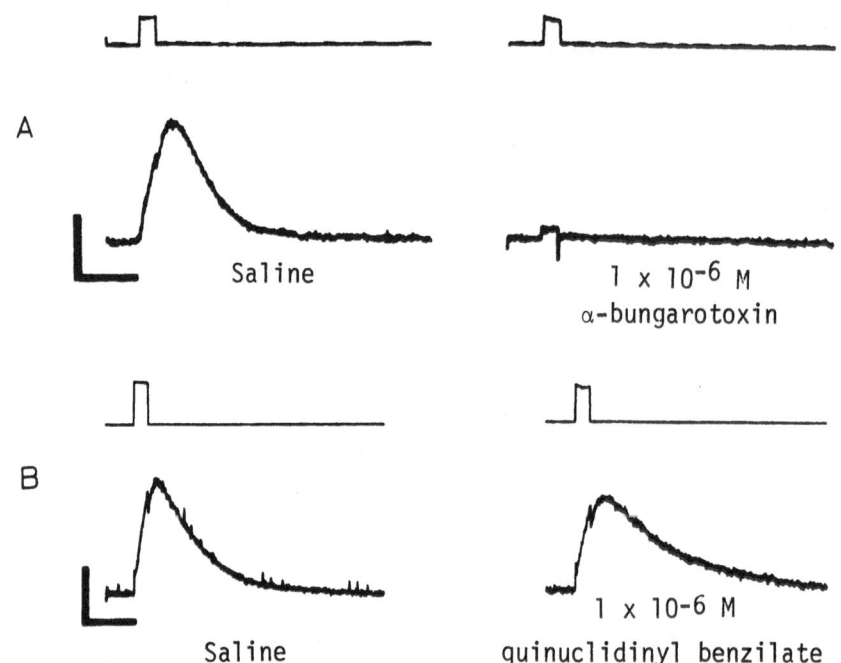

Fig. 8.   Effects of α-bungarotoxin and quinuclidinyl benzilate on the response to ionophoretically-applied acetylcholine of the cell body membrane of GI 2.   (A) 1.0 x 10^{-6}M α-bungarotoxin (60 min), ionophoretic dose 25 nC;   (B) 1.0 x 10^{-6}M quinuclidinyl benzilate (60 min), ionophoretic dose 115 nC.   Calibration: (A) vertical (upper) 120 nA, (lower) 5 mV, horizontal 2 s;   (B) vertical (upper) 300 nA, (lower) 2.5 mV, horizontal 2 s.   Modified from Harrow & Sattelle (1983).

acetylcholine was reduced (Harrow et al., 1982; David & Sattelle, 1983).   Complete block of the acetylcholine-induced depolarizing response of D_f was achieved at a concentration of 5.0 x 10^{-8}M after exposure to the toxin for 120 min.   By contrast quinuclidinyl benzilate was without effect on the acetylcholine response at concentrations of 1.0 x 10^{-5}M (60 min exposure).   At higher concentrations of this muscarinic receptor ligand a significant block was observed, though 1.0 x 10^{-3}M was required to completely block the acetylcholine response.   Decamethonium was without effect on the local application of acetylcholine to the cell body membrane of D_f after a 60 min exposure at concentrations up to 1.0 x 10^{-3}M.   At higher concentrations a blocking action was detected (David & Sattelle, 1983).

     Thus of the three molecular probes used so far in binding studies to characterize putative cholinergic receptors in insects, α-bungarotoxin was the most effective on the cell body receptors of D_f.

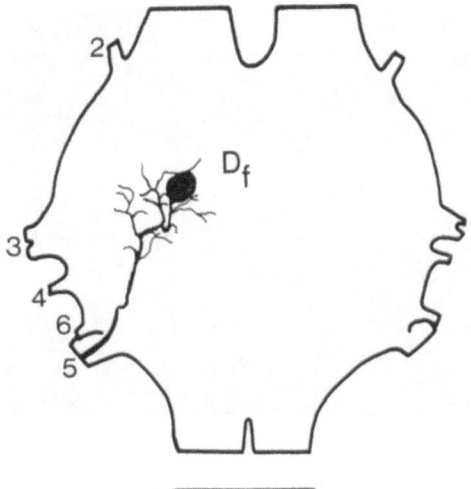

Fig. 9. Morphology of the fast coxal depressor motoneurone (D_f) of the metathoracic ganglion (T3) of <u>Periplaneta americana</u>. Camera lucida representation of the cell injected with cobalt via a microelectrode placed in the cell body. The nerve trunks are enumerated. Scale bar represents 500 μm.

Consistent with the observation that the nicotinic cholinergic antagonist (α-bungarotoxin), was highly effective on D_f, the parallel dose-response curves obtained for the depolarizing actions of four bath-applied agonists showed the following order of effectiveness: nicotine > acetylcholine (in the presence of 1.0 x 10^{-7}M neostigmine) > carbamylcholine > tetramethylammonium. By contrast dimethyl-4-phenyl piperazinium, suberyldicholine, D,L-muscarine, oxotremorine, acetyl-β-methylcholine and sebacinylcholine were practically ineffective (David & Sattelle, 1983).

A number of cholinergic antagonists were compared for (a) relative effectiveness and (b) voltage-dependence in blocking the response of D_f to acetylcholine ionophoresis. Dose-response curves for inhibition of the acetylcholine response were constructed. Clearly α-bungarotoxin, α-cobratoxin, mecamylamine, dihydro-β-erythroidine and benzoquinonium were highly effective. Less potent and almost equally effective were atropine, d-tubocurarine, pancuronium and quinuclidinyl benzilate. Even less effective were hexamethonium, gallamine, decamethonium and succinylcholine, all requiring concentrations of 1.0 x 10^{-3}M and higher to produce a sigificant block of the acetylcholine response.

With the cell body of D_f voltage-clamped, the acetylcholine-

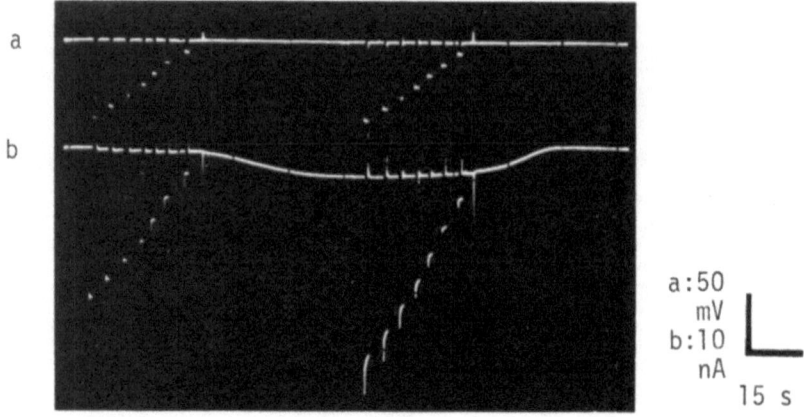

a:50
   mV
b:10
   nA
      15 s

Fig. 10. Measurement of the acetylcholine-induced current, recorded from the voltage-clamped cell body membrane of D_f at a series of membrane potentials. Upper trace, clamp potential; lower trace, clamp current. The membrane potential was held at -50 mV and jumped to a series of more negative potentials (-120 mV to -70 mV) in 10 mV steps both in the absence and presence of ionophoretically-applied acetylcholine.

induced current was obtained at a series of membrane potentials between -120 mV and -60 mV. From an initial holding potential of -60 mV, the membrane was jumped to the different potentials (each jump lasting 700 ms) both before and during acetylcholine application. In this way (Fig. 10) the degree of antagonism of the acetylcholine response at different membrane potentials was assessed. The results of such experiments are compared in Table 2. Whereas the blocking actions of α-bungarotoxin , dihydro-β-erythroidine and mecamylamine were independent of membrane potential in the range -120 to -60 mV, those of d-tubocurarine and atropine were strongly dependent on membrane potential. Clearly the actions of the voltage-independent blockers could be accounted for in terms of an interaction with the closed receptor/ion channel. D-tubocurarine and atropine were strongly voltage-dependent in their blocking actions and therefore probably interact with the open receptor/ion channel complex. d-Tubocurarine also induces a voltage-dependent reduction of acetylcholine responses of Aplysia neurones (Marty et al., 1976) and frog neuromuscular junctions (Katz & Miledi, 1978; Colquhoun et al., 1979). As in the case of D_f, atropine also appears to block the ion channel at vertebrate peripheral nicotinic acetylcholine receptors (Feltz et al., 1977) and cell body receptors in Aplysia neurones (Ascher et al., 1978).

Table 2. Voltage-dependence of antagonist suppression of acetylcholine-induced current in the cell body membrane of D_f.

Concentration of Antagonist	Voltage-independent block of acetyl-choline response	Voltage-dependent block of acetyl-choline response
5.0×10^{-7}M	α-Bungarotoxin	
1.0×10^{-6}M	Dihydro-β-erythroidine	
5.0×10^{-6}M	Mecamylamine	
1.0×10^{-4}M		d-Tubocurarine
1.0×10^{-5}M		Atropine

Thus acetylcholine regulates a voltage-dependent sodium channel (though a contribution from other ions cannot be ruled out at this stage). The pharmacological profile of this CNS acetylcholine receptor/ion channel complex is nicotinic.

A GABA receptor/ion channel complex

In vertebrate CNS tissues two classes of GABA receptors: bicuculline-sensitive $GABA_A$-receptors (Olsen et al., 1981) and baclofen-sensitive $GABA_B$ receptors (Bowery et al., 1981) have been identified and characterized but there has been comparatively little work to date on the pharmacology of insect CNS GABA receptors. Bicuculline which in vertebrates appears to bind to the recognition site of the GABA receptor (Mohler & Okada, 1978) appeared to reversibly block GABA induced hyperpolarization and ipsps in unidentified dorsal midline cells of the cockroach A6 ganglion (Walker et al., 1971), albeit at rather high concentrations. Recently Walker et al. (1981) found that bicuculline methochloride was ineffective on cockroach GABA receptors. It is however effective on D_f blocking GABA-induced currents at 1.0×10^{-5}M concentrations and at similar concentrations ^{3}H-GABA binding to cockroach CNS extracts is largely inhibited (Lummis et al., 1982). Using chloride selective electrodes changes in intracellular chloride were observed accompanying bath application of GABA to D_f.

Thus GABA appears to regulate a voltage-dependent chloride channel. Pharmacological studies indicate that bicuculline-sensitive GABA receptors are present in the cockroach CNS.

CONCLUSIONS

Uniquely identifiable neurones of known function in the insect CNS are accessible to the application of voltage-clamp techniques, thereby facilitating studies on CNS neurotransmitter receptors and ion channels in these animals. When combined with the results of radiolabelled ligand binding investigations performed on the same tissue, the role of specific neurotransmitter receptors and ion channels in pathways of known function can be determined. Neuropharmacological studies of this type are also improving our understanding of the mode of action of insecticides, many of which modify the functions of neurotransmitter receptors and ion channels. Such actions may be amplified by the resulting changes in the pattern of neurotransmitter and neurohormone release.

ACKNOWLEGEMENTS

The author is indebted to his colleagues M. Pelhate, S.C.R. Lummis, J.A. David and I.D. Harrow each of whom contributed to aspects of the work surveyed here.

REFERENCES

Armstrong, C.M., 1971, Interaction of tetraethylammonium derivatives with the potassium channels of giant axons. J. gen. Physiol., 58: 413–437.

Ascher, P., Marty, A. and Neild, T.D., 1978, The mode of action of antagonists of the excitatory responses to acetylcholine in Aplysia neurones. J. Physiol. (Lond.), 278:207–235.

Bowery, N.G., Doble, A., Hill, D.R., Hudson, A.L., Shaw, J.S., Turnbull, M.J. and Warrinton, R. 1981, Bicuculline-insensitive GABA receptors on peripheral autonomic nerve terminals. Eur. J. Pharmacol. 71, 53–70.

Callec, J.J. 1974, Synaptic transmission in the central nervous system of insects. In Insect Neurobiology (ed. J.E. Treherne). pp. 119–178 North Holland, American Elsevier: Amsterdam & New York.

Callec, J.J., Boistel, J. 1967, Les effets de l'acétylcholine aux niveaux synaptique et somatique dans le cas du dernier ganglion abdominal de la Blatte, Periplaneta americana. C.R. Séances Soc. Biol. (Paris). 161, 442–446.

Callec, J.J. & Boistel, J. 1971a, Further evidence for ACh transmission in the cockroach central nervous system studied at the unitary level. Proc. XXV Int. Congr. IUPS, Munich 9, 95.

Callec, J.J., and Boistel, 1971b, Role possible du GABA comme mediateur inhibiteur du système de fibres géantes chez la Blatte (Periplaneta americana). J. Physiol. (Paris), 63, 119A.

Callec, J.J. and Sattelle, D.B., 1973, A simple technique for monitoring the synaptic actions of pharmacological agents J. exp. Biol. 59, 725-738.

Colquhoun, D., Dreyer, F. and Sheridan, R.E. 1979, The action of d-tubocurarine at the frog neuromuscular junction. J. Physiol. (Lond), 293, 247-284.

David, J.A. and Pitman, R.M. 1982, The effects of axotomy upon the extrasynaptic acetylcholine sensitivity of an identified motoneurone in the cockroach Periplaneta americana. J. exp. Biol. 98, 329-341.

David, J.A., Lummis, S.C.R., and Sattelle, D.B.. 1983, Receptors for GABA and acetylcholine in the central nervous system of an insect. Abstr. Br. Pharmacol. Soc. C8 (July, 1983).

Duclohier, H. and Georgescauld, D. 1979, The effects of the insecticide decamethrin on action potential and voltage-clamp currents of Myxicola giant axon. Comp. Biochem. Physiol. 62C, 217-223.

David, J.A. and Sattelle, D.B. 1983, Actions of cholinergic pharmacological agents on the cell body membrane of the fast coxal depressor motoneurone of the cockroach (Periplaneta americana). J. exp. Biol. 107 in press

Dudai, Y. 1982, Genetic approaches to insect neurochemistry. In Neuropharmacology of Insects (Ciba Foundation Symposium 88) pp.199-206 Pitman:London.

Eldefrawi, A.T., Mansour, N.A. and Eldefrawi, M.E. 1982, Insecticides affecting acetylcholine receptor interactions. Pharmac. Ther. 16, 45-65.

Feltz, A., Large, W.A. and Trautmann, A. 1977, Analysis of atropine action at the frog neuromuscular junction. J. Physiol (Lond.) 269, 109-130.

Gepner, J.I., Hall, L.M., Sattelle, D.B. 1978, Insect acetylcholine receptors as a site of insecticide action. Nature (Lond.) 276, 188-190.

Gitschier, J., Strichartz, G.R. and Hall, L.M. 1980, Saxitoxin binding to sodium channels in head extracts from wild-type and tetrodotoxin-sensitive strains of Drosophila melanogaster. Biochim. Biophys. Acta. 595, 291-303.

Goodman, C.S. and Spitzer, N.C. 1980, Embryonic development of neurotransmitter receptors in grasshoppers. In Receptors for Neurotransmitters, Hormones and Pheromones in Insects (eds. D.B. Sattelle, L.M. Hall, and J.G. Hildebrand) pp.195-207. Elsevier/North-Holland Biomedical Press: Amsterdam.

Hall, L.M., Wilson, S.D., Gitschier, J., Martinez, N. and Strichartz, G.R. 1982, Identification of Drosophila melanogaster mutant that affects the saxitoxin receptor of the voltage-sensitive sodium channel. In: Neuropharmacology of Insects (Ciba Foundation Symposium 88) pp207-220. Pitman:London.

Hall, Z.W., Hildebrand, J.G. and Kravitz, E.A. 1975, The Chemistry of Synaptic Transmission. Chiron Press: Newton, Massachusetts, U.S.A.

Harrow, I.D., David, J.A. and Sattelle, D.B. 1982, Acetylcholine receptors of identified insect neurons. In: Neuropharmacology of Insects (Ciba Foundation Symposium 88) pp.12-31. Pitman:London.

Harrow, I.D., Hue, B., Pelhate, M. and Sattelle, D.B., 1980, Cockroach giant interneurones stained by cobalt-backfilling of dissected axons. J. exp. Biol, 84, 341-343.

Harrow, I.D. and Sattelle, D.B. 1983, Acetylcholine receptors on the cell body membrane of giant interneurone 2 in the cockroach, Periplaneta americana. J. exp. Biol. 105, 339-350.

Hildebrand, J.G. 1980, Development of putative acetylcholine receptors in normal and deafferented antennal lobes during the metamorphosis of Manduca sexta. In: Receptors for Neurotransmitters, Hormones and Pheromones in Insects (eds. D.B. Sattelle, L.M. Hall, and J.G. Hildebrand) pp.209-220. Elsevier/North-Holland Biomedical Press: Amsterdam.

Hildebrand, J.G. 1982, Chemical signalling in the insect nervous system. In: Neuropharmacology of Insects (Ciba Foundation Symposium 88) pp5-11. Pitman:London.

Katz, B. and Miledi, R. 1978, A re-examination of curare action at the motor end plate. Proc. Roy. Soc. B. 203, 119-133.

Kerkut, G.A., Newton, L.C., Pitman, R.M., Walker, R.J. and Woodruff, G.N. 1970, Acetylcholine receptors of invertebrate neurones. Br. J. Pharmacol. 40, 586p.

Kerkut, G.A., Pitman, R.M. and Walker, R.J. 1968, Electrical activity in insect nerve cell bodies. Life. Sci. 7, 605-607.

Kerkut, G.A., Pitman, R.M. and Walker, R.J. 1969a, Sensitivity of neurons of the insect central nervous system to iontophoretically applied acetylcholine or GABA. Nature, (Lond). 222, 1075-1076.

Kerkut, G.A., Pitman, R.M. and Walker, R.J. 1969b, Iontophoretic application of acetylcholine and GABA onto insect central neurones. Comp. Biochem. Physiol. 31, 611-633.

Laufer, J., Roche, M., Pelhate, M. Elliott, M., Janes, N.F. and Sattelle, D.B. 1983, Pyrethroid insecticides: actions of deltamethrin and related compounds on insect axonal sodium channels. J. Insect. Physiol. (in press)

Lees, G.V., Pichon, Y. and Poussart, D. 1981, Spectrum analysis on membrane current fluctuations in voltage-clamped axons of the cockroach. J. Physiol. (Lond.). 319, 29pp.

Lund, A.E. and Narahashi, T. 1981, Kinetics of sodium channel modification by the insecticide tetramethrin in squid axon membranes. J. Pharmacol. Exp. Ther. 219 464-473.

Marty, A., Neild, T.O. and Ascher, P. 1976, Voltage sensitivity of acetylcholine currents in Aplysia neurones in the presence of curare. Nature (Lond.), 261, 501-503.

Meves, H. and Pichon, Y. 1977a, The effect of internal and external 4-aminopyridine on the potassium currents in intracellularly perfused squid giant axons. J. Physiol. (Lond.). 268, 511-532.

Meves, H. and Pichon, Y. 1977b, Modèle d'action de la 4-aminopyridine au niveau des 'canaux' potassium de l'axone geant de Calmar (Loligo forbesi L.). C.R. Acad. Sci. (Paris) 284, 1325-1328.

Mohler, H. and Okada, T. 1978, Properties of GABA receptor binding with (^{3}H)-bicuculline methiodide in rat cerebellum. Mol. Pharmacol. 14, 256-265.

Narahashi, T. 1974, Chemicals as tools in the study of excitable membranes. Physiol. Rev. 54, 813-889.

Narahashi, T. 1975, Toxins as tools in the study of ionic channels of nerve membranes. Proc. 6th Int. Congr. Pharmacol. Helsinki, Finland. pp.97-108.

Olsen, R.W., Bergman, M.O., Van Ness, P.C., Lummis, S.C., Watkins, A.E. Napias, C. and Greenlee, D.V. 1981, Gamma aminobutyric acid receptor binding in mammalian brain: heterogeneity of binding sites. Mol. Pharmacol. 19, 217-227.

Pelhate, M., Hue, B. and Chanelet, J. 1972, Effets de la 4-aminopyridine sur le système nerveux d'un Insecte: la Blatte (Periplaneta americana L.). C.R.Soc. Biol. (Paris) 166, 1598-1605.

Pelhate, M., Hue, B. and Chanelet, J. 1974a, Modifications, par la 4-amino-pyridine, des charactéristiques électriques de l'axone géant isolé d'un insecte, la Blatte (Periplaneta americana L.) C.R. Soc. Biol. (Paris) 168, 27-34

Pelhate, M., Hue, B., Pichon, Y. and Chanelet, J. 1974b, Actions de la 4-aminopyridine sur la membrane de l'axone isolé d'Insecte. C.R. Acad. Sci. (Paris) 278, 2807-2809.

Pelhate, M., Hue, B., Pichon, Y. and Chanelet, J. 1982, Interactions of aminopyridines and related compounds with ionic channels in the isolated cockroach axon. In: Effect of Aminopyridines and Similarly Acting Drugs on Nerves, Muscles and Synapses (eds. W.C. Bowman, P. Lechat and S. Thesleff). Pergamon Press: Oxford.

Pelhate, M., Hue, B. and Sattelle, D.B. 1980, Pharmacological properties of axonal sodium channels in the cockroach Periplaneta americana L. II. Slowing of sodium current turn off by Condylactis toxin. J. exp. Biol. 83, 49-58.

Pelhate, M., Hue, B., and Sattelle, D.B., 1980, Actions of natural and synthetic toxins on the axonal sodium channels of the cockroach. In: Insect Neurobiology and Pesticide Action (Neurotox 79) pp. 65-71, Society of Chemical Industry: London.

Pelhate, M., Mony, L., Hue, B. and Chanelet, J. 1975, Action de la 4-aminopyridine (4-AP) sur les courants ioniques membranaires; cas de l'axone géant isolé de la Blatte (Periplaneta americana) J. Physiol. (Paris) 71, 306A.

Pelhate, M., Mony, L., Hue, B., and Chanelet, J. 1976, Action de la 9-aminoacridine sur la membrane axonale de la Blatte (Periplaneta americana L.) C.R. Soc. Biol. (Paris) 140, 1182-1187.

Pelhate, M. and Pichon, Y. 1974, Selective inhibition of potassium current in the giant axon of the cockroach. J.Physiol. (Lond.) 242, 90-91p.

Pelhate, M. and Sattelle, D.B. 1978, Synthetic saxitoxin selectively inhibits sodium currents in the cockroach giant axon. J. Physiol (Lond). 284, 89-90p.

Pelhate, M. and Sattelle, D.B. 1982, Pharmacological properties of insect axons: a review. J. Insect Physiol. 28, 889-903.

Pelhate, M. and Zlotkin, E. 1982, Actions of insect toxin and other toxins derived from the venom of the scorpion, Androctonus australis on the isolated giant axons of the cockroach (Periplaneta americana). J. exp. Biol. 97, 67-77.

Pichon, Y. 1969 Aspects électriques et ioniques du functionnement nerveux chez les insects. Cas particulier de la chaine nerveuse abdominale d'une blatte Periplaneta americana. L. Thèse d'Etat, Université de Rennes (France).

Pichon, Y. 1974a, Axonal conduction in insects. In: Insect Neurobiology (ed, J.E. Treherne). pp73-117 North Holland-Elsevier: Amsterdam.

Pichon, Y. 1974b, The pharmacology of the insect nervous system. In: The Physiology of Insecta 2nd ed. (Ed. M. Rockstein) pp. 101-174. Academic Press: New York.

Pichon, Y. and Boistel, J. 1966, Etude de la fibre nerveuse isolee d'insecte. Enregistrements des potentiels de membrane et des potentiels d'action de la fibre géante de blatte (Periplaneta americana) C.R. Soc. Biol. (Paris) 160, 1948-1954.

Pichon, Y. and Boistel, J. 1967, Current-voltage relations in the isolated giant axon of the cockroach under voltage-clamp conditions. J. exp. Biol. 47, 343-355.

Pichon, Y. and Boistel, J. 1969, Effets compares du D.D.T. et de la vératrine sur la perméabilité ionique de la membrane nerveuse d'insecte. J. Physiol. (Paris) 61, suppl.2, 373-374.

Pichon, Y., Meves, H. and Pelhate, M. 1982, Effects of aminopyridines on ionic currents and ionic channel noise in unmyelinated axons. In: Effects of Aminopyridines and Similarly Acting Drugs (eds. W.C. Bowman, P. Lechat and S. Thesleff) Pergamon Press: Oxford.

Pitman, R.M. and Kerkut, G.A. 1970, Comparison of the actions of iontophoretically applied acetylcholine and gamma aminobutyric acid with the EPSP and IPSP in cockroach central neurons. Comp. gen. Pharmacol. 1, 221–230.

Sattelle, D.B. 1978, The insect central nervous system as a site of action of neurotoxicants. In: Pesticide and Venom Neurotoxicity (eds. D.L. Shankland, R.M., Hollingworth, and T. Smyth. Jr.) pp.7–26. Plenum Press: New York.

Sattelle, D.B. 1980, Acetylcholine receptors of Insects. Adv. Insect. Physiol. 15, 215–315.

Sattelle, D.B., David, J.A., Harrow, I.D. & Hue, B. 1980, Actions of α-bungarotoxin on identified central neurones. In: Receptors for Neurotransmitters Hormones and Pheromones in Insects (eds. D.B. Sattelle, L.M. Hall and J.G. Hildebrand) pp.125–139 Elsevier/North Holland Biomedical Press: Amsterdam.

Sattelle, D.B., Hall, L.M. and Hildebrand, J.G. (eds) 1980, Receptors for Neurotransmitters Hormones and Pheromones in Insects pp 1–310. Elsevier/North Holland Biomedical Press: Amsterdam.

Sattelle, D.B., Harrow, I.D., Hue, B., Pelhate, M., Gepner, J.I. & Hall, L.M. 1983, α-Bungarotoxin blocks excitatory synaptic transmission between cercal sensory neurones and giant interneurone 2 of the cockroach, Periplaneta americana. J. exp. Biol. 107, in press.

Sattelle, D.B., McClay, A.S., Dowson, R.J., and Callec, J-J. 1976, The pharmacology of an insect ganglion: actions of carbamylcholine and acetylcholine. J. exp. Biol. 64, 13–23.

Sattelle, D.B., Pelhate, M. and Hue, B. 1979, Pharmacological properties of axonal sodium channels in the cockroach Periplaneta americana L. I. Selective block by synthetic saxitoxin. J. exp. Biol. 83, 41–48.

Sauviat, M.P. and Pichon, Y. 1976, Modifications de la perméabilité ionique de la membrane axonale de Periplaneta americana L. sous l'effect de la palytoxine. J. Physiol. (Paris) 73, 56p.

Shain, W. and Carpeneter, D.O. 1981, Mechanisms of synaptic modulation. Int. Rev. Neurobiol. 22, 205–250.

Shankland, D.L., Rose, J.A. and Donniger, C. 1971, The cholinergic nature of the cercal nerve-giant fibre synapse in the sixth abdominal ganglion of the American cockroach, Periplaneta americana (L.). J. Neurobiol. 2, 247–262.

Smith, D.S. and Treherne, J.E. 1965, the electron microscopic localization of cholinesterase activity in the central nervous system of an insect Periplaneta americana. J. Cell Biol., 26, 445–465.

Vijverberg, H.P.M. and van den Bercken, J. 1979, Frequency dependent effects of the pyrethroid insecticide decamethrin in frog myelinated nerve fibres. Eur. J. Pharmacol. 58, 501-504.

Walker, R.J., Crossman, A.R., Woodruff, G.N. and Kerkut, G.A. 1971, The effect of bicuculline on the gama-aminobutyric acid (GABA) receptors of neurones of Periplaneta americana and Helix aspersa. Brain. Res. 33, 75-82.

Walker, R.J., James, V.A., Roberts, C.J. and Kerkut, G.A. 1981, Studies on amino acid receptors of Hirudo, Helix, Limulus and Periplaneta. In: Neurotransmitters in Invertebrates (ed. K.S. Rozsa). Vol. 22 pp.161-190. Pergamon Press: Oxford.

ISOLATION AND CHARACTERISATION OF NEUROHORMONES FROM LOCUSTS

W. Mordue and P. J. Morgan

Department of Zoology
University of Aberdeen
Tillydrone Avenue, Aberdeen, U.K.

Our knowledge of the chemistry of insect neurohormones is still very limited. The principal reasons for this paucity of information on the structure and biosynthesis of neurosecretory and other peptides are consequences of both lack of starting material and low resolution chromatographic and insensitive detection techniques. The deficiencies in amounts of material, especially if the active principle is labile, are difficult to overcome but the development of High Performance Liquid Chromatographic (HPLC) separation methods has however improved greatly the ability to resolve and detect small amounts of peptides. It is to be expected that in the next few years, or perhaps even during this symposium, that the major impact of HPLC on insect neurohormone research will be felt. The next phase of significant advances in our understanding of the role(s) of insect peptides will be possible only when more is known of their chemistry. However, for many peptides their true endocrine, modulatory or transmitter functions have not been established with certainty. For some peptides it is a real possibility that the biological event(s) which form the basis of their bioassay may be a pharmacological rather than a physiological effect. Improvements in chemical methods will have to be complemented with rigorous adherence to criteria for establishing candidate peptides as true hormones or transmitters - or both.

For many peptides precise information on circulatory titres will be possible only after chemically purified peptides have been isolated. The perfection of RIA (Radio-immunoassay) methods for peptides and cataloguing of chromatographic indices is possible only when quantitative and routine chemical methods are established. It is unfortunate that all of this wide repertoire of chemical and physiological techniques has not been applied to any particular

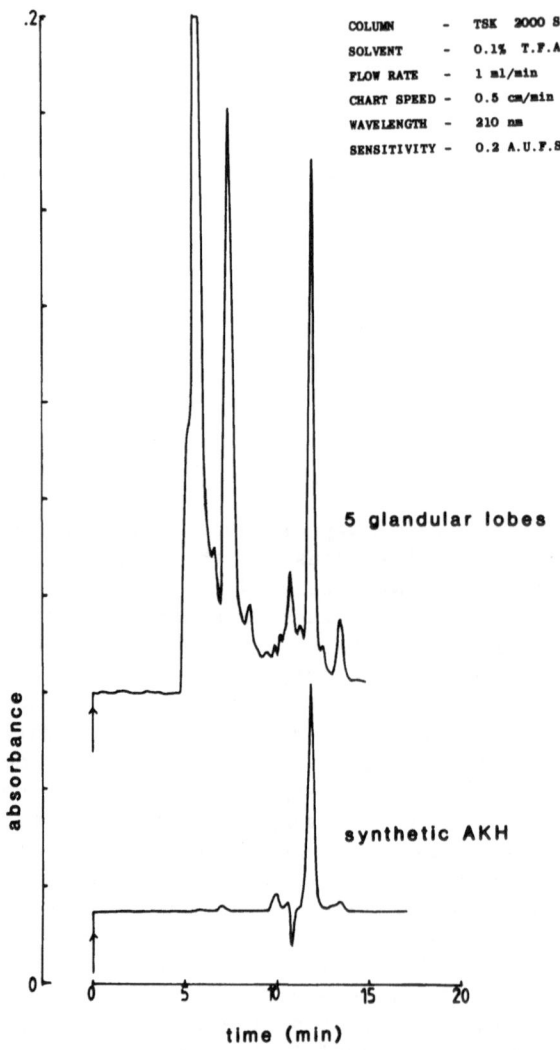

Fig. 1.   Upper trace: separation of
          methanol extract of 5 glandular
          lobes of <u>Locusta</u> by high per-
          formance size-exclusion chroma-
          tography using a 30 cm TSK 2000
          SW column.
          Lower trace: elution profile
          for synthetic AKH chromato-
          graphed in an identical manner.
          The retention time is <u>c</u>. 12 min.

insect peptide.   The production of RIA methods and receptor binding
studies are also impossible until radio-labelled peptides of high
specific activity are available.   Until recently the only radiolabelled invertebrate peptides available have been produced using
radio-iodination.   This is suitable for some studies but the large
size of the iodine moiety relative to the mass of a small peptide
limits the usefulness of iodinated compounds in cellular investigations and binding studies.   However, Hardy et al. 1983 have now
reported the synthesis of tritiated AKH with a specific activity of
115 Ci mmol^{-1}.   This compound should prove invaluable in receptor
binding and studies of peptide degradation within the intact insect.

ADIPOKINETIC HORMONE (AKH) AND RELATED PEPTIDES

Adipokinetic hormone

     We must remember that at present there are still only two neuro-
secretory peptides, AKH and proctolin, which are characterised fully
and available routinely.   Proctolin will be considered in detail
elsewhere in this symposium.   AKH is a blocked decapeptide of the
following structure:-

     PCA - Leu - Asn - Phe - Thr - Pro - Asn - Trp - Gly - Thr - NH$_2$.

AKH is released from the glandular lobes of locust corpora cardiaca
in response to flight to effect changes in metabolism, particularly
mobilisation of lipid, enabling long term flight to occur.   There
is a considerable body of information concerning the isolation,
characterisation, roles and modes of action of this peptide (see
Stone and Mordue 1980).

     AKH is methanol soluble which allowed methanolic extracts of
dissected glandular lobes to be used as starting material for the
isolation procedures.   Chromatographic separation was carried out
using either aqueous or organic solvents on columns of controlled-
pore glass beads or Sephadex LH-20.   Both methods provided reliable
and relatively rapid mechanisms for effecting an initial purification
of AKH.   However, more recently we have been developing techniques
utilising HPSEC (High Performance size-exclusion chromatography) on
a TSK 2000 SW column.   This high performance size-exclusion step
applied to the separation of a methanol extract of glandular lobes
from Locusta is shown in Fig. 1.   The adipokinetic activity of the
gland extract co-chromatographs exactly with synthetic AKH.   The
advantage of such an initial purification step using a HPSEC column
is that separation is achieved in minutes rather than hours;   this
time factor is not especially critical for AKH but is of paramount
importance when dealing with labile materials.   This first stage
separation is now used routinely in our laboratory and has been
applied with success to the isolation and purification of locust
diuretic hormone (DH) - see Morgan and Mordue this symposium.   The

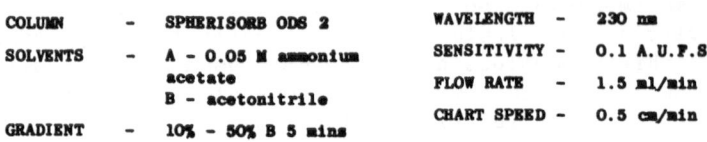

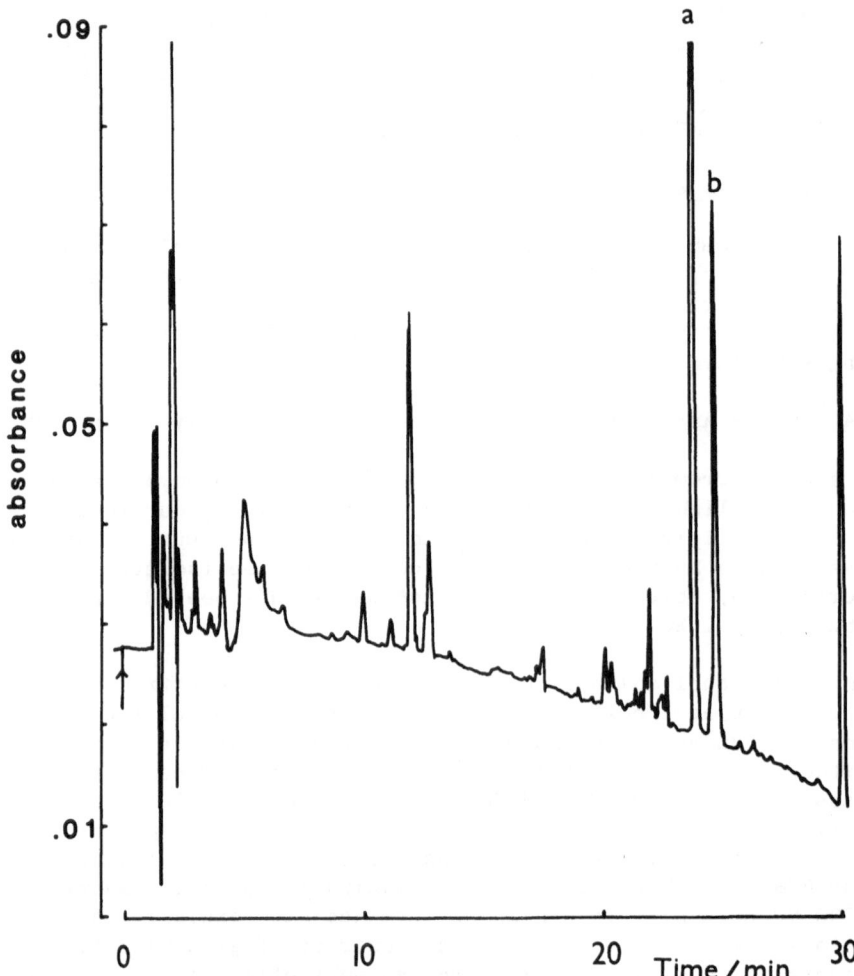

Fig. 2.   Separation of methanol extract of 25 glandular
           lobes of <u>Locusta</u> by reversed-phase HPLC, following
           the method of Morgan and Mordue (1983).   Both
           peaks (a) and (b) possess substantial AKH activity;
           (a) corresponds to AKH and (b) is possibly AKH II.

resolving properties are such that although AKH, DH and other small
peptides do co-elute they are nevertheless resolved from a considerable number of other components within the corpora cardiaca.

Material from the TSK chromatographic step (after lyophilization) or methanolic extracts of corpora cardiaca have been subject
to reversed-phase HPLC.   A typical elution profile and conditions
used are shown in Fig. 2.   The technique is sufficiently sensitive
to allow AKH to be detected in extracts from single pairs of corpora
cardiaca.   Peak (a) in Fig. 2 is AKH and peak (b) with a longer
retention time also possesses significant adipokinetic activity;   we
suspect this second peak to be AKH II (Carlsen et al. 1979).   Current
work is endeavouring to confirm the identity of this and other adipokinetic peaks detected in glandular lobe extracts.   Such methods
may prove invaluable in ascertaining the possible release of AKH and
AKH II into the haemolymph under a variety of physiological conditions.

Related Peptides

Corpora cardiaca from many insect species contain factors -
often peptidergic in nature - which appear to be important in controlling metabolism.   The concentrations of haemolymph metabolites,
especially carbohydrates and lipids, and key metabolic pathways in
certain tissues are altered by injections of corpus cardiacum extracts.   Such reports, particularly those concerned with hyperglycaemic activity are too numerous to list but examples of this
phenomenon are found in most major orders of insects.   AKH in
addition to being hyperlipaemic can exert hyperglycaemic actions in
other insects.   Thus it has been intriguing to investigate whether
the hyperglycaemic factors well documented for Periplaneta and other
insects are related to AKH.   However, in spite of the considerable
efforts of a number of laboratories only incomplete and rather un-
satisfactory composition data have been published (see Mordue 1981).
Similarly a potent hyperglycaemic and hyperlipaemic factor from
Carausius has also eluded full characterisation (Mordue and Gäde,
unpublished observations;   Mordue 1981).   However, in Fig. 3 the
resolution of a methanolic extract of stick insect corpora cardiaca
on reversed-phase HPLC is shown;   a very clear peak at retention
time (R_T) of 23.6 min is shown.   This peak possesses potent biological activity and it is worthy of note that under identical
chromatographic conditions the R_T for AKH is the same.   The resolution of the stick insect peptide on HPLC is so vastly superior
to previous resolution that significant progress as to its composition is now expected.

From the published data there are a number of amino acid residues in common between AKH, AKH II, RPCH (red-pigment concentrating
hormone), roach, stick insect and honey bee peptides.   Such similarities in structure can be deduced also from the biological activities of the different peptides.   All these peptides cross-react

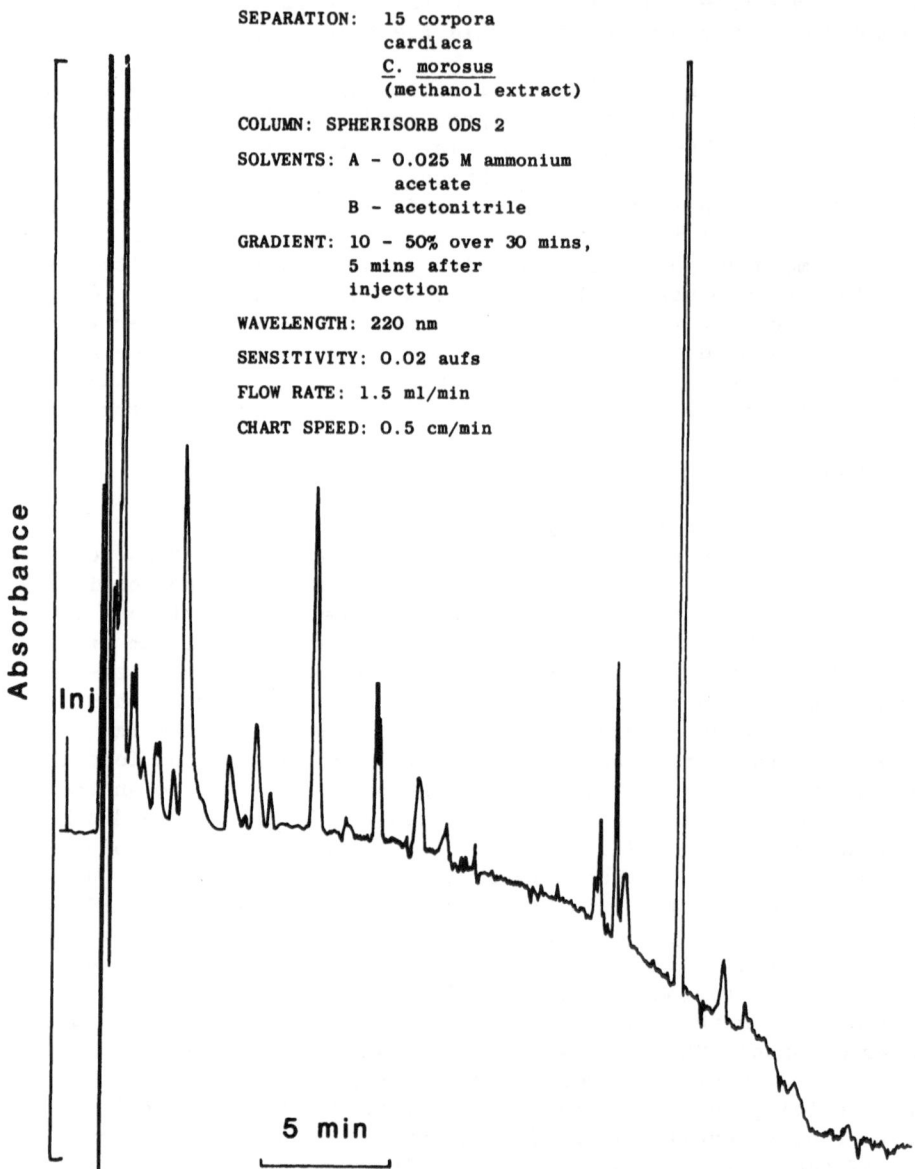

SEPARATION: 15 corpora
              cardiaca
              C. morosus
              (methanol extract)

COLUMN: SPHERISORB ODS 2

SOLVENTS: A - 0.025 M ammonium
              acetate
          B - acetonitrile

GRADIENT: 10 - 50% over 30 mins,
          5 mins after
          injection

WAVELENGTH: 220 nm

SENSITIVITY: 0.02 aufs

FLOW RATE: 1.5 ml/min

CHART SPEED: 0.5 cm/min

Absorbance

Inj

5 min

Fig. 3.   Separation of a methanolic extractof 15 corpora cardiaca
           by reversed-phase HPLC on a Spherisorb ODS 2 column,
           following the method of Morgan and Mordue (1983).   The
           peak at R_T 23.6 min possesses substantial adipokinetic
           and hyperglycaemic activity.

in different bioassays, such as the hyperglycaemic and hyperlipaemic
assays and the prawn erythrophore assay; these are important observations but unfortunately do not clarify whether or not a particular
peptide(s) eluted is pure or not.   Routine application of HPLC
methods should again alleviate the confusion concerning biological
activities by ensuring that only homogenous peptides are assayed.
It is worth re-emphasising that progress on an endocrine front is
also essential.   In the stick insect and the colorado beetle factors
are present within the corpus cardiacum which are potent when assayed
in locusts and roaches but are more difficult to detect when assayed
on host animals.   In the beetle the effect of the peptide(s) is to
influence metabolism of proline which is the primary energy source
for flight.   Significant changes in lipid and carbohydrate levels
are not detectable.   Thus the endocrine or other role(s) played by
a peptide must be ascertained as complementary part of the isolation
and characterisation programme - what is hyperlipaemic in one insect
may well have a very different role in another.

The scenario so far indicates that we are likely to be dealing
with a family of structurally related peptides in insects and crust-
acea.   These peptides may stem from some ancestral molecule.   However, there has been considerable evolution to produce the diversity
in biological activities not only in the peptides but there must
also have been an evolution of the receptor systems which occurred
in parallel.   Such hypothesis can be tested best through structure
activity analyses of peptides of known structures.

STRUCTURE ACTIVITY RELATIONSHIPS OF AKH

These studies were initiated once the fully active hormone had
been synthesised.   The attempts to determine which residues and
regions of the peptide are essential for full adipokinetic activity
have, in part, been reported previously (Stone et al. 1978;   Mordue
1981) and in consequence will be dealt with only briefly.

Peptides related structurally to AKH are shown in Table 1;
these have been assayed for their adipokinetic activity.   From these
compounds and numerous other studies a number of key factors relating
structure with function can be elucidated.   Peptides with fewer than
eight amino acid residues (whether blocked or possessing a net
charge) all lack adipokinetic activity.   The octa-, nona- and decapeptides tested which possessed detectable activity were all uncharged.   Compounds 11 and 13 (Table 1) indicate the minimum
requirements: chain length of a minimum of eight residues, no net
charge, N-terminal pyroglutamate and C-terminal amide.

The requirements for the terminal residues are precise.   The
N-terminal residue of AKH or its agonists must be the L-enantiomer
of pyroglutamic acid: compare the activities of compounds 11 and 12
and compound 17 with 18.   For both epimer pairs the presence of

Table 1.   Adipokinetic activities of adipokinetic hormone and structurally related compounds.

Compound	Residues 1	2	3	4	5	6	7	8	9	10	Relative agonist activity*
1								H - Asn	Trp	Thr - NH$_2$	< 0.3
2							Ac - Asn	Trp	Gly	Thr - NH$_2$	< 0.3
3					H - Thr	Pro	Asn	Trp	Gly	Thr - NH$_2$	< 0.3
4				H - Phe	Thr	Pro	Asn	Trp	Gly	Thr - NH$_2$	< 0.3
5	PCA - OH										< 0.3
6	PCA	Leu	Asn								< 0.3
7	PCA	Leu	Asn	Phe							< 0.3
8	PCA	Leu	Asn	Phe	Thr	Pro					< 0.3
9	PCA	Leu	Asn	Phe	Thr	Pro - NH$_2$					< 0.3
10	PCA	Leu	Asn	Phe	Thr	Pro	Trp - NH$_2$				< 0.3
11	PCA	Leu	Asn	Phe	Thr	Pro	Asn	Trp - NH$_2$			20
12	D - PCA	Leu	Asn	Phe	Thr	Pro	Asn	Trp - NH$_2$			8
13	PCA	Leu	Asn	Phe	Ser	Pro	Gly	Trp - NH$_2$ (RPCH)			20
14	PCA	Leu	Asn	Phe	Thr	Pro	Asn	Trp	Gly - NH$_2$		7
15	PCA	Leu	Asn	Phe	Thr	Pro	Gly	Trp	Gly	Thr - NH$_2$	30
16	PCA	Leu	Asn	Phe	Thr	Pro	Asn	Trp	Thr	Gly - NH$_2$	3
17	D - PCA	Leu	Asn	Phe	Thr	Pro	Asn	Trp	Gly	Thr - NH$_2$	5
18	PCA	Leu	Asn	Phe	Gly	Pro	Asn	Trp	Gly	Thr - NH$_2$	33
19	PCA	Leu	Asn	Phe	Ser	Pro	Asn	Trp	Gly	Thr - NH$_2$	< 0.3
20		PCA	Asn	Phe	Thr	Pro	Asn	Trp	Gly	Thr - NH$_2$	< 0.3
21	PCA	Leu	Asn	Phe	Thr	Pro	Asn	Trp	Gly	Thr - NH$_2$ (AKH)	100

* The activity of adipokinetic hormone is defined as 100. Other preparations were compared with hormone standard on a molar basis. ED$_{50}$ of adipokinetic hormone is 3.0–4.5 pmol per locust. Activities shown as < 0.3 indicate that the maximum dose tested (200–250 pmol per locust) did not produce a significant adipokinetic response.

D-pyroglutamate reduces activity markedly.   The natural sequence of
Gly-Thr-NH$_2$ at the C-terminus is essential for maximal activity:
additions to compound 11 demonstrate this.   Addition of Gly-NH$_2$ to
the partially active octapeptide-compound 11 to produce a nonapeptide compound 14 reduces activity markedly.   Whereas addition
of Gly-Thr-NH$_2$ to produce AKH provides maximal activity.   If the
addition to compound 11 is Thr-Gly-NH$_2$ then the resulting decapeptide has very low activity.   Analogues with eight residues can
exhibit activity but nonapeptides all possess low activity.   The
precise requirements above eight residues are stressed further by
the very low activity exhibited by another nonapeptide compound 20
(des-Leu AKH).

The rigorous constraints for substitution and deletion within
AKH have lead to the suggestion that there may be a preferred configuration for AKH (Stone et al. 1978).   Substitution within the
molecule for Asn with Gly at position 7 does not affect the biological activity of octapeptides (as in compounds 11 and 13) but does
reduce markedly the activity of the decapeptide-compound 15.   From
this information and conformational analysis the existence of a β-
bend configuration for AKH has been suggested.   Stability for such
a configuration would be conferred by hydrogen bonding between residues 3 and 10 and between 5 and 8.   Such a possibility is supported
further by the marked reduction in activity which follows substitution of Thr at residue 5 by either Gly or Ser (compounds 18 and
19).   Moreover, the dehydro-analogue (4, 5-dehydro Leu) AKH has
activity identical with AKH but (3, 4-dehydro Pro) AKH has much
reduced activity.   Dehydrogenation is a subtle structural variation
which does not affect the activity of the Leu-residue but does affect
the agonist activity (perhaps β-bend formation) of the Pro-residue
(see also Hardy and Sheppard 1983).

DIURETIC HORMONE (DH)

The corpora cardiaca of locust comprise two more or less discrete lobes.   The more ventral storage lobes contain material which
originates predominantly, but not exclusively, from the cerebral
neurosecretory cells.   Storage lobe extracts are known to contain a
number of biologically active factors (see Goldsworthy and Mordue
1974).   In this paper we shall restrict our discussions to diuretic
hormone.   Our recent progress in the isolation and characterisation
of DH from locusts is discussed elsewhere in this symposium - see
Morgan and Mordue.   Here we shall restrict ourselves to the relationships between DH and other peptides.

During the course of our work on the isolation and characterisation of locust DH a method has been developed using reversed-phase
HPLC, by which it is possible to separate DH from methanol extracts
of storage lobe material (Morgan and Mordue 1983).   Consistently
two peaks have been found to contain diuretic activity, which are

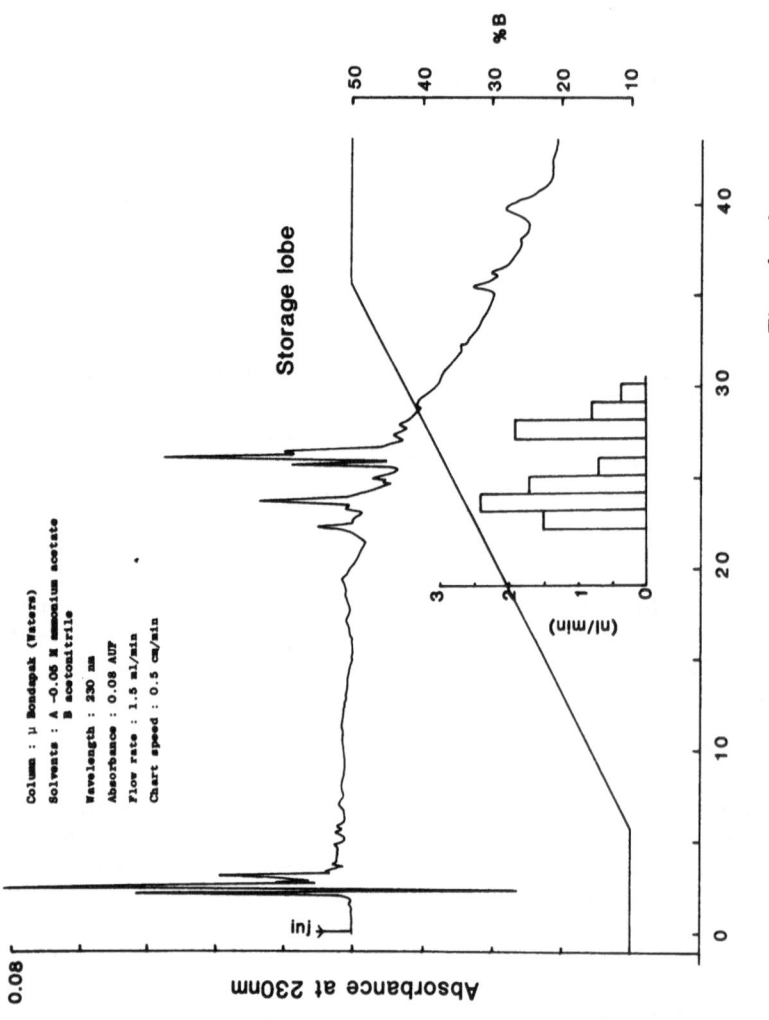

Fig. 4. Reversed-phase HPLC of a methanol extract of 25 storage lobes of Locusta on a Waters μ-Bondapak column, following the method of Morgan and Mordue (1983). Inset shows distribution of diuretic hormone activity.

distinctly separated from one another (Fig. 4).   As the peptides
were separated from methanol extracts of storage lobes, each peak
contains a peptide of relatively small molecular mass;   they are
therefore unlike the 'high' and 'low' molecular weight forms of DH
that have been isolated from the thoracic ganglia of Rhodnius (Aston
and White 1974).

    The locust diuretic peptides are isolated from the corpora
cardiaca, and our recent work indicates that they are chemically dissimilar to either vasopressin or the vasopressin-like diuretic factor
identified in the sub-oesophageal ganglion (SOG) by Proux and colleagues (Proux and Rougon-Rapuzzi 1980;   Proux et al. 1982;   Morgan
and Mordue, in preparation).   The evidence to support this view is
as follows: we have compared the elution characteristics of DH from
corpora cardiaca with five peptides from the posterior lobes of
vertebrate pituitaries using reversed-phase HPLC.   Fig. 5 shows the
elution profile of a mixture of arg-vasopressin, lys-vasopressin,
isotocin, vasotocin and oxytocin using a Waters μ-Bondapak column
employing chromatographic conditions identical to those used for the
isolation of DH from corpora cardiaca (Morgan and Mordue 1983).
Even though the five peptides differ only slightly in structure they
are readily resolved from each other into five discrete peaks.   The
retention times for four of the peptides are distinctly shorter than
both peaks of DH: oxytocin, however, chromatographs almost identi-
cally with the first peak of DH (DH I).   Oxytocin has been tested
in a variety of ways but even concentrations as high as 10^{-5}M it has
no diuretic effect on locust Malpighian tubules.   (It is worth
mentioning that arg-vasopressin also lacks any discernible diuretic
activity in locusts).   Estimates for the molecular mass of DH I
(Morgan and Mordue, this symposium) predict a size of c. 1000;   this
molecule differs from that isolated by Proux and colleagues.   The
latter peptide from the SOG, although showing some immunocytochemical
affinity with arg-vasopressin, has a molecular mass of c. 2500 (Cupo
and Proux 1983).   In view of the differences in chromatographic
properties and molecular mass it seems unwarranted to describe the
diuretic peptide isolated by Cupo and Proux (1983) as vasopressin-
like.   It shows only some common antigenic determinants at the C-
terminus with arg-vasopressin but differs substantially in many other
chemical and physiological properties.   Notwithstanding these
comments both DH I and DH II as indicated in Fig. 4 are very distinct
from the 'vasopressin-like' peptide.   Therefore from chromatographic
and biological data we can conclude that DH is not arg-vasopressin
and from biological data it is not oxytocin.   The immunocytochemical
and other evidence suggesting a similarity between mammalian vasopressin and DH may have to be re-examined.

DISTRIBUTION OF PEPTIDES WITHIN THE CNS

    There are a number of recent reports which describe the distri-
bution of peptides, both of insect origin and vertebrate-like,

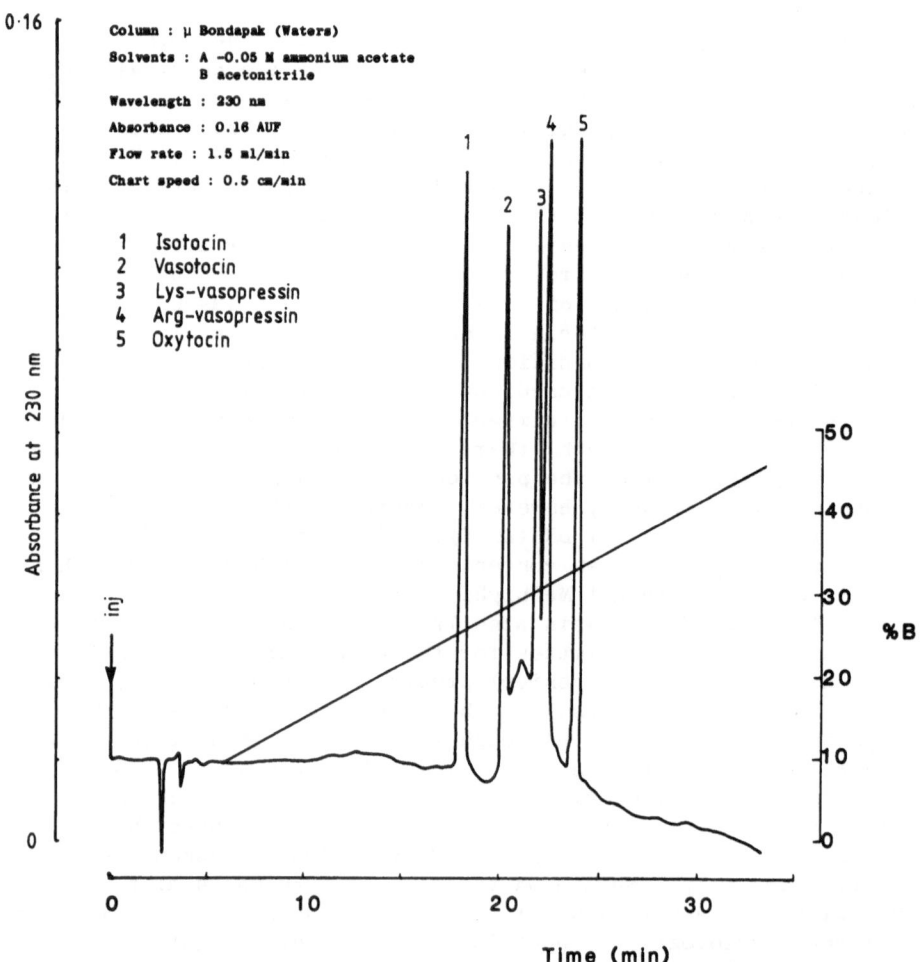

Fig. 5.   Reversed-phase HPLC of a mixture of vasopressin and
related peptides (8 µg of each) on a Waters µ-Bondapak
column, following the method of Morgan and Mordue (1983).

throughout the insect CNS.   For the greater part such reports depend
on immunocytochemical evidence.   For small peptides especially, but
also for many other peptides, the number of antigenic determinants
will be few.   Therefore the possibility of even specific anti-sera
against small peptides reacting with antigenic sites on a variety of
peptidergic cellular inclusions must be high.   Indeed from an extreme
standpoint it may be said that in a number of instances the immunocytochemical description 'hormone-like' is merely another way of
confirming that the reactive material is peptidergic!

Diuretic Hormone

Proux et al. (1982) demonstrated that the SOG contains significant levels of diuretic hormone, and although we might dispute that
this diuretic factor is a vasopressin-like molecule, our results
confirm that the SOG does contain significant levels of a DH.
Indeed dose-response curves for a diuretic hormone in the brain, SOG
and thoracic ganglia have been produced (Morgan and Mordue, in
preparation), and one of the intriguing findings has been the similarity of the levels of diuretic activity in the various parts of
the CNS.   An exception is the terminal abdominal ganglion, where
no discernible diuretic activity could be detected.

Brain, SOG and thoracic ganglia were extracted in methanol and
subsequently delipidated with dichloromethane.   These extracts were
then chromatographed on a Waters µ-Bondapak reversed-phase column
using the method of Morgan and Mordue (1983) (see also Fig. 4) and
for each tissue a peak corresponding to DH I, possessing biological
activity, was eluted.   This result strongly supports the view that
the diuretic activity extractable from the ventral nerve cord is
chemically similar to DH I extractable from corpora cardiaca.

Other Peptides

In addition to DH, proctolin, bursicon, AKH and a number of
vertebrate-like peptides are distributed similarly throughout the
CNS.   Many of the peptidergic neurones are found in segmental and
symmetrical arrangements within the CNS and such a distribution may
reflect the ontogenic development of the CNS.   The function of
such peptides throughout the CNS is obscure;   some may be relics
of evolutionary progress and indicate merely the genetic potential
which other neurones possess to produce a particular peptide in
large amounts;   for others they may have an important neurotransmitter or neuromodulatory role.   However, with the exception of DH,
proctolin and pancreatic polypeptide in <u>Calliphora</u> (see also Thorpe
and Duve, this symposium) little is known of the chemical properties
of these peptides in the CNS.   For many peptides their distribution
has been inferred from immunocytochemical reactivity.   Until more
rigorous confirmation of their presence and assessment of the quantities present is made we should resist speculation concerning the

roles for the plethora of peptides with such apparent widespread
distribution.

ACKNOWLEDGEMENTS

      Much of this work was supported by the ARC (Agricultural
Research Council).   We thank Miss R. Bisset for expert technical
assistance.   Investigations using Carausius material were carried
out with Dr. G. Gäde, University of Bonn and supported by NATO.

REFERENCES

ASTON R. J. and WHITE A. F. (1974) Isolation and purification of the
     diuretic hormone from Rhodnius prolixus. J. Insect Physiol. 20,
     1673-1682.
CARLSEN J., HERMAN W. S., CHRISTENSEN M. and JOSEFSSON L. (1979)
     Characterisation of a second peptide with adipokinetic and
     red-pigment concentrating activity from the locust corpora
     cardiaca. Insect Biochem. 9, 497-501.
CUPO A. and PROUX J. (1983) Biochemical characterisation of a vaso-
     pressin-like neuropeptide in Locusta migratoria. Evidence of
     high molecular weight protein encoding vasopressin sequence.
     Neuropeptides 3, 309-318.
GOLDSWORTHY G. J. and MORDUE W. (1974) Neurosecretory hormones in
     insects. J. Endocr. 60, 529-558.
HARDY P. M. and SHEPPARD P. W. (1983) Synthesis of (4, 5-dehydro
     Leu^2) - and (3, 4-dehydro Pro^6) - locust adipokinetic hormone.
     J. Chem. Soc. Perkin Trans. I, 723-729.
HARDY P. M., SHEPPARD P. W., BRUNDISH D. E. and WADE R. (1983)
     Tritiated Peptides. Part 13. Synthesis of (4, 5-^{3}H-Leu^2) - and
     (3, 4-^{3}H-Pro^6) - Locust Adipokinetic Hormone. J. Chem. Soc.
     Perkin Trans. I, 731-734.
MORDUE W. (1981) Adipokinetic hormone and related peptides. In
     Neurosecretion: molecules, cells, systems. Ed. D. S. Farner
     and K. Lederis. Plenum Press, New York.
MORGAN P. J. and MORDUE W. (1983) Separation and characteristics of
     diuretic hormone from the corpus cardiacum of Locusta. Comp.
     Biochem. Physiol. 75B(1), 75-80.
PROUX J. and ROUGON-RAPUZZI G. (1980) Evidence of a vasopressin-like
     molecule in migratory locust. Radioimmunological measurements
     in different tissues: correlation with various states of
     hydration. Gen. comp. Endocrinol. 42, 378-383.
PROUX J., ROUGON G. and CUPO A. (1982) Enhancement of dye excretion
     across locust Malpighian tubules by a diuretic vasopressin-like
     hormone. Gen. comp. Endocrinol. 47, 449-457.
STONE J. V. and MORDUE W. (1980) Adipokinetic Hormone. In Neuro-
     hormonal Techniques in Insects. Ed. T. A. Miller. Chap. 2.
     p.31-80. Springer-Verlag, New York.
STONE J. V., MORDUE W., BROOMFIELD C. E. and HARDY P. M. (1978)
     Structure-activity relationship for the lipid-mobilising action

of locust adipokinetic hormone. Synthesis and activity of a series of hormone analogues. Eur. J. Biochem. 89, 195-202.

NEUROENDOCRINE CONTROLS ON INSECT REPRODUCTION

K. G. Davey

Department of Biology, York University
Downsview, Ontario, M3J 1P3 Canada

INTRODUCTION

This paper presents an overview of the involvement of "neuro-secretion" in the control of reproduction in the adult insect. It will be preoccupied with the female system, largely because the control of reproduction in the male has not received sufficient attention from researchers. It does not attempt to constitute an exhaustive review, but focuses on recent experimental evidence.

It is first necessary to understand what is implied by the term "neurosecretion". All neurones secrete, but neurosecretory cells have customarily been defined on histological and ultrastructural grounds as those neurons with a more or less prominent content of membrane bound granules and in which the cytoplasm exhibits an affinity for one or more of a variety of stains. The two principal staining techniques are chrome-hematoxylinphloxin and paradelhyde-fuschin, and a very extensive literature exists describing cycles of staining intensity in such cells (see Raabe, 1982). The products of such cells are typically peptides, although aminergic cells are commonly referred to as neurosecretory. Because many of these stainable or granule-containing cells terminate in neurohemal structures, the term neurosecretory has always carried a strong implication of hormonal function. However, the realisation that peptides may be acting entirely within the nervous system as possible neurotransmitters has required the modification of this view. "Neuroendocrine cells", therefore, are nerve cells which produce and release

hormones. In anatomical terms, as they apply to insects, neuroendocrine cells are nerve cells which discharge their secretion into the
hemolymph at neurohemal sites, or, at the very least, which release
their secretion into spaces which exceed the dimensions of the typical synaptic space. This paper will be concerned with neuroendocrine
cells.

    Many neuroendocrine cells exhibit staining properties charac-
teristic of neorosecretory cells, and there is an extensive literature which attempts to relate changes in the histological properties
of such cells to various reproductive events. Thus, in Schistocerca
gregaria neurosecretory cells in the pars intercerebralis of the
brain contain little paraldehyde fuschin-positive material while the
terminal oocytes are in rapid vitellogenesis, but accumulations of
stainable material are noted when growth of the oocytes is complete
or when their growth is inhibited (Highnam and Lusis, 1962). In
the tsetse fly Glossina austeni, a neurosecretory outflow from cells
in the pars intercerebralis has been deduced on the basis of changes
in staining intensity in the perikarya and corpus cardiacum. This
outflow is correlated with the cycle of pregnancy (Ejezie and Davey,
1974). Studies similar to these have led to the generally accepted
view that cells with rather little content of stainable material
were likely to be actively releasing products, while those which
stain more prominently are not. Good support for this view comes
from the observation that histologically "empty" cells exhibit a
higher rate of incorporation of ^{35}S-cystine than more heavily stained
cells (Highnam and Mordue, 1970; Mordue and Highnam, 1973).

    Similar cycles of staining in neurosecretory cells have been
correlated with vitellogenesis for cells occurring in the suboesophageal ganglia in Locusta (Fréon, 1964), in Gryllus domesticus
(Huignard, 1964) and in the suboesophageal ganglia and ventral nerve
cord of the ovoviviparous cockroach, Leucophaea maderae (de Bessé
1965).

    The occurrence of such cycles in adult insects is intriguing,
but purely histological evidence, unsupported by direct experimental
intervention, can only be suggestive. Even when correlations are
firmly established, as they have been in a number of the examples
cited above, it is possible that the observed neurosecretory cycles
may be correlated with unsuspected physiological events not directly
related to reproduction. For example, although cycles may be more
closely correlated with feeding rather than with any reproductive
cycle (Grossman, 1977; Grossman and Davey, 1978). The remainder
of this paper, then, will be devoted to examining events for which
a neurosecretory control has been established by more directly experimental means.

NEUROENDOCRINE CONTROL OF ENDOCRINE ORGANS

One of the classical concepts to emerge from studies on
vertebrates is that of neuroendocrine hormones as "master" hormones,
controlling the synthesis and/or release of hormones from other
endocrine organs. While there is much suggestive evidence
in the literatue, concrete evidence which demonstrates a direct
link between neuroendocrine factors and the activity of other
endocrine organs in reproduction is rare for insects. In only
two systems, the corpus allatum and the ovary, is there good
direct evidence.

In the majority of insects, the corpus allatum is essential
for expression of the normal egg production: allatectomy stops or
reduces very markedly egg production. For most insects, juvenile
hormone, the product of the corpus allatum,i s the gonadotropin,
regulating the process of vitellogenesis. It is also involved in
other phenomena, such as behaviour, the control of accessory glands
and early events in the gonads. Given that the precise involvement
of juvenile hormone in these processes may vary among insects and
given that neuroendocrine factors may themselves intervene directly
in some of these processes, including vitellogenesis, then neuroendocrine control of the corpus allatum becomes difficult to
establish.

There is a close anatomical association between the corpus
allatum and elements of the neurosecretory system, and this has
been extensively studied in various orthopteroids. The allatum is
innervated by the nervi corporis allati (nca) I, from the brain
and corpora cardiaca, and the nca II, from the suboesophageal
ganglion (Willey, 1961). The fibres in nca II appear to be
primarily neurosecretory, emanating from a small number of neurosecretory cells in the suboesophageal ganglion (Mason, 1973;
Pipa and Novak, 1979). The situation in nca I is more complex.
The nerve has both neurosecretory and non-neurosecretory axones
(Scharrer, 1964) which originate in the brain and which travel
to the allatum via the nervi corporis cardiacii (ncc), which also
supply the corpus cardiacum. While it is difficult to be precise,
some, at least, of the axones have their origins in neurosecretory
cells of the contralateral pars intercerebralis and the ipsilateral pars lateralis (Mason, 1973; Pipa and Novak, 1978). It is
worth noting that the ncc I, some fibres of which contribute to the
nca I, receives fibres from many parts of the brain (Willey, 1961).
There is no doubt that there are neurohaemal structures in the
corpus allatum (Scharrer, 1952), but it is also important to note
that there may be neurohemal organs along the length of nca II
(Weber and Gaude, 1971).

While these important anatomical facts are suggestive of neurosecretory control of the corpus allatum, it is important to
recognise that they demonstrate only that neurosecretory cells
terminate in or very close to the allatum. They do not in themselves demonstrate that the neurosecretory products have
anything to do with the control of the allatum. Indeed, there
is evidence which suggests that the corpus allatum is the
release site for prothoracicotropic hormone in Lepidoptera
(Raabe, 1982). The occurrence of non-neurosecretory axons
in nca I further complicates matters, so that interpretation
of experiments based exclusively on surgical manipulation such
as nerve section becomes very difficult indeed.

The existence of a radiochemical assay which directly monitors
the capacity of the corpus allatum to synthesise juvenile hormone
in vitro (Tobe and Pratt, 1976) has presented opportunities to
explore the control of the corpus allatum. The ovoviviparous
cockroach, Diploptera punctata, has provided excellent experimental
material. In this insect, eggs are retained by the mother, and the
embryos develop and grow in a special brood chamber where they are
supplied with a secretion from "milk glands" (Stay and Coop, 1974).
In Diploptera, virgin females do not mature eggs, and mating trig-
gers a bout of vitellogenesis which is juvenile hormone (JH)-
dependant. Similarly, during pregnancy vitellogenesis is inhibited
and pre-mature removal of the embryos is followed by a bout of
JH-dependant vitellogenesis. It has been known for many years
that egg production can be inhibited in pregnant or virgin females
of another ovoviviparous cockroach, Leucophaea maderae, by
denervating the allatum, and careful electrocoagulation studies
led Engelmann and Luscher (1957) to the conclusion that this activation was governed by non-neurosecretory fibres travelling in ncc I.

Against this background, the question of controls on the
corpus allatum has been examined in more detail in Diploptera by
Stay and Tobe and their colleagues. Initial studies confirmed
that the corpus allatum exhibited a cycle of capability to
synthesize C16 JH in vitro which was closely correlated with the
gonotrophic cycle (Tobe and Stay, 1977) and with changes in the
volume, cell number and cytoplasmic/nuclear ratio in the gland
itself (Szibbo and Tobe, 1981). Thus, the removal of the gland
and the assessment of its synthetic activity in vitro is a
reliable measure of its synthetic activity in vivo just before
its removal. Such studies also confirmed that innervation is
important, for denervation of the inactive allatum in a virgin
leads to a detectable increase in its synthetic capacity in vitro
within four hours of the denervation. More significant in the
present context is the observation that the denervated allatum
in the virgin female undergoes the same cycle of increase and

decrease of synthetic activity as an intact allatum in a normal
mated female. Clearly, humoral factors which do not depend on
an intact innervation are influencing the synthetic activity of
the allatum (Tobe and Stay, 1977).

Among the humoral factors which influence the rate of
biosynthesis of JH is JH itself which appears to stimulate
biosynthesis at low concentrations and inhibit at higher concentrations (Tobe and Stay, 1979, 1980). This effect appears
to be mediated by the brain via humoral factors (Tobe, 1980).
The ovary itself also plays an important regulatory role: the
synthetic capability of ovariectomised mated females remains low
and the synthetic capability can be re-established by implanting
an undeveloped ovary. The young ovary is thus a source of a
stimulatory factor. The ovary is also the source of an inhibitory
factor. Transplanting a corpus allatum from a recently mated
female into a male results in the normal increase in synthetic
activity, but that activity remains high: it does not exhibit
the decrease characteristic of the female. If an ovary is trans-
planted with the corpus allatum, the ovary grows in the normal way,
and the activity of the corpus allatum decreases after reaching its
normal peak. This inhibitory effect of a mature ovary is mimicked
by the injection of edcysterone, and it is probably reasonable to
assume that the mature ovary signals the corpora allata via ecdysone
(Stay et al., 1980). Ecdysterone added to the incubation medium
does not suppress synthesis in isolated corpora allata (Tobe, 1980),
suggesting that the effect of ecdysterone may not be direct. All
that is clear at present is that the allatum in Diploptera is
controlled by both inhibitory and stimulatory humoral factors.
The fact that these factors are as diverse as JH and ecdysterone
might perhaps argue for some integrative centre. Additionally,
ecdysterone is known to have effects on neurosecretory cells in
the brain of Rhodnius (Ruegg et al., 1982).

Is there evidence of humoral effects on the corpora allata
emanating from the brain? Good evidence for a trophic effect is
coming from studies on Diploptera. It has already been noted that
implanting corpora allata from mated females early in the gonotrophic cycle into males results in the normal increase in synthetic
capability over the ensuing five days (Stay et al., 1980). If
however, the pars intercerebralis of the recipient male has been
destroyed by radio frequency cautery, the rate of synthesis of the
implanted corpora allata is greatly reduced compared to those
implanted into males with portions of the optic lobes destroyed.
The demonstration that a trophic humoral influence is involved rests
with transplanting a brain from a female along with the allatum into
a male with the pars intercerebralis destroyed. In such cases the
inhibition of biosynthesis observed without the brain transplant is

partially reversed (Tobe et al., 1981). The effect of this
allatotropin is only apparent when the allatum is free of its
nervous connections, suggesting that the nervous inhibitory pathway
can override the trophic influence of the brain.

Perhaps the most important realisation to emerge from these
studies is the enormous complexity of controls in the allatum.
These controls apparently involve inhibitory control via nca I,
feedback effects from hemolymph JH, a stimulatory factor from the
ovary, an inhibitory factor, probably an ecdysteroid, from the
mature ovary and a humoral trophic factor from the central nervous
system. In the present context it is the last effect which is of
greatest interest. At present the role of this apparent neurosecretory influence would appear to be a rather minor one. However,
the importance of this role would alter dramatically if the effects
of JH and the ovarian factors were determined to operate via the
brain. The Diploptera system is clearly a useful one, and should
contribute in a major way to our understanding of the controls on
the allatum.

In the cockroach Nauphoeta cinerea, the synthetic activity of
quiescent corpora allata can be stimulated by implanting them into
females near the beginning of the gonotrophic cycle. However, this
stimulation does not occur if the recipients are decapitated
(Lanzrein et al., 1978).

Other reports, involving less direct evidence, support the view
that the brain exerts a humoral influence on the allatum in the adult
female. In the orthopterans Anacridium aegyptium and Locusta,
denervation of the allatum does not affect vitellogenesis, but
electrical stimulation of the pars intercerebralis leads to ovarian
growth (Girardie, 1966; Girardie et al., 1974). Reproductive
diapause in Leptinotarsa decemlineata is dependant upon photoperiod
acting via JH. Implanting a pair of allata from a non-diapausing
female into a diapausing female will not in itself result in egg
production in the recipient. However, if the recipient is
allatectomised, ovarian growth ensues (De Wilde and De Boer, 1961;
1969). These results have been interpreted as revealing humoral
influences on the allatum (De Kort and Granger, 1981). In the
linden bug, Pyrrhocoris apterus, experiments suggest an inhibitory
humoral influence from the brain (Hodkova, 1979). Finally, although
the influence is not humoral, the fact that the corpus allatum
receives neurosecretory fibres via nca II from the suboesophageal
ganglion should be mentioned. Severing this nerve leads to
vitellogenesis in Leucophaea, but has no effect in Locusta (Raabe,
1982) and Schistocerca (Strong, 1965). It is important to emphasize,
however, that these experiments rely on vitellogenesis as an index
of allatum activity. Such experiments require great caution in
their interpretation, for there may be a direct neuroendocrine
control of vitellogenesis.

The other major study involving the neuroendocrine control
of an endocrine gland involves the ovary, which has been demonstrated to secrete ecdysone in response to a neurosecretory
hormone in mosquitoes. It is important to understand in at
least a cursory way the endocrine background, most of which
has been derived from studies on Aedes aegypti. The corpus
allatum is essential for egg development only during previtellogenic growth of the follicle, where it governs the
development of existing follicles up to the immediately
previtellogenic "resting stage" (Gwadz and Spielman, 1973).
Ecdysone has been demonstrated to have two influences. It
promotes the development of secondary follicles up to the
point of their separation as identifiable follicles (Beckemeyer
and Lea, 1980). Ecdysone is also the gonadotropin in mosquitos
in that it controls the synthesis of vitellogenin by the fat
body (Fallon et al., 1974; Hagedorn, 1983). The third hormone
known to influence egg production is a neurosecretory factor
from the head. It was known that the head is required for
vitellogenesis for about 8 hours after feeding (Gillet, 1957).
This influence was traced to neurosecretory cells in the pars
intercerebralis which released their product through the corpus
cardiacum in response to a blood meal; the hormone was given
the name "egg development neurosecretory hormone" or EDNH (Lea,
1967, 1972).

This factor was eventually shown to promote ecdysone synthesis by the ovary and to be essential for the appearance of
ecdysteroid in the hemolymph (Hanoaka and Hagedorn, 1980). The
fact that EDNH stimulates ecdysone production by ovaries in vitro
(Hagedorn et al., 1979; Hanoaka and Hagedorn, 1980) presents
opportunities for the characterisation of the peptide and the
determination of titres in the hemolymph.

On the basis of currently available information, it is
possible to draw some tentative conclusions concerning the release
of EDNH. Since ecdysteroids do not appear in the hemolymph of
Aedes aegypti until after feeding (Hanoaka and Hagedorn, 1980),
since the head is required for vitellogenesis for no more than 8
hours after feeding (Gillet, 1957), and since ecdysteroid titres
fall during the second day post feeding (Hanoaka and Hagedorn, 1980),
it is reasonable to assume that feeding initiates the release of
the neurosecretory product. The careful surgical experiments of
Lea (1972) suggest that a humoral factor released after feeding
acts directly on the corpus cardiacum to bring about release of EDNH.

The source and number of such factors remain uncertain. The
inginous experiments of Chang and Judson (1977) demonstrated that
blood-borne stimuli originating from the digestion of the blood meal,
and mediated via the head, were essential for egg development. An

ovarian factor is also required for the release of EDNH (Borovsky, 1982; Lea and van Handel, 1982). The relationship between the ovarian factor and the factor resulting from the digestion of blood is unclear. The effect of the ovary on EDNH release parallels the effect of the young ovary in stimulating the synthesis of JH by the allatum in <u>Diploptera</u> (see above).

The evidence thus far available suggests, but does not prove, that EDNH disappears from the hemolymph after the first day post-feeding. It this is so, it may be that rising titres of ecdysone inhibit release of EDNH, as has been suggested for prothoracico-tropic hormone, which controls the synthesis of ecydsone by the prothoracic glands (Steel and Davey, 1983).

NEUROENDOCRINE CONTROL OF PROTEIN SYNTHESIS

The direct involvement of neurosecretion in vitellogenesis has been postulated by many authors, but concrete evidence has yet to be provided. There is a large number of studies which relate abla-tion of the pars intercerebralis to a failure in vitellogenesis. The probability that a neuroendocrine control of corpus allatum activity exists renders suspect those investigations which do not explore the effects of JH therapy on vitellogenesis in such operated insects. Indeed, JH or implantation of corpora allata appear able to repair the defect caused by ablation of the pars intercerebralis in a number of insects (Raabe, 1982). Similarly, for mosquitoes, the results of decapitation or ablation of neurosecretory cells are fully explicable on the basis of the effect of EDNH on vitellogenesis via ecdysone synthesis. A similar explanation may also be valid for other Diptera, although that possibility has been insufficiently explored.

Nevertheless, there remains a group of observations which strongly suggest a direct involvement of neuroendocrine factors in vitellogenesis. <u>Locusta</u> <u>migratoria</u> has been the object of a good deal of attention in this respect. Several reports (Girardie, 1966; Minks, 1967) emphasize that both the corpora allata and the pars intercerebralis are essential for the full expression of egg pro-duction. McCafferey (1976) showed that electrocoagulation of the cerebral median neurosecretory cells prevents vitellogenesis. Implanting corpora allata into such females does not result in the formation of complete eggs, although a few follicles deposit yolk. Similar, but less dramatic, observations were made by Luscher (1968), who showed that while decapitated females of <u>Schistocerca gregaria</u> into which corpora allata had been implanted were capable of complet-ing egg formation in an apparently normal fashion, the process was enhanced at least slightly the presence of a corpus cardiacum.

These studies, of course, deal with vitellogenesis rather than with protein synthesis itself. A more direct demonstration of the involvement of neurosecretion with protein synthesis is provided by recent work on the control of the accessory glands of the adult male of Rhodnius prolixus. The transparent accessory glands of the male are responsible for forming the spermatophore. The gland consists of three finger-like lobes on either side of the male, and a simple epithelium surrounds a large central reservoir containing the proteinaceous secretion (Davey, 1959). The accumulation of protein by the gland is sensitive to both allatectomy and ablation of the pars intercerebralis. JH therapy will restore the protein levels in the glands of allatectomised males to near normal levels. JH treatment also increased the protein content of glands from males lacking their neurosecretory cells, but failed to increase it to normal levels (Barker and Davey, 1981). It is possible to achieve synthesis of the proteins in the secretion by incubation of the gland in vitro (Barker and Davey, 1982) and a polypeptide has been partially characterised from the brain and corpus cardiacum which stimulated protein synthesis in such preparations (Barker and Davey, 1983). While it has not yet been possible to examine the role of JH in vitro in these preparations, there is no doubt that JH stimulates protein accumulation in vivo. There are thus some parallels between the interaction of JH and neuroendocrine factors in male Rhodnius and in some female insects. Both are essential, but JH can replace to some degree the neurosecretory influence. While the unravelling of this enigma will have to await additional data, it is tempting to speculate that JH directs the synthesis of specific proteins, at least partly by governing the development of the appropriate cellular machinery, while the neurosecretory factor governs the general level of protein synthesis, as originally suggested for the locust by Hill (1962), who showed that electrocautery of the median neurosecretory cells of the brain lowered the concentration of haemolymph proteins, while injections of homogenates of corpora cardiaca restored the levels. Injections of extracts of corpora cardiaca and pars intercerebralis into female desert locusts results in an increase in protein synthesis in the fat body (Osborne et al., 1968). These experiments, particularly the latter study, are both open to the objection that the factor may be acting indirectly. However, Carlisle and Loughton (1979) have reported two factors from the corpus cardiacum of locusts which affect protein synthesis in vitro in the fat body of males. One, from the glandular lobe, is adipokinetic hormone and suppresses protein synthesis, the other from the storage lobe, stimulates protein synthesis. While adipokinetic hormone, which also stimulates lipid mobilisation, is presumably released during flight, no information is yet available on the possible relationship of the second factor to reproduction.

There is strong circumstantial evidence that the general level
of protein synthesis may be under neuroendocrine control. If
definitive proof of such controls emerges, then the role of such
controls in governing the synthesis of vitellogenesis will need to
be investigated. It is, of course, already clear that JH directs
in a specific way the synthesis of vitellogenins. Perhaps a
neuroendocrine control of the general level of protein synthesis
might act in a synergistic or permissive way with JH.

OVULATION-OVIPOSITION

There are many reports in the literature which suggest a
neuroendocrine control of ovulation and/or oviposition. It should
be noted that ovulation, the release of the egg from the ovary, is
difficult to separate from oviposition, the deposition of the egg.
They are obviously closely related in that the latter cannot occur
in the absence of the former. Ovulation is difficult to detect in
the living female, and oviposition is usually assessed. However,
while ovulation may be entirely dependant on humoral factors, since
the ovary is not innervated, oviposition will almost certainly
involve neural factors as well.

Endocrine control of ovulation/oviposition has been suggested
by the induction of egg laying by injection of hemolymph from ovi-
positing females (Nayar, 1958; Okelo, 1971; Chaudbury and Dhadialla,
1976). These studies suffer because they use endocrinologically
competent recipients. Extracts from various parts of the central
nervous system have been shown to stimulate contractions in isolated
oviducts (Koller, 1954; Enders, 1956; Davey, 1967; Girardie and
Lafon-Cazal, 1972). Removal of the pars intercerebralis inhibits
ovulation/oviposition (Davey, 1967; Furtado, 1971; Foster, 1974;
Lazarovici and Pener, 1978). These facts are all suggestive of
neuroendocrine control of ovulation, but recent work in this
laboratory has established neuroendocrine control of ovulation in
Rhodnius in a more definitive way.

Rhodnius females take a large blood meal, which results in the
production of a batch of eggs. Virgin females do not ovulate, but
retain their eggs in the ovary.

Mating leads to ovulation (and oviposition), and matedness involves
a humoral factor from the spermathecae of mated females. Removal
of the neurosecretory cells of the pars intercerebralis inhibits
ovulation in mated females, and application of material from
neurosecretory cells to exposed reproductive ducts initiates powerful contractions (Davey, 1965, 1967).

By inserting a plasic window into the abdomens of females, it was possible to observe the contractions of the muscles of the ovary. Two peaks were observed in mated females: one occurs immediately after feeding, and the second coincides with the period of ovulation/ oviposition. The second peak is absent in virgin females, and in females from which the neurosecretory cells were removed after feeding (Kriger and Davey, 1982).

An extract of ten identifiable large neurosecretory cells from the pars intercerebralis increases the power of the contractions when added to an isolated ovary preparation. Similar myotropic activity is not present in extracts of the ocellar nerve. The activity is heat stable and trypsin-sensitive, suggesting that a peptide is involved. It is important to note that the ten cells were dissected free of other tissue before the extract was prepared (Kriger and Davey, in preparation).

Using the isolated ovary preparation as an assay, it has been possible to measure the myotropic activity of the hemolymph. Two peaks of activity are detected in the hemolymph of mated females, and only one in virgin females, coinciding with the peaks of ovarian motility observed in vivo (Kriger, 1981). Injections of extracts of the ten identifiable neurosecretory cells into females lacking the neurosecretory cells results in ovulation (Kriger and Davey, 1983).

It is clear that ovulation in Rhodnius is stimulated by a peptide neurohormone originating in ten identifiable neurosecretory cells in the pars intercerebralis and that the hormone is released on the one hand by feeding and on the other by mating. Nothing is known about the feeding stimulus, but the stimuli involved in the mating-induced release of the myotropin leading to the second peak have been investigated. While matedness in Rhodnius involves a spermathecal factor released at mating (Davey, 1965), this does not constitute the only stimulus, since mating precedes ovulation by some days. A second stimulus is necessary, and that stimulus is provided by ecdysteroids from the ovary.

The titre of ecdysteroid in the hemolymph varies in an identical way in both virgin and mated females, such that a peak occurs on about the fifth day after feeding, just prior to the peak of myotropic activity in the hemolymph of mated females. The peak of ecdysteroid is absent in ovariectomised females, as is the peak of myotropic activity. Injections of physiological doses of ecdysterone into ovariectomised females results in an increase in myotropic activity of the hemolymph in mated females, but not in virgins (Ruegg et al., 1981).

It was possible to associate bursts of action potentials recorded from the corpus cardiacum at the time of ovulation with the ten large neurosecretory cells which are the source of the myotropin. Exposure of isolated brain-retrocerebral complexes from mated ovariectomised females to ecdysterone in vitro initiates the bursts of action potentials characteristic of release (Ruegg et al., 1982). The action of the ecdysterone on the cells is not direct, but involves an aminergic pathway (Orchard et al., 1983). Aminergic cells are known to occur in the pars intercerebralis of Rhodnius (Hales and Davey, unpublished observations).

Ovulation in Rhodnius is thus under the control of a peptide neurohormone originating in the pars intercerebralis and released from the cardiacum. The hormone is released in mated females when an increase in ecdysteroid titre in the hemolymph signals that mature eggs are present in the ovary.

A similar mechanism may operate in tsetse flies. Removal of the neurosecretory cells of the pars intercerebralis (Foster, 1974) or of the corpus cardiacum (Chaudhury and Dhadialla, 1976) prevents ovulation in Glossina austeni. In virgin females, which do not ovulate, the outflow of neurosecretion detected histologically in mated females is absent (Ejezie and Davey, 1974, 1977). Ejezie and Davey (1976) have suggested that two stimuli - mating and a signal from the ovary - are required for ovulation in Glossina. In this respect, it is worth noting that injection of ecydsone leads to premature larviposition in tsetse flies (Denlinger, 1975).

OTHER NEUROENDOCRINE CONTROLS

In Rhodnius, abdominal neurosecretory organs are the source of an antigonadoptropin which antagonises the action of JH on the follicle cells. JH causes the follicle cells to shrink, thereby opening up large extracellular spaces between them (Abu-Hakima and Davey, 1977). This action of JH is antagonised by an antigonado-tropin, a peptide orginally thought to emanate from the ovary (Huebner and Davey, 1973; Liu and Davey, 1974). More recent studies have shown that the peptide originates in at least four pairs of abdominal neurosecretory organs which occur as diaphanous strips of tissue running from the ventral to the dorsal body wall at the antero-lateral borders of each of segments II, III, IV and V (Davey and Kuster, 1981; Kuster and Davey, 1981). These structures are innervated from the central nervous system, but an additional humoral input, perhaps from the ovaries, has been hypothesized as controlling the release of this neuroendocrine factor (Davey, 1982).

Abdominal neurosecretory centres, usually located in the ventral ganglionic chain, have not received the experimental attention that they perhaps deserve. There are many studies which document histological changes during the gonotrophic cycle in neurosecretory cells of the ventral nerve cord of females of a wide variety of insects (Raabe, 1982). There are some scattered experimental results. In Schistocerca, removal of the last abdominal ganglion prevents egg development, while severing all of the nervous connections of the ganglion does not. Implanting a terminal ganglion from a mature female into a deganglionated female of the appropriate age re-establishes egg production (Delphin, 1963).

How these putative neurohormones from the ventral nerve cord exert their action remains a matter for conjecture. A neglected possibility involves the chain of events between mating and egg production and oviposition. For many insects matedness in the female involves one or more factors transmitted by the male to the female during mating, and in some cases this factor has been shown to enter the circulation of the female (Gillot and Friedel, 1977; Davey, 1983). In most cases, these factors lead to oviposition, and it has been assumed that they act in a more or less direct way on the brain (Friedel and Gillot, 1976). This is possibly true in a number of cases, but an alternate explanation is possible.

In Rhodnius, for example, there is no doubt that a factor which emanates from mated spermathecae stimulates the central nervous system (Davey, 1965; Dumser, 1969; Ruegg, 1981). Although elements from the accessory glands of the male were demonstrated to be important in the response of the female, they were ineffective when injected into the hemocoel (Ruegg, 1981). Neurosecretory fibres have been observed to be associated with the spermathecae (unpublished observations).

In Hyalophora cecropia, the male transfers to the female a substance which stimulates the bursa copulatrix to secrete a factor which in turn acts to bring about the release of an oviposition inducing hormone from the corpus cardiacum (Truman and Riddiford, 1971; Riddiford and Ashenhurst, 1973). Nothing is known of the substance from the bursa. Perhaps the most suggestive evidence comes from work on the mating response in the ovoviviparous cockroach, Leucophaea maderae. In this cockroach, mating leads to activation of the ovaries via the removal of nervous inhibition on the corpora allata. Severing the nerve cord immediately after mating prevents the activation of the allatum. However, the nerve cord must be intact for two days after mating in order for mating to be fully effective (Engelmann, 1970).

Similarly, the inhibition of the corpus allatum which obtains during pregancy is relieved by severing the ventral nerve cord, and replacing the egg case in the brood chamber by a glass bead maintains the inhibition. This suggests that the inhibition is a purely mechanical event, transmitted to the allatum via the nerve cord (Engelmann, 1970). However, even if all of the nerves connecting the genitalia to the last abdominal ganglion are severed, inhibition of the allatum is nevertheless maintained. Moreover, when the nerve cord is severed at different levels, activation of the allatum occurs more rapidly in animals in which the cord is severed more anteriorly (Engelmann, 1970). Neurosecretory cells in the ventral ganglionic chain of Leucophaea have been shown to undergo histological changes during the gonotrophic cycle (de Besse, 1965). These facts concerning Leucophaea are at least provocative, suggesting, perhaps, that neurohormones from the ventral ganglionic chain may play a role in informing the central nervous system about the status of matedness or pregancy in Leucophaea.

In a few cases, evidence has been brought forward which suggests that neuroendocrine factors may influence in a direct way the functioning of the gonad. In particular, it is worth directing attention to the remarkable studies of Naisse (1969) on sexual differentiation in the glow-worm Lampyris noctiluca. In the early larval instars, the development of the male and female gonad is identical. Near the end of the fourth instar, specialised apical tissue differentiates from the mesoderm in the male. This secretory material disappears by the end of the fifth instar, when spermato-genesis begins. Careful surgical manipulations, including extirpations transplantations and parabioses, have led to the conclusion that the development of the apical tissue is dependant on a neurosecretory hormone emanating from the pars intercerebralis of males. Gonads from females can be induced to form the apical tissue by transplanting them into castrated males or by transplanting brains from males into female larvae.

In the triatomid bug Panstrongylus megistus, mitotic activity of the oogonia has been claimed to be under the direct control of a neuroendocrine factor from the brain, while differentiation of the ovary and meiotic divisions of the oocytes were found to be dependant on the secretion of ecdysone under the control of PTTH (Furtado, 1979). This form of control is at variance with what is known in the closely related Rhodnius prolixus, where pre-meiotic divisions in the testis are under the control of ecdysterone and JH (Dumser and Davey, 1975). In the female of Rhodnius, a mitosis which differentiates the tropho-cytes from the oogonia occurs on the first day of the fifth larval instar (Case, 1970), when ecdysteroid titres are low (Steel et al., 1982). However, the oocytes in Rhodnius do not undergo further divisio the nucleus forming the familiar "germinal vesicle", until fertilisatio initiates meiosis, as in most other insects.

In <u>Tenebrio</u> <u>molitor</u>, cautery of the pars intercerebralis decreases the previtellogenic growth of the ovary (Mordue, 1965). In <u>vitro</u> studies on <u>Tenebrio</u> suggest that this effect may be direct (Laverdure, 1972). It is clear that more thorough investigation of the possible effect of neuroendocrine factors on gonadal development should be productive.

CONCLUSION

This overview reveals that while there are rather few examples in which a neuroendocrine control has been demonstrated in reproductive processes, there is an abundance of evidence which suggests that neuro-secretory cells may play a central role in the control of reproduction. As evidence accumulates, the control of reproduction in insects is revealed as increasingly complex. For example, in <u>Rhodnius</u>, no fewer than six hormones: JH, ecdysterone, ovulation hormone, the antigonado-tropin, and the spermathecal factor, are now known to impinge on egg production. As this complexity becomes more apparent, attention will have to be focussed on the integrative action of the central nervous system, particularly its neurosecretory elements.

ACKNOWLEDGEMENTS

Research in the author's laboratory is supported by grants from the Natural Sciences and Engineering Research Council of Canada.

REFERENCES

Abu-Hakima R. and Davey K.G. (1977) The action of juvenile hormone on follicle cells of <u>Rhodnius</u> <u>prolixus</u> in vitro: The effect of colchicine and cytochalasin B. <u>Gen</u>. <u>Comp</u>. <u>Endocrinol</u>.32, 360-370.

Barker J.F. and Davey K.G. (1981) Neuroendocrine regulation of protein accumulation by the transparent accessory reproductive gland of male <u>Rhodnius</u> <u>prolixus</u>. <u>Internat</u>. <u>J</u>. <u>Invert</u>. <u>Reprod</u>. 3, 291-296.

Barker J.F. and Davey K.G. (1982) Intraglandular synthesis of protein in the transparent accessory reproductive gland in the male <u>Rhodnius</u> <u>prolixus</u>, <u>Insect</u> <u>Biochem</u>. 12, 157-159.

Barker J.F. and Davey K.G. (1983) A polypeptide from the brain and corpus cardiacum of male <u>Rhodnius</u> <u>prolixus</u> which stimulates <u>in</u> <u>vitro</u> protein synthesis in the transparent accessory reproductive gland. <u>Insect Biochem</u>. 13, 7-10.

Beckemeyer E.F. and Lea A.O. (1980) Induction of follicle separation in the mosquito by physiological amounts of ecdysterone. <u>Science</u>. 209, 819-821.

Borovsky D. (1982) Release of egg development neurosecretory hormone in <u>Aedes</u> <u>aegypti</u> and <u>Aedes</u> <u>taeniorhynchus</u> induced by an ovarian factor, <u>J</u>. <u>Insect Physiol</u>. 28, 311-316.

Carlisle J.A. and Loughton B.G. (1979) Adipokinetic hormone inhibits protein synthesis in <u>Locusta</u>. <u>Nature</u>. 282, 420-421.

Case D.C. (1970) <u>Postembyronic deveopment of the ovary of Rhodnius prolixus Stal</u>. M.Sc. Thesis, McGill University, Montreal.

Chang Y.Y.H. and Judson C.L. (1977) The role of isoleucine in differential egg production by the mosquito <u>Aedes aegypti</u> following feeding on human or guinea pig blood. <u>Comp</u>. <u>Biochem</u>. <u>Physiol</u>. 57, 23-28.

Chaudhury M.F.B. and Dhadialla T.S. (1976) Evidence of hormonal control of ovulation in tsetse flies. <u>Nature</u>. 260, 243-244.

Davey K.G. (1959) Spermatophore production in <u>Rhodnius prolixus</u>. <u>Quart</u>. <u>J</u>. <u>Microscop</u>. <u>Sci</u>. 100, 221-230.

Davey K.G. (1965) Corpulation and egg production in <u>Rhodnius prolixus</u>: the role of the spermathecae. <u>J</u>. <u>Exp</u>. <u>Biol</u>. 42, 373-378.

Davey K.G. (1967) Some consequences of copulation in <u>Rhodnius prolixus</u>. <u>J</u>. <u>Insect</u> <u>Physiol</u>. 13, 1629-1636.

Davey K.G. (1982) The effect of severing abdominal nerves on egg production in <u>Rhodnius prolixus</u> Stal. <u>J</u>. <u>Insect</u> <u>Physiol</u>. 28, 509-512.

Davey K.G. (1983) The female system. In <u>Comprehensive Insect Physiology, Biochemistry and Pharmacology</u>, (Ed. by Gilbert L.I. and Kerkut G.A.), Pergamon Press, New York (in press).

Davey K.G. and Kuster J.E. (1981) The source of an antigonado-tropin in the female of <u>Rhodnius prolixus</u> Stal. <u>Can</u>. <u>J</u>. <u>Zool</u>. 59, 761-764.

de Bessé N. (1965) Recherches histophysiologiques sur la neuro-sécrétion dans la chaine nerveuse ventrale d'une blatte, <u>Leucophaea maderae</u> (F.). <u>C</u>. <u>R</u>. <u>Acad</u>. <u>Sci</u>. 260, 7014-7017.

de Kort C.A.D. and Granger N.A. (1981) Regulation of the juvenile hormone titer. <u>Ann</u>. <u>Rev</u>. <u>Entomol</u>. 26, 1-128.

Delphi n F. (1963) Histology and possible functions of neuro-secretory cells in the ventral ganglia of <u>Schistocerca gregaria</u> Forsk. <u>Nature</u> 200, 913-915.

Denlinger D.L. (1975) Insect hormones as tsetse abortifacients. <u>Nature</u> 253, 347-348.

De Wilde J. and De Boer J.A. (1961) Physiology of diapause in the adult Colorado beetle: II. Diapause as a case of pseudo-allatectomy. <u>J</u>. <u>Insect Physiol</u>. 6, 152-161.

De Wilde J. and De Boer J.A. (1969) Humoral and nervous pathways in photoperiodic induction of diapause in <u>Leptinotarsa decemlineata</u>. <u>J</u>. <u>Insect Physiol</u>. 15, 661-675.

Dumser J.B. (1969) <u>Evidence For a Spermathecal Hormone in Rhodnius prolixus (Stal)</u>. M.Sc. Thesis, McGill University, Montreal.

Dumser J.B. and Davey K.G. (1975) The <u>Rhodnius</u> testis: Hormonal effects on germ cell division. <u>Can</u>. <u>J</u>. <u>Zool</u>. 53, 1683-1689.

Ejezie G.C. and Davey K.G. (1974) Changes in the neurosecretory cells, corpus cardiacum, and corpus allatum during pregnancy in <u>Glossina austeni</u> Newst. (Diptera, Glossinidae) <u>Bull</u>. <u>Entomol</u>. <u>Res</u>. 64, 247-256.

Ejezie G.C. and Davey K.G. (1976) Some effects of allatectomy
    in the female tsetse, Glossina austeni. J. Insect Physiol. 22,
    1743-1749.
Ejezie G.C. and Davey K.G. (1977) Some effects of mating in
    female tsetse, Glossina austeni Newst. J. Exp. Zool. 200, 303-
    310.
Enders E. (1956) Die hormonale steuerung rhythmischer bewegungen
    von insekten-ovidukten. Verh. Dtsch. Zool. Ges. 19, 113-116.
Engelmann F. (1970) The Physiology of Insect Reproduction,
    Pergamon Press, Oxford, England.
Engelmann F. and Luscher M. (1957) Die hemmende wirkung des
    gehirns auf die corpora allata bei Leucophaea maderae (Orthop-
    tera). Verhandl. Deut. Zool. Ges. Hamburg. 1956, 215-220.
Fallon A.M., Hagedorn H.H., Wyatt G.R. and Kaufer H. (1974)
    Activation of vitellogenin synthesis in the mosquito Aedes
    aegypti by ecdysone. J. Insect Physiol. 20, 1815-1823.
Foster W.A. (1974) Surgical inhibition of ovulation and gestation
    in the tsetse fly Glossina austeni Newst. Bull. Entomol. Res.
    63, 483-493.
Fréon G. (1964) Contribution a l'étude de la neurosécrétion dans
    la chaine nerveuse ventrale du criquet migrateur, Locusta
    migratoria (L.) Bull. Soc. Zool. Fr. 89, 819-830.
Friedel T. and Gillot C. (1976) Male accessory gland substance of
    Melanoplus sanguinipes: an oviposition stimulant under the control
    of the corpus allatum. J. Insect Phyiol. 22, 489-495.
Furtado A.E. (1971) Recherches sur le controle endocrine cérébral
    de la vitellogenese et de la parturition chez une Punaise
    vivipare, Stilbocoris natalenis (Heteropteres, Lygeides).
    C. R. Acad. Sci. 272, 2468-2471.
Furtado A.E. (1979) The hormonal control of mitosis and meiosis
    during oogenesis in a blood-sucking bug Panstrongylus megitus.
    J. Insect Physiol. 25: 561-570.
Gillett J.D. (1957) Variation in the time of release of the
    ovarian development hormone in Aedes aeqypti. Nature 180,
    656-657.
Gillot C. and Friedel T. (1977) Fecundity-enhancing and
    receptivity-inhibiting substances produced by male insects: A
    review. Adv. Invert. Repro. 1, 199-218.
Girardie A. (1966) Controle de l'activite genitale chez Locusta
    migratoria. Mise en evidence d'un facteur gonadotrope et d'un
    facteur allatrope dans la pars intercerebralis. Bull. Soc.
    Zool. Fr. 91, 423-431.
Girardie A. and Lafon-Cazal M. (1972) Controle endocrine des
    contractions de l'oviducte isolé de Locusta migratoria
    migratorioides (R. et F.). C. R. Acad. Sci. 274, 2208-2210.
Girardie A., Moulins M. and Girardie J. (1974) Rupture de la
    diapause ovarienne d'Anacridium aegyptium par stimulation
    electrique des cellules neurosecretrices medians de la pars
    intercerebralis. J. Insect Physiol. 20, 2261-2275.

Grossman M. (1977) Neurosecretion in the Adult Male Tsetse Fly, Glossina austeni Newstead. M.Sc. Thesis, York University, Toronto.

Grossman M. and Davey K.G. (1979) The effect of feeding on the cerebral neurosecretory system of the adult male tsetse, Glossina austeni Newst. Can. J. Zool. 56, 1988-1992.

Gwadz R.W. and Spielman A. (1973) Corpus allatum control of ovarian development in Aedes aegypti. J. Insect Physiol. 19, 1441-1448.

Hagedorn H.H. (1983) The role of ecdysteroids in the adult insect. In: Insect Endocrinology (Ed. by Laufer H. and Downer R.). Alan R. Liss, Inc., New York in press.

Hagedorn H.H., Shapiro J.P. and Hanoaka K. (1979) Ovarian ecdysone secretion is controlled by a brain hormone in an adult mosquito. Nature 282, 92-94.

Hanaoka K. and Hagedorn H.H. (1980) Brain hormone control of ecdysone secretion by the ovary in the mosquito. In: Progress in Ecdysone Research, (Ed. by Hoffman J.A.), pp. 467-480, Elsevier/North Holland, Amsterdam.

Highnam K.C. and Lusis O. (1962) The influence of mature males on the neurosecretory control of ovarian development in the desert locust, Quart. J. Micro. Sci. 103, 73-81.

Highnam K.C. and Mordue A.J. (1970) Estimates of neurosecretory activity by an autoradiographic method in adult female Schistocerea gregaria. Gen. Comp. Endocrinol. 15, 31-38.

Hill L. (1962) Neurosecretory control of haemolymph protein concentration during ovarian development in the desert locust. J. Insect Physiol. 8, 609-619.

Hodkova M. (1979) Hormonal and nervous inhibition of reproduction by brain in diapausing females of Pyrrhocoris apterus (L.) (Hemiptera). Zool. J. Physiol. 83, 126-136.

Huebner E. and Davey K.G. (1973) An antigonadotropin from the ovaries of Rhodnius prolixus Stal. Can. J. Zool. 51, 113-120.

Huignard J. (1964) Recherches histophysiologiques sur le controle hormonal de l'ovogenese chez Gryllus domesticus L. C. R. Acad. Sci. 259, 1557-1560.

Koller G. (1954) Zur Frage der hormonalen Steuerung bei rhythmischer Eingeweide bewegungen von Insekten, Verh. Dtsch. Zool. Ges. 27, 417-422.

Kriger F.L. (1981) Ovulation in Rhodnius prolixus (Stal). Ph.D. Thesis, York University, Toronto.

Kriger F.L. and Davey K.G. (1982) Ovarian motility in mated Rhodnius prolixus requires an intact cerebral neurosecretory system. Gen. Comp. Endocrinol. 48, 130-134.

Kriger F.L. and Davey K.G. (1983) Ovulation in Rhodnius prolixus is induced by an extract of neurosecretory cells, Can. J. Zool. 61, 684-686.

Kuster J.E. and Davey K.G. (1981) Fine structure of the abdominal neurosecretory organs of Rhodnius prolixus Stal. Can. J. Zool. 59, 765-770.

Lanzrein B., Gentinetta V., Fehr R. and Luscher M. (1978)
    Correlation between hemolymph juvenile hormone titer, corpus
    allatum volume and corpus allatum in vivo and in vitro activity
    during oocyte maturation in a cockroach (Nauphoeta cinerea).
    Gen. Comp. Endocrinol. 36, 339-345.
Laverdure A.M. (1972) L'évolution de l'ovaire chez la femelle
    adulte de Tenebrio molitor: La prévitellogenèse. J. Insect
    Physiol. 18, 1477-1491.
Lazarovici P. and Pener M.P. (1978) The relations of the pars
    intercerebralis, corpora allata and juvenile hormone to oocyte
    development and oviposition in the African migratory locust.
    Gen. Comp. Endocrinol. 35, 375-386.
Lea A.O. (1967) The medial neurosecretory cells and egg maturation
    in mosquitoes, J. Insect Physiol. 13, 419-429.
Lea A.O. (1972) Regulation of egg maturation in the mosquito by
    the neurosecretory system: the role of the corpus cardiacum.
    Gen. Comp. Endocrinol. Suppl. 3, 602-608.
Lea A.O. and Van Handel E. (1982) A neurosecretory hormone-
    releasing factor from ovaries of mosquitoes fed blood. J.
    Insect Physiol. 28, 503-508.
Liu T.P. and Davey K.G. (1974) Partial characterization of
    a proposed antigonadotropin from the ovaries of the insect
    Rhodnius prolixus Stal. Gen. Comp. Endocrinol. 24, 405-408.
Luscher M. (1968) Hormonal control of resopiration and protein
    synthesis in the fat body of the cockroach Nauphoeta cinerea
    during oocyte growth, J. Insect Physiol. 14, 499-511.
Mason C.A. (1973) New features of the brain-retrocerebral
    neuroendocrine complex of the locust Schistocerca vaga (Scudder)
    Z. Zellforsch. 141, 19-32.
McCaffery A.R. (1976) Effects of electrocoagulation of cerebral
    neurosecretory cells and implantation of corpora allata on
    oocyte development in Locusta migratoria. J. Insect Physiol.
    22, 1081-1092.
Minks A.K. (1967) Biochemical aspects of juvenile hormone action
    in the adult Locusta migratoria. Arch. neerl. Zool. 17,
    175-257.
Mordue A.J. and Highnam K.C. (1973) Incorporation of cysteine
    into the cerebral neurosecretory system of adult desert
    locusts. Gen. Comp. Endocrinol. 20, 351-357.
Mordue W. (1965) Neuroendocrine factors in the control of oocyte
    production in Tenebrio molitor (L.). J. Insect Physiol. 11,
    617-629.
Naisse J. (1969) Rôle des neurohormones dans la differenciation
    sexuelle de Lampyris nocticula. J. Insect Physiol. 15, 877-892.
Nayar K.K. (1958) Studies on the neurosecretory system of Iphita
    limbata Stal. Part V. Probable endocrine basis of ovi-
    position in the female insect. Proc. Indian Acad. Sci. B., 47:
    233-251.

Okelo O. (1971) Physiological control of oviposition in the female
    desert locust, Schistocerca gregaria Forsk. Can. J. Zool. 49,
    969-974.
Orchard I., Ruegg R.P. and Davey K.G. (1983) The role of
    central aminergic neurons in the action of 20-hydroxy ecdysone
    on neurosecretory cells of Rhodnius prolixus. J. Insect
    Physiol., in press.
Osborne D.J., Carlisle D.B. and Ellis P.E. (1968) Protein
    synthesis in the fatbody of the female desert locust,
    Schistocerca gregaria Forsk., in relation to maturation.
    Gen. Comp. Endocrinol. 11, 347-354.
Pipa R.L. and Novak F.I. (1979) Pathways and fine structure of
    neurons forming the nervi coporis allati II of the cockroach
    Periplaneta americana (L.). Cell Tissue Res. 201, 227-237.
Raabe M. (1982) Insect Neurohormones. Plenum Press, New York.
Riddiford L.M. and Ashenhurst J.B. (1973) The switchover from
    virgin to mated behaviour in female cecropia moths: the role
    of the bursa copulatrix. Biol. Bull. Woods Hole. 144, 162-171.
Ruegg R.P. (1981) Factors Influencing Reproduction in Rhodnius
    prolixus. PhD Thesis, York University, Toronto.
Ruegg R.P., Kriger, F.L., Davey, K.G. and Steel, C.G.H. (1981)
    Ovarian ecdysone elicits release of a myotropic ovulation
    hormone in Rhodnius (Insecta: Hemiptera). Internat. J. Invert.
    Reprod. 3, 357-361.
Ruegg R.P., Orchard I. and Davey K.G. (1982) 20-hydroxyecdysone
    as a modulator of electrical activity in neurosecretory cells of
    Rhodnius prolixus. J. Insect Physiol.3, 243-248.
Scharrer B. (1952) Neurosecretion XI. The effects of nerve section
    on the intercerebralis-cardiacum-allatum system of the
    insect Leucophaea maderae. Biol Bull. (Woods Hole) 102,
    261-272.
Scharrer B. (1964) Histophysiological stiduies on the corpus
    allatum of Leucophaea maderae. IV. Ultrastructure during
    normal activity cycle. Z. Zellforsch. 62, 125-148.
Stay B. and Coop A.C. (1974) "Milk" secretion for embryogenesis
    in a viviparous cockroach. Tissue and Cell 6, 669-693
Stay B., Friedel T., Tobe S.S. and Mundall E.C. (1980) Feed-
    back control of juvenile hormone synthesis in cockroaches:
    possible role for ecdysterone. Science 207, 898-900.
Steel C.G.H. and Davey K.G. (1983) Integration in the insect
    endocrine system. In Comprehensive Insect Physiology,
    Biochemistry and Pharmacology (Ed. by Gilber L.I. and Kerkut
    G.A.) Pergamon Press, New York (in press).
Strong L. (1965) The relationships between the brain , corpora
    allata and oocyte growth in the Central American locust
    Schistocerca sp. II. The innervation of the corpora allata,
    the lateral neurosecretory complex and oocyte growth. J.
    Insect Physiol. 11, 271-282.
Szibbo C.M. and Tobe S.S. (1981) Cellular and volumetric changes
    in relation to the activity cycle in the corpora allata of
    Diploptera punctata. J. Insect Physiol. 27,655-665.

Tobe S.S. (1980) Regulation of the corpora allata in adult female
    insects. In Insect Biology in the Future (Ed. by Locke, M.,
    and Smith D.S.) pp. 345-367. Academic Press, New York.
Tobe S.S. and Pratt G.E. (1976) Farnesenic acid stimulation of
    juvenile hormone biosynthsis as an experimental problem in
    corpus allatum physiology. In The Juvenile Hormones (Ed. by
    Gilbert, L.I.) pp. 147-163. Plenum Press, New York.
Tobe S.S. and Stay B. (1977) Corpus allatum activity in vitro
    during the reproductive cycle of the viviparous cockroach,
    Diploptera punctata (Eschscholz) Gen. Comp. Endocrinol. 31,
    138-147.
Tobe S.S. and Stay B. (1979) Modulation of juvenile hormone
    synthesis by an analogue in the cockroach. Nature. 281, 481-
    482.
Tobe S.S. and Stay B. (1980) Control of juvenile hormone
    biosynthesis during the reproductive cycle of the viviparous
    cockroach. III. Effects of denervation and age on compensation
    with unilateral allatectomy and supernumerary corpora allata.
    Gen. Comp. Endocrinol. 40, 89-98.
Tobe S.S., Stay B., Friedel T., Feyereisen R. and Paulson C.
    (1981) The role of the brain in regulation of the corpora
    allata in female Diploptera punctata. In Juvenile Hormone
    Biochemistry (Ed. by Pratt, G.E. and Brooks, G.T.) pp.161-174.
    Elsevier/North Holland, Amsterdam.
Truman J.W. and Riddiford L.M. (1971) Role of the corpora
    cardiaca in the behaviour of saturniid moths. II. Oviposition.
    Biol.Bull. (Woods Hole) 140, 8-14.
Weber W. and Gaude H. (1971) Ultrastruktur des neurohaemalorgans
    im nervus corporis allati II von Acheta domesticus. Z.
    Zellforsch. 121, 561-572.
Willey R.B. (1961) The morphology of the stomodeal nervous system
    in Periplaneta americana and other Blattaria. J. Morph. 108,
    219-262.

THE ROLE OF BIOGENIC AMINES IN THE REGULATION OF

PEPTIDERGIC NEUROSECRETORY CELLS

Ian Orchard

University of Toronto, Dept. of Zoology
Toronto, Ontario
Canada, M5S 1A1

INTRODUCTION

The morphological analogy between the pars-intercerebralis-corpus-cardiacum complex of insects and the hypothalamo-pituitary complex of vertebrates was appreciated early in the development of neurosecretion (SCHARRER and SCHARRER, 1944; HANSTROM, 1953). Both systems are composed of groups of neurosecretory cells located in specific areas of the brain, which are linked to a neurohaemal organ containing both extrinsic neurosecretory axons and intrinsic cells (Fig. 1). Cells and fibers containing neurosecretory material of peptidergic nature and others containing biogenic amines are common to both systems (BAUMGARTEN et al., 1972;BJORKLUND, 1968; KLEMM and AXELSSON, 1973; OKSCHE, 1976). Functional similarities have also been found with active factors controlling metabolic and developmental processes, and it has been postulated that the mechanisms of integration of the two systems may be similar (SCHARRER and SCHARRER, 1944; WARTON and DUTKOWSKI, 1978).

In vertebrates, it is now well established that biogenic amines regulate the release of peptidergic hormones from the hypothalamo-pituitary complex (see McCANN et al., 1972; WEINER and GANONG, 1978). These peptidergic hormones include oxytocin released from the posterior pituitary, and hypothalamic hypophysiotropic hormones (releasing factors) which in turn regulate the release of the anterior pituitary hormones. Thus, the secretion of the six established anterior pituitary hormones is controlled largely by releasing factors that reach this portion of the gland via the hypophyseal-portal blood vessels. There are at least seven releasing factors: corticotropin releasing hormone (CRH) which increases the secretion of adrenocorticotropic hormones (ACTH); growth hormone releasing hormone (GHRH) and somatostatin which increase and

115

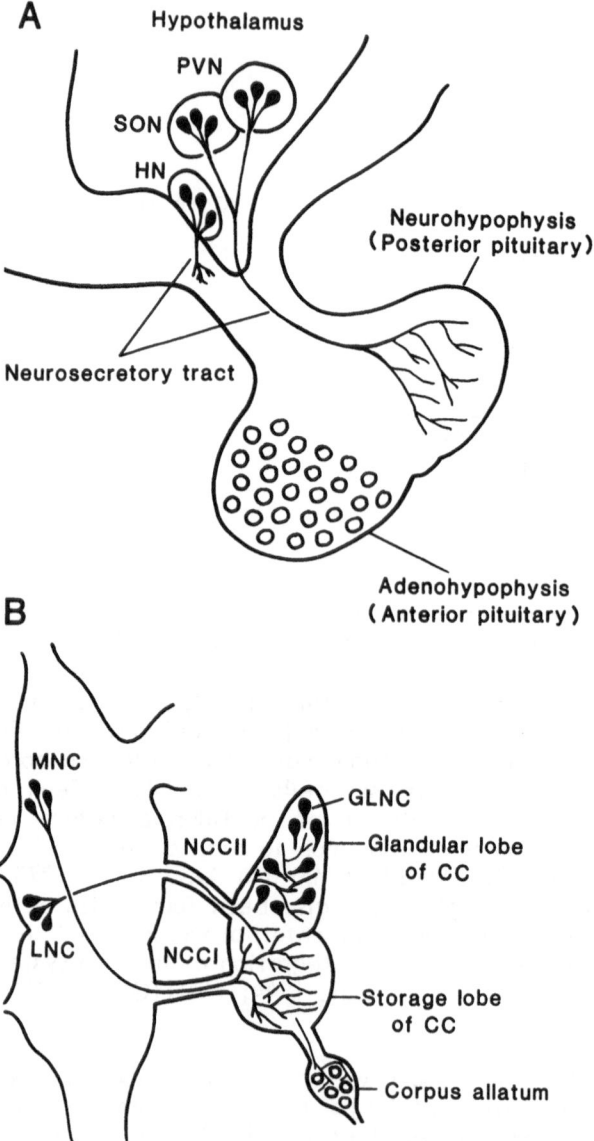

FIGURE 1. Diagram to show structural similarities between A, hypo-
thalomo-pituitary complex of mammals and B, brain-corpus cardiacum
complex of insects, as represented by the locust.
CC, corpus cardiacum; GLNC, glandular lobe neurosecretory cells;
HN, hypothalamic nuclei; LNC, lateral neurosecretory cells; MNC,
medial neurosecretory cells; NCCI and NCCII, nervi corporis cardiaci
I and II; PVN, paraventricular nucleus; SON, supraoptic nucleus.
Neurosecretory cells shaded, gland cells clear.

decrease respectively the secretion of growth hormone (GH); thyrotropin-releasing hormone (TRH) which stimulates the secretion of thyroid stimulating hormone (TSH); luteinising hormone releasing hormone (LHRH) which stimulates the secretion of luteinising hormone (LH) and follicle stimulating hormone (FSH); prolactin-releasing hormone (PRH) and prolactin-inhibiting hormone (PIH) which stimulate and inhibit respectively the secretion of prolactin. Except for PIH these releasing factors are believed to be peptides (see WEINER and GANONG, 1978). The hypothalamus is innervated in part by neurons that secrete dopamine, nor-epinephrine, epinephrine and serotonin (5-hydroxytryptamine, 5-HT). Some of the dopaminergic neurons secrete dopamine into the hypophyseal portal blood where it is transported to and acts directly upon the anterior pituitary to inhibit prolactin secretion (ie. dopamine is the PIH). The remainder of the amines appear to function as neurotransmitters involved in controlling the hypothalamic hormone-secreting neurons which secrete releasing factors. There is evidence to show that these aminergic neurons make synaptic contact with the hypothalamic hormone-secreting neurons. While there is still much work to be done to fully characterise these systems (for full review see WEINER and GANONG, 1978) it appears that nor-epinephrine acts in an excitatory way upon CRH-secreting neurons and TRH-secreting neurons, and also possibly upon GHRH-secreting neurons. Dopaminergic neurons influence the activities of LHRH-secreting neurons but the effect is dependent upon gonadal steroids and may be excitatory or inhibitory. Dopamine also appears to increase GH release but its action is outside of the blood-brain barrier and therefore is not directly upon the GHRH-secreting cells. 5-HT may stimulate the release of PRH thereby increasing prolactin release (CLEMENS et al., 1978). In addition there is evidence that the peptidergic hormones released from the posterior pituitary may also be under aminergic control. Thus, there appears to be a central aminergic component in the milk ejection reflex of rats (TRIBOLLET et al., 1978). Nor-adrenergic nerve terminals terminate on neurosecretory neurons (CARLSSON et al., 1962) and nor-epinephrine, dopamine and 5-HT applied iontophoretically influence the firing rate of identified neurosecretory cells (BARKER et al., 1971).

In comparison to the enormous body of information on the hypothalamo-pituitary complex of vertebrates, relatively little is known about the control of neurosecretory cells in insects (in particular the involvement of amines). While the association of biogenic amines with the peptidergic neurosecretory systems of insects and other invertebrates has been stressed by several authors (see EVANS, 1980; COOKE and GOLDSTONE, 1970) experiments on a functional relationship have been rare. This of course is due to a lack of basic information on the identity of neurohormones, the function they control and the specific neuronal pathways through which they are activated. However, in recent years evidence has been forthcoming for a physiological relationship between aminergic and peptidergic neurons in insects and there is now sufficient knowledge to make some comparisons with vertebrates. This review is aimed at collecting together the evidence obtained over the last few years which suggests a similar functional link between aminergic and peptidergic cells of insects and those of vertebrates. The review will begin with a brief survey of the close anatomical association between the two types of neurons and lead on to the current physiological evidence which links the two.

DISTRIBUTION OF AMINES IN THE INSECT NEUROSECRETORY SYSTEM

The presence of the catecholamines, dopamine and nor-epinephrine, the indolalkylamine, 5-hydroxytryptamine (5-HT) and the phenolamine, octopamine has been demonstrated in the insect nervous system (for complete review see EVANS, 1980). Especially pertinent to the present review is the distribution of these amines within the neurosecretory system.

Amine-containing cell bodies (revealed by fluorescent histochemistry) are located among the neurosecretory cells of the anterior pars-intercerebralis of several insect species including silver fish (Lepisima saccharina), house crickets (Acheta domesticus), locusts (Schistocerca gregaria and Locusta migratoria), dragonflies (Aeschna virdis and Aeschna cyanae)(see KLEMM and FALCK, 1978), and cockroaches (Blaberus craniifer, Byrsotria fumigata and Periplaneta americana, KLEMM, 1983). In locusts KLEMM and FALCK (1978) found several hundred green-fluorescent cell bodies within the anterior pars-intercerebralis which probably contained dopamine, although a few may have contained nor-epinephrine. One yellow fluorescent cell body was also found which contained an unidentifed indolalkylamine. These monoamine containing cell bodies were not identical to the peptidergic neurosecretory cells as revealed by double staining for monoamines and A and B peptidergic cells.

In addition, biogenic amines are also associated with neurohaemal organs. The neurosecretory cells in the brain provide the axons which constitute the nervi corporis cardiaci (NCC) passing to the corpus cardiacum (the neurohaemal organ). In some insects there are a pair of nerves, the NCCI and NCCII passing to the corpus cardiacum. In others these nerves are fused. In insects with separate NCCI and NCCII (see Fig. 1) it has been shown that the neurosecretory cells of the anterior pars intercerebralis contribute axons to the NCCI, while lateral neurosecretory cells contribute axons to NCCII. Such is the case in locusts. KLEMM and FALCK (1978) found that the tract of NCCI which lay within the brain was devoid of aminergic fluorescence except for small amounts of green and yellow droplets of fluorescent material or occasional fine green fluorescent varicose fibers at the edges. However at the periphery of the brain the NCCI became highly fluorescent as it passed to the corpus cardiacum. No fluorescent fibers have been detected in the NCCII. Fluorescence histochemistry has demonstrated the presence of dopamine, an unidentified yellow fluorescent indolalkylamine and an unidentified fluorophore (cysteinyldopa?) in the corpus cardiacum of the locusts Schistocerca gregaria and Locusta migratoria (KLEMM and FALCK, 1978; LAFON-CAZAL, 1981; LAFON-CAZAL and ARLUISON, 1976) and the cockroach

Blaberus craniifer (GERSCH et al., 1974). The presence of 5-HT has been indicated using biochemical techniques in Periplaneta americana (GERSCH et al., 1961; MIGLIORI-NATALIZI et al., 1970). Radioenzymatic assays have now demonstrated the presence of dopamine, nor-epinephrine and octopamine in the corpus cardiacum of Periplaneta americana, Locusta migratoria and Schistocerca gregaria (DYMOND and EVANS, 1979; EVANS, 1978; LAFON-CAZAL, 1981; ORCHARD and LOUGHTON, 1981a; DAVID and LAFON-CAZAL, 1979).

Within the locust the corpus cardiacum is separated into the storage lobe (consisting of terminals of neurosecretory cells located within the brain) and glandular lobe (consisting of intrinsic neurosecretory cells). Fluorescence histochemistry reveals that the catecholamines and indolalkylamines are restricted to the storage lobe, with only a few fibers passing into the glandular lobe (KLEMM and FALCK, 1978; LAFON-CAZAL, 1981). The glandular lobe itself is apparently devoid of catecholamines or indolalkylamines. Radioenzymatic studies have, however, revealed the presence of octopamine in both glandular and storage lobe (DAVID and LAFON-CAZAL, 1979; ORCHARD and LOUGHTON, 1981a; LAFON-CAZAL, 1981; GOOSEY and CANDY, 1982).

Autoradiographical studies performed at the ultrastructural level have recently been performed. These studies make use of the re-uptake properties of aminergic neurons such that incubation of aminergic neurons in small amounts of amine results in the uptake of the amine into the cell. LAFON-CAZAL (1981) found that ^{3}H-5HT was taken up into storage lobe axons containing electron-dense granules of 100nm diameter. ^{3}H-dopamine accumulated in 200 nm diameter electron-dense granules. The axons terminating in the glandular lobe contained 100 nm diameter electron-dense granules which were slightly labelled with 5×10^{-6} M of ^{3}H-dopamine, nor-epinephrine and 5-HT and in the absence of fluorescence these were considered octopaminergic. Some interesting ultracytochemical results were obtained using the proteolytic enzymes, pronase and pepsine (LAFON-CAZAL, 1981). One hour incubations in 5% solutions of these enzymes resulted in the digestion of the dense granules in some axons (presumably those containing peptides) while others were resistant (presumably the aminergic axons which accumulated the tritiated amines). This technique may prove a useful one for distinquishing peptidergic and aminergic axons at the ultrastructural level.

It was once believed that the brain and corpus cardiacum were the only sites of neurosecretory cells and neurohaemal organs in insects. RAABE (1965) was the first to recognise the existence of neurosecretory cells in the ventral ganglia and to locate neurohaemal organs on the median unpaired nervous system of phasmids. These perisympathetic organs are structurally identical to the corpus cardiacum and have been found in 15 orders of insects (RAABE et al., 1971; GRILLOT, 1972). Biogenic amines have been found to be associated with these neurohaemal organs. Thus, radioenzymatic assays have shown octopamine to be present in the perisympathetic organs of Periplaneta and Schistocerca (EVANS, 1978) and the stick insect Carausius morosus (ORCHARD, unpublished observation) while dopamine and small amounts of nor-epinephrine are also present in Periplaneta (DYMOND and EVANS, 1979). SMALLEY (1970) reported the uptake of ^{3}H-dopamine by the abdominal perisympathetic organs and the ventral ganglia of cockroaches but RAABE (1971) found no evidence for amines using fluorescent histochemistry. Octopamine is not a fluorescent amine and so it is possible that dopamine was accumulated in octopaminergic neurons. Since these neurohaemal organs contain more than one type of axon as judged by granule content it would appear that the aminergic fibers probably lie amongst classical peptidergic fibers. Cobalt backfilling has also been used to trace neurons which project into the perisympathetic organs of Periplaneta (ALI and PIPA, 1978), Schistocerca (EVANS, 1980) and Carausius (see ORCHARD, 1983). Groups of filled cell bodies occur at various levels adjacent to the mid-saggital plane, and also in the lateral aspects of the ganglia. These cells presumably represent in part the neurosecretory cells which terminate in the perisympathetic organs. Some of the cells in Schistocerca and Periplaneta are similar in size and position to the presumed aminergic neurons revealed by neutral red staining (EVANS and O'SHEA, 1978; DYMOND and EVANS, 1979; see EVANS, 1980).

ROLE OF AMINES IN THE INSECT NEUROSECRETORY SYSTEM

It is clear from the previous account that biogenic amines are located within the insect neurosecretory system. Physiologically there may be two reasons for such a close apposition. The first is that these amines may in themselves be neurohormones and that the most efficient site for their release into the haemolymph is from a neurohaemal organ. Secondly, amines may control the release of peptidergic hormones by acting either as conventional neurotransmitters, regulating the activities of peptidergic cells, or acting as neuromodulators, modifying the hormonal output from the peptidergic neurosecretory terminals (see Fig. 2).

There is now convincing evidence for the role of octopamine as a neurohormone in insects (review by ORCHARD, 1982). Octopamine is present in the haemolymph of Locusta (DAVID and LAFON-CAZAL, 1979; ORCHARD et al., 1981), Schistocerca (GOOSEY and CANDY, 1980) and Periplaneta (BAILEY, B.A. personal communication), where it has been shown to increase in concentration following flight or following excitation due to handling. Neurohormonal functions for octopamine include its ability to stimulate the release of lipid or trehalose from fat body; increase the activity of glycogen phosphorylase and stimulate glycogenolysis in ventral nerve cord; increase the rate of glucose oxidation in flight muscle; and increase sodium-dependent respiration in nerve cord (for references see ORCHARD, 1982). As a result of these effects, octopamine has been considered to be the effector of a generalized response to excitation leading to the 'fight or flight' response in insects. The analogy with the vertebrate sympathetic nervous system has been pointed out previously (HOYLE, 1975; GOLE and DOWNER, 1979; DOWNER, 1980). The source of circulating octopamine is at present unknown although it is tempting to attribute its presence in the haemolymph to release from neurohaemal organs. Clearly the corpus cardiacum and perisympathetic organs are in an ideal position to quickly flood the haemolymph with octopamine but it has not been demonstrated that octopamine is released from these structures.

To date there is no evidence for a neurohormonal function for any other amines in insects. Dopamine, nor-epinephrine, and 5-HT have yet to be shown present in the haemolymph. However, there are tissues which appear to possess 5-HT receptors, but that do not receive any 5-HT innervation (eg. malphigian tubules, see MADDRELL and NORDMANN, 1979). However, the physiological function for these receptors is yet to be demonstrated.

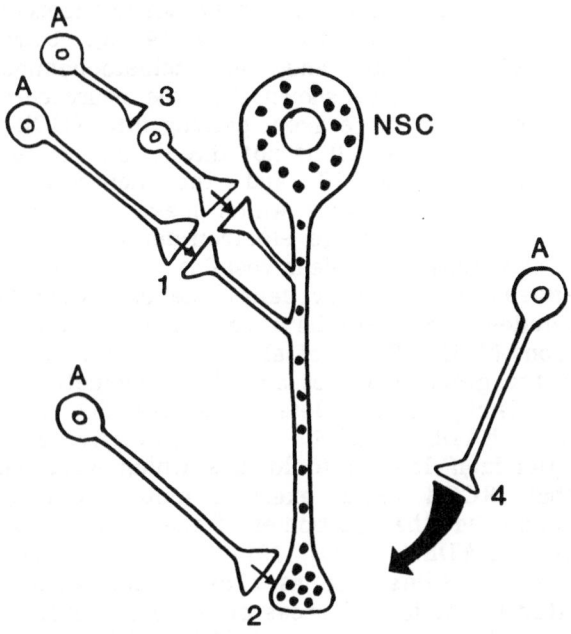

FIGURE 2. Possible sites at which aminergic neurons (A) could act to affect secretion from peptidergic neurosecretory cells (NSC). Aminergic neurons could synapse with the collaterals (1) or terminals (2) of the NSC. They could act via one or many neurons in pathway converging onto NSC (3) or they could release amine close to the terminal or into the haemolymph which may influence terminal (4).

Recent studies have now established a role for biogenic amines in the regulation of peptidergic neurosecretory cells. The most convincing evidence comes from the control of release of adipokinetic hormones in Locusta. The so-called glandular cells of the glandular lobe of the corpus cardiacum of Locusta are the source of two hyperlipaemic hormones represented by adipokinetic hormone I (AKH I) which has been characterised, sequenced and synthesised (STONE et al., 1976; BROOMFIELD and HARDY, 1977) and, to a lesser extent , by a more recently discovered second peptide, AKH II (CARLSEN et al., 1979). The glandular cells possess short axon-like processes as revealed following dissection (KROGH, 1973) or following staining with AKH-antibody (SCHOONEVELD et al., 1983), and so may be considered neuronal and hence are neurosecretory cells (see KROGH and NORMANN, 1977). These neurosecretory cells release AKH I and AKH II in response to flight (CHEESEMAN and GOLDSWORTHY, 1979; ORCHARD and LANGE, 1983a,b; see also LANGE and ORCHARD, this volume). These hormones mobilise stored lipid which is then used as a fuel for flight. ORCHARD and LOUGHTON (1981a,b) and ORCHARD and LANGE (1983a) found that the release of AKH's from the glandular lobe was under the immediate control of the nervi corporis cardiaci II (NCCII). Electrical stimulation of NCCII resulted in the release of AKH's. Interestingly, axons in NCCI appeared to modulate the effects of NCCII stimulation but did not in themselves induce release of hormone (ORCHARD and LOUGHTON, 1981b). Ultrastructural studies (RADEMAKERS, 1977a,b) and immunohistochemical studies (SCHOONEVELD et al., 1983) have shown that the neurosecretory cells within the glandular lobe do not have axons passing into NCCII. Thus the control of hormone release by NCCII is not direct, but in fact involves synapses between axons of NCCII and the neurosecretory cells (RADEMAKERS, 1977a,b). The axons which make these synapses are characterised by the presence of large electron-dense granules of about 100nm diameter and numerous small clear vesicles. There is only one type of axon present between the glandular cells. The characteristics of the electron-dense granules are similar to those from identified octopaminergic neurons (SCHAEFFER et al., 1978; HOYLE et al.,1980) in that the density of the granules varies and the substructure of rod-like tubular subunits is visible in the less-dense granules (KROGH and NORMANN, 1977; RADEMAKERS and BEENAKKERS, 1977). Transplantation of glandular lobes into locusts which were subsequently flown suggested that NCCII axons exert a motor control over the neurosecretory cells and that the cell bodies of these axons lay outside of the corpus cardiacum (RADEMAKERS, 1977b). Cell bodies containing secretory granules of the same size as those found within the axon terminals were located in the lateral areas of the protocerebrum and the axons of these cell bodies were shown by cobalt backfilling to enter NCCII and pass to the glandular lobe . It seems most likely that these cells are the ones responsible for the NCCII mediated synaptic activation of the neurosecretory cells in the glandular lobe.

The earliest suggestion that these neurosecretory cells may be under aminergic control came from the work of SAMARANAYAKA (1976).

SAMARANAYAKA (1976) found that insecticide poisoning of locusts resulted in an elevation of haemolymph lipid, which she attributed to the release of adipokinetic hormone. Aminergic antagonists prevented the elevation of lipid (release of adipokinetic hormone?) in insecticide-treated insects, leading to the suggestion that the synapses between NCCII and the neurosecretory cells were aminergic. However these experiments were conducted on the intact insect and so the precise site of action of these drugs was impossible to determine.

Recently DAVID and LAFON-CAZAL (1979) and LAFON-CAZAL (1981) reported the presence of octopamine in the glandular lobe and also traces of dopamine. The NCCII and the neurosecretory cells of the glandular lobe did not contain dopamine or 5-HT or, apparently any other amine fluorescing with the Falck-Hillarp test (the trace amount of dopamine was probably contamination from fibers which pass from the storage lobe). On the basis of the presence of octopamine, plus the observation that NCCII axons took up tritiated amines without showing fluorescence DAVID and LAFON-CAZAL (1979) and LAFON-CAZAL (1981) suggested that octopamine could be acting as the transmitter regulating the release of the AKH's. Physiological data now indicates this suggestion to be true. As mentioned earlier, electrical stimulation of NCCII results in the release of the AKH's. In addition, it also results in an elevation in cyclic AMP in the glandular lobe (ORCHARD et al., 1983a) and it has been argued that the increase in cyclic AMP is coupled to the process of synaptic transmission and that the increase occurs in the intrinsic neurosecretory cells. Thus there are two criteria upon which to assess synaptic activation of the glandular cells; release of hormone and elevation in cyclic AMP. Corpora cardiaca, removed from locusts which had been pre-treated with reserpine, failed to release AKH's or to elevate cyclic AMP in response to NCCII stimulation (ORCHARD and LOUGHTON, 1981a; ORCHARD et al., 1983a). The simplest interpretation of this response is that reserpine acted in its role as depletor of aminergic stores in presynaptic terminals, and indeed, the levels of octopamine in control gland (0.62 pmol) were depleted to 0.3 pmol by reserpine treatment (ORCHARD and LOUGHTON, 1981a). Thus an amine was implicated as the transmitter in this system. The neurally-evoked release of hormone and elevation in cyclic AMP were blocked by the α-adrenergic blockers phenoxybenzamine and phentolamine but not blocked by the β- adrenergic blocker propranolol. Surprisingly propranolol actually enhanced the release of hormone but the mechanism leading to this enhancement is not understood. These experiments indicate that stimulation of the presynaptic axons in NCCII leads to the release of an aminergic transmitter which subsequently results in both an elevation in cyclic AMP of the post-synaptic neurosecretory cells and the release of adipokinetic hormones. Exogenous application of the natural transmitter of this system should mimic the effects of nerve stimulation. Initial experiments revealed that application of octopamine at low concentrations, could indeed induce the release of hormone (ORCHARD and LOUGHTON, 1981a) and elevate cyclic AMP levels (ORCHARD et al., 1983a). Both of these effects were abolished in the presence of α-adrenergic receptor antagonists. Recent studies have extended these observations (ORCHARD, GOLE and DOWNER, unpublished), and the effectiveness of several putative aminergic neurotransmitters and antagonists in mimicking the neurophysiological response has been examined. The receptors mediating the release of the AKH's exhibited a specificity for the monophenolic amines octopamine and synephrine (Table 1). At concentrations of 10^{-7} M only octopamine and synephrine were capable of stimulating significant release of hormone from incubated glandular lobes. Nor-epinephrine and tyramine, which only differ from octopamine by a single hydroxyl group, and dopamine which differs by two hydroxyl groups, did not stimulate the release of hormone at 10^{-7} M. 5-HT was also without effect. Similar results were obtained from cyclic AMP studies (Table 1). Octopamine and synephrine induced the largest elevation in cyclic AMP. The response to octopamine was dose-dependent (Fig. 3) with half-maximal stimulation at 5×10^{-6} M and maximal stimulation, representing an alsmost ten-fold

TABLE 1. Effects of amines upon hormone release (assayed as an increase in haemolymph lipid of bioassayed locusts) and upon cyclic AMP levels of glandular lobe.

Treatment	Increase in lipid[a] (µg per µl)	Cyclic AMP[b] (pmol per gland)
Synephrine	$4.6 \pm 1.0^{*}$	$1.43 \pm 0.30^{+}$
Octopamine	$4.2 \pm 0.6^{*}$	$1.58 \pm 0.17^{+}$
5-HT	-0.4 ± 0.3	$0.88 \pm 0.07^{+}$
Dopamine	1.0 ± 0.5	0.29 ± 0.08
Nor-epinephrine	1.1 ± 0.6	0.45 ± 0.09
Metanephrine	-0.7 ± 0.7	0.53 ± 0.12
Tyramine	-0.3 ± 0.1	0.29 ± 0.08
Epinephrine	0.8 ± 0.6	0.27 ± 0.05

[a] 2 glands incubated for 20 min in Ringers followed by 20 min in 10^{-7} M amine. Perfusate bioassayed for adipokinetic activity.

[b] 10 min incubation in 5×10^{-6} M amine in the presence of 0.5mM IBMX.

* Denotes significant increase at $P < 0.05$; + Denotes significant difference from baseline value ($P < 0.05$). Data from ORCHARD, GOLE and DOWNER (unpublished).

increase over basal levels, at 10^{-4}M. 5-HT also stimulated a significant elevation in cyclic AMP, although at 5×10^{-6} M the response was much lower than that of octopamine. However the 5-HT stimulated increase in cyclic AMP was not blocked by phentolamine whereas the octopamine-stimulated increase was (Table 2).

The results of all the above experiments are consistent with octopamine being the natural transmitter within this neurosecretory system. In summary, electrical stimulation of NCCII results in the release of adipokinetic hormones (ORCHARD and LOUGHTON, 1981a,b) and an elevation in cyclic AMP (ORCHARD et al., 1983a). Both of these effects are prevented by prior treatment with reserpine , and antagonised by α-adrenergic receptor antagonists. Octopamine is present in the glandular

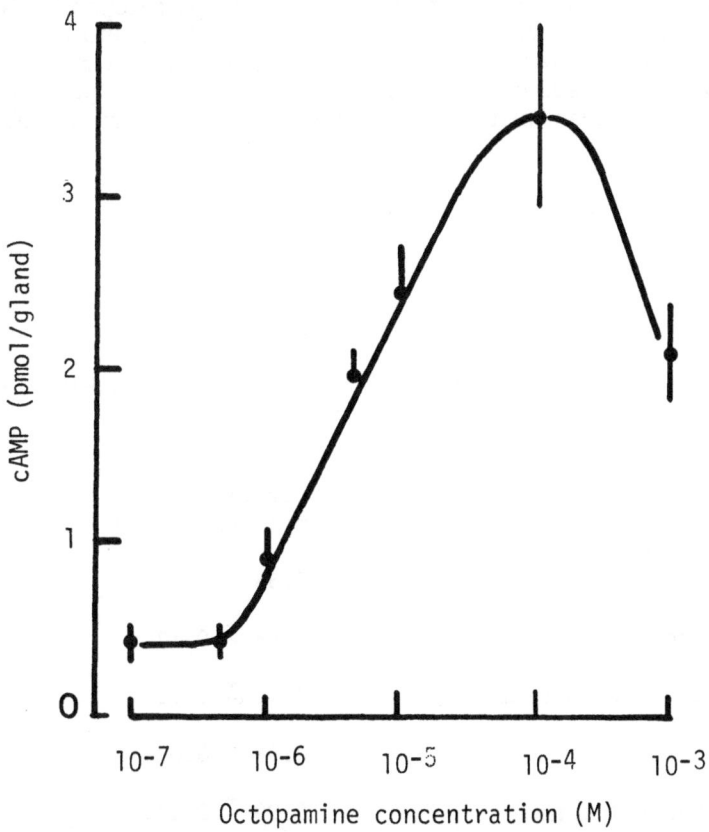

FIGURE 3. Dose-response curve for the action of DL-octopamine on cAMP content of glandular lobes (10 min incubations in the presence of IBMX). Data from ORCHARD, GOLE and DOWNER, unpublished).

lobe (ORCHARD and LOUGHTON, 1981a) whereas there is no evidence for any fluorogenic amine (KLEMM and FALCK, 1978; LAFON-CAZAL, 1981). The most potent amines at mimicking both of the synaptically activated events are the monophenolic amines octopamine and synephrine (ORCHARD, GOLE and DOWNER, unpublished), and in the absence of any demonstration of naturally occurring synephrine, octopamine must be the prime candidate as the natural transmitter (we can discount 5-HT as the natural transmitter since the 5-HT induced elevation in cyclic AMP is not blocked by phentolamine, a known antagonist of synaptic activation in this system and 5-HT did not induce release of hormone).

An interesting facet of the octopaminergic control of release of AKH's is that flight induces a short term elevation in haemolymph octopamine levels (GOOSEY and CANDY, 1980). Octopamine can act as a neurohormone (see earlier, also ORCHARD, 1982) and in the locusts acts upon the fat body to cause the release of lipid (ORCHARD et al., 1981, 1982). An intriguing possibility is that the haemolymph octopamine levels may reach a sufficient magnitude to potentiate the neurally evoked (octopamine-mediated) release of AKH's. Thus there may be dual control of the neurosecretory cells of the glandular lobe; synaptic input utilising octopamine as the transmitter and hormonal input from circulating octopamine.

TABLE 2. Effects of antagonists on amine-stimulating elevations in cyclic AMP of the glandular lobe.

Treatment	Cyclic AMP (pmol per gland)
Ringers	0.33 ± 0.07
Octopamine	1.58 ± 0.17
Octopamine + Phentolamine	0.40 ± 0.11
5-HT	0.88 ± 0.07
5-HT + Phentolamine	0.99 ± 0.10

All concentrations, 5×10^{-6}M.
Phentolamine blocks the action of octopamine but not that of 5-HT.
Data from ORCHARD, GOLE and DOWNER (unpublished).

Similar evidence for an involvement of octopamine in regulating hormone release has recently been obtained in Periplaneta (ORR, DOWNER and ORCHARD, unpublished; see also DOWNER et al., this volume). In Periplaneta a hyperglycaemic hormone is released from the corpus cardiacum. Electrical stimulation of NCCII resulted in the release of a hyperglycaemic hormone (GERSCH, 1972) and also induced an elevation in cyclic AMP. Octopamine, applied to corpora cardiaca in vitro mimicked both of these neurally evoked events, and the action of antagonists was consistent with there being α-adrenergic receptors involved with the control of release of hyperglycaemic hormone (ORR, DOWNER and ORCHARD, unpublished). However, hyperglycaemic hormone is believed to be manufactured in cell bodies within the brain, and transported along axons to the corpus cardiacum for storage and subsequent release. The action of octopamine appears therefore to be on the axons or terminals within the corpus cardiacum. Octopamine may therefore be playing a neuromodulatory role in this system, regulating release at or close to the release sites for the hormone.

Recent experiments have provided evidence for aminergic regulation of certain medial neurosecretory cells of adult females of the blood sucking bug Rhodnius prolixus (ORCHARD et al., 1983b). In this insect the timing and rate of oviposition in mature females is governed by a myotropic peptidergic ovulation hormone derived from ten large neurosecretory cells in the brain (KRIGER, 1981). In mated females, the electrical activity of these median neurosecretory cells and their axons, and the associated release of ovulation hormone, were stimulated by the appearance of ecdysteroid in the haemolymph (RUEGG et al., 1981,1982). It is believed that the presence of the ovaries are required for ecdysteroid to appear in the haemolymph. Thus in ovariectomised mated females, the peak of ecdysteroid was absent and so too was the characteristic pattern of electrical activity in the corpus cardiacum. However, injection of 20-hydroxyecdysone into ovariectomised mated females restored the patterning of electrical activity and the resultant release of ovulation hormone (RUEGG et. al., 1981,1982). The enhancement of electrical activity of the neurosecretory cells in the brain and corpus cardiacum induced by 20-hydroxyecdysone has been used as a means of examining the possible role of aminergic neurons in this reflex (ORCHARD et al.,1983b). The response of the brain and corpus cardiacum, from mated ovariectomised females, to 20-hydroxyecdysone was blocked by the α-adrenergic receptor antagonists phentolamine and phenoxybenzamine, but not by the β-adrenergic receptor antagonist propranolol. Treatment of insects with reserpine was also effective at abolishing the response to 20-hydroxyecdysone, and dopamine (10^{-7}M) was capable of mimicking the action of 20-hydroxyecdysone in mated ovariectomised females as well as in reserpine-treated ovariectomised females. The action of 20-hydroxyecdysone on the neurosecretory cells in Rhodnius does not therefore appear to be direct, but involves aminergic interneurons located within the brain. The simplest interpretation would have aminergic neurons sensitive to the ecdysteroid, and synaptically linked to the neurosecretory cells. However these experiments cannot distinguish the number of pathways involved within this reflex.

An interesting result was obtained using preparations of virgin Rhodnius (ORCHARD et al., 1983b). Brains from virgin Rhodnius did not respond to 20-hydroxyecdysone in the same manner as brains from mated Rhodnius (RUEGG et al., 1982), and it had been previously suggested that a second input, probably from the 'spermathecal factor' (DAVEY, 1965) was required in addition to 20-hydroxyecdysone. While virgin brains did not repond to 20-hydroxyecdysone or to 10^{-7} M dopamine, they did respond to 10^{-5} M dopamine (ORCHARD et al., 1983b). This may suggest that the mechanism of action of the 'spermathecal factor' could be to enhance the response of the neurosecretory cells or other neurons involved in the reflex to amines. Maybe the 'spermathecal factor' results in an unmasking of aminergic receptors or a stimulation in synthesis of new receptors.

Some evidence for a relationship between aminergic and peptidergic cells has been obtained from histological and ultrastructural studies. Tenebrio molitor fed with a 1% reserpine diet demonstrated conspicuous changes in the median neurosecretory cells, with an accumulation of paraldehyde fuchsin material in the cell bodies and axons (MASNER et al., 1970). These results were interpreted as the result of a reduction in the release of neurohormone, suggestive of an inhibitory adrenergic control over neurosecretory activity. More recent studies examined the ultrastructure of neurosecretory cells following administration of nor-epinephrine (WARTON and DUTKOWSKI. 1978), reserpine (WARTON and DUTKOWSKI, 1977) or disulfiram, an inhibitor of nor-adrenaline synthesis (WARTON, 1981). Injection of nor-epinephrine into the haemocoel of VII instar larvae of Galleria mellonella resulted in an accumulation of resorcin-fuchsin-positive material within the cell bodies of some of the median neurosecretory cells. Ultrastructural studies confirmed that an accumulation of electron-dense granules occurred in the cytoplasm of Type I cells containing granules of 180-200nm diameter, and to a lesser extent Type III cells containing granules of 250-270 nm diameter. The other neurosecretory cells were apparently not affected. Since there was no evidence of an effect on the formation of secretory granules from the Golgi complexes, the accumulation of granules was interpreted as being due to an inhibition of release, rather than stimulation of synthesis. Injection of reserpine into Galleria (WARTON and DUTKOWSKI, 1978) reduced the number and electron opacity of dense-core vesicles (60-80nm diameter) found in aminergic cells in the vicinity of median neurosecretory cells. At the same time there was an accumulation of electron-dense granules in the cell bodies of Type I, II (granule diameter 160-170 nm) and III neurosecretory cells. Again, the results are suggestive of an inhibition over release of neurohormones, with the assumption that reserpine resulted in the release of an amine. Finally injection of disulfiram resulted in a lowering of electron density of dense-core vesicles and a decrease in the amount of neurosecretory granules in the Type I neurosecretory cells (WARTON, 1981) Disulfiram apparently inhibits dopamine-β-hydroxylase which may therefore inhibit the production of nor-epinephrine. In the absence of nor-epinephrine the neurosecretory cells apparently release neurohormones. Again these results are suggestive of an inhibitory control over neurosecretory cells.

SCHOONEVELD (1974) described synaptic contacts between neurosecretory axons and axons containing dense-core vesicles of a size to be expected in aminergic cells in Leptinotarsa decemlineata. Thus there is ultrastructural evidence that median neurosecretory cells may be controlled synaptically by aminergic neurons.

CONCLUSION

As has been reported in vertebrates there is also a close association between aminergic cells and peptidergic neurosecretory cells in insects. Neurosecretory cells and aminergic cells are grouped together within the central nervous system, and neurohaemal organs contain axons belonging to both classes of cells. Octopamine, dopamine and nor-epinephrine have been identified within neurohaemal organs using radioenzymatic assays and fluorescent histochemistry.

The evidence in this review makes it abundantly clear that the close physiological relationship which is known to exist between aminergic and peptidergic neurosecretory cells in vertebrates (see WEINER and GANONG, 1978) is also found within insects. It seems likely that multiple aminergic neuronal systems are involved in the control of secretion of a number of neurohormones. Thus octopamine acts as an excitatory transmitter stimulating the release of AKH's in Locusta (ORCHARD and LOUGHTON, 1981a; ORCHARD, 1982). It is also an intriguing possibility that circulating levels of octopamine may potentiate the release of hormone at a critical time during the early stages of flight. Octopamine may act as an excitatory neuromodulator, stimulating the release of hyperglycaemic hormones from terminals in Periplaneta (ORR, DOWNER and ORCHARD, unpublished). The action of ecdysteroid in stimulating the median neurosecretory cells in Rhodnius is mediated via aminergic interneurons with dopamine being capable of mimicking the effect of 20-hydroxyecdysone(ORCHARD et al., 1983b). Nor-epinephrine may be an inhibitory transmitter on some of the neurosecretory cells in Galleria, inhibiting the activity of these cells, such that hormone is not released(WARTON and DUTKOWSKI, 1977, 1978; WARTON, 1981). Aminergic cells appear to make synaptic contact with peptidergic neurosecretory cells in Leptinotarsa (SCHOONEVELD, 1974).

ACKNOWLEDGEMENT

I am most grateful to Angela B. Lange for helpful suggestions and for the preparation of the manuscript; and to Barry G. Loughton for reading an early draft of the manuscript.

REFERENCES

ALI Z.I. and PIPA R. (1978) The abdominal perisympathetic neurohaemal organs of the cockroach Periplaneta americana: Innervation revealed by cobalt chloride diffusion. Gen. Comp. Endocrinol. 36: 396-401.

BARKER J. L., CRAYTON J. W. and NICOLL R. A. (1971) Noradrenaline and acetylcholine responses of supraoptic neurosecretory cells. J. Physiol. 218: 19-32.

BAUMGARTEN H. G., BJORKLUND A., HOLSTEIN A. F. and NOBIN A. (1972) organization and ultra-structural identification of catecholamine nerve terminals in the neuronal lobe and pars intermedia of the rat pituitary. Z. Zellforsch. 126: 483-517.

BJORKLUND A.(1968) Monoamine-containing fibres in the pituitary neurointermediate lobe of the pig and rat. Z. Zellforsch. 89: 573-589.

BROOMFIELD C. E. and HARDY P. M. (1977) The synthesis of locust adipokinetic hormone. Tetrahedron Lett. 25: 2201-2204.

CANDY D.J. (1978) The regulation of locust flight muscle metabolism by octopamine and other compounds. Insect Biochem. 8: 177-181.

CARLSEN J., HERMAN W. S., CHRISTENSEN, M. and JOSEFSSON L. (1979) Characterisation of a second peptide with adipokinetic and red pigment-concentrating activity from the locust corpora cardiaca. Insect Biochem. 9: 497-501.

CARLSSON A., FALCK B. and HILLARP N. A. (1962) Cellular localization of brain monoamines. Acta Physiol. Scand. 56:1-28.

CHEESEMAN P. and GOLDSWORTHY G. J. (1979) The release of adipokinetic hormone during flight and starvation in Locusta. Gen. Comp. Endocrinol. 37: 35-43.

CLEMENS J.A., ROUSH M.E. and FULLER R.W. (1978) Evidence that serotonin neurons stimulate secretion of prolactin releasing factor. Life Sci. 22: 2209-2214.

COOKE I. M. and GOLDSTONE M. W. (1970) Fluorescence localization of monoamines in crab neurosecretory structures. J. Exp. Biol. 53: 651-668.

DAVID J.-C. and LAFON-CAZAL M. (1979) Octopamine distribution in the Locusta migratoria nervous and nonnervous systems. Comp. Biochem. Physiol. 64C: 161-164.

DOWNER R. G. H. (1980) Short term trehalosemia by excitation in Periplaneta americana. J. Insect Physiol. 25: 59-63.

DYMOND G. R. and EVANS P. D. (1979) Biogenic amines in the nervous system of the cockroach, Periplaneta americana: association of octopamine with mushroom bodies and dorsal unpaired median (DUM) neurones. Insect Biochem. 9: 535-545.

EVANS P. D. (1978) Octopamine distribution in the insect nervous system. J. Neurochem. 30: 1009-1013.

EVANS P. D. (1980) Biogenic amines in the insect nervous system. Adv. Insect Physiol. 15:317-474.

EVANS P.D. and O'SHEA M. (1978) The identification of an octopaminergic neurone and the modulation of a myogenic rhythm in the locust. J. Exp. Biol. 73: 235-260.

GERSCH M. (1972) Experimentelle untersuchungen zum freis etz ungsmechanismus von neurohormonen nach elektrischer reizung der corpora cardiaca von Periplaneta americana in vitro. J. Insect Physiol. 18: 2425-2439.

GERSCH M., HENTSCHEL E. and UDE J. (1974) Aminergic substanzen in lateralen herznerven und im stomatogastrischen nervensystem der schabe Blaberus craniifer burm. Zool. Jb. Physiol. 78: 1-15.

GERSCH M., FISCHER F., UNGER H. and KABITZA W. (1961) Vorkommen von seratonin im nervensystem von Periplaneta americana L. (Insecta). Z. Naturforsch. 16B: 351-352.

GLOWACKA S.K. (1982) Disturbances in the development of the egg chambers of Galleria mellonella L. (Lepidoptera) after reserpine administration. J. Insect Physiol. 28: 249-256.

GOLE J.W.D. and DOWNER R.G.H. (1979) Elevation of adenosine 3',5'-monophosphate by octopamine in fat body of the american cockroach, Periplaneta americana L. Comp. Biochem. Physiol. 64C: 223-226.

GOOSEY M. W. and CANDY D. J. (1980) The D-octopamine content of the haemolymph of the locust, Schistocerca americana gregaria and its elevation during flight. Insect Biochem. 10: 393-397.

GOOSEY M. W. and CANDY D. J. (1982) The release and removal of octopamine by tissues of the locust Schistocerca americana gregaria. Insect Biochem. 12:681-685.

GRILLOT J. -P. (1972) Les organes périsympathetique lieux de stockage et de diffusion de produits de neurosécrétion chez les insectes. Ann. Univ. Brazzaville. 8(C): 107-112.

HANSTROM, B. (1953) Neurosecretory pathways in the head of crustaceans, insects and vertebrates. Nature 171: 72-73.

HOYLE, G. (1975) Evidence that insect dorsal unpaired median (DUM) neurones are octopaminergic. J. Exp. Zool. 193: 425-431.

HOYLE G., COLQUHOUN W. and WILLIAMS M. (1980) Fine structure of an octopaminergic neuron and its terminals. J. Neurobiol. 11: 103-126.

KLEMM, N. (1983) Monoamine-containing neurons and their projections in the brain (supraoesophageal ganglion) of cockroaches. Cell Tiss. Res. 229: 379-402.

KLEMM N. and AXELSSON S. (1973) Determination of dopamine, noradrenaline and 5-hydroxytryptamine in the brain of the desert locust, Schistocerca gregaria Forsk. (Insecta, Orthoptera). Brain Res. 57: 289-298.

KLEMM N. and FALCK B. (1978) Monoamines in the pars intercerebralis-corpus cardiacum complex of locusts. Gen. Comp. Endocrinol. 34: 180-192.

KRIGER F.L. (1981) Neuroendocrine Regulation of Ovulation in an insect, Rhodnius prolixus Stal. Ph.D. Thesis. York University. Toronto, Canada.

KROGH I. M. (1973) Light microscopy of living neurosecretory cells of the corpus cardiacum of Schistocerca gregaria. Acta Zool. 54: 73-80.

KROGH I.M. and NORMANN T. C. (1977) The corpus cardiacum neurosecretory cells of Schistocerca gregaria. Electron microscopy of resting and secreting cells. Acta. Zool. 58: 69-78.

LAFON-CAZAL M. (1981) Monoamines in the corpora cardiaca of locusts.
    Adv. Physiol. Sci. 22: 255-267.
LAFON-CAZAL M. and ARLUISON M. (1976) Localization of monoamines in
    the corpora cardiaca and the hypocerebral ganglion of locusts. Cell
    Tiss. Res. 172: 517-527.
MADDRELL S. H. P. and NORDMANN J.J. (1979) "Neurosecretion". New
    York, Toronto, John Wiley and Sons.
MASNER P., HUOT L., CORRIVAULT G.-W. and PRUDHOMME J.C. (1970)
    Effect of reserpine on the function of the gonads and its neuro-
    endocrine regulation in tenebrionid beetles. J. Insect Physiol. 16:
    2327-2344.
McCANN S. M., KALRA P.S., DONOSO A.O., BISHOP W., SCHNEIDER
    H.P.G., FAWCETT C.P. and KRULICH L. (1972) The role of
    monoamines in the control of gonado-tropin and prolactin secretion.
    In: Brain-Endocrine Interaction. Median Eminence: Structure and
    Function. Int. Symp. Munich 1971, pp224-235 (Karger, Basel)
MIGLIORI-NATALIZI G., PANSA M.C., D'AJELLO V., CASAGLIA O.,
    BETTINI S. and FRONTALI N. (1970) Physiologically active factors
    from corpora cardiaca of Periplaneta americana. J. Insect Physiol.
    16:1827-1836.
OKSCHE A. (1976) The neuroanatomical basis of comparative
    neuroendocrinology. Gen. Comp. Endocrinol. 29: 225-239.
ORCHARD I. (1982) Octopamine in insects: neurotransmitter,
    neurohormone, and neuromodulator. Can. J. Zool. 60: 659-669.
ORCHARD I. (1983) Neurosecretion: morphology and physiology. In: Insect
    Endocrinology. DOWNER R.G.H. and LAUFER H. (eds) Alan R. Liss
    Inc, New York (IN PRESS).
ORCHARD I. and LOUGHTON B.G. (1981a) Is octopamine a transmitter
    mediating hormone release in insects? J. Neurobiol. 12: 143-153.
ORCHARD I. and LOUGHTON B.G. (1981b) The neural control of release of
    hyperlipaemic hormone from the corpus cardiacum of Locusta
    migratoria. Comp. Biochem. Physiol. 68A: 25-30.
ORCHARD I. and LANGE A.B. (1983a) Release of identified adipokinetic
    hormones during flight and following neural stimulation in Locusta
    migratoria. J. Insect Physiol. 29: 425-429.
ORCHARD I. and LANGE A.B. (1983b) The hormonal control of haemolymph
    lipid during flight in Locusta migratoria. J. Insect Physiol. (IN
    PRESS).
ORCHARD I., LOUGHTON B. G. and WEBB R. A. (1981) Octopamine and
    short-term hyperlipaemia in the locust. Gen. Comp. Endocriol. 45:
    175-180.
ORCHARD I., CARLISLE J. C., LOUGHTON B.G., GOLE J. W. D. and
    DOWNER R. G. H. (1982) In vitro studies on the effect of
    octopamine on locust fat body. Gen. Comp. Endocrinol. 98: 7-13.
ORCHARD I., LOUGHTON B.G., GOLE J.W.D. and DOWNER R.G.H.
    (1983a) Synaptic transmission elevates adenosine 3',5'-monophosphate
    (cyclic AMP) in locust neurosecretory cells. Brain Res. 258: 152-155.
ORCHARD I., RUEGG R.P. and DAVEY K. G. (1983b) The role of central
    aminergic neurons in the action of 20-hydroxyecdysone on

neurosecretory cells of Rhodnius prolixus. J. Insect Physiol. 29: 387-391.

RAABE M. (1965) Recherches sur la neurosécrétion dans la chaîne nerveuse ventrale du Phasme, Clitumnus extradentatus: Les elements neurosécréteurs. C.R. Acad. Sci. D Paris. 260: 6710-6713.

RAABE, M. (1971) Neurosécrétion dans la chaîne nerveuse ventrale des insectes et organes neurohémaux métamériques. Arch. Zool. exp. gen. 112: 679-694.

RAABE M., BAUDRY N., GRILLOT J.-P. and PROVANSAL A. (1971) Les organes perisympathiques des Insects Ptérygotes. Distribution. Caractères générau. C.R. Acad. Sci. D Paris. 273: 2324-2327.

RADEMAKERS L. H. P. M. (1977a) Effects of isolation and transplantation of the corpus cardiacum on hormone release from its glandular cells after flight in Locusta migratoria. Cell Tiss. Res. 184: 213-224.

RADEMAKERS L. H. P. M. (1977b) Identification of a secretomotor centre in the brain of Locusta migratoria, controlling the secretory activity of the adipokinetic hormone producing cells of the corpus cardiacum. Cell Tiss. Res. 184: 381-395.

RADEMAKERS L.H.P.M. and BEENAKKERS A.M.Th. (1977) Changes in the secretory activity of the glandular lobe of the corpus cardiacum of Locusta migratoria induced by flight. Cell Tiss. Res. 180: 155-171.

RUEGG R.P., KRIGER F. L., DAVEY K.G. and STEEL C.G.H. (1981) Ovarian ecdysone elicits release of a myotropic ovulation hormone in Rhodnius (Insecta: Hemiptera). Int. J. Invert. Reprod. 3: 357-361.

RUEGG R.P., ORCHARD I. and DAVEY K.G. (1982) 20-Hydroxyecdysone as a modulator of electrical activity in neurosecretory cells of Rhodnius prolixus. J. Insect Physiol. 28: 243-248.

SAMANARAYAKA M. (1976) Possible involvement of monoamines in the release of adipokinetic hormone in the locust, Schistocerca gregaria. J. Exp. Biol. 65: 415-425.

SCHAEFFER S.S., LIVINGSTONE M. and KRAVITZ E. A. (1978) Octopamine and serotonin nerve endings in the lobster. 8th Annual Meeting, Society of Neuroscience, Abstract 1033.

SCHARRER B. and SCHARRER E. (1944) Neurosecretion. VI. A Comparison between the intercerebralis cardiacum-allata system of insects and the hypothalamo-hypophysial system of the vertebrates. Biol. Bull. 87: 242-251.

SCHOONEVELD H. (1974) Ultrastructure of the neurosecretory system of the Colorado potato beetle, Leptinotarsa decemlineata (Say). II. Pathways of axonal secretion, transport and innervation of neurosecretory cells. Cell Tiss. Res. 154: 289-301.

SCHOONEVELD H., TESSER G. I., VEENSTRA J. A. and ROMBERG-PRIVEE H.M. (1983) Adipokinetic hormone and AKH-like peptide demonstrated in the corpora cardiaca and nervous system of Locusta migratoria by immunocytochemistry. Cell Tiss. Res. 230: 67-76.

SMALLEY K.A. (1970) Median nerve neurosecretory cells in the abdominal ganglion of the cockroach, Periplaneta americana. J. Insect Physiol. 16: 241-250.

STONE J.V., MORDUE W., BATLEY K. E. and MORRIS H. R. (1976) Structure of locust adipokinetic hormone, a neurohormone that regulates lipid utilization during flight. Nature 263: 207–211.

TRIBOLLET E., CLARKE G., DREIFUSS J.J. and LINCOLN D. W. (1978) The role of central adrenergic receptors in the reflex release of oxytocin. Brain Res. 142: 69–84.

WARTOŃ S. (1981) Effect of disulfiram on the ultrastructure of the peptidergic and aminergic cells in the pars intercerebralis of Galleria mellonella (Lepidoptera). Cell Tiss. Res. 215: 417–424.

WARTOŃ S. and DUTKOWSKI A. B. (1977) Ultrastrucure of the neurosecretory cells of the pars intercerebralis of Galleria mellonella (Lepidoptera) after noradrenaline administration. Gen. Comp. Endocrinol. 33: 179–186,

WARTOŃ S. and DUTKOWSKI A. B. (1978) Ultrastrucural analysis of the action of reserpine on the brain neuroendocrine system of the waxmoth, Galleria mellonella L., Lepidoptera. Cell Tiss. Res. 192: 143–155.

WEINER R. I. and GANONG W. F. (1978) Role of brain monoamines and histamine in regulation of anterior pituitary secretion. Physiol. Rev. 58: 905–976.

WEINER R. I., GORSKI R. A. and SAWYER C. H. (1972) Hypothalamic catecholamines and pituitary gonadotropic function. In: Brain-Endocrine Interaction. Median Eminence: Structure and Function. Int. Symp. Munich 1971. pp. 236–244. Karger, Basel.

CYCLIC AMP AND PROTEIN PHOSPHORYLATION AS A TRANSDUCING

MECHANISM FOR CERTAIN NEUROHORMONES AND NEUROTRANSMITTERS

James A. Nathanson

Department of Neurology, Harvard Medical School
Neuropharmacology Research Laboratory
Massachusetts General Hospital, Boston, MA 02114

INTRODUCTION

During the past decade, much progress has been made in our understanding of the role of membrane receptors in mediating the initial actions of many neurohormones and neurotransmitters. Less is known, however, about the subsequent intracellular biochemical mechanisms which translate membrane binding into an alteration in intracellular physiology. In vertebrate nervous systems, much attention has been focused, in recent years, on the role of protein phosphorylation as an intracellular translational process. As envisioned by Greengard and colleagues, phosphorylation represents a central mechanism by which various second messengers (e.g., calcium, cyclic AMP, cyclic GMP) exert their effects, after having being affected by a large variety of first messengers (e.g., neurotransmitters or membrane depolarization) (GREENGARD, 1978). This article will summarize briefly some of the salient features of second messenger/protein phosphorylation systems and will present some data which indicate their existence in the insect nervous system.

SECOND MESSENGERS

Although certain neurotransmitter receptors (e.g., the nicotinic cholinergic) appear to exert direct conformational changes in membrane proteins, a variety of other receptors act through the production of, or alteration in, the intracellular concentration of a second messenger, such as cyclic AMP, cyclic GMP or calcium ion. In the case of calcium, changes in intracellular concentration are usually dependent upon influx of extracellular calcium (e.g., through voltage-dependent channels opened by membrane depolarization) or mobilization of intracellular stores. In the case of the cyclic nucleotides, production occurs through direct or indirect activation of a nucleotide cyclase (adenylate cyclase or guanylate cyclase), with the resultant intracellular synthesis of cyclic AMP from ATP or cyclic GMP from GTP. A good example of an insect neurohormone and neurotransmitter that exerts many of its effects through cyclic nucleotide production is the phenylethylamine, octopamine. Since this amine appears to be an important first messenger, which occurs primarily in invertebrates, it may be useful to discuss the receptor actions of octopamine as an illustration of the second messenger concept.

Specific membrane receptors for octopamine, detected by activation of octopamine-sensitive adenylate cyclase, were first described by Nathanson and Greengard (1973) (Fig. 1). On the basis of the enrichment of such receptors in segmental ganglia as compared to connectives and on the known presence of octopamine in invertebrate nerve tissue, these investigators postulated that octopamine might function as a true neurotransmitter in insects (NATHANSON and GREENGARD, 1973, 1974; NATHANSON, 1976). Although octopamine-sensitive adenylate cyclase was first reported to be present in insect nerve tissue, subsequent studies have indicated that octopamine receptors exist in both neuronal and non-neuronal tissue of a number of classes of invertebrates (see EVANS, 1980; LINGLE et al, 1982 for reviews). The presence of octopamine receptors in invertebrates is in marked contrast to the failure (at least to date) to detect specific octopamine receptors or octopamine-sensitive adenylate cyclase in vertebrates.

Since the discovery of octopamine receptors, anatomical, physiological, and biochemical studies from a number of laboratories have shown that octopamine-containing neurons exist, not only in insects, but in a large number of invertebrate species (see EVANS, 1980; LINGLE et al, 1982 for reviews). From such studies, it appears that, in invertebrates, octopamine (and, possibly, its N-methyl derivative, synephrine) may function, physiologically, in a fashion similar to that by which norepinephrine and epinephrine function in vertebrates. In particular, there is evidence that octopamine may act as a circulating neurohormone (affecting cardiac function, fat and carbohydrate metabolism), as a peripheral neuromodulator (affecting muscular contraction), and as a centrally-acting neurotransmitter (affecting various motor and behavioral activities). Octopamine may also be involved in the release of hormones from neurosecretory cells, such as in the corpus cardiacum (ORCHARD et al, 1983). As with norepinephrine and epinephrine, a number of the above actions of octopamine appear to be mediated by the second messenger, cyclic AMP.

In those cases in which octopamine receptors are associated with adenylate cyclase, receptor binding of octopamine, in the presence of magnesium and GTP, causes a marked enhancement of cyclic AMP synthesis by the catalytic protein of adenylate cyclase. The GTP requirement and the sensitivity of the enzyme to cholera toxin (NATHANSON and HUNNICUTT, 1979; 1981; NATHANSON, unpublished observations) suggest that, as with many vertebrate hormones, receptor-mediated activation of the catalytic subunit requires the presence of a GTP-binding protein (G-subunit).

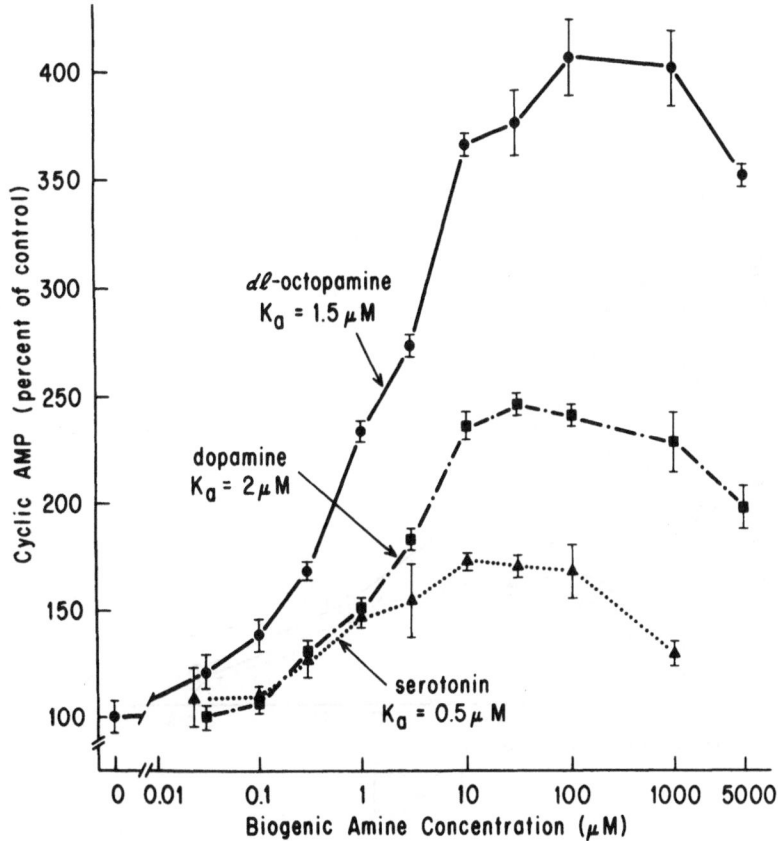

Fig. 1.   Demonstration, in broken cell preparations of cockroach thoracic ganglia, of adenylate cyclase stimulation by octopamine, dopamine and serotonin. Additivity and antagonist experiments (not shown) indicate that each of these three hormones is acting at a distinct receptor (from Nathanson and Greengard, copyright by AAAS, 1973).

However, calcium does not seem to be required for hormone
activation, and, in fact, higher concentrations of this divalent
ion are inhibitory (Fig. 2).

The presence of numerous hormone-stimulated adenylate
cyclases in insect nerve tissue has made it difficult to separate
and examine the pharmacological and biochemical characteristics
of octopamine- versus other hormone-sensitive adenylate cyclases.
Recently, however, Nathanson has reported (NATHANSON, 1979;
NATHANSON AND HUNNICUTT, 1981) that an extremely active
octopamine-sensitive adenylate cyclase exists in the firefly

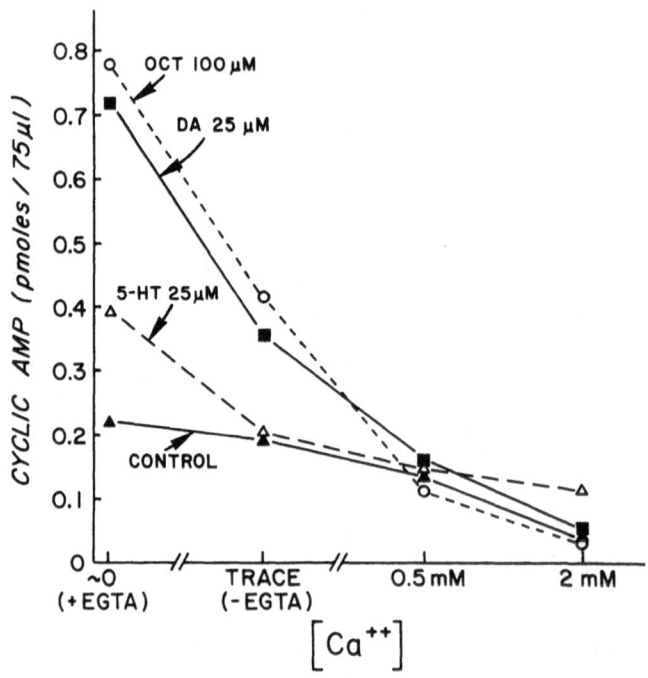

Fig. 2. Effects of calcium and EGTA on control and hormone-
stimulated adenylate cyclase activity of cockroach
ganglion homogenates. In the absence of added EGTA,
calcium ion concentration was determined by the
endogenous calcium (usually 10-50 micromolar) present in
the tissue homogenate.

light organ (Fig. 3).   This tissue appears to lack other
amine-stimulated adenylate cyclases.   Also, the activity of the
octopamine-stimulated enzyme is greater than that reported for
any other adenylate cyclase present in an excitable tissue
(NATHANSON, 1979 and in preparation).   Octopamine-containing DUM
cells appear to innervate the light organ (CHRISTENSEN AND
CARLSON, 1982), and the cyclic AMP produced by this enzyme upon
hormonal stimulation appears to be associated with the chain of
events involved in the regulation of light production in the
light organ (NATHANSON, 1983 and in preparation).   It is hoped
that further study of this tissue will aid in understanding the
mechanism of action of second messengers in the insect.

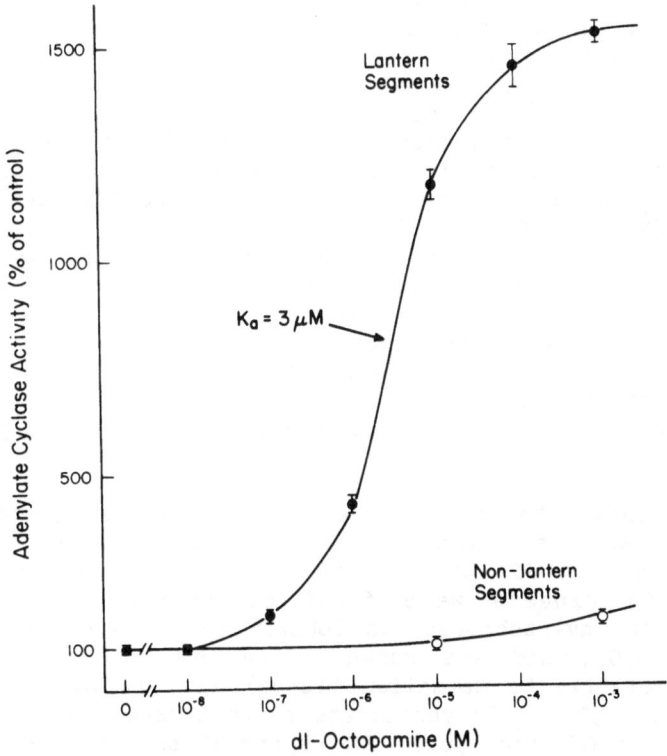

Fig. 3. Effect of octopamine on adenylate cyclase activity in
         broken cell preparations of firefly abdominal segments
         that contain light organs and in segments which lack
         light organs.   These and other experiments have revealed
         that the firefly light organ contains the most active
         octopamine-sensitive adenylate cyclase yet described
         (from Nathanson, copyright by AAAS, 1979a).

PROTEIN KINASES

        Evidence accumulated during the past decade has indicated
that, in virtually all eukaryotic organisms, protein kinases
appear to mediate the intracellular actions of cyclic AMP
(GREENGARD, 1982). More recent studies suggest that at least two
other second messengers - cyclic GMP and calcium - also appear to
utilize protein kinases to carry out many of their actions.

        Cyclic AMP-dependent protein kinases catalyze the phos-
phorylation, by ATP (usually in the presence of magnesion ion),
of protein substrates, according to the reaction:

        ATP + protein -----> ADP + phosphoprotein

This reaction is stimulated in the presence of cyclic AMP (see
BEAVO and MUMBY, 1982 for review). Cyclic AMP-dependent protein
kinases are composed of a regulatory (inhibitory) subunit and a
catalytic subunit (each subunit probably existing as a dimer).
Binding of cyclic AMP to the regulatory subunit(s) leads to a
dissociation of the enzyme, resulting in activation of the
catalytic subunit(s) and an increase in the rate (and usually
degree) of phosphorylation of the appropriate substrate.
Phosphorylation can cause a change in the charge or configuration
of the protein, thereby affecting its function. Phosphate may be
removed from the protein (thereby reversing the reaction) by a
class of enzymes known as phosphoprotein phosphatases. This
change in the level of intracellular phosphorylation of certain
proteins by protein kinases and phosphatases appears to be the
mechanism by which cyclic AMP (stimulated by an extracellular
first messenger) exerts its action on intracellular physiology.
As will be illustrated below, cyclic AMP-dependent phosphorylation is prominent in insect nerve tissue.

        Cyclic GMP appears to exert its effects through a special
group of protein kinases activated by cyclic GMP (for recent
review, see KUO and SHOJI, 1982). Of interest, cyclic GMP-
dependent protein kinases were first discovered, not in
vertebrates, but in invertebrates (in lobster tail muscle) (KUO
and GREENGARD, 1970), and are known to be present in high
concentrations (relative to mammalian tissue) in silk worm fat
bodies (KUO et al, 1971). In vertebrate nerve tissue, evidence
suggests that cyclic GMP and cyclic GMP-dependent protein kinase
may mediate some of the actions of acetylcholine at muscarinic
receptors. However, less is known about the physiological
actions of cyclic GMP-dependent phosphorylation in insects.
Compared with the cyclic AMP system, the cyclic GMP system
appears to be less associated with the plasma membrane. Also,
calcium may play a more important role in cyclic GMP production
than in cyclic AMP production.

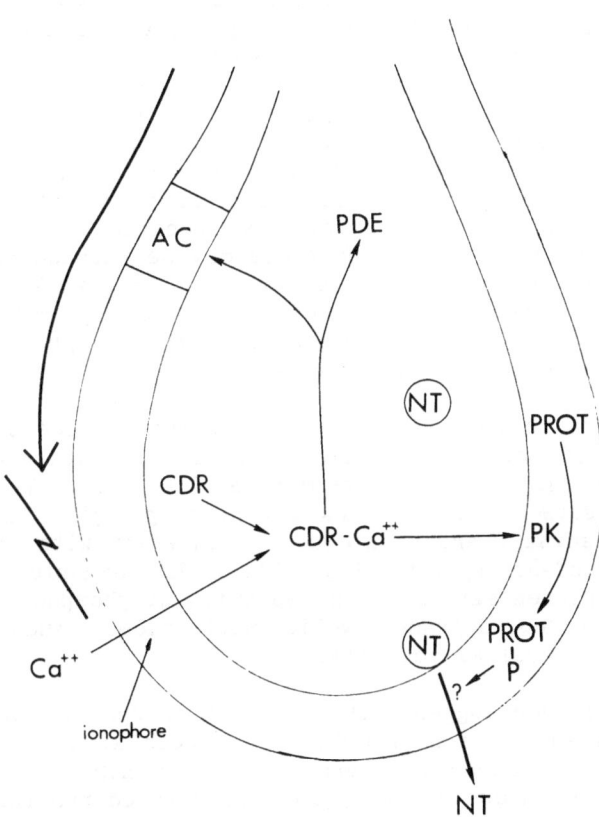

Fig. 4. Postulated interaction, in synaptic terminals, between calcium and cyclic AMP second messenger systems and protein phosphorylation. Experimental evidence suggests that depolarization of synaptic terminals leads to an influx of calcium ion which then combines with a calcium-dependent regulator protein ("CDR" or "calmodulin"). The calcium-calmodulin complex is capable of activating a specific protein kinase (PK) which phosphorylates certain membrane-associated proteins.
In some cases, the same protein may be phosphorylated, independently, by a cyclic AMP-dependent protein kinase. Calcium-calmodulin may interact, also, with other components of the cAMP 2nd messenger system, including adenylate cyclase (AC) and phosphodiesterase (PDE).

A great deal of evidence now suggests that many primary actions of calcium (independent of cyclic GMP) may be mediated through stimulation of a group of calcium-dependent protein kinases (for review, see SCHULMAN, 1982; CHEUNG and STORM, 1982). These enzymes require, also, the presence of the calcium-binding protein, calmodulin (also known as calcium-dependent regulator or CDR). Calcium-dependent phosphorylation may be stimulated when, for example, depolarization allows an influx of extracellular calcium. This calcium combines with intracellular calmodulin, and, in turn, the calcium-calmodulin complex activates the protein kinase which phosphorylates a specific substrate. In some cases, the same protein substrate may be phosphorylated (at different sites) by both a calcium-dependent protein kinase and a cyclic AMP-dependent protein kinase, thereby raising the possiblility of dual regulation of certain physiological processes by these two second messengers.

Of interest to the study of neurosecretory processes is the fact that one of the major sites of action of calcium-dependent phosphorylation is at nerve terminals (see Fig. 4). Other evidence suggests that calcium-dependent phosphorylation may provide a mechanism whereby calcium can interact with the cyclic nucleotide-dependent system (Fig. 4). For example, calcium-calmodulin complexes may alter the activity of phosphodiesterases (the enzymes which degrade cyclic nucleotides) and may also affect adenylate cyclase activity.

Recent evidence suggests that certain phospholipids may also play a significant role in regulating protein kinases, and that there may exist a separate class of protein kinases (so-called protein kinase C) whose activity is primarily controlled by such compounds as phosphatidylserine (FLOCKHART and CORBIN, 1982). As yet, the physiological role of such kinases in insects has not been investigated.

Nature of Protein Substrates and Cellular Localization of Phosphorylation Mechanisms

The nature of the protein substrates phosphorylated by the various protein kinases mentioned above varies considerably. The protein may be a soluble enzyme (e.g., phosphorylase b kinase in mammalian liver), in which case phosphorylation may stimulate or inhibit its activity. The protein may be membrane-bound (e.g. a receptor-associated ion channel), in which case phosphorylation may change its conformation, opening or closing the channel. The substrate protein may be a nuclear protein (e.g., histone), in which case phosphorylation may alter RNA synthesis. Figure 5 summarizes some of the possible sites of action of second messenger-regulated phosphorylations, indicating the diverse effects that such reactions can have on cellular metabolism.

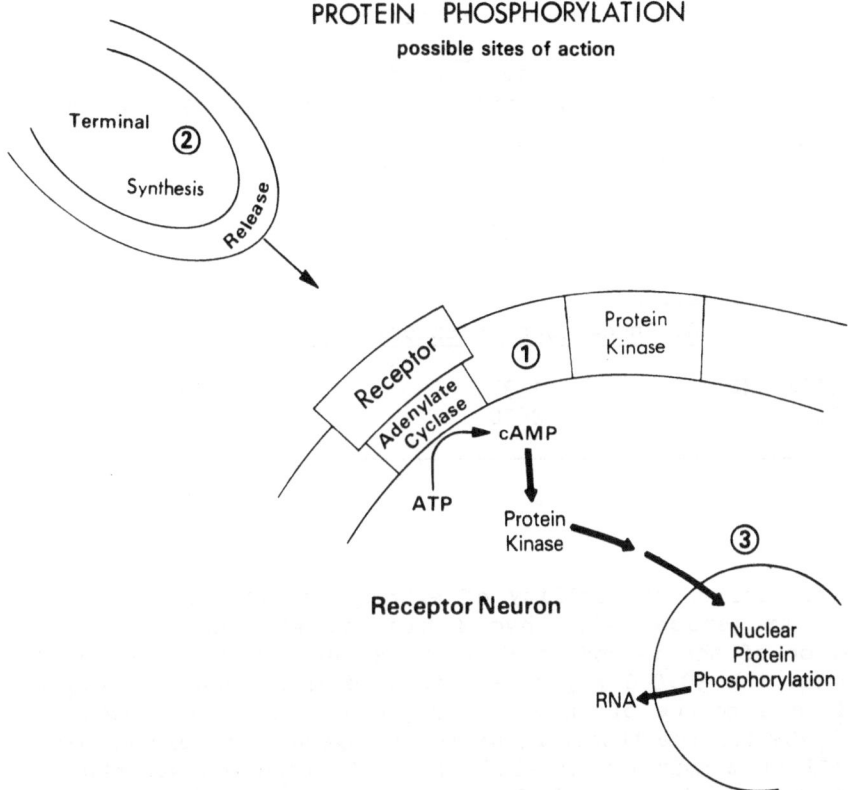

PROTEIN PHOSPHORYLATION
possible sites of action

Fig. 5.   Possible cellular sites of action of second messenger-
          activated protein phosphorylation systems: 1) the
          post-synaptic membrane or neighboring cytoplasm (e.g.,
          phosphorylation of an ion channel protein); 2) the
          presynaptic terminal (e.g., phosphorylation and
          activation of a synthetic enzyme such as tyrosine
          hydroxylase; 3) the cell nucleus (e.g., phosphorylation
          of a nuclear regulatory protein with resultant change in
          RNA synthesis.

CYCLIC-AMP-DEPENDENT PHOSPHORYLATION IN INSECT NERVE TISSUE

     In order to illustrate some of the above mechanisms, the
following sections describe experiments, which utilize Periplaneta americana thoracic ganglia, to demonstrate the presence
of an active cyclic AMP-dependent protein phosphorylation and
dephosphorylation system in this species.

Table 1. Protein Kinase Activity

Condition	CPM incorporated into protein	pmole ^{32}P transferred per mg enzyme protein
A. Histone Phosphorylation		
minus cyclic AMP	1400	126
plus cyclic AMP	4575	404
B. Endogenous Phosphorylation		
minus cyclic AMP	1337	20
plus cyclic AMP	2630	40

Specifically, the ability of extracts of ganglia to phosphorylate exogenous calf thymus histone was assayed by a modification of the method of Miyamoto et al. (1969). In these experiments, the 30,000 x g supernatant of metathoracic ganglia was used as a source of enzyme. Exogenous histone was used as the substrate for the transfer, by the endogenous protein kinase, of the ^{32}P of terminally labeled ATP. The reaction was run in the presence or absence of 5 micromolar cyclic AMP, and the amount of ^{32}P incorporated into total TCA precipitable protein was measured at the end of the incubation. Table 1A shows the resultant histone kinase activity. In the presence of cyclic AMP, there was more than a three-fold stimulation of ^{32}P incorporation into histone protein.

In other experiments, the ability of endogenous ganglion proteins to serve as substrates for the cyclic AMP-dependent protein kinase was measured. In these experiments, both enzyme and protein substrates were obtained from a 150,000 x g pellet of homogenized pro-, meso-, and metathoracic ganglia. As before, the incorporation of ^{32}P into TCA-precipitable protein was used as a measure of phosphorylation. Table 1B shows that 5 micromolar cyclic AMP stimulated phosphorylation approximately 2-fold, thus indicating the presence of endogenous substrates for the endogenous cyclic AMP-dependent protein kinase.

It was of interest to determine whether the cyclic AMP-stimulated incorporation of ^{32}P into endogenous protein was a generalized phenomenon, occurring on all cellular proteins, or whether it was limited to only a small number of proteins. To

measure ^{32}P incorporation into individual proteins, SDS-polyacrylamide gel electrophoresis was used to separate the various
endogenous proteins present in the ganglia. In this method, the
phosphorylation reaction described above was stopped by
the addition of a sulfhydryl reducing agent and sodium dodecyl
sulfate (SDS), which binds to proteins and causes a disruption of
secondary and tertiary structure. Because SDS binds in a

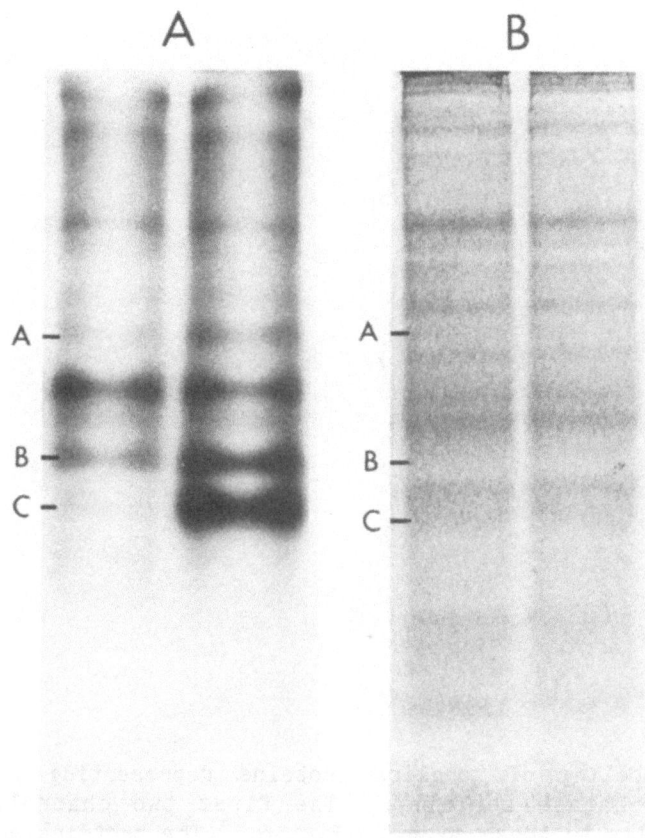

Fig. 6. Effect of cyclic AMP on phosphorylation of endogenous
ganglion proteins (see text for details of procedure).
In part A, the two lanes show the autoradiograms of
phosphorylated proteins labeled with ^{32}P and separated
by SDS-polyacrylamide electrophoresis. The first lane
had been incubated in the absence of cyclic AMP, while
the second had been incubated in 5 micromolar cyclic
AMP. "A", "B", and "C" mark the three bands whose
phosphorylation was cyclic AMP-dependent. The two lanes
on the right show the cooresponding protein staining
patterns of the two gels.

constant ratio according to molecular weight, the proteins can be
separated on the basis of molecular weight by electrophoresis,
applied through a polymerized acrylamide gel.   After electrophoresis, the gel is stained for protein, dried, and exposed to
X-ray film to form an autoradiogram of the ^{32}P which was
incorporated into the protein bands of the gel.

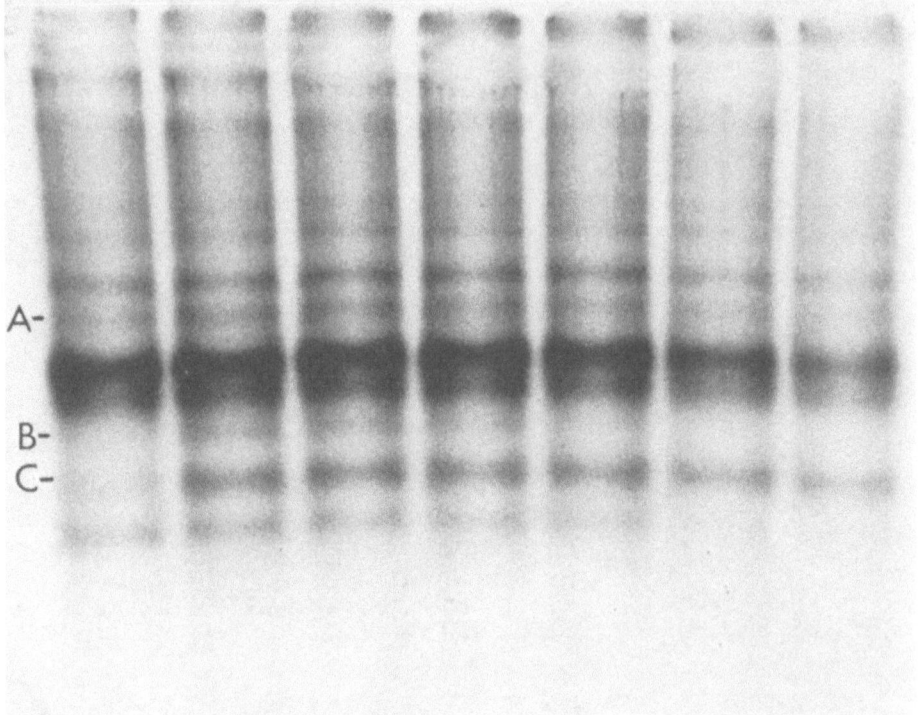

Fig. 7. Unlabeling of ganglion proteins representing protein
         phosphatase activity.   The first two channels are
         similar to those shown in Fig. 6. (The proteins applied
         to the first channel had been incubated without cyclic
         AMP present; the proteins applied to the second, with 5
         micromolar cyclic AMP.)   The remaining 5 channels are
         similar to the second channel, except that the proteins
         had been post-incubated in an excess of cold ATP for
         increasing periods of time.   The cold ATP stopped,
         effectively, any further incorporation of labeled
         phosphate by the protein kinase, and allowed the protein
         phosphatase to unlabel the proteins.   The post-incuba-
         tion times (channels 3-7) were: 15 sec, 45 sec, 2 min,
         10 min, and 30 min.

Figure 6A shows the results of two gels run side by side. In both gels, the material applied to the origin (top) was obtained from incubations in which the 150,000 x g pellet from thoracic ganglia homogenates had been used both as a source of protein kinase and of substrate. In the left-hand gel, no cyclic AMP had been added during the incubation; on the right, 5 micromolar cyclic AMP had been present. It can be seen that the phosphorylation of most protein bands was not greatly influenced by the presence of cyclic AMP. However, three bands (labeled "A", "B" and "C") were quite cyclic AMP-dependent, one ("A") strikingly so. The corresponding protein staining patterns in Fig. 6B show that the three cyclic AMP-dependent bands represent only a small fraction of the total proteins present.

It was also of interest to know if the thoracic ganglia contained a phosphoprotein phosphatase capable of removing the phosphate from endogenous proteins which previously had been phosphorylated by cyclic AMP-dependent protein kinase. Figure 7 shows an experiment which illustrates the presence of such a phosphatase. In this figure, the first two gel channels are similar to those in Fig. 6A. The proteins shown in channel 1 had been incubated for 45 seconds in the absence of cyclic AMP and those in channel 2, in the presence of cyclic AMP. All of the proteins shown in the remaining 5 channels had been incubated in the presence of cyclic AMP (like channel 2) except that, after 45 seconds, a large excess (200-fold) of unlabeled ATP had been added in order to dilute the specific activity of the labeled ATP already present. This effectively stopped any further incorporation of labeled ATP into protein. The proteins in the five right-hand gels had then been post-incubated for increasing periods of time to allow the protein to unlabel. This unlabeling is presumed to represent phosphoprotein phosphatase activity, since (as shown by other experiments) non-enzymatic phosphate removal was minimal over the time periods studied.

Figure 7 demonstrates that phosphatase activity was present, and that it showed a rather narrow substrate specificity. Cyclic AMP-dependent protein band B was rapidly unlabeled, whereas band B was not. In addition, a single non-cyclic AMP-dependent band (located below C) was rapidly unlabeled. The remaining bands, both cyclic AMP and non-cyclic AMP-dependent, were unlabeled much more slowly. The specificity of the rapid-acting phosphatase is quite interesting, for it supplies a possible means, together with the cyclic AMP-dependent protein kinases, for hormone regulation of specific intracellular processes. Of course, in all the experiments described, it must be remembered that the tissue used contains several cell types. Thus, it is likely that some of the labeled proteins are present in one cell type, while others are present in other cell types.

ACKNOWLEDGEMENT

This work was supported, in part, by a grant from the McKnight Foundation.

REFERENCES

BEAVO J.A. and MUMBY M.C. (1982) Cyclic AMP-dependent Protein Phosphorylation. In Cyclic Nucleotides I (Ed. by Nathanson, J. and Kebabian, J.) pp. 363-392. Springer Verlag, New York.

CHEUNG W.Y. and STORM D.R. (1982) Calmodulin Regulation of Cyclic AMP Metabolism. In Cyclic Nucleotides I (Ed. by Nathanson, J. and Kebabian, J.) pp. 301-323. Springer Verlag, New York.

CHRISTENSEN T.A. and CARLSON A.D. (1982) The neurophysiologgy of larval firefly luminescence: direct activaiton through four bifurcating (DUM) neurons. (1982) J. Comp. Physiol. 148, 503-514.

EVANS P.D. (1980) Biogenic amines in the insect nervous system. Adv. Insect Physiol. 15, 317-473.

GREENGARD P. (1978) Cyclic Nucleotides, Phosphorylated Proteins, and Neuronal Function. Raven Press, New York.

KUO J.F. and GREENGARD P. (1970) Cyclic nucleotide-dependent protein kinases. VI. Isolation and partial purification of a protein kinase activated by guanosine 3',5'-monophosphate. J. Biol. Chem. 245, 2493-2498.

KUO J.F. and SHOJI M. (1982) Cyclic GMP-dependent Protein Phosphorylation. In Cyclic Nucleotides I (Ed. by Nathanson, J. and Kebabian, J.) pp. 393-424. Springer-Verlag, New York.

KUO J.F., WYATT G.R., and GREENGARD, P. (1971) Cyclic nucleotide-dependent protein kinases. IX. Partial purification and some properties of guanosine 3',5'-monophosphate-dependent and adenosine 3'.5'-monophosphate-dependent protein kinases from various tissues and species of arthropoda. J. Biol. Chem. 246, 7159-7167.

LINGLE C.J., MARDER E. and NATHANSON J.A. (1982) The Role of Cyclic Nucleotides in Invertebrates. In Cyclic Nucleotides II (Ed. by Kebabian, J. and Nathanson, J.), pp. 787-845. Springer Verlag, New York.

NATHANSON J.A. (1976) Octopamine-sensitive adenylate cyclase and its possible relationship to the octopamine receptor. In Trace Amines and the Brain (Ed. by Usdin, E. and Sandler, M.), pp. 161-190. Marcel Dekker, New York.

NATHANSON J.A. (1979) Octopamine receptors, adenosine 3',5'-monophosphate, and neural control of firefly flashing. Science 203, 65-68.

NATHANSON J.A. (1983) The firefly light organ: an intriguing neuroeffector system for investigating aminergic synapses. Winter Conf. on Brain Research, Abstr., 38.

NATHANSON J.A. and GREENGARD P. (1973) Octopamine-sensitive
    adenylate cyclase: evidence for a biological role of octopamine
    in nervous tissue. Science 180, 308-310.
NATHANSON J.A. and GREENGARD P. (1974) Serotonin-sensitive
    adenylate cyclase in neural tissue and its similarity to the
    serotonin receptor: a possible site of action of lysergic acid
    diethylamide. Proc. Natl. Acad. Sci. 71, 797-801.
NATHANSON J.A. and HUNNICUTT E.J. (1979a) Neural control of light
    emission   in   Photuris   larvae: identification   of
    octopamine-sensitive   adenylate   cyclase. J. exp. Zool.   208,
    255-262.
NATHANSON J.A. AND HUNNICUTT E.J. (1979b) Octopamine-sensitive
    adenylate cyclase: properties and pharmacological characteri-
    zation. Soc. Neurosci. Abstr. 5, 346.
NATHANSON J.A. and HUNNICUTT E.J. (1981) N-demethylchlordimeform:
    a potent partial agonist of octopamine-sensitive adenylate
    cyclase. Mol. Pharmacol. 20, 68-75.
ORCHARD I., LOUGHTON B.G., GOLE J.W.D. and DOWNER R.G.H. (1983)
    Synaptic   transmission   elevates   cyclic   AMP   in   locust
    neurosecretory cells. Brain Res. 258, 152-155.
SCHULMAN H. (1982) Calcium-dependent Protein Phosphorylation. In
    Cyclic Nucleotides I (Ed. by Nathanson J. and Kebabian J.) pp.
    425-478. Springer-Verlag, New York.

EXTRINSIC CONTROL OF PHYSIOLOGICAL PROCESSES MEDIATED BY THE

INSECT NEUROENDOCRINE SYSTEM

J. de Wilde

Department of Entomology
Agricultural University
Wageningen, The Netherlands

1. Neuro-endocrine integration and homeostasis

Neurosecretory systems in the central nervous system and peripheral ganglia of insects are now known to control a large number of elements of development and reproduction: moulting, differentiation, eclosion, sclerotization, vitellogenesis and ovulation, to name only a few. These effects are partly exerted directly on the target tissues by neurosecretory hormones and partly indirectly by the adenotropic effect of these hormones on subordinated endocrine glands such as the corpora allata and the prothoracic glands.

A major function of the neuroendocrine system of the brain is the harmonious control of these subordinated mediator and effector systems in such a way that the homeostatic levels of hormones in the body fluids are adequately maintained. Next to internal coordination and regulation, this implies the adaptation of the insect to various ecological situations. To quote E. and B. Scharrer (1963) "Channels must exist through which these modifying influences enter the control system and become integrated with or supersede the feedback mechanism."

It should be realized that the very existence of homeostasis excludes the direct and specific interference of ecological factors with the internal milieu. And nevertheless, such factors deeply influence the physiological state of an insect. Even factors derived from cosmic constellations such as photoperiod can be shown to result in specific protein syntheses (de Loof & de Wilde 1970, de Kort 1981). The transduction of external into internal information by the neurosecretory system guarantees the integrity

151

of the organism and at the same time allows it to adapt to
environmental changes and to meet a number of conditions belonging
to the "ecological niche" which is the basis of the life cycle
strategy of the insect.

As stated before (de Wilde 1981), neuroendocrine integration
is the solution of a problem posed by the continuous adjustment of
internal relations to external relations, of the rigid homeostatic
regulation of the body fluids on the one hand, as stated by Claude
Bernard (1878) and the almost perfect fitting of an animal into its
environment on the other hand, as stated by Von Uexküll (1909).
It was Richet (1900) who saw first that the facts that internal
relations lead to stability, whereas the external relations lead
to adaptability, needed reconciliation. This was subsequently done
by the Scharrers (1963) in the formulation quoted above.

2. Seasonal stages – long-term synchronization

a. General

In many insects that have solved the problem of resistance to
adverse conditions, such as extreme freezing temperatures or lack
of free water, the life cycle is homodynamic and the succession of
generations is directly determined by the favourable or adverse
state of the environmental conditions. They freeze or desiccate
with the substrate in which they live. The larvae of tropical and
polar Chironomids belong to this category. They live in
environments where food is present as soon as the insect regains
activity. But if the ecological niche necessitates the
synchronization of the life cycle with ecological requisites such
as food plant, host or prey, the insect cannot merely rely on
resistance to undercooling or desiccation. It necessitates a
specific resting state, development or reproduction being triggered
by token stimuli assuring rest and activity at a proper (mostly
seasonal) time (Fig. 1).

Time measurement is most precise in photoperiodic responses,
which can be properly corrected by temperature. The ecological
requisite itself can emanate tokens which have their effect during
host feeding or consuming physiologically aged hostplants, assuring
a final correction and acting as a "veto" factor in seasonal
timing.

In this way, synchronization may occur with the astronomical
season, the climatic season, and, e.g., the hostplant season.

Such systems often include forecasting elements, by which
summer conditions may induce a process leading to rest in the
subsequent autumn. Delayed photoperiodic responses are an example,
depending on different receptive and responsive phases in the life
cycle (Fig. 1).

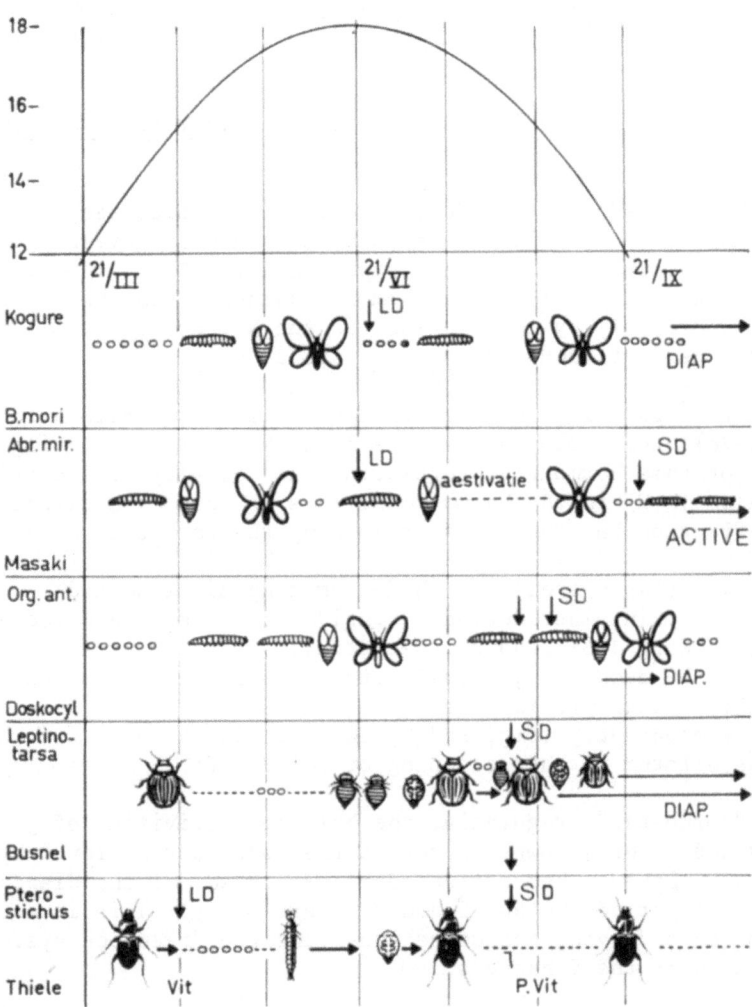

Fig. 1. Receptive and responsive phases with respect to photoperiod, and the effect on seasonal timing of diapause and reproduction in five insect species.

Embryonic, larval, pupal or adult diapause are syndromes, the elements of which may vary greatly between species. There are probably not two species with a completely identical diapause syndrome. Similarly, break of diapause may be caused by specific temperatures, photoperiodic and humidity conditions, acting during variable periods of time, according to the life cycle strategy of the species.

b. Case study - the Colorado beetle (<u>Leptinotarsa decemlineata</u> Say)

In the Colorado beetle adult, the states of diapause and reproduction are controlled by the brain, mainly by governing the activity of a master gland - the corpora allata (Fig. 2). This control is exerted partly by the humoral pathway, via the corpora cardiaca, and partly by innervation via the Nn. corporis allati (de Wilde & de Boer 1969, Khan <u>et al</u>. 1982a, b).

Next to regulating the rate of synthesis of juvenile hormone (Kramer 1978), the brain furthermore contributes to regulating the titre of this hormone by regulating the activity of specific jh esterase (jh E) in the haemolymph. This is done by a direct humoral effect on the synthesis of jh E by the fatbody.

The two "state-levels" of jh determining diapause and oogenesis are maintained during a considerable time (de Wilde <u>et al</u>. 1968; de Kort 1981). Jh E is especially significant in determining the diapause jh level, its activity increasing very considerably during prediapause (Kramer & de Kort 1976). This activity is apparently triggered by the initial release of jh by the corpus allatum in the beginning of adult life.

The "hormostat", regulating the balanced activities of jh synthesis and jh breakdown, is set at the reproductive level by photoperiods above 15 hours (long-day). It is set at the diapause level by photoperiods below 15 hours (short - day). This is a clear example of a "modifying influence" entering the "feedback system", as postulated by the Scharrers.

As we have not been able to observe any effect of varying jh titres on "<u>in vitro</u>" synthesis of jh by the corpus allatum (Khan <u>et al</u>. 1982), the "hormostat" clearly resides in the brain. Its location and nature are subject to further studies, but the lateral NSC are good candidates (Khan 1983) (Fig. 3).

The two "state-levels" of jh have a profound influence on a series of subcellular processes characterizing the syndromes of reproduction and diapause. The effect on reproduction is not so much in the phase of mating and courtship (Thibout 1982) but very outspoken in the phase of vitellogenesis, both as regards the

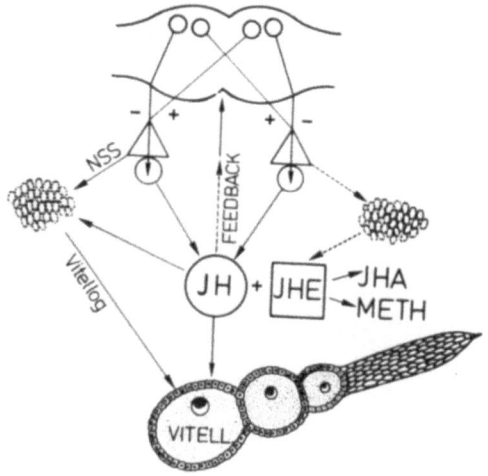

Fig. 2. Regulation of JH titre in the Colorado potato beetle and
        its effect on vitellogenesis. Inhibitory nervous stimuli
        emerge from the lateral, activating NS stimuli from the
        medial NSC and the corpora cardiaca.
        NSS = neurosecretory substance
        JHE = JH esterase
        JHA = JH acid
        Meth = Methanol
        Feedback by JH titre takes place via the brain.

synthesis and release of vitellogenin by the fatbody (Dortland
1979) and the uptake of this protein by the oocyte (de Loof &
Lagasse, 1972).

    The effect on diapause covers all the syndrome elements known
by us at present, and especially flight muscle atrophy (de Kort
1969), burrowing behaviour (de Wilde & de Boer 1969), the synthesis
and release of a "diapause protein" by the fatbody (de Loof & de
Wilde 1970, Dortland 1979), and programmed breakdown of yolk by
the oocyte (de Loof and Lagasse, 1970).

    Histologically, neurosecretory activity in the brain under

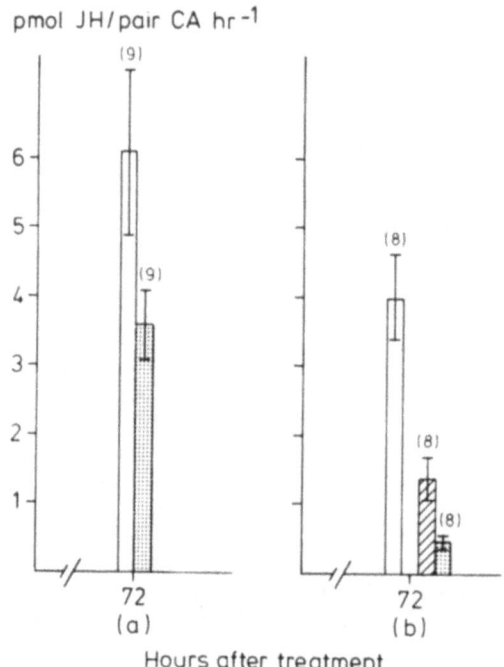

pmol JH/pair CA hr^{-1}

Hours after treatment

Fig. 3. a. Effect of elevated JH titre after denervation of
corpora allata, and rate of JH synthesis. Open columns:
Controls. Shaded columns: JH treated.
b. Effect of elevated JH titre on host (shaded columns)
and implanted (striped columns) corpora allata in the
same insects on the rate of JH synthesis. Donors as well
as recipients were 2 days old females. Open columns:
Controls. JH synthesis was measured in vitro 72 hours
after treatment. (after Khan et al. 1982).

various photoperiods and as a function of hostplant condition, has
been extensively studied by Schooneveld (Schooneveld 1970, 1974;
de Wilde et al. 1969). These histological patterns of neurosecretory activity mainly concern the complex of cerebral neurosecretory cells as a whole; up till now, specific effects on
certain types of NSC have hardly been detected.

3. Castes as ecomorphs

a. General

The environment sometimes induces insects to enter into irreversible morphogenetic states such as observed in seasonal dimorphism or in alternating generations of sexual and asexual forms.

It has recently become clear that castes in social insects belong to this category, for which I have proposed the term ecomorphs (de Wilde 1975).

Larval life of social insects is confined to the colony, either in a relatively free state or enclosed in a cell. Larvae of either sex in the termites, or female larvae of social Hymenoptera may be induced to follow different lines of development, leading to specific adult morphs such as reproductives, soldiers and workers. Some of these lines may be blastogenic as recently found for reproductive castes of termites, but other lines of development are induced during larval development by conditions in the colony. As was first detected in the termite Calotermes flavicollis, soldier formation is induced by an increased titre of juvenile hormone (Lüscher 1958). The same hormone inhibits the formation of reproductives (Lüscher 1957). The exact nature of the environmental stimuli leading to these endocrine effects has not yet been elucidated; it is supposed that pheromonal messages within the colony are involved. Similarly, soldier formation in the ant Pheidole bicarinata is induced by an increased titre of juvenile hormone during a critical period in the last larval instar (Wheeler & Nijhout 1981). The precise stimuli leading to this effect are unknown.

b. Caste formation in the honey bee Apis mellifera L.

We are considerably better informed of the causal factors in caste formation in the honey bee where painstaking research during more than 20 years has led us to exclude a number of possibilities and on the other hand to render a good deal of probability to other potential factors.

In the honey bee, sex determination is by haplo-diploidy. The diploid is the female sex, occurring in two castes: workers and queens (Fig. 4).

Some investigators have approached the problem starting from the idea that special factors are needed to induce the formation of queens. In this line of thought, a Queen Determinator is present in royal jelly, a glandular secretion deposited exclusively in queen cells. Already the nomenclature applied to this

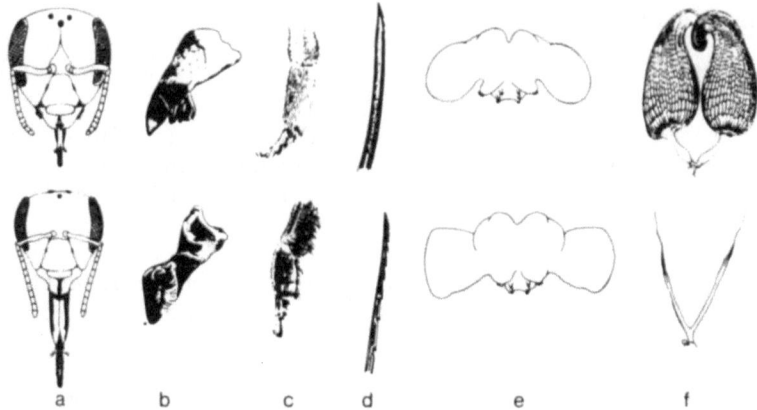

Fig. 4. Some essential caste features in the worker and queen
        honey bee, which are used in the evaluation of caste
        differentiation (after de Wilde and Beetsma 1982).

    (a) Shape of head capsule and size of proboscis;

    (b) mandible;

    (c) tibia and basitarsus of hind leg;

    (d) sting;

    (e) brain;

    (f) ovary.

substance has rendered it a magical reputation, and in pharma-
ceutical products as well as "health foods" it has found a market
among those seeking anti-senescence and euphory. More than twenty
years of intensive chemical investigation (Butenandt 1955,
Rembold 1976) have failed to isolate and identify the hypothetic
factor in royal jelly responsible for queen differentiation.

    During this intensive chemical research, an alternative line of
thought has been overlooked, which had been proposed nearly thirty
years ago by Lukoschus (1956). It departs from the logical line of
reasoning that in the evolution of social insects the novelty is the
mass production of one or more sterile castes rather than reproductives. Otherwise than in the Osoptera, in the family of Apidae the pre-

ponderant number of species is solitary, and we have thereby ample
occasion for further comparison. We then observe that the normal
way of provisioning larvae in the Apidae is by mass provisioning,
and that the piecemeal feeding of larvae leading to the worker
caste of bees is apparently the novel factor in social evolution.
This, consequently, leads to the hypothesis that the active
process is the suppression of reproductive potentialities and the
switch in the developmental program leading to the remarkable
facilities characteristic for the worker caste. In this line of
thought, the trophic milieu of mass provisioning activates the
corpora allata, while the reduced feeding of the worker larva
during the critical phase of caste programming inhibits the
activity of these glands. The genes for queen programming are
activated by a high JH titre, while at a low JH titre these genes
are repressed and the genes for worker programming are switched on.

The likeliness of this line of thought is beautifully
confirmed by the endocrine situation. In the honey bee, the
differentiation of cerebral neurosecretory cells is remarkably
retarded during the early larval instars, and this retardation is
supported by the treatment given to the presumptive worker larvae
(Ulrich 1979). In both queen and worker larvae, up till the fourth
day of larval life a normal number of 160 presumptive ovarioles
is present in the ovarian "anlagen". In the worker larva, after
this day an ovarian regression sets in, during which all but two
or three presumptive ovarioles are maintained and undergo further
growth and differentiation (de Wilde & Beetsma 1982).

As found by Wirtz (1973) the corpora allata of larvae transferred from worker cells to queen cells rapidly increase in size
and activity, and this is apparently caused by the acceleration
of the retarded neuroendocrine activity in these larvae. The
juvenile hormone titre in presumptive queen larvae during the
critical phase of caste determination is a factor 10 higher than
is observed in worker larvae (Wirtz 1973). Application of juvenile
hormone during the third day of larval life determines the program
of development in the direction of queen formation. All
characteristic features of queen bees are induced by this
treatment, including the prevention of ovarian regression leading
to a normal number of queen ovarioles. The mandibular glands are
activated to produce "queen substance" (Fig. 5).

A most interesting confirmation of juvenile hormone impact on
caste behaviour as already demonstrated in the larval stage, has
come from the experiments made by Ebert (1980). She could show
that the vertical position taken by the pupating queen larva is
not merely determined by the vertical position of the queen cell,
but is a result of positive geotaxis exhibited at the end of
larval life. Larvae in worker combs live in horizontal cells, but
when they are treated by jh during the caste determining period

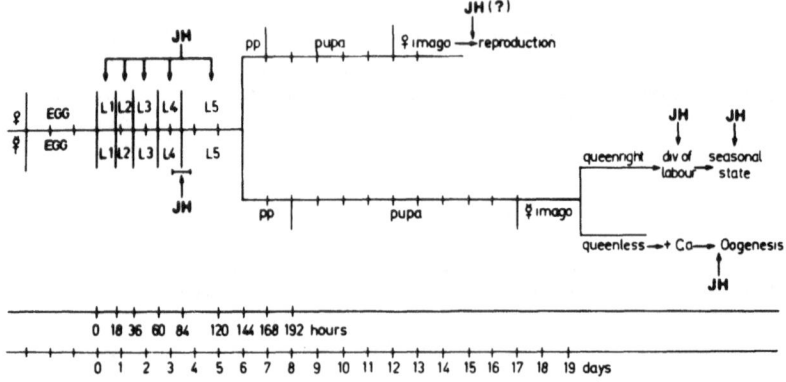

Fig. 5. Rate of caste development in the honey bee and JH-
        sensitive phases for instar, stage and caste programming,
        oogenesis and worker polyethism (after de Wilde and
        Beetsma 1982).

they tend to take a vertical position, head downwards, in the beginning of the prepupal phase!

As regards the causal environmental factors leading to the
onswitch of the worker developmental program, it has recently been
found by Brouwers (1983) in our institute that until the fourth
day of larval life, the frequency as well as the duration of
feeding of worker larvae by the nurse bees are remarkably reduced,
causing larval food to be both rare and stale. After the caste-
determining period the rate of feeding increases most remarkably
(Fig. 6). Differences in the composition of the nutrients before
the caste determining period have not been found (Brouwers 1982)
contrary to some data in literature suggesting a higher sugar
content of queen jelly (Asencot and Lensky 1976).

For a more detailed discussion, readers are referred to de
Wilde & Beetsma (1982).

Our colleagues from the Max Planck Institute for Chemistry
in Munich are now actively supporting the endocrine approach. Some
of their findings are shown in Fig. 7 (Rembold 1981). If these
data are confirmed, it is apparent that the course of jh and
ecdysone titres during larval development in the honey bee
deviates considerably from those found in solitary insects. This
is especially true for the jh titre, and this fact alone shows the
enormous impact of caste programming during the early instars.

Similar effects of jh on caste determination as described
above have in meantime been found in stingless bees (Velthuis &

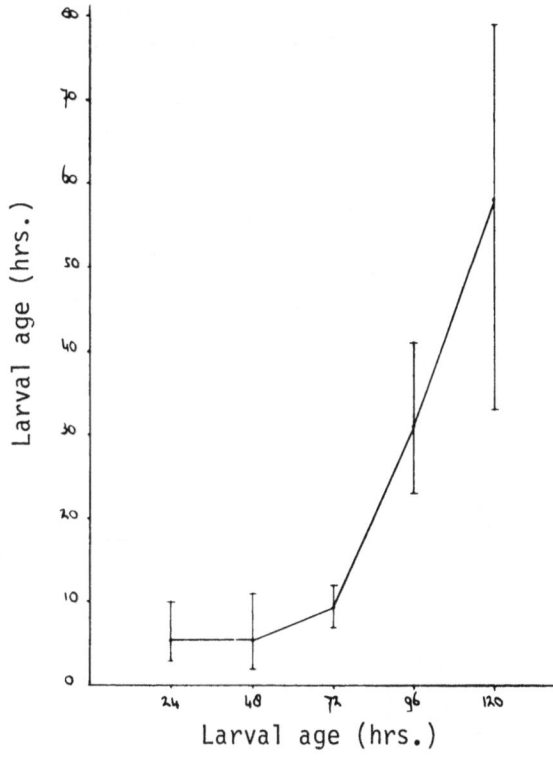

Fig. 6. Feeding frequency of worker larvae in the honey bee, by the nursing bees.
Note the considerable increase of feeding frequency after the critical period for caste programming (after Brouwers 1983).

Larval age (hrs.)

Velthuis-Klüppell 1975) and in bumble bees (Röseler 1976).

c. Endocrine impact on worker functions

It has been reported that in the adult worker of the honey bee, social functions such as cleaning, nursing, building and foraging succeed each other and are age-dependent (Rösch 1925, 1930). It has also been found by Lindauer (1952) that this succession is not merely a question of age, but that the "labour market" in the bee hive profoundly influences these types of activity, albeit that a tendency towards a sequence in the above-mentioned sense is observed with ageing. It seems that the "de facto" choice of activities, within the cadre set by age, is made during "wandering periods" of the worker bee within the colony, formerly considered as "empty" or "resting" periods (Lindauer 1952). It now appears that the determining imprints which the worker receives during her strolls around the colony govern the activity of the corpora allata, obviously by the neuroendocrine pathway. High titres of jh inhibit the activity of the hypopharyngeal food glands and the wax glands, and nursing and building behaviour of young worker bees is thereby inhibited. On the other hand, foraging behaviour and aggressive behaviour as demonstrated by field bees and guard bees are promoted by such high JH titres (Beetsma and ten Houten, 1975; Rutz et al., 1976; Breed, 1979; Fluir et al., 1982) (Fig. 5).

It has finally been found that the suppressive effect of the presence of the queen in ovarian activity in worker bees is brought about by inhibition of the corpora allata (Gast, 1967). This is a result of the release by the queen of the pheromone 9-oxo decenoic acid.

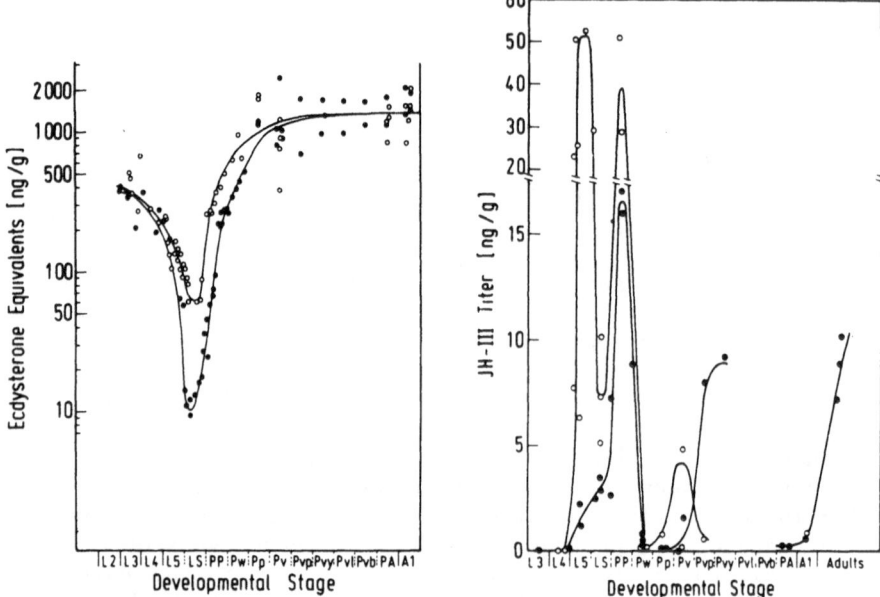

Fig. 7. Titres of Ecdysterone and JH in developmental stages of worker and queen honey bees.
Left: Ecdysterone equivalents
Right: JH titres
● worker o queen
L2-L5 = 2nd till 5th larval instar
LS,PP = spinning larva and prepupa
Pw-Pvb = stages in eye differentiation
PA, A1 = pharate, resp. 1 day old adult.
(after Rembold and Hagenguth 1981).

4. Endocrine impact in the regulation of population density

As I mentioned in an earlier paper (de Wilde, 1981), a fascinating field of study is where population regulation meets the homeostatic processes in the individual. There is in the first place the synchronization of the life cycle of parasites with their hosts. Generally, the parasite is dominated by the endocrine situation prevailing in the host. As pupal diapause is frequently occurring during hibernation, and as this type of diapause is caused by the absence of ecdysone, the increasing activity of the prothoracic glands during break of diapause stimulates the immature stages of endoparasites, which have hibernated within their host to follow its development and to assure a synchronous emergence (Syrphidae and Diplazon sp., Schneider 1950; Bupalus and Eucarcelia, Schoonhoven 1962.

In other cases, the parasite destroys or inhibits the host's neuroendocrine system and thereby causes a standstill in or modification of development or, in adults, sterility (Diplazon pectoratorius and Syrphids, Schneider 1950; Mermis and Bombus queens, Palm 1948, Stylops and solitary bees, Salt 1927, 1931, Brandenburg 1956.

But most interesting (and surprisingly little studied) is the endocrine impact in intraspecific density effects known as functional responses (Solomon 1949). It is a generally observed fact in population ecology that population densities above the so-called equilibrium density result in an increase of mortality and a decrease in the rate of reproduction. Partly this is due to the effect of natural enemies, but there are many species where mutual interference between the elements of a population results in a degree of self-regulation. Thus, in aphids, the formation of alate migrants has this effect and is due to crowding (Bonnemaison 1951), and in locusts, the formation of the gregarious phase and the subsequent migratory activity reduces the population in the breeding grounds (Uvarov 1966). In both cases, there is little doubt that the signals are transduced by the neuroendocrine system. In Locusta migratoria Staal (1961) has shown this to be the case for a number of phase features.

As regards the density effect on reproduction, the mechanisms vary in a considerable degree between species. Competition for food can directly affect the adults, but can also have its effect via larval growth and the resulting adult size, as in the grey larch moth, Zeiraphera diniana (Benz 1974).

Even more intriguing are cases where the meeting between individuals in a population has a direct sensory effect resulting in abrupt changes in the programming of larval development and a reduced adult size. In a recent paper (de Wilde 1981) I have

quoted the results of Gruys (1970) who has described dramatic
effects of this kind in Bupalus piniarius. The existence of
sensitive stages in these intraspecific effects and the nature
of the resulting phenomena, being of a similar kind as those
induced by photoperiod and temperature, point to a neuroendocrine involvement comparable with what we observe in caste
induction and similar ecomorphoses.

5. Freedom in internal and external relations

Most of our work on diapause in Leptinotarsa has been performed with newly emerged beetles shortly after the adult moult.
We have recently started to extend our experiments to beetles
which have nearly or totally completed diapause development.

It was found by Dortland (1979) that the Colorado beetle
female, allatectomized at the end of diapause when the jh-titre
was already almost zero, and before feeding had been resumed, is
capable of vitellogenin synthesis and, after subsequent feeding,
of a reduced rate of oviposition.

Already in the early beginning of our investigations (de
Wilde et al. 1959) it was found that post-diapause beetles no
longer respond to photoperiod. This is illustrated by an experiment
in which it is shown that transfer of post-diapause beetles from
long-day to short-day and vice versa, does not result in any
photoperiodic signal-effect but merely in a slight quantitative
change in the rate of oviposition, corresponding with the change
in the duration of feeding in this diurnal insect. During her
postgraduate work in our laboratory, it was found by Miss K.
Lefevre that the medial neurosecretory cells of the brain become
activated during break of diapause, and that the corpora allata
resume their activity in a matter of days.

Allatectomy immediately after break of diapause does not
result in a second diapause, but after a maximum of 5 days the
corpora allata have regained control. It is, however, no longer
influenced by daylength and the histological picture of the
central MNSC and the postcerebral glands shows activity at long
as well as short photoperiods.

Freedom is also observed in the comparison between species.
In Rhodnius (Wigglesworth 1936) and in Locusta (Wyatt et al. 1976)
the fatbody synthesizes vitellogenin in response to jh alone,
while in Leptinotarsa (de Loof & de Wilde 1970) and in Sarcophaga
bullata (Wilkens 1969) it needs the combined action of jh and
brain hormone. And the same system may even respond to different
hormones in different species, as is demonstrated by mosquitoes
where the fatbody appears to synthesize vitellogenin under the
influence of α-ecdysone (Hagedorn 1974).

At the level of sequestration of vitellogenins by the oocyte, differences also appear. While this sequestration generally requires jh, in Sarcophaga bullata it appears to occur under the influence of brain hormone (Wilkens 1969). Diapause in Leptinotarsa requires the atrophy of the flight muscles. In Scolytid beetles, which oviposit while embedded in plant tissues, atrophy of the flight muscles occurs during reproduction. In Leptinotarsa, flight muscle atrophy is induced by near-zero jh titres, while in Scolytids, high titres of jh have the same effect.

Final instructions on hormonal effects are apparently provided by the hormone receptors present in the subordinated effector cells. The ecological niche appears to have profound consequences in this respect, as I have emphasized on former occasions (de Wilde 1970, 1982). In the mealworm, Tenebrio molitor, the reproductive state does not depend on photoperiod. In the pear weevil, Anthonomus cinctus, which depends on the buds of pear trees prevailing during the winter season, reproduction is a short-day effect.

There are even species with racial differences in photoperiodic relations. The short-day race of Ceutorrhynchus pleurostigma oviposits in autumn, while the photoperiod-independent, monovoltine race oviposits in spring after adult diapause (Ankersmit 1964). Apparently, the ecological niche has for every species its profound consequences for the ensuing life cycle strategy and the underlying extrinsinc and intrinsic control of endocrine systems.

References

Ankersmit G.W. (1964) Voltinism and its determination in two beetles of cruciferous crops. Meded. Landb. Hogesch. Wageningen 64-8, 60 pp.

Asencot M. and Lensky Y. (1976) The effect of sugars and juvenile hormone on the differentiation of the female honey bee larva (Apis mellifera L.) to queens. Life Sci. 18, 693-700.

Beetsma J. and ten Houten, A. (1975) Effects of juvenile hormone analogues in the food of honey bee colonies (Apis mellifera L.). Z. angew. Ent. 77, 292-300.

Benz G. (1974) Negative Rückkoppelung durch Raum- und Nahrungskonkurrenz sowie zyklische Veränderung der Nahrungsgrundlage als Regelprinzip in der Populationsdynamik des grauen Lärchenwicklers, Zeiroptera diniana. Z. angew. Entomol. 76, 196-228.

Bernard C. (1878) Leçons sur les phénomènes de la vie communs aux animaux et aux dégétaux. Baillière, Paris.

Bonnemaison L. (1951) Contribution à l'étude des facteurs provoquant l'apparition des formes ailées et sexuéés chez les aphidinae. Thèses Fac. Sc. Univ. Paris, 380 pp.

Brandenburg J. (1956) Das endokrine System des Kopfes von
    Andrena vaga Pz. (Ins. Hymenopt.) und Wirkung der
    Stylopization (Stylops, Ins. Strepsipt). Z. Morph. Oek. der
    Tiere 45, 343-364.
Breed M.D. (1979) Does juvenile hormone regulate dominance and
    aggression in honey bee? Proc. 16th Int. ethol. Conf.
    Vancouver.
Brouwers E.V.M. (1983) Personal communication.
Butenandt A. (1955) Wirkstoffe des Insektenreiches. Nova Acta
    Leopoldina (N.F.) 17, 445-471.
de Kort C.A.D. (1969) Hormones and the structural and biochemical
    properties of the flight muscles in the Colorado beetle.
    Meded. Landb. Hogeschool Wageningen 69 (2), 1-63.
de Kort C.A.D. (1981) Hormonal and metabolic regulation of adult
    diapause in Leptinotarsa decemlineata. Entomol. Generalis
    7, 261-271.
de Loof A. and de Wilde J. (1970) The relation between haemolymph
    proteins and vitellogenesis in the Colorado beetle,
    Leptinotarsa decemlineata Say. J. Insect Physiol. 16,
    157-169.
de Loof A. and Lagasse A. (1970) Resorption of the terminal oöcyte
    in the allatectomized Colorado beetle, Leptinotarsa
    decemlineata Say. Proceedings, Kon. Ned. Ak. v. Wet. C73,
    284-297.
de Loof A. and Lagasse A. (1972) Proteid yolk formation in the
    Colorado beetle, with special reference to the mechanism of
    the selective uptake of haemolymph proteins. Proceedings,
    Kon. Ned. Ak. v. Wet. C75, 127-143.
de Wilde J. (1975) As endocrine view of metamorphosis, polymorphism
    and diapause in insects. Am. Zool. suppl. 15, 13-27.
de Wilde J. (1981) Ecological reflections on insect neuroendocrine
    activity. Amer. Zool. 21, 625-630.
de Wilde J. and Beetsma J. (1982) The physiology of caste develop-
    ment in social insects. Adv. Insect Physiol. 16, 167-246.
de Wilde J. and de Boer J.A. (1969) Humoral and nervous pathways
    in photoperiodic induction of diapause in Leptinotarsa
    decemlineata. J. Insect Physiol. 15, 661-675.
de Wilde J., Bongers W and Schooneveld H. (1969) Effects of host
    plant age on phytophagous insects. Ent. exp. et appl. 12,
    714-720.
de Wilde J., Staal G.B., de Kort C.A.D., de Loof A. and Baard G.
    (1968) Juvenile hormone titre in the haemolymph as a function
    of photoperiodic treatment in the adult Colorado beetle
    (Leptinotarsa decemlineata Say). Proc. Kon. Ned. Ak. v. Wet.
    C71. 321-326.
Dortland J.F. (1979) Hormonal control of protein synthesis in the
    fat body of Leptinotarsa decemlineata. Thesis, Wageningen
    87 pp.
Ebert R. (1980) Influence of juvenile hormone on gravity
    orientation in the female honey bee larva (Apis mellifera L.).

J. comp. Physiol. A137, 7-16.

Fluri P., Lüscher M., Wille H. and Gerig L. (1982) Changes in weight of the pharyngeal gland and hemolymph titre of juvenile hormone, protein and vitellogenin in worker bees. J. Insect Physiol. 28, 61-68.

Gast R. (1967) Untersuchungen ueber den Einfluss der Königinnen-substanz auf die Entwicklung der endokrinen Drüsen bei der Arbeiterin der Honigbiene (Apis mellifica). Ins. Soc. 14, 1-12.

Gruys P. (1970) Growth in Bupalus piniarius (Lepidoptera, Geometridae) in relation to larval population density. Centre for Agric. Publicat. and Document., Wageningen.

Hagedorn H.H. (1974) The control of vitellogenesis in the mosquito, Aedes aegypti. Amer. Zool. 14, 1207-1217.

Khan M.A., Koopmanschap A.B., Privee H. and de Kort C.A.D. (1982a) The mode of regulation of the corpus allatum activity during starvation in adult females of the Colorado potato beetle, Leptinotarsa decemlineata (Say). J. Insect Physiol. 28, 791-796.

Khan M.A., Koopmanschap A.B. and de Kort C.A.D. (1982b) The effects of juvenile hormone, 20-hydroecdysone and precocene II on activity of corpora allata and the mode of negative-feedback regulation of these glands in the adult Colorado potato beetle. J. Insect Physiol. 28, 995-1001.

Khan M.A. (1983) Thesis, Wageningen.

Kramer S.J. (1978) Regulation of the juvenile hormone titre in the Colorado potato beetle. Thesis, Wageningen, 72 pp.

Kramer S.J. and de Kort C.A.D. (1976) Age-dependent changes in juvenile hormone esterase and general carboxy-esterase activity in the haemolymph of the Colorado potato beetle, Leptinotarsa decemlineata. Mol. Cell. Endocr. 4, 43-53.

Lindauer M. (1952) Ein Beitrag zur Arbeitsteilung im Bienenstaat. Z. Vergl. Physiol. 34, 299-345.

Lüscher M. (1958) Experimentelle Erzeugung von Soldaten bei der Termite Kalotermus flavicollis (Fabr.). Naturwiss. 45, 1-2.

Lüscher M. (1957) Ersatzgeschlechtstiere bei den Termiten und die Beeinflüssung ihrer Entstehung durch die corpora allata. Verh. Ber. dtsch. Ges. angew. Ent. 14, 144-150.

Lukoschus H. (1956) Zur Kastendetermination bei der Honigbiene. Z. Bienenforsch. 3, 190-199.

Palm N.B. (1948) Normal and pathological histology of the ovaries in Bombus Latr. (Hymenoptera). Opuscula Ent., Lund., Suppl. VII.

Rembold H. (1976) The role of determination in caste formation in the honey bee. In: Phase and Caste determination in Insects (M. Lüscher, Ed.). Pergamon Press, Oxford, 21-34.

Rembold H. and Hagenguth H. (1981) Modulation of hormone pools during postembryonic development of the female honey bee castes. Regulation of Insect Development and Behaviour, Wroclaw, 427-440.

Rösch G.A. (1925) Untersuchungen ueber die Arbeitsteilung im
    Bienenstaat 1. Z. Vergl. Physiol. 2, 571-631.
Rösch G.A. (1930) Untersuchungen ueber die Arbeitsteilung im
    Bienenstaat 2. Z. Vergl. Physiol. 12, 1-71.
Röseler P.F. (1976) Juvenile hormone and queen rearing in bumble-
    bees. In: Phase and Caste determination in Social insects
    (M. Lüscher, ed.). Pergamon Press, Oxford, 55-61.
Rutz W., Gerig L., Wille H. and Lüscher M. (1976) The function of
    juvenile hormone in adult worker honeybees, Apis mellifera.
    J. Insect Physiol. 22, 1485-1491.
Rutz W., Imboden H., Jaycox E.R., Wille H., Gerig L. and Lüscher
    M. (1977) Juvenile hormone and polyethism in worker honey-
    bees (Apis mellifera). Proc. 8th Int. Congr. IUSSI,
    Wageningen, 26-27.
Salt G. (1972) The effects of stylopization on aculeate
    Hymenoptera. J. exp. Zool. 48, 223-319.
Salt G. (1931) A further study of the effects of stylopization on
    wasps. J. exp. Zool. 59, 133-163.
Scharrer E. and Scharrer B. (1963) Neuroendocrinology. Columbia
    University Press, Nw York.
Schneider F. (1951) Einige physiologische Beziehungen zwischen
    Syrphidenlarven und ihren Parasiten. Z. angew. Ent. 33,
    150-162.
Schooneveld H. (1970) Structural aspects of neurosecretory and
    corpus allatum activity in the adult Colorado beetle,
    Leptinotarsa decemlineata Say. Neth. J. Zool. 20, 151-237.
Schooneveld H. (1974) Ultrastructure of the neurosecretory system
    of the Colorado potato beetle, Leptinotarsa decemlineata
    (Say). In: Characterization of the protocerebral neuro-
    secretory cells. Cell & Tissue Res. 154, 275-288.
Schoonhoven L. (1962) Diapause and the physiology of host-
    parasite synchronization in Bupalus piniarius L. and
    Encarcelia rutilla. Arch. néerl. Zool. 15, 111-174.
Staal G.B. (1961) Studies on the physiology of phase induction in
    Locusta migratoria migratorioides (R and F). Publ. Fonds
    Landb. Export Bur. 40, 1-125.
Solomon M.E. (1949) The natural control of animal populations. J.
    animal Ecol. 18, 1-35.
Thibout E. (1982) Le comportement sexuel du doryphore, Leptinotarsa
    decemlineata Say, et son possible controle par l'hormone
    juvenile et les corps allates. Behaviour 80, 199-217.
Ulrich G. (1979) Histologische und biochemische Untersuchungen zur
    endokrinen steuerung der Kastenwerkmale bei der Honigbiene
    (Apis mellifera). Ph.D. Thesis, Köln.
Uvarov B.P. (1966) Grasshoppers and locusts. Cambridge University
    Press.
Velthuis H.H.W. and Velthuis-Klüppell F.M. (1975) Caste
    differentiation in a stingless bee, Melipona quadrifasciata
    Lep., influenced by juvenile hormone application.
    Proceedings, K. Ned. Akad. Wet. C78, 81-94.

Von Uexküll J. (1909) Umwelt und Innenwelt der Tiere. Julius
    Springer, Berlin.
Wheeler D.E. and Nijhout H.F. (1981) Soldier determination in ants:
    New role for juvenile hormone. Science 213, 361-363.
Wigglesworth V.B. (1936) The function of the corpus allatum in the
    growth and reproduction of Rhodnius prolixus (Hemiptera).
    Quart. J. Micr. Sci. 79, 91-121.
Wilkens J.L. (1969) The endocrine control of protein metabolism as
    related to reproduction in the fleshfly, Sarcophaga bullata.
    J. Insect Physiol. 15, 1015-1024.
Wirtz P. (1973) Differentiation in the honey bee larva. Meded.
    Landb. Hogesch. Wageningen 73-5, 1-155.
Wyatt G.R., Chen T.T. and Couble P. (1976) Juvenile hormone
    induced vitellogenin synthesis in locust fat body in vitro.
    In: Invertebrate Tissue Culture, ed. E. Kurstak, K.
    Maramorosch. Academic Press, Nw. York, 195-202.

COMPARATIVE ASPECTS OF INSECT-VERTEBRATE NEUROHORMONES

H. Duve and A. Thorpe

School of Biological Sciences
Queen Mary College, University of London
London E1 4NS.   U.K.

INTRODUCTION

Studies on the endocrine mechanisms of insects have mainly been concerned with the control of physiological processes such as growth, moulting, differentiation, diuresis and reproduction. In comparison, relatively little interest has been paid to the hormonal influences on metabolism. The role of the adipokinetic hormone in the control of lipid metabolism is a notable exception and one of the few other metabolic regulatory factors to attract attention is the hyperglycemic factor, first identified by Steele (1961). He demonstrated the presence of a substance in extracts of the corpora cardiaca (CC) of *Periplaneta* that elevates the haemolymph trehalose content markedly. These observations were later supported by others from Ralph (1962) and Ralph & McCarthy (1964) who showed that the CC extract acted by causing a breakdown of fatbody glycogen with a resultant release of trehalose. Two hormones are thought to be involved, both having a similar effect in converting the enzyme phosphorylase in the fatbody from its inactive to its active form (Steele, 1963). There was also an effect on the glycogen content of some other tissues such as the nerve cord but very little effect on the glycogen of muscle or the gut. The hyperglycemic factors from the CC of *Periplaneta* were partially purified using Sephadex and Biogel column chromatography and shown to be peptides by Natalizi and Frontali (1966). The hyperglycemic factor(s) have subsequently been identified and studied in several other insect species. Thus, in the locust, two chromatographically distinct carbohydrate mobilizing factors are present in the CC, one potent factor localised in the glandular lobes, and one, less potent, in the storage lobes (Goldsworthy, 1969; Mordue & Goldsworthy, 1969). Both have the effect of increasing the level of active phosphorylase in the fatbody.

Evidence for the existence of a hyperglycemic factor with a possible hormonal regulatory effect on the trehalose metabolism of the blowfly *Phormia regina* was presented by Friedman (1967) and Chen & Friedman (1977). Furthermore, it was shown by Normann and Duve (1969) that the CC may be activated via the brain to release the hyperglycemic hormone. However, it was not possible to state whether the hormone was produced within the CC itself, or whether it was produced within the brain and transported from there to the CC.

Recently, the possibility that the hyperglycemic hormone of the lepidopteran, *Manduca sexta* is identical with or very similar to vertebrate glucagon has been studied by Tager et al (1975, 1976), Kramer (1979) and Kramer et al. (1980).

Studies on the existence of a hypoglycemic factor, presumably acting in concert with the hyperglycemic hormone are considerably fewer, but the presence of such a factor was suggested early on by Dixit and Patel (1964). More recently Normann (1975) reported the existence of a hypoglycemic hormone in the blowfly, *Calliphora*, possibly located in the median neurosecretory cells. This idea gained further support from the study of Chen and Friedman (1977) who demonstrated a hormone in the adult blowfly *Phormia regina*, able to depress hemolymph trehalose levels.

It has of course long been established in vertebrates, that the principal hyperglycemic hormone, glucagon, acts together with the dominant hypoglycemic hormone, insulin, to control carbohydrate metabolism and it is not surprising that, as well as associating insect hyperglycemic factors with glucagon, attempts have been made to link the hypoglycemic factor with insulin. Thus, Ishay et al. (1976) have reported the presence of insulin in tissues of various hymenopterans by a rather crude radioimmunoassay technique and subsequently, Tager et al. (1976) and Kramer et al. (1977) have found an insulin-like substance in extracts of the neurosecretory system and in the hemolymph of the tobacco hornworm moth, *Manduca sexta.* Using a physiological approach, Duve (1978) has shown that the perikarya of the median neurosecretory cells of the blowfly *Calliphora* produce a hormone with at least one biological similarity to insulin. Thus, homogenates of MNC or acid alcohol extracts of heads when injected into hyperglycemic flies cause a significant hypoglycemia.

It is perhaps reasonable to assume that if the biochemical and metabolic processes that exist in insects are the same or very similar to those in vertebrates, then the control or regulating molecules will also be the same or similar. We attempted to study this intriguing problem in *Calliphora* by examining firstly the

relationship of the hypoglycemic factor to insulin and the hyperglycemic factor to glucagon. We then extended the study by looking
for other members of the gastroenteropancreatic series of biologically active peptides.   Finally we have explored the possibility
that *Calliphora* possesses vertebrate-type neuropeptides of the
endorphin/enkephalin series.

Several approaches have been used.

(a) We have used region specific antisera directed against
mammalian peptides in a series of immunocytochemical studies
designed to localise the peptides within the insect nervous system.

(b) In parallel, we have carried out biochemical studies aimed at
the extraction and purification of the peptides.

(c) We have established as far as possible the immunological and
physicochemical characteristics of the peptides, with the ultimate
goal being the amino acid composition and sequence.

(d) Finally, we have examined some of the biological properties
of the molecules both within *Calliphora* itself and also in appropriate mammalian bioassay systems.

INSULIN-LIKE PEPTIDE

Cellular localisation
─────────────────────

A variety of different antisera directed against both porcine
and bovine insulins have been used in immunocytochemical studies
of the brain and, in particular, the median neurosecretory cells
(MNC) of *Calliphora* (Duve &/Thorpe, 1979, 1983a). As most of the
antisera to both porcine and bovine insulin are produced in guinea
pigs it has not been possible to use the more sensitive peroxidase
anti-peroxidase (PAP) immunocytochemical method.   Instead we have
relied on the indirect immunofluorescent technique (see Thorpe &
Duve, this volume). The results of these studies have supported the
earlier suggestions that the insect hypoglycemic hormone is indeed
insulin-like.   Thus, 6-8 cells within the MNC of *Calliphora*
contain material immunoreactive towards the insulin antisera.
Quite often, we have observed only relatively small amounts of
insulin-like peptide material within the cell bodies (Fig. 1), and
this fact, together with the necessity of using the less sensitive
method has made it impossible for us, as yet, to trace the pathways
of the insulin-like immunoreactive material leading away from the
cell bodies. We assume, however, that the material is passed along
the fibres within the median bundle to be deposited in the corpus
cardiacum prior to its release into the hemolymph (Duve & Thorpe,
1983b). Subsequently, there have been other reports confirming that

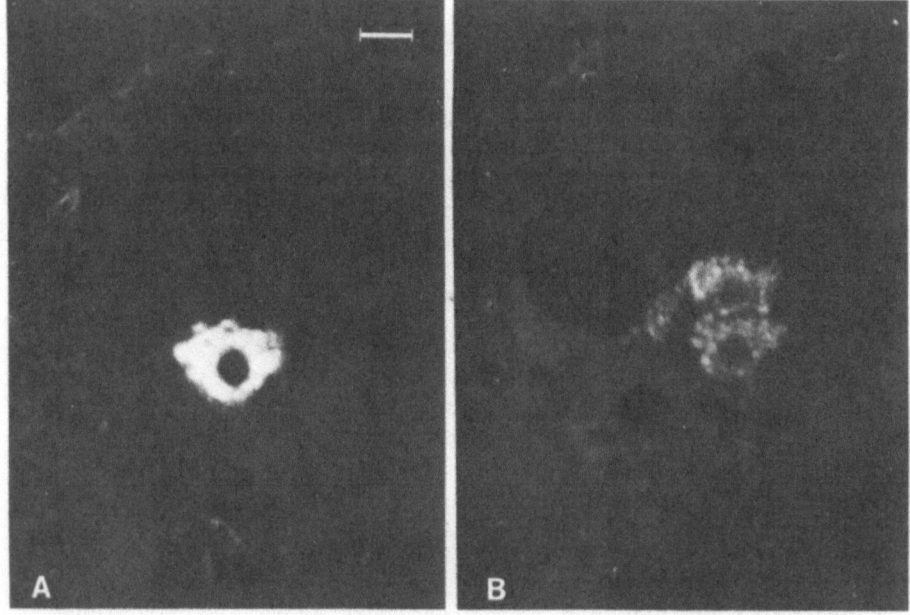

Fig. 1.   A and B. Median neurosecretory cells of pars intercerebralis of *Calliphora* showing insulin immunoreactive
cells. (Anti-bovine insulin antiserum; indirect immunofluorescence technique). The variability in the amounts
of neurosecretory material visible is characteristic of
these cells.   Bar = 10 μm.

an insulin-like material is present in neurones and/or fibres of
Bombyx mori (Yui et al., 1980), *Eristalis aeneus* (El-Salhy et al.,
1980), the locust (Orchard and Loughton, 1980), *Drosophila* (LeRoith
et al., 1981) and *Manduca sexta* (El-Salhy et al., 1983).

EXTRACTION AND PURIFICATION

    For extraction of the insulin-like peptide of the blowfly, the
well known acid alcohol extraction procedure as used on the vertebrate pancreas was employed. Four separate purifications have now
been completed, the major one of these involving the use of more
than a million flies. As the insulin-containing cells are few
and, as far as is known, located in the brain, only heads were used.
We hoped by this to avoid the problem of contamination from other
major protein sources such as those of the flight muscles, fatbodies
and the bulk of the cuticular proteins. The extraction fluid (70
litres) was removed in a cyclical distillation apparatus and the
residue, suitably prepared by centrifugation and filtration, was
applied to a large gel-filtration column (14 x 100 cm) for the
initial purification stages. The media employed were Sephadex
G-15 followed by G-50 superfine, eluted with 3 M acetic acid.
Further purification was accomplished by a combination of some of
the following steps: ion-exchange chromatography on CM-52 cellulose,
eluted with a gradient of 0 - 0.3 M NaCl (Duve et al., 1979).
QAE-Sephadex, eluted with 0.06 M HCl, and DE-52 DEAE cellulose,
eluted with a 20 - 400 mM gradient of ammonium acetate in the
presence of 20% acetonitrile (Duve et al., 1982).

    The insulin immunoreactive peaks after purification on a
DEAE-cellulose column were pooled and purified by high performance
liquid chromatography (HPLC) with the successive use of Ultrasphere
ODS (octadecyl silane) and Cyano (cyano propyl) columns. The two
peak fractions containing insulin immunoreactivity were used
separately for amino acid analysis (Table 1: Fractions 5 & 6;
Fig. 2A, C).

    The extracted peptides were monitoried through the entire
purification procedure by means of radioimmunoassays employing
homologous mammalian (bovine or porcine) antisera, standard and
tracers. Furthermore, the immunoreactive fractions were
frequently tested for linearity of dilution and parallelism when
compared with the corresponding standard curves (Fig. 2C inset).
The yield of the insulin-like peptide from the original starting
material (5½ kg of heads) was approximately 30 µg, but since
nothing is known of the % yield from the original extraction, or of
the losses incurred during some of the original filtration stages
it is difficult to assess the degree to which final yield reflects
the amount of insulin-like material per fly. The preliminary
results from the amino acid analysis of the insulin-like peptide

suggest a high degree of similarity to typical mammalian insulin, but further studies are required before a definitive composition can be given.

CHARACTERISATION

The molecular weight of the blowfly peptide, determined by means of Sephadex-gel filtration is very similar to that of bovine insulin and on polyacrylamide-gel electrophoresis the blowfly peptide has an almost identical R_f-value. In immunocytochemical assays and in radioimmunoassays, the substance cross-reacts with bovine/porcine insulin antibodies. Furthermore, in two distinct mammalian bioassays, the incorporation of glucose into lipid in the isolated rat fat-cell and the displacement of ^{125}I-labelled insulin from liver plasma membrane receptors (Fig. 3C), the material behaves in an insulin-like manner. In this respect, however, it is important to note that the slight differences in the response curves of the *Calliphora* insulin-like material as compared with bovine insulin, suggest that the molecule probably has a slightly different conformation. A further bioassay was carried out on *Calliphora* itself, where it was shown that the hyperglycemia induced by MNC extirpation could be normalised by injection of the partially purified insulin-like material (Fig. 3A, B) (Duve et al., 1979). This result, together with the fact that it is possible to measure insulin immunoreactivity in pooled hemolymph samples (Duve & Thorpe, 1983b), is important in suggesting a physiological role in carbohydrate metabolism for the peptide in *Calliphora* (Fig. 3D).

On the other hand, in experiments where the insulin-like peptide was iodinated with ^{125}Iodine, run on polyacrylamide gel electrophoresis, and the eluted iodinated fraction subjected to sulphitolysis, followed by paper electrophoresis, it was only possible to label the tyrosine in the B-chain (Fig. 4). These results, together with the results from the amino acid composition, where a smaller number of tyrosine residues were observed, suggest a certain change in the structure of the molecule, particularly in respect of the A chain.

Thus, in summary, the purification and characterisation studies have enabled us to conclude that an insulin-like peptide is present in the brain of *Calliphora* with striking similarities at least to the typical mammalian insulin counterpart.

The results of these studies confirm the findings of Tager et al. (1976) in respect of the *Manduca* insulin-like peptide and the subsequent report of LeRoith et al. (1981) on the *Drosophila* insulin-like peptide. The inevitable conclusion is that an insulin-like molecule has been in existence over a very much longer evolutionary period than hitherto supposed.

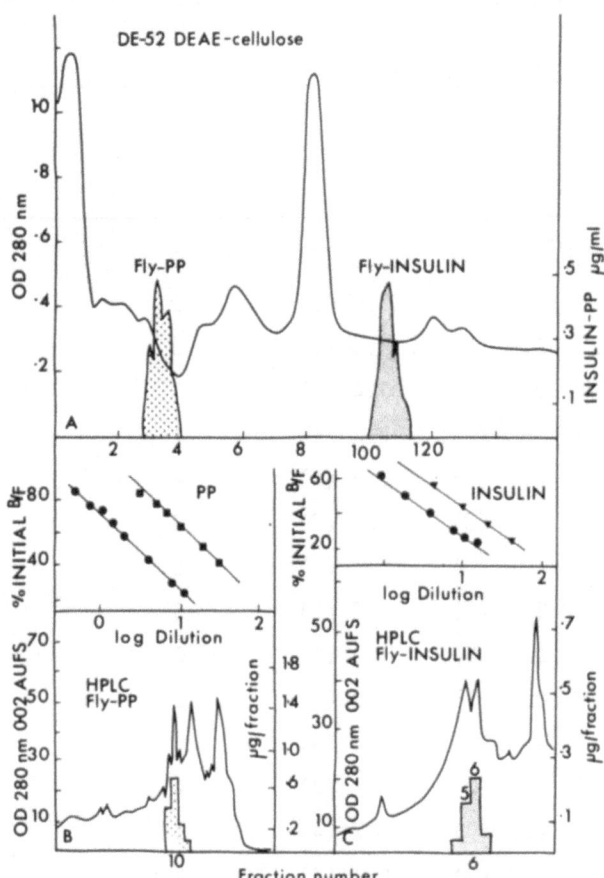

Fig. 2.   DE-52-DEAE-cellulose chromatography of reconstituted
freeze-dried material from *Calliphora* brain (Column size
1.5 cm x 30 cm; gradient 20-400mM ammonium acetate, pH 8.5,
in the presence of 20% acetonitrile; fractions, 4 ml).
Samples were immunoassayed for insulin and PP. Dotted
graphs refer to immunoreactivity and the continuous line
to A_{280}. Final analytical HPLC of PP (B) and insulin (C)
from the immunoreactive peaks from DEAE. Ultrasphere ODS;
0.4 cm x 25 cm; gradient 40-80% methanol/1% TFA. Fraction
10 was used for amino acid analysis of PP and both fractions
5 & 6 for insulin analysis. The insets show the parallelism
of the dilution curves and suggests similarity of antigenic
determinants.

TABLE 1

Amino acid composition of a *Calliphora* insulin-like neuropeptide
and vertebrate insulins.

	Calliphora			Bovine*	Porcine*
	I-5		I-6		
Asx	7.3	5.0 (6)	6.3 (6)	3	3
Thr	3.0	2.0 (3)	2.9 (3)	1	2
Ser	2.9	2.9 (3)	2.8 (3)	3	3
Glx	7.1	7.2 (7)	7.4 (7)	7	7
Pro	0.5	− (1)	1.2 (1)	1	1
Gly	4.1	3.7 (4)	3.8 (4)	4	4
Ala	3.0	3.0 (3)	3.0 (3)	3	2
Cys	−	− (−)	− (−)	6	6
Val	3.9	2.5 (3)	2.7 (3)	5	4
Met	3.3	1.8 (2)	0.0 (0)	0	0
Ile	0.7	0.9 (1)	0.8 (1)	1	2
Leu	5.7	4.6 (5)	5.0 (5)	6	6
Tyr	2.5	2.2 (2)	2.5 (2)	4	4
Phe	3.9	2.5 (3)	3.7 (4)	3	3
His	2.4	1.3 (2)	1.9 (2)	2	2
Lys	2.8	2.9 (3)	2.4 (2)	1	1
Arg	3.7	− (4)	3.9 (4)	1	1

Purified *Calliphora* neuropeptide was analysed by means of an automatic ion exchange analyser (Joel) with both O-phthaldialdehyde
(column 1, I-5) and ninhydrin (column 2, I-5 and column 3, I-6).
Two separate analyses for I-5 and a single one for I-6 were
obtained from 24 hr acid hydrolysates of approx. 100 pmol each.
Values for serine and threonine have been corrected for an
estimated 30% loss during hydrolysis. Cysteine and tryptophan were
not determined. Results from the analyses are given with the
nearest whole integers in parentheses. Integer values were
normalised to 3 residues of alanine.

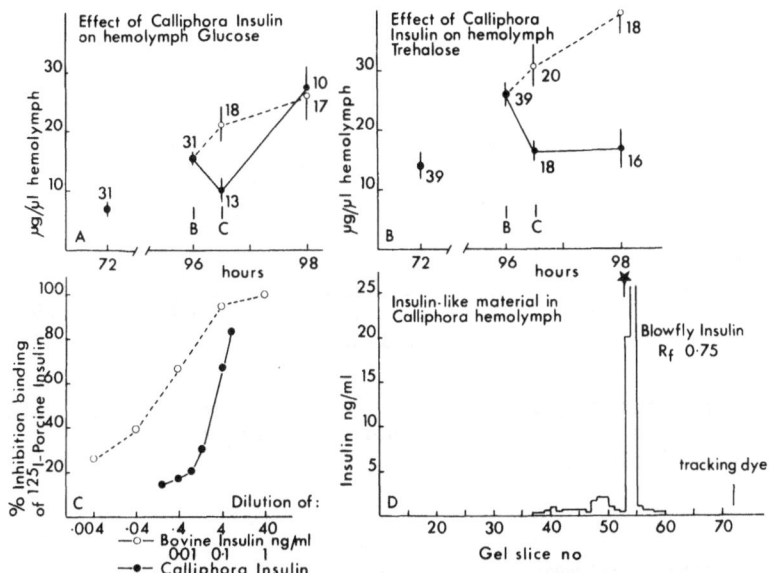

Fig. 3.   Bioassays for blowfly insulin-like peptide. Time course of
changes in hemolymph concentrations of (A) glucose and (B)
trehalose after extirpation of median neurosecretory cells
(MNC) and injection of fly insulin-like material into blow-
flies. MNC were removed 72 h after adult ecdysis. Fly
insulin-like material (•) or carrier solvent (o) was
injected at 96 h when the flies were hyperglycaemic. A
return to normal concentration in both sugars was observed
after 30 min.   (C) Displacement of binding of ^{125}I-labelled
porcine monocomponent insulin from rat liver plasma
membrane insulin receptors by fly insulin-like material (•)
and bovine insulin (o). (D) Profile of insulin-immuno-
reactive material from 200 μl pooled *Calliphora* hemolymph
after polyacrylamide gel electrophoresis. 1.5 mm gel
slices were extracted in ammonium bicarbonate and immunoassayed. The R_f value of porcine insulin is starred.
(A, B and C redrawn after Duve et al., 1979).

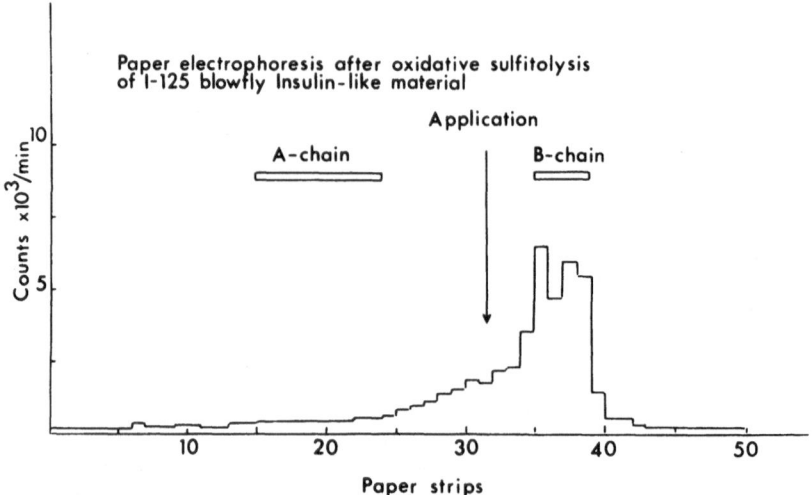

Fig. 4. Paper electrophoresis of *Calliphora* insulin-like material
         labelled with [125]Iodine and following oxidative
         sulphitolysis.

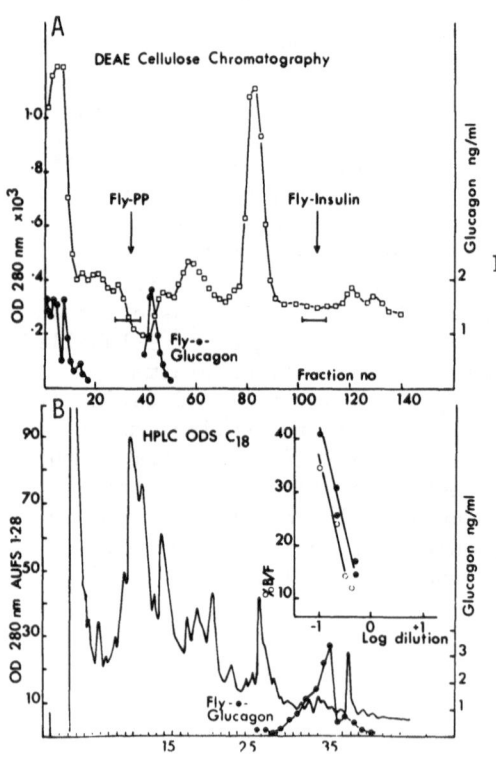

Fig. 5. (A) DE-52-DEAE cellulose
         chromatography of freeze-
         dried material from
         Calliphora brain to show
         elution position of the
         glucagon-like material.
         Details as for Fig. 2.
         (B) HPLC of pooled
         glucagon immunoreactive
         material from (A).
         Details as for Fig. 2.

GLUCAGON-LIKE PEPTIDE

The possibility that the hyperglycemic hormone of insects referred to by several authors (Steele, 1961, 1963; Bowers & Friedman, 1963; Wiens & Gilbert, 1967; Goldsworthy, 1969; Normann & Duve, 1969; Natalizi et al., 1970; Chen & Friedman, 1977; Loughton & Orchard, 1981) is related either closely or distantly to glucagon has recently attracted considerable interest.   Thus, in a series of biochemical studies on the tobacco hornworm moth, *Manduca sexta*, a glucagon-like peptide has been identified from extracts of the CC/CA complex (Kramer et al., 1980; Tager et al., 1976) and also of the midgut (Tager et al., 1980).   The cellular localisation of this peptide has been demonstrated in the brain of *Manduca* by El-Salhy et al. (1983) and also in the brain of the hoverfly, *Eristalis aeneus* (El-Salhy et al., 1980).   There is also a report on the occurrence of glucagon-immunoreactive cells in the midgut of the cockroach (Iwanaga et al., 1981).   In *Calliphora*, attempts by the authors to identify glucagon-immunoreactive material in the corpus cardiacum by a variety of glucagon antisera directed against the mid-core, and the NH_2 and COOH termini, as well as by antisera to glicentin, have proved unsuccessful, and at the moment we must conclude, on the basis of this evidence, that the hyperglycemic hormone of the CC of *Calliphora* is not an easily identifiable glucagon molecule.   In the brain, the suboesophageal ganglion and the abdominal ganglion, however, there are a limited number of glucagon-immunoreactive cells (unpublished data).   Furthermore, in the same extracts from which the insulin-like (and PP-like) materials were obtained, a glucagon-like peptide is evidenced by means of radioimmunoassays (Fig. 5).   Preliminary studies show that the material dilutes in parallel to the mammalian glucagon standards but the true nature of this immunoreactive material remains to be determined.

PANCREATIC POLYPEPTIDE (PP)-LIKE SUBSTANCE

Cellular localisation.

The localisation of a mammalian type pancreatic polypeptide immunoreactive material in both the central nervous system (CNS) and the autonomic nervous system of *Calliphora* has been thoroughly documented in two previous publications (Duve & Thorpe, 1980, 1982). Using antisera directed towards the COOH-terminal region of the bovine PP molecule as well as antisera considered to read the whole molecule, immunoreactive cells of the brain were shown to be more numerous than those which contained the insulin-like peptide. Some of the cells corresponded to neurones previously identified tentatively as neurosecretory and some were totally unknown from previous neurosecretory studies (Fig. 6).   Furthermore, immunoreac-

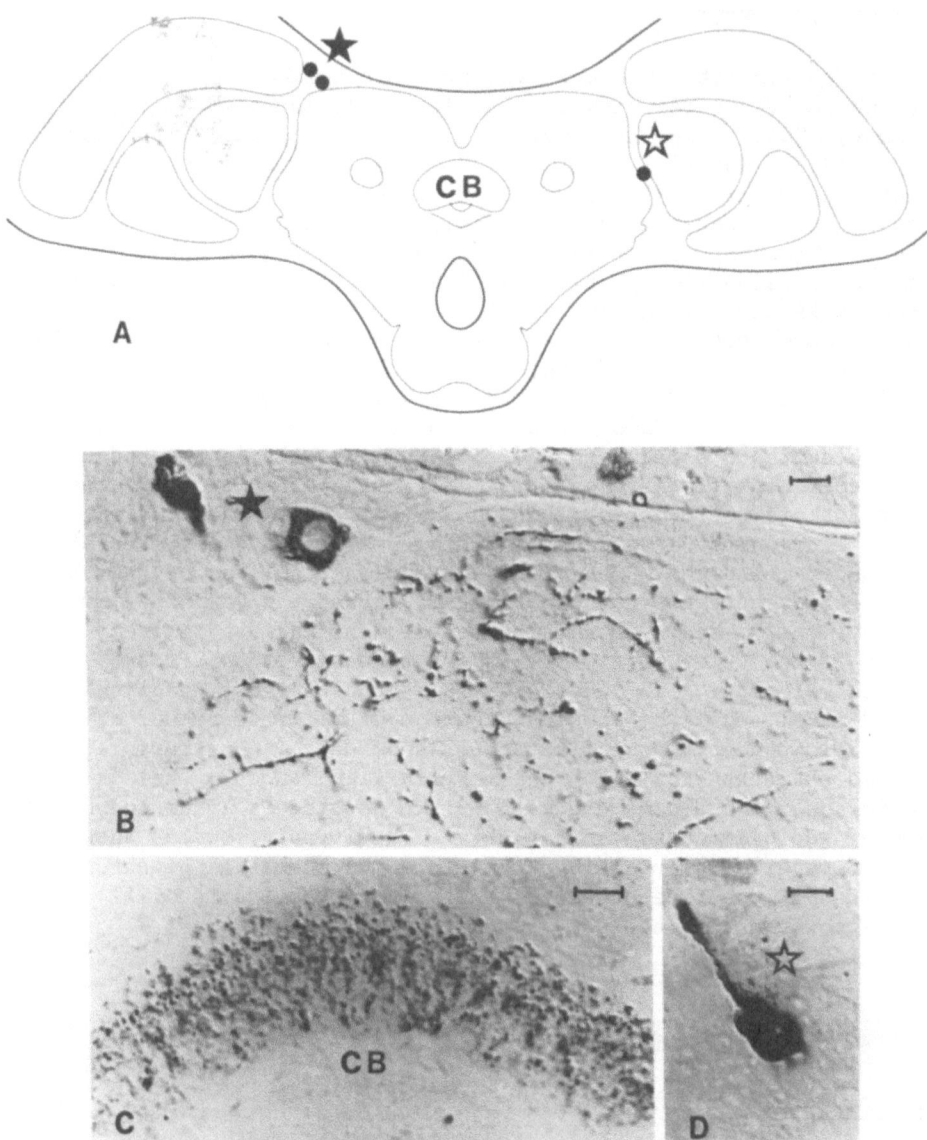

Fig. 6.   Drawing of a transverse section of the brain of *Calliphora*
           demonstrating regions which contain cells and axons immuno-
           reactive to bovine pancreatic polypeptide (BPP) antiserum.
           (B-D) Transverse sections of the brain treated with an
           antiserum to BPP (PAP technique, antiserum dilution 1:2500).
           The position of the BPP immunoreactive cells in (B) is
           indicated by a black star on the drawing in (A) and in
           (D) with a white star.   (C) The central body (CB) showing
           BPP-like immunoreactivity.   Bar = 10 μm.

tive neurones occurred in the suboesophageal ganglion, the thoracic ganglion and abdominal ganglionic components all belonging to the CNS.   Besides it occurrence in the CNS, PP immunoreactivity was also demonstrated in certain of the cells in the hypocerebral ganglion, an organ belonging to the stomatogastric nervous system. With the use of the PAP technique, it has been possible to visualise material in axons within the neuropil of the various ganglions and to follow their pathways.   It has also been possible to localise a putative neurohaemal organ/area within the dorsal sheath of the thoracic ganglion.   Apart from elements of the nervous system which have been shown to contain PP, certain flask-shaped cells of the midgut epithelium of *Calliphora* also contain PP immunoreactive material.   These cells almost certainly belong to the granular type as described by Priester (1971) in an ultrastructural study of the *Calliphora* midgut epithelium.   This finding confirms the recent demonstration of PP immunoreactive cells in the cockroach midgut by Nishiitsutsuji-Uwo & Endo (1981).   Further supporting evidence for the presence of PP-like material in insects has come from immunocytochemical studies on *Bombyx mori* (Yui et al., 1980), *Eristalis aeneus* (El-Salhy et al., 1980) and *Manduca sexta* (El-Salhy et al., 1983).

EXTRACTION, PURIFICATION AND CHARACTERISATION

The acid alcohol extraction procedure is known from vertebrate pancreas studies to be suitable for the extraction of pancreatic polypeptide (PP) and glucagon as well as insulin. The specific objective of the large-scale purification of *Calliphora* heads was to isolate all three peptides.   Radioimmunoassays employing homologous mammalian antisera, standards and tracers carried out on eluates from the various Sephadex G-15, 25 and 50 purification steps, showed immunoreactivity for PP in the same fractions as those in which insulin was isolated (Duve et al., 1981). A complete separation of the PP and insulin immunoreactive materials was first achieved without any losses of the two peptides, only by use of DE-52 DEAE cellulose chromatography, and from then onwards, the purification procedures for PP and insulin were carried out separately, although in parallel.   Thus, the fractions containing PP immunoreactivity were pooled and purified by HPLC with successive use of Ultrasphere ODS and Cyano columns and the immunoreactive fractions from the final purification step were used for amino acid analysis (Fig. 2A, B; Table II) (Duve et al., 1982).

The following data from these experiments suggest close homology with the known vertebrate species of PP. (I) The total number of amino acid residues present is 36 (based on the assumption that the molecular weight is approximately that of bovine PP, i.e. 4,200). (II) The amino acids are present in proportions similar to the known vertebrate species. (III) There is a lack of

TABLE 2

Amino acid composition of *Calliphora* neuropeptide and vertebrate pancreatic polypeptides.

	Calliphora			Bovine	Porcine	Ovine	Human	Avian
Asx	3.4	3.1	(3)	3	3	3	4	6
Thr	2.4	2.5	(2)	2	2	2	2	2
Ser	1.3	1.2	(1)	0	0	1	0	1
Glx	5.7	6.1	(6)	6	5	6	4	4
Pro	3.8	4.2	(4)	5	5	4	5	4
Gly	1.9	1.9	(2)	1	1	1	1	2
Ala	3.9	3.7	(4)	5	5	5	5	1
Cys	–	–	–	0	0	0	0	0
Val	1.2	1.4	(1)	0	1	0	1	3
Met	1.4	1.6	(1)	2	2	2	2	0
Ile	2.2	1.6	(2)	1	1	1	1	1
Leu	2.5	2.8	(3)	3	3	3	3	3
Tyr	4.0	4.0	(4)	4	4	4	4	4
Phe	0	0	(0)	0	0	0	0	1
His	0	0	(0)	0	0	0	0	1
Lys	0	0	(0)	0	0	0	0	0
Arg	3.1	3.3	(3)	4	4	4	4	3
Trp	–	–	–	0	0	0	0	0

Purified *Calliphora* neuropeptide was analysed by means of an automatic ion exchange analyser (Joel) with both ninhydrin and o-phthaldialdehyde monitoring channels. Two separate analyses were obtained from 24 hr acid hydrolysates of approx. 100 pmol each. Values for serine and threonine have been corrected for an estimated 30% loss during hydrolysis. Cyteine values were recorded as zero as compared with 0.5 nmol standards. Tryptophan was not determined although its absence is predicted on the basis of a general lack of absorbance of the peptide. Results from the two analyses given are ninhydrin detection, with the nearest whole integers in parentheses. Integer values were normalised to 4 residues of tryosine.

phenylalanine, histidine and lysine residues, three amino acids
which are not found in any of the known PP molecules so far analysed.
(IV) The material eluted in the expected position of PP on HPLC.
(V) The fly PP peptide dilutes in parallel with bovine PP in RIA
(Fig. 2B inset).   These results strongly suggest, therefore, that
the primary structure of *Calliphora* PP is very similar to the known
mammalian PP molecules.   It is hoped that it will be possible to
obtain the amino acid sequence from a new batch of *Calliphora* PP
which is currently being purified. As no function has been ascribed
with certainty to PP in mammals or other vertebrates, physiological
and metabolic studies have been omitted for the time being for the
insect PP-like material.

     One obvious potential source of difficulty in purification
studies of this type is presented by the threat of exogenous peptide
contamination during certain stages of the purification. For this
reason extreme precautions were taken. The work was carried out in
special insulin-, glucagon- and PP-free areas and new glassware,
chemicals and gel-filtration media were used throughout.

GASTRIN/CHOLECYSTOKININ (CCK)-LIKE PEPTIDES

Cellular localisation

     The presence of gastrin/cholecystokinin-like immunoreactive
material in neurones of the central nervous system and the neurosecretory (endocrine) system of *Calliphora* has been demonstrated by
means of both PAP and fluorescence immunocytochemical techniques
(Duve & Thorpe, 1981, 1983c).   A range of 5 antisera specific
for the COOH terminus of both gastrin and CCK, a single antiserum
directed against the NH_2 terminus of human gastrin G34 and an antiserum directed against the NH_2 terminus of CCK-8 have been used.
The results show that the COOH-terminal region is recognised in a
number of neurones in the brain, thoracic ganglion and corpus
cardiacum by the specific antisera, whereas the NH_2 terminus of
both gastrin G34 and CCK-8 seem not to be recognised by any cells,
at least in young flies that are less than 7 days old.

     Several points of interest have arisen from this work. (i) It
was possible to distinguish several separate populations of immunoreactive cells based on shape and size differences, both in the
brain and in the fused thoracic ganglion. The two most frequently
occurring cell types are within the size ranges 6-12 μm and 18-24 μm,
but another, even larger type appears in the median neurosecretory
cell groups of flies older than 21 days and maintained on a high
protein diet (Fig. 7). (ii) Studies on the larger cells (the so-
called giant neurons) within the MNC, using a combined immunocytochemical and cobalt backfilling technique, have enabled us to trace
the complete axonal pathway of the gastrin/CCK-like neuropeptide

Fig. 7.   Neurones of the brain of *Calliphora* (4 weeks old, fed
          sugar, ovaltine & water), showing gastrin/CCK-like
          immunoreactivity (PAP technique; antiserum L-112
          specific for COOH terminus of gastrin and CCK, diluted
          1:2000).   (A) Sagittal section through the brain and
          suboesophageal ganglion in the planeof the MNC.
          Bar = 100 µm.
          (C) Showing the MNC in detail. (B) Transverse section
          through the MNC showing the difference in content of the
          immunoreactive material, black star; cells containing no
          immunoreactivity, white star; highly immunoreactive, and
          ★; moderately immunoreactive   . (D) and (E) Transverse
          sections through MNC, showing (D) gastrin/CCK immuno-
          reactivity (fluorescence technique).   (E) The same
          section as (D), but here showing a cobalt-silver deposit
          from a cobalt backfilling of the neurones at the position
          of the corpus cardiacum. Two of the cells are clearly
          both backfilled and immunoreactive. Bar = 10 µm.
          (MNC) median neurosecretory cells; (cb) central body;
          (br) brain; (SO) suboesophageal ganglion.

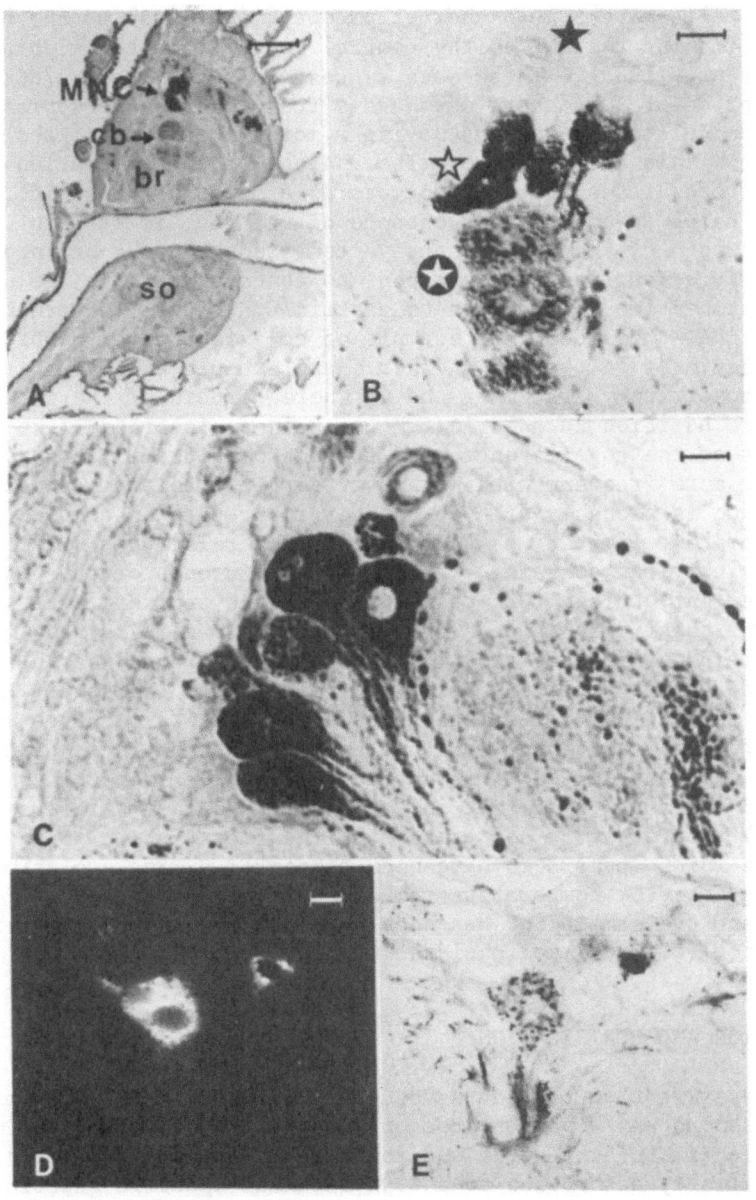

out of the brain via the long neurosecretory neurones to a neurohaemal site usually within the corpus cardiacum, but possibly even
extending in some fibres beyond the CC to a region or regions as
yet unidentified. With the same technique (except for using cobalt
frontfilling) it has been possible to demonstrate that certain of the
smaller cells have a much shorter pathway, intrinsic to the brain
itself, linking the MNC to the central body within the midbrain
(Duve et al., 1983). These data suggest that the gastrin/CCK-
like peptide of *Calliphora* may have a function as a neurotransmitter
or neuromodulator as well as having a more widespread metabolic
function within the organism. The analogy of this situation to
that in vertebrates where CCK occurs as a major neuropeptide, as
well as being an endocrine substance of the gut, is striking.
(iii) Certain of the immunoreactive cells both in the brain and in
the thoracic ganglionic mass appear to contain at least one other
different peptide in addition to gastrin/CCK, namely pancreatic
polypeptide.   Besides this example of co-existence of peptides
within neurones in the CNS another example could be demonstrated in
the neurosecretory system of the CC where certain of the glandular
cells, in addition to gastrin/CCK, contained a secretin-like peptide.
Examples of the co-existence of different peptides in neurones are
becoming more frequent within the vertebrate literature and the
phenomenon is probably of even greater importance in complex
invertebrates, limited, as they are, by the relatively small number
of cells they can contain.   (iv) That the neurones of the blowfly
responded only to the COOH-termainal-directed antisera, illustrates
the point that immunocytochemistry, dependent as it is upon selective
antigenic determinants, requires as many antisera as possible,
directed against different parts of the molecule (Larsson & Rehfeld,
1977a).   These so-called region-specific antisera enable a more
detailed appraisal of the molecule under investigation and in the
case of the gastrin/CCK-like substance it can be seen that whereas
the COOH terminus is present in *Calliphora*, the NH_2 terminus (of
gastrin) apparently is not present.   The fact that it is the COOH
terminus which appears to have been conserved during evolution has
been suggested by comparative vertebrate studies (Larsson & Rehfeld,
1977b) and the same point has been demonstrated in other invertebrates too (c.f. Grimmelikhuijzen et al., 1980).

EXTRACTION AND PURIFICATION

The extraction procedure used for the gastrin/CCK-like material
of *Calliphora* was a variant commonly used in vertebrate studies.
Extraction in boiling water was followed by purification using
immunoaffinity adsorption to a COOH-terminal-specific gastrin antiserum.   The peptide was eluted from the column with 15% formic
acid and subsequently freeze dried.   Gastrin/CCK immunoreactive
material eluted in two peaks after separation by gel filtration on
Sephadex G-50.   Both peaks reacted in radioimmunoassays using

antisera specific for the COOH terminus of the gastrin/CCK molecule
and did not react with antisera specific for the NH$_2$ terminus or
intact gastrin G17.   The immunoaffinity-purified material diluted
in parallel with standard curves using 3 different COOH-terminal
antisera.   The elution characteristics of the fly peptide from
the G-50 did not, however, correspond with any previously characterised forms of mammalian gastrin or CCK (Dockray et al., 1981).
The results of this work confirm that peptides with immunochemical
properties resembling those of the COOH-terminal fragments of
mammalian gastrin and CCK occur in *Calliphora*.   This suggestion
has been made earlier for *Manduca* by Kramer et al. (1977) who
demonstrated a gastrin-like peptide in the brain-CC complex.

ENDORPHIN AND ENKEPHALIN-LIKE PEPTIDES

Cellular localisation

     In a study of the mammalian-type opiate-like peptides in
Calliphora the most interesting results have come from work with a
series of antisera directed against α-endorphin.   This peptide is
the portion 61-76 of pro-opiocortin, the common precursor molecule
of the cells of the pars intermedia and the cortiocotrophin cells
of the pars distalis of mammals, from which a wide variety of
peptides, including the endorphin family are produced.

     The immunoreactive cells have been demonstrated by means of
the PAP and fluorescence techniques to be present in a restricted
number (12-14) of the median neurosecretory cells of the pars
intercerebralis of the brain and nowhere else.   The cells are
clearly distinguishable from the insulin- and gastrin/CCK
immunoreactive cells of this region and they stain strongly with
paraldehyde fuchsin (see Thorpe & Duve, this volume).   Furthermore,
the α-endorphin immunoreactive material can be followed in axons
of the median bundle, terminating either in the corpus cardiacum
or passing through this structure in the cardiac-recurrent nerve
(Duve & Thorpe, 1983a).   α-Endorphin-like substances have also
been demonstrated in a few earlier immunocytochemical reports in
the suboesophageal ganglion of two larval lepidopterans *Bombyx
mori* and *Thaumetopoea pityocampa* (Rémy et al., 1978) and, in more
recent studies, enkephalin-related peptide receptors have been
demonstrated in the cerebral ganglion and midgut of *Leucophaea
maderae* by Stefano & Scharrer (1981) and Stefano et a. (1982).

     Studies with β-endorphin and met- and leu-enkephalin antisera
in *Calliphora* have revealed immunoreactive cells, particularly in
the suboesophageal ganglion but also in the fused thoracic ganglionic
mass.   This work is at a preliminary stage.

EXTRACTION AND PURIFICATION

For extraction of the α-endorphin-like peptide in *Calliphora*, whole bodies have been used. In a typical experiment, 50g of tissue was homogenized and extracted in 0.025 N HCl. The extract was centrifuged at 70,000g and the supernatant containing the peptide was freeze dried and subjected to Sephadex G-50 chromatography followed by reverse phase HPLC using C_{18} and Cyano columns (see Thorpe & Duve, this volume) (Fig. 7). The various chromatographic stemps were monitored by RIA with two antisera highly specific for α-endorphin (cross-reactivity with, for example, β-endorphin is only 0.07%). The α-endorphin immunoreactive peptide of *Calliphora* diluted in parallel with the standard curve. It is not known how closely related chemically this material is to the mammalian α-endorphin or whether it is part of a bigger parent molecule, but it seems reasonable to assume on the basis of the evidence to hand that peptides at least very similar to α-endorphin, do, indeed, occur in the nervous tissues of *Calliphora*.

CONCLUSION

Results to date suggest that the nervous tissues and to some extent the gut of *Calliphora* are rich sources of peptides, similar in varying degrees to known vertebrate hormonal peptides. Thus, insulin-, pancreatic polypeptide (PP)-, glucagon- and gastrin/CCK-like peptides, as representatives of the gastroenteropancreatic (GEP) peptides, and the endorphin/enkephalin-like peptides of the more orthodox type of neuropeptides, can all be identified by immunocytochemical and biochemical/immunochamical methods. They occur in specific neurones and axonal pathways in both the central and autonomic nervous systems. In experiments with combined antibody and cobalt labelling of entire neurones, it has been possible to trace specific peptidergic pathways, from which something of the possible physiology of the peptides may be inferred.

From purification and characterisation studies of the insulin- and PP-like peptides a close similarity to the corresponding mammalian molecules is suggested by their behaviour in radioimmunoassays, and by their chromatographic and electrophoretic properties. The insulin-like peptide has been shown, in three different bioassays, to have at least some of the characteristic properties of insulin, although a preliminary amino acid composition of the purified peptide suggest several substitutions as compared with the bovine insulin molecule. For the purified PP-like peptide, a much closer similarity to the known mammalian molecules has been predicted from amino acid analysis studies. Investigations on the gastrin/CCK-like peptide(s) appear to be in line with other invertebrate studies in suggesting that the resemblance to mammalian peptides of this type is restricted to the COOH terminus.

The data from the glucagon and endorphin studies are as yet incomplete. From the preliminary immunochemical studies on the partially purified glucagon a close similarity to mammalian glucagon is suggested, since the fly peptide exhibits specific immunoreactivity towards both NH_2- and COOH-terminal specific mammalian antibodies to glucagon.

For one of the peptides belonging to the opiate class, namely α-endorphin, both immunocytochemical and biochemical/immunochemical studies suggest a close similarity to the corresponding mammalian molecule.

From these results it is concluded that molecular evolution of this series of biologically active peptides did not begin at the level of vertebrates, but much further back in time. Parallel studies of the type described here for *Calliphora*, on other species of insects confirm this view, and studies on more primitive organisms, even down to the level of unicellular organisms (LeRoith et al., 1982), suggest that the evolutionary origins of these peptides are very much earlier than previously supposed.

ACKNOWLEDGMENTS

It is a great pleasure to express our sincere appreciation of the collaboration of our co-workers in various aspects of the original research reviewed in this article.

In particular, we wish to thank our colleague Dr. N. R. Lazarus of the Wellcome Foundation, Beckenham, Kent, U.K., for his continuing stimulating and active involvement in the project at all levels and for allowing us to discuss certain of our previously unpublished data on the *Calliphora* insulin- and glucagon-like neuropeptides in this article.

We also wish to thank our other collaborators, Mr. R. Neville and Dr. D. Stone of the Wellcome Foundation and Dr. P. J. Lowry of the Pituitary Hormone Laboratory, St. Bartholomew's Hospital, London, and Dr. Suzanne Linde of the Hagedoorn Research Laboratory, Gentofte, Copenhagen, Denmark, who participated in various aspects of the biochemical studies on insulin-, glucagon-, and PP-like substances; Professor G. J. Dockray of the Department of Physiology, University of Liverpool, for our collaboration on the gastrin/CCK-like material and Mr. A. Scott of the School of Biological Sciences, Queen Mary College for collaboration on the α-endorphin-like peptide.

We also gratefully acknowledge the support of the Science and Engineering Research Council of Great Britain in the form of grants to AT (GR/B88136 and GR/C20789).

REFERENCES

Bowers W.S. and Friedman S. (1963) Mobilization of fat body
    glycogen by an extract of corpus cardiacum. Nature, Lond.
    198, 685.
Chen A.C. and Friedman S. (1977) Hormonal regulation of trehalose
    metabolism in the blowfly, *Phormia regina*: Interaction between
    hypertrehalosemic and hypotrehalosemic hormones. J.Insect
    Physiol. 23, 1223-1232.
Dixit P.K. and Patel N.G. (1964) Insulin-like activity of larval
    foods of the honeybee. Nature (London) 202, 189-190.
Dockray G.J., Duve H. and Thorpe A. (1981) Immunochemical charac-
    terisation of gastrin/cholecystokinin-like peptides in the
    brain of the blowfly, *Calliphora vomitoria*. Gen.Comp.Endocrinol.
    45, 491-496
Duve H. (1978) The presence of a hypoglycemic and hypotrehalocemic
    hormone in the neurosecretory system of the blowfly, *Calliphora
    erythrocephala*. Gen.Comp.Endocrinol. 36, 102-110.
Duve H. and Thorpe, A. (1979) Immunofluorescent localization of
    insulin-like material in the median neurosecretory cells of the
    blowfly, *Calliphora vomitoria* (Diptera). Cell Tissue Res.
    200, 187-191.
Duve H. and Thorpe A. (1980) Localisation of pancreatic polypeptide
    (PP)-like immoreactive material in neurones of the brain of the
    blowfly, *Calliphora erythrocephala* (Diptera). Cell Tissue Res.
    210, 101-109.
Duve H. and Thorpe A. (1981) Gastrin/cholestokinin (CCK)-like
    immunoreactive neurones in the brain of the blowfly, *Calliphora
    erythrocephala* (Diptera). Gen.Comp.Endocrinol. 43, 381-391.
Duve H. and Thorpe A. (1982) The distribution of pancreatic poly-
    peptide in the nervous system and gut of the blowfly,
    Calliphora vomitoria (Diptera). Cell Tissue Res. 227, 67-77.
Duve H. and Thorpe A. (1983a) Immunocytochemical identification of
    α-endorphin-like material in neurones of the brain and corpus
    cardiacum of the blowfly, *Calliphora vomitoria*. (Diptera).
    Cell Tissue Res. (In press).
Duve H. and Thorpe A. (1983b) Vertebrate-type brain/gut peptides in
    neurones of the blowfly *Calliphora*. In: Symposium Proceedings
    of the Ninth International Symposium on Comparative Endocrino-
    logy. (Lofts B. and Chan D.K.O., eds). Hong Kong University
    Press.(In press).
Duve H. and Thorpe A. (1983c) Immunochemical identification of
    brain/gut peptides of vertebrate type in insect nerve cells.
    In: Experimental Entomology: Neuroanatimical Techniques. Vol. 2.
    (Strausfeld N., ed.). Springer-Verlag, Heidelberg. (In press).
Duve H., Thorpe A. and Lazarus, N.R. (1979) Isolation of material
    displaying insulin-like immunological and biological activity
    from the brain of the blowfly, *Calliphora vomitoria*. Biochem.J.
    184, 221-227.

Duve H., Thorpe A. and Strausfeld N.J. (1983) Cobalt-immunocyto-
    hemical identification of peptidergic neurons in *Calliphora*
    innervating central and peripheral targets. J.Neurocytol.
    (In press).

Duve H., Thorpe A., Lazarus N.R. and Lowry P.J. (1982) A neuro-
    peptide of the blowfly *Calliphora vomitoria* with an amino acid
    composition homologous with vertebrate pancreatic polypeptide.
    Biochem.J. 201, 429-432.

Duve H., Thorpe A., Neville R., Lazarus N.R. (1981) Isolation and
    partial characterization of pancreatic polypeptide-like
    material in the brain of the blowfly *Calliphora vomitoria*.
    Biochem.J. 197, 767-770.

El-Salhy M., Abou-El-Ela R., Falkmer S., Grimelius L., Wilander E.
    (1980) Immunohistochemical evidence of gastro-entero-pancreatic
    neurohormonal peptides of vertebrate type in the nervous system
    of the larva of a dipteran insect, the hoverfly, *Eristalis
    aeneus*. Regulatory Peptides 1, 187-204.

El-Salhy M., Falkmer S., Kramer K.J. and Speirs R.D. (1983)
    Immunohistochemical investigations of neuropeptides in the
    brain, corpora cardiaca and corpora allata of an adult
    lepidopteran insect, *Manduca sexta* (L.). Cell Tissue Res.
    (In press).

Friedman S. (1967) The control of trehalose synthesis in the
    blowfly, *Phormia regina* Meig. J.Insect Physiol. 13, 397-405.

Goldsworthy G.J. (1969) Hyperglycaemic factors from the corpus
    cardiacum of *Locusta migratoria*. J.Insect Physiol. 15, 2131-
    2140.

Grimmelikhuijzen G.J.P., Sundler F. and Rehfeld J.F. (1980)
    Gastrin/CCK-like immunoreactivity in the nervous system of
    coelenterates. Histochemistry 69, 61-68.

Ishay J., Gitter S., Galun R., Doron M. and Laron Z. (1976)
    The presence of insulin in and some effects of exogenous
    insulin on Hymenoptera tissues and body fluids. Comp.Biochem.
    Physiol. 54A, 203-206.

Iwanaga T., Fujita T., Nishiitsutsuji-Uwo J. and Endo Y. (1981)
    Immunohistochemical demonstration of PP-, somatostatin-,
    enteroglucagon- and VIP-like immunoreactivities in the cock-
    roach midgut. Biomed.Res. 2, 202-207.

Kramer K.J. (1979) Insulin-like and glucagon-like hormones in
    insects. In: Experimental Entomology: Insect Neurohormones
    (Miller T.A., ed.) pp. 116-136. Springer-Verlag, New York,
    Heidelberg, Berlin.

Kramer K.J., Speirs R.D. and Childs C.N. (1977) Immunochemical
    evidence for a gastrin-like peptide in insect neuroendocrine
    system. Gen.Comp.Endocrinol. 32, 423-426.

Kramer K.J., Tager H.S. and Childs C.N. (1980) Insulin-like and
    glucagon-like peptides in insect hemolymph. Insect Biochem.
    10, 179-182.

Kramer K.J., Tager H.S., Childs C.N. and Speirs R.D. (1977)
    Insulin-like hypoglycemic and immunological activities in

honey-bee royal jelly. J.Insect Physiol. 23, 293-295.

Larsson L.-I. and Rehfeld J.F. (1977a) Characterization of antral gastrin cells with region-specific antisera. J.Histochem. Cytochem. 25, 1317-1321.

Larsson L.-I. and Rehfeld J.F. (1977b) Evidence for a common evolutionary origin of gastrin and cholecystokinin. Nature (London) 269, 335-338.

LeRoith D., Lesniak M.A. and Roth J. (1981) Insulin in insects and annelids. Diabetes 30, 70-76

LeRoith D., Liotta A.S., Roth J., Shiloach J., Lewis M.E., Pert C.B. and Krieger D.T. (1982) Corticotrophin and β-endorphin-like materials are native to unicellular organisms. Proc. Natl. Acad.Sci. 79, 2086-2090.

Loughton B.G. and Orchard I. (1981) The nature of the hyperglycaemic factor from the glandular lobe of the corpus cardiacum of Locusta migratoria. J.Insect Physiol. 27, 383-385.

Mordue W. and Goldsworthy G.J. (1969) The physiological effects of corpus cardiacum extracts in locusts. Gen.Comp.Endocrinol. 12, 36)-369.

Natalizi G.M. and Frontali N. (1966) Purification of insect hyper-glycaemic and heart accelerating hormones. J.Insect Physiol. 12, 1279-1287.

Nishiitsutsuji-Uwo J. and Endo Y. (1981) Gut endocrine cells in insects: the ultrastructure of the endocrine cells in the cockroach midgut. Biomed.Res. 2, 30-44.

Normann T.C. (1975) Neurosecretory cells in insect brain and pro-duction of hypoglycaemic hormone. Nature (London) 254, 259-261.

Normann T.C. and Duve H. (1969) Experimentally induced release of a neurohormone influencing hemolymph trehalose level in Calliphora erythrocephala (Diptera). Gen.Comp.Endocrinol. 12, 449-459.

Orchard I. and Loughton B.C. (1980) A hypolipaemic factor from the corpus cardiacum of locusts. Nature (London) 286, 494-496.

Priester W. de (1971) Ultrastructure of the midgut epithelial cells in the fly Calliphora erythrocephala. J.Ultrastruct.Res. 36, 783-805.

Ralph, C.L. (1962) Action of extracts of cockroach nervous system on fat bodies in vitro. Am.Zool. 2, 550.

Ralph C.L. and McCarthy R. (1964) Effects of brain and corpus cardiacum extracts on haemolymph trehalose of the cockroach, Periplaneta americana. Nature 203, 1195-1196.

Rémy C., Girardie J. and Dubois M.P. (1978) Présence dans le ganglion sous oesophagien de la chenille processionaire du Pin (Thaumetopoea pityocampa Schiff) de cellules révélées en immunofluorescence par un anticorps, anti-α-endorphine. C.R.Acad.Sci. 286, 651-653.

Steele J.E. (1961) Occurrence of a hyperglycaemic factor in the corpus cardiacum of an insect. Nature (London) 192, 680-681.

Steele J.E. (1963) The site of action of insect hyperglycaemic hormone. Gen.Comp.Endocrinol. 3, 46-52.

Stefano G.B. and Scharrer B. (1981) High affinity binding of an
    enkephalin analog in the cerebral ganglion of the insect
    Leucophaea maderae (Blattaria). Brain Res. 225, 107-114.
Stefano G.B., Scharrer B. and Assanah P. (1982) Demonstration,
    characterization and localisation of opioid binding sites in
    the midgut of the insect *Leucophaea maderae* (Blattaria).
    Brain Res. 253, 205-212.
Tager H.S. and Kramer K.J. (1980) Insect glucagon-like peptides:
    evidence for high molecular weight form in midgut from *Manduca
    sexta* (L.). Insect Biochem. 10, 617-619.
Tager H.S., Markese J., Spiers R.D. and Kramer K.J. (1975) Glucagon-
    like immunoreactivity in insect corpus cardiacum. Nature
    (London) 254, 707-708.
Tager H.S., Markese J., Kramer K.J., Spiers R.D. and Childs C.N.
    (1976) Glucagon-like and insulin-like hormones of the insect
    neurosecretory system. Biochem.J. 156, 515-520.
Wiens A.W. and Gilbert L.I. (1967) Regulation of carbohydrate
    mobilization and utilization in *Leucophaea maderae*. J.Insect
    Physiol. 13, 779-794.
Yui R., Fujita T. and Ito S. (1980) Insulin-, gastrin-, pancreatic
    polypeptide-like immunoreactive neurones in the brain of the
    silkworm, *Bombyx mori*. Biomed.Res. 1, 42-46,

IMMUNOCHEMICAL APPLICATIONS IN THE STUDY OF INSECT NEUROPEPTIDES

WITH SPECIAL EMPHASIS ON THE PEPTIDES OF VERTEBRATE TYPE

A. Thorpe and H. Duve

School of Biological Sciences
Queen Mary College, University of London
Mile End Road, London E1 4NS. U.K.

INTRODUCTION

Many biologically active peptides previously known only from
the brain and gastroentero-pancreatic system of vertebrates have now
been shown to be present in insects and other invertebrates.   In
the preceding paper within this volume we have discussed the localisation and characterisation of several of these peptides present
within the blowfly, *Calliphora*.   The aim of the present paper is
(a) to give a more specific account of the immunochemical techniques
which are available for insect peptide studies and (b) to relate
them more especially to the *Calliphora* neuropeptides of vertebrate
type with which we are currently working.

The immunocytochemical identification of the active substances
within neurons is a useful starting point in many neuropeptide
studies and the neuroscience literature has been inundated with
immunocytochemical papers in recent years.   There is no doubt that
this powerful technique has been the force behind a revolution in
neuroanatomy and neurochemistry.   The older, non-selective staining
procedures, such as the Golgi technique, permit study of the
structural organisation of the brain but they do nothing or very
little to differentiate between the widely heterogeneous component
cells and tissues.   The important classical neurosecretory
staining techniques give a greater insight into the chemical nature
of particular neurons but even these methods are somewhat empirical
and depend on relatively simple and common chemical bonds such as
the presence of sulphydryl groups. Immunocytochemistry on the other
hand has provided the means of chemical microdissection of these
components. At a simple level it has permitted the identification
of a very much greater number and variety of neurosecretory

peptides than was possible by means of the traditional neurosecretory methods. When used optimally the method provides precise
information on highly specific neuropeptide processing, packaging,
transport and release. It also allows precise neuroendocrine
circuitry to be worked out which in turn is providing us with
important clues as to the function of the neuropeptides.

As will be shown later in the insect studies to be described,
immunocytochemistry is usually specific only to a certain degree.
To illustrate this point, it is possible using an anti-bovine
insulin serum to identify the insulin-producing cells of a
codfish as well as those of the ox. In other words, the technique
is specific in the sense that it will recognise only insulin cells,
but it is not sufficiently specific, in this instance, to
distinguish between the two insulin molecules which have variations
in more than 30% of the amino acid residues.

It is essential, therefore, to undertake the complete chemical
characterisation of the peptides identified by immunocytochemistry.

The extraction and purification procedures are completely
reliant, at all stages, on an accurate identification of the
peptides. This is achieved preferably by a combination of bioassay and radioimmunoassay (RIA) but more often than not by only
one of these assay methods. The second part of this paper will
deal exclusively with the application of RIA to neuropeptide
studies in insects.

IMMUNOCYTOCHEMISTRY

Many reviews on the technique of immunocytochemistry have
appeared in recent years (c.f. Sternberger, 1979; Forsmann et al.,
1981; Larsson, 1981; Van Noorden & Polak, 1981) and it is not
the purpose of the present paper to add to these by dealing
esoterically with the subject. Rather, we shall concentrate on
some of the essential and more interesting aspects of the subject
as it relates to the study of insect neuropeptides.

(a) Tissue preparation

The specific objectives of tissue preparation are the preservation of the immunoreactive sites (the antigenic determinants) of
the peptide under study and the preservation of a near-normal
tissue structure. Only when this is achieved is it possible to
correlate the localisation of the peptide within the nervous
tissue accurately. Unfortunately, or fortunately, there are no
hard and fast rules to follow and the best approach is often
empirical and varies considerably from tissue to tissue and from
peptide to peptide (c.f. Brandtzaeg, 1982). Quite often a balance

must be struck between good, but not perfect, overall tissue preservation and satisfactory peptide structure preservation. This is especially true for EM studies.

The essence of immunocytochemistry is to bring the antibody into direct contact with the antigen and in insect studies this has usually been achieved by the use of thin (4-10 μm) chemically fixed tissue sections subjected to dilute solutions of antibody. In our own studies we have found that perfusion with aqueous Bouin's fluid, followed by wax embedding has given the most satisfactory results. The structure of the tissues is preserved reasonably well and, as far as we have been able to judge, the peptides we have worked with have retained their structure and, accordingly, their immunoreactivity. In working with vertebrate-type peptides we have a definite advantage in that mammalian control tissues can be run in parallel. Whilst it does not necessarily follow that the insect peptides will behave exactly as their vertebrate counterparts, it is probably a useful guide. One other useful feature of Bouin fixation is that the yellow colour of the picric acid allows accurate orientation of the small and delicate tissues during the embedding procedure. This in turn has enabled us, for example, to section the whole nervous system of *Calliphora* in the sagittal plane and to follow neuropeptide distribution within complete axonal pathways such as the route from the median neurosecretory cells via the median bundle to the corpus cardiacum (Fig. 1). Although Bouin's fluid has been shown to be satisfactory for most peptide hormones, some others e.g. vasoactive intestinal peptide and secretin may give a better result if a fixative containing parabenzoquinone is used (Larsson, 1980a). Although aldehyde fixatures are generally to be preferred on account of their preservation of structure and antigenicity, we have found that glutaraldehyde and paraformaldehyde used either separately or in combination, have proved impossible to work with in association with wax embedding, resulting in dense, extremely hard tissues which are impossible to cut.

Other workers with insect neuropeptides of vertebrate type have also mainly used Bouin fixed, wax embedded preparations (El-Salhy et al., 1980, 1983; Yui et al., 1980; Hansen et al., 1982). There is certainly scope for a greater exploration of the methods of tissue handling, including fixation.

(b) Antibodies

The primary antibody is the key component in successful immunocytochemistry and high affinity/high avidity antibodies are much in demand. Most of the studies conducted on insect neuro-peptides have made use of antisera containing antibodies of poly-clonal origin. These have certain limitiations by virtue of their heterogeneity and the possibility of contaminating antibodies.

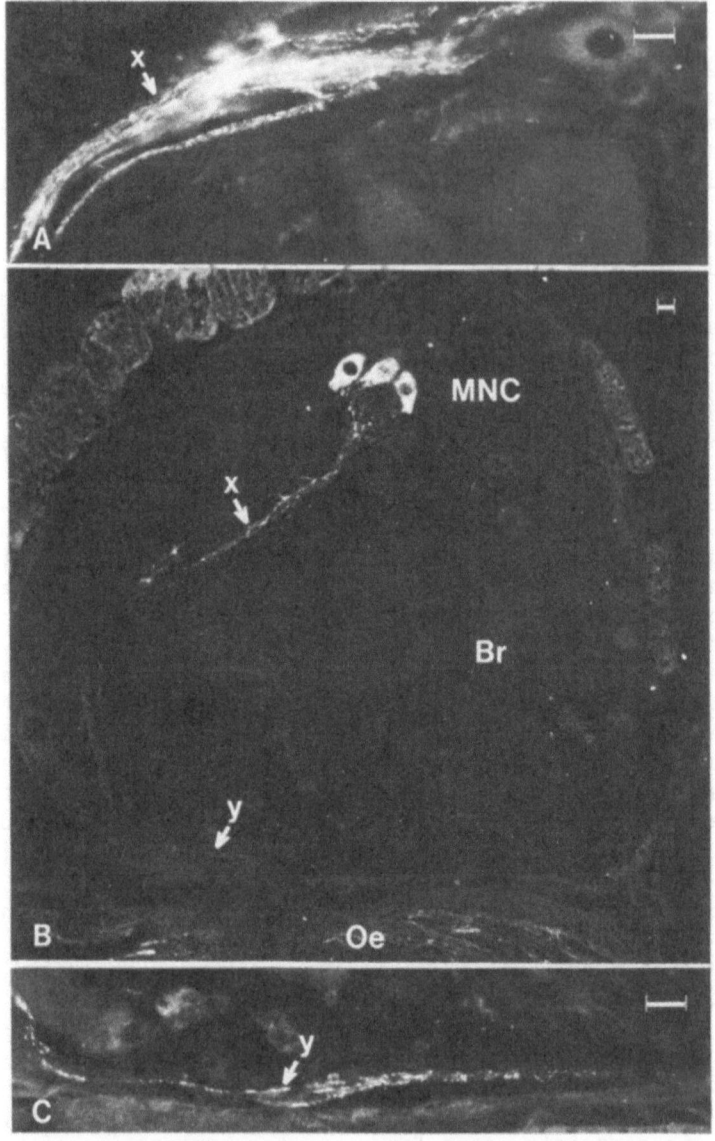

Fig. 1.   Sagittal sections through the brain of *Calliphora*;
          (B) α-Endorphin immunoreactive material in neurones of
          MNC.   (Antiserum to synthetic α-endorphin, α-E1(7) 1:2000 -
          secondary antibody, rhodamine conjugated sheep antirabbit).
          (A) & (C) details of immunoreactive material in the median
          bundle.   Bar = 10 µm.
          (B) Reproduced from Cell & Tissue Research, Duve &
          Thorpe (1983), in press.

In the future these could be overcome by the use of monoclonal antibodies (Eisenbarth & Jackson, 1982).

The use of several region-specific antisera all directed against one peptide is extremely valuable both in terms of over-coming the possible non-specific effects caused by the cross-reactivity of one of the minority antibody populations within a single antiserum and also in helping to give a more complete breakdown on the peptide structure (c.f. Larsson & Rehfeld, 1977). Our studies on the gastrin/cholecystokinin-like peptides of *Calliphora* exemplify this approach. Thus, the use of 6 or more antisera has revealed a typically mammalian COOH-terminus specificity (Fig. 2) and has shown that the NH_2 terminus must be rather different from that of the mammalian gastrin G17 forms. The immunoreactive peptide cannot be very similar to mammalian CCK in that one of the antisera, specific for a mid-core region of the molecule, gives no cross-reactivity.

(c) Specificity

Much has been written about the specificity of immunocyto-chemical procedures and it is indeed important that both method-specificity and serum-specificity controls should be carried out as extensively as possible (for recent review see Pool et al., 1982). The conventional approach towards achieving specificity in an immunocytochemical procedure is to effect a positive staining by means of an antiserum raised against, for example, gastrin, on a known gastrin-containing tissue. The antiserum is then adsorbed with the immunogen, in this instance, gastrin or C- or N-terminal fragments of gastrin, after which no reaction should be observed in the control tissue. The same method and the same materials are then switched to the unknown (insect) tissue suspected of containing the same antigen. If the same results are obtained, the conclusion is that the positive reaction is due to a tissue peptide or protein having at least some degree of likeness to the antigen used in the adsorption test, i.e. the sharing of common antigenic determinants. This type of specificity has been called method specificity and is commonly taken as the ultimate proof for specificity in immunocytochemistry (Petrusz et al., 1976). However, the adsorption test as described above depends heavily upon the fact that the adsorbing antigen is pure. Also, as stated above, it gives only a little information as to the true identity of the immunoreactive compound(s) occurring in the tissue. Other criteria are required, therefore, to be assured of immunocytochemical specificity. One very useful test we have carried out in some of our vertebrate-type peptide studies is the serum specificity test ELISA which makes use of isolated antigens (Engvall & Perlmann, 1971). ELISA (enzyme-linked immunosorbent assay) involves the adsorption of antigens to the surface of a polyethylene tube followed by application of the antiserum, immunoperoxidase

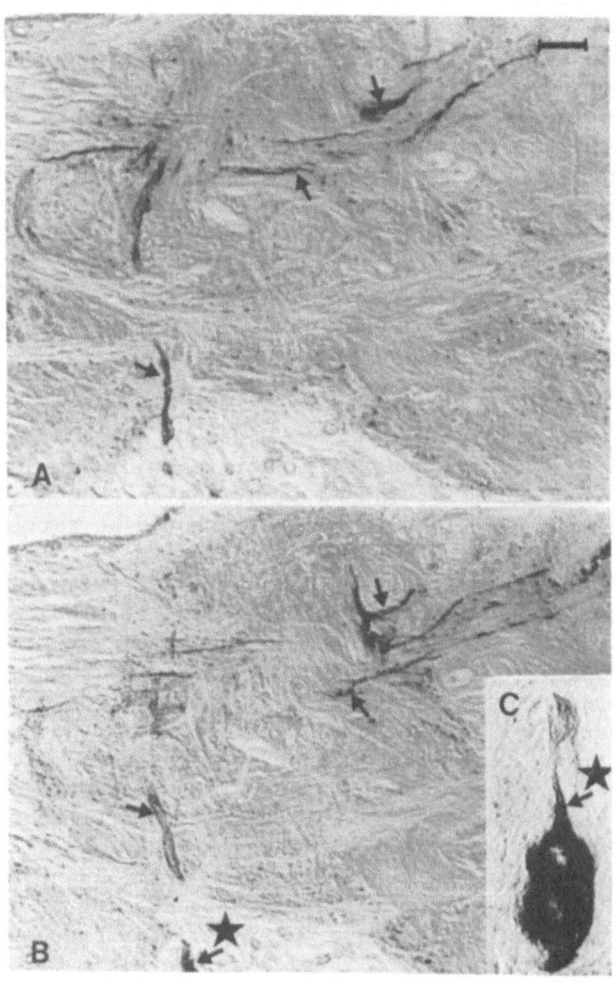

Fig. 2.   Cells (C) and axons (A) and (B), arrows, showing immuno-
           reactivity towards a gastrin/CCK antiserum in a longitu-
           dinal section of the thoracic ganglion of *Calliphora* (PAP)
           technique;   antiserum L-112 specific for COOH terminus of
           gastrin and CCK, diluted 1:2500).   Inset (C) (arrowed)
           shows complete cells indicated in B.   Bar = 10 μm.

staining and spectrophotometer readings with a soluble chromogen
(O'Beirne & Cooper, 1979). The ELISA technique allows for precise
information on the reactivity of the antiserum towards known
compounds believed to be in the tissue. As an example, the
α-endorphin antiserum we have been working with on *Calliphora*
tissue (Duve & Thorpe, 1983) (Fig. 3) has been shown to be highly
specific for the α-endorphin molecule itself, but unreactive
towards other α-endorphin-related peptides (S. Jackson, personal
communication).

One very important point, however, is that although ELISA
accurately defines the immunoreactive properties of the antiserum
it only defines the tissue antigens indirectly, i.e. by comparison
with known peptides. Positive identification of antigens actually
present in the tissue requires the separation of the tissue
components followed by a test on their reactivity in an immunocytochemical staining procedure, as in GEDELISA (SDS gel-electrophoresis-
derived ELISA) (Lutz et al., 1979). Here the tissues are solubilized by means of SDS (sodium dodecyl sulphate) and/or urea, and
the peptides/proteins separated by means of SDS-electrophoresis
and tested in the ELISA technique.

The other alternative is to accept that with immunocytochemistry
it is only possible to gain a limited amount of information regarding
the identity of the immunoreactive material within the tissue and to
work towards its extraction, purification and characterization using
other means such as RIA, bioassay and ultimately amino acid sequencing. This will be discussed later.

(d)   Immunocytochemical procedure

As with all aspects of immunocytochemistry much has been
written about the actual techniques employed and there are many
combinations and varieties of markers and procedures (see Bullock
& Petrusz, 1982). Briefly, the range of markers includes fluorescent compounds (c.f. Coons, 1978), enzymes such as peroxidase,
colloidal gold and, for EM, ferritin. Varieties of the procedure
include direct techniques where the marker is directly coupled to
an antibody. This method could increase in importance with the
introduction of monoclonal antibodies since they can be easily
adsorbed to colloidal gold (De Mey et al., 1982). The indirect, or
sandwich, techniques in which the primary antibody is linked to a
secondary molecule which can be either a secondary antibody conjugated to a marker, protein A or avidin, are more widely used.
The most commonly used methods, however, and the ones favoured by
almost all insect neuropeptide workers are those involving more
than 2 steps. The PAP method is perhaps the best known (see
Vandesande, 1975; Burns, 1982). The primary antibody (usually
raised in a rabbit) binds to the tissue antigens. The second,
bridging antibody is swine anti-rabbit IgG in excess which may or

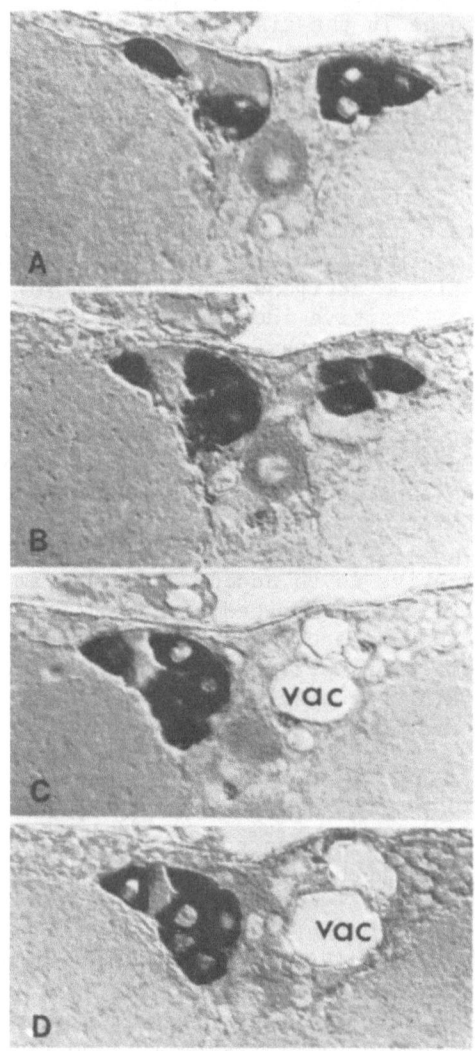

Fig. 3.   A – D.   Serial transverse sections of MNC of the pars
            intercerebralis of *Calliphora*.   Certain of the cells show
            immunoreactivity towards an α-endorphin antiserum (PAP
            technique, antiserum α-E1(7), diluted 1:2500) whereas
            others are unstained.   Vac: vacuolar cells.   Bar = 10 μm.

may not be conjugated with peroxidase.   Finally, the PAP complex
(a soluble peroxidase rabbit anti-peroxidase antibody) completes
the bridge (Sternberger et al., 1970). The peroxidase molecules
are rendered visible by standard peroxidase substrates such as
3,3'-diaminobenzidine hydrochloride (DAB) (no longer believed to
be carcinogenic according to Heydermann - quoted by Burns, 1982)
(see Heydermann, 1980), 3-amino-9-ethyl-carbazole, 4-Cl-naphthol.
or Hanker-Yates reagent.   Sections may be counterstained in
haematoxylin and mounted in DPX for the DAB and Hanker-Yates
reactions or in glycerine gelatine for carbazole or 4-Cl-naphthol.

The PAP method is usually used with paraffin embedded material
and immunofluorescence techniques given over to frozen tissue
material despite the fact that both methods are applicable to both
types of prepared material.   The PAP method intensifies immunological staining over and above that seen in, for example, direct and
indirect immunoenzymic bridge methods.   A further intensification
has been demonstrated by Vacca et al. (1980) by means of a double
bridge PAP application in which a second bridge IgG antibody is
applied in excess and followed by a second application of PAP
serum.

Among the many other methods available, but untried by insect
neurophysiologists and neuroendocrinologists, are those in which the
marker is not an antibody or linking molecule but the antigen itself.
Two such techniques make use of radiolabelling (Larsson & Schwartz,
1977) and gold labelling (Larsson, 1979, 1981c).   The latter, gold
labelled antigen detection (GLAD) technique, may be used at the
ultrastructural level and because it is possible to obtain gold
granules in different sizes it is possible to use the method to
demonstrate different peptides simultaneously.

(e)   Immunoelectronmicroscopy

The problem of the simultaneous preservation of structural
integrity and the overlying antigenicity of the peptides is even
more acute at the ultrastructural level than it is in light
microscopy.   Fixatives that have been successfully used include
Bouin's fluid and glutaraldehyde and paraformaldehyde either
separately or as a mixture.   The expoy resins so widely used in
normal electron microscopy have been shown, in some instances, to
interact with tissue components during polymerization in a way that
could interfere with antigenicity.   For this reason methacrylates
have been preferred in some instances.   For convenience, nickel
grids (not copper) are processed tissue side down on drops of the
various incubating fluids placed on parafilm-covered microscopy
slides.

We have used a postembedding unlabelled antibody (PAP)
technique on tissues of *Calliphora* to reveal α-endorphin immuno-

reactive granules in axons leading from the median neurosecretory
cells (Fig. 4). The method is little changed from that used in
light microscopy (Duve & Thorpe, 1983) and largely follows the
procedure adopted by Van Leeuwen and Swaab (1977). Material was
fixed in either 2.5% glutaraldehyde or 4% paraformaldehyde in
phosphate buffer (0.1 M) + 5% sucrose and embedded in araldite.
Ultrathin sections were collected on nickel slot-grids coated with
2% pioloform in chloroform. The sections were etched in 6%
H_2O_2 for 6 mins and then processed according to standard PAP
procedure. Osmium tetroxide is often applied to the sections after
the DAB incubation is complete (Van Leeuwen & Swaab, 1977) but it
can also be used in the normal post-fixation manner.

(f) Specific applications of immunocytochemistry to insect studies

    Neuropeptide mapping. Probably the most important question
regarding insect neuropeptides relates to their physiological role
and as a prelude to answering this difficult question it is essential
to plot their distribution within the complete nervous system.
Immunocytochemical studies using permanent PAP/DAB preparations
have proved extremely useful in this connection. In our studies of
vertebrate-type peptides we have made considerable use of immunocytochemical staining in Bouin fixed, wax embedded serial sections
in order to reconstruct the peptide distribution (Duve & Thorpe,
1979, 1980, 1981, 1982, 1983) as have other workers with vertebrate-
type peptides in insects (El-Salhy et al., 1980, 1983; Yui et al.,
1980,; Iwanaga et al., 1981). Likewise the distribution of one
of the specific insect neuropeptides, AKH, has recently been
demonstrated by Schooneveld et al. (1983) using the post-embedding
technique.

    Our preliminary studies on the staining of whole-mounts prior
to embedding, either by means of the fluorescence or PAP technique,
have given extremely encouraging results (see Duve et al., 1983).
In order to obtain the fine detail of material within the neuropils
it is necessary to solubilize the tissues for several days by
soaking them in buffer containing 2.5% saponin. The whole-mount
procedure has been very successfully applied to the mapping of
proctolin-like immunoreactive material in the CNS of the cockroach
Periplanata americana by Bishop et al. (1981) and O'Shea (1982)
who point out the degree to which this procedure simplifies and
speeds the mapping of immunoreactive neurons.

    One of the interesting points to emerge from mapping studies
of both types is the widespread occurrence of particular peptides
within the nervous system suggesting according to Bishop et al.
(1981) "diverse central and peripheral functions". We have
commented on the symmetry of the pattern of distribution of pancreatic polypeptide within the central nervous system arguing
that it probably relates to the ancestral, segmented nature of

organisms which gave rise to the insects (Duve & Thorpe, 1982).
Concerning the significance of the widespread and symmetrical
distribution of another peptide, the gastrin/CCK-like peptide of
Calliphora we have recently provided anatomical evidence to
support the idea of a dual central and peripheral function. Thus,
we have shown by a combined immunocytochemical/cobalt labelling
technique at least one complete pathway intrinsic to the brain and
another in which fibres containing this particular peptide actually
leave the CNS to supply the neurohaemal organ, the corpus cardiacum
(Duve et al., 1983).   We regard this as a possible evolutionary
stage in the development of complete duality of occurrence (and
function?) of peptides such as CCK which occur in both the CNS and
gut of vertebrates.

The double labelling technique has been successfully employed
in vertebrate neuropeptide studies (Reaves & Hayward, 1979, 1980;
Van der Kooy and Steinbusch, 1980) but has not been greatly used in
insect studies (Eckert & Ude, 1983). The essential strategy of the
technique as we have used it in *Calliphora* is to employ a CO^{2+}
backfilling procedure with subsequent silver intensification,
combined with immunocytochemistry on the filled cells. By means of
this technique it is possible to trace complete peptidergic pathways
which are not always visible when using immunocytochemistry alone.
The technique is not altogether simple or straightforward but it is
of undoubted value in providing clues as to possible function.
Thus, the peptides are likely to function as neurohormones in the
strict sense if they can be seen to be released at a neurohaemal
organ for general distribution to the body tissues. If, on the
other hand, pathways intrinsic to the CNS are reconstructed, then
a neuromodulator/neurotransmitter function is indicated.

A general point, worth emphasising here is the importance of
the particular state and stage of development of the insect under
study. It is, of course, notoriously difficult to make generalised
statements about insects from different taxonomic groupings but it
is also apparent from our studies on *Calliphora* that considerable
individual variation exists. Thus, the manufacture and distribution
of neuropeptides appears to be highly dependent upon natural
agencies, such as age and sex, and also upon environmental influences
such as diet. It is interesting in this connection that Levine and
Truman (1982) have recently demonstrated changes in the morphology
and synaptic interactions of identified neurones during the metamorphosis of *Manduca* which allows the cells to perform adult as
opposed to larval functions. Although we have not yet compared
larval with adult *Calliphora* neuropeptide-producing cells, we have
observed during the adult phase of the life cycle pronounced changes
in both the number and peptidergic content of particular cells.
This indicates quite strongly that mechanisms operate to switch on
or off particular biosynthetic peptidergic pathways within these
cells and confirms that there is no such animal as a 'Standard'
further experimentation.

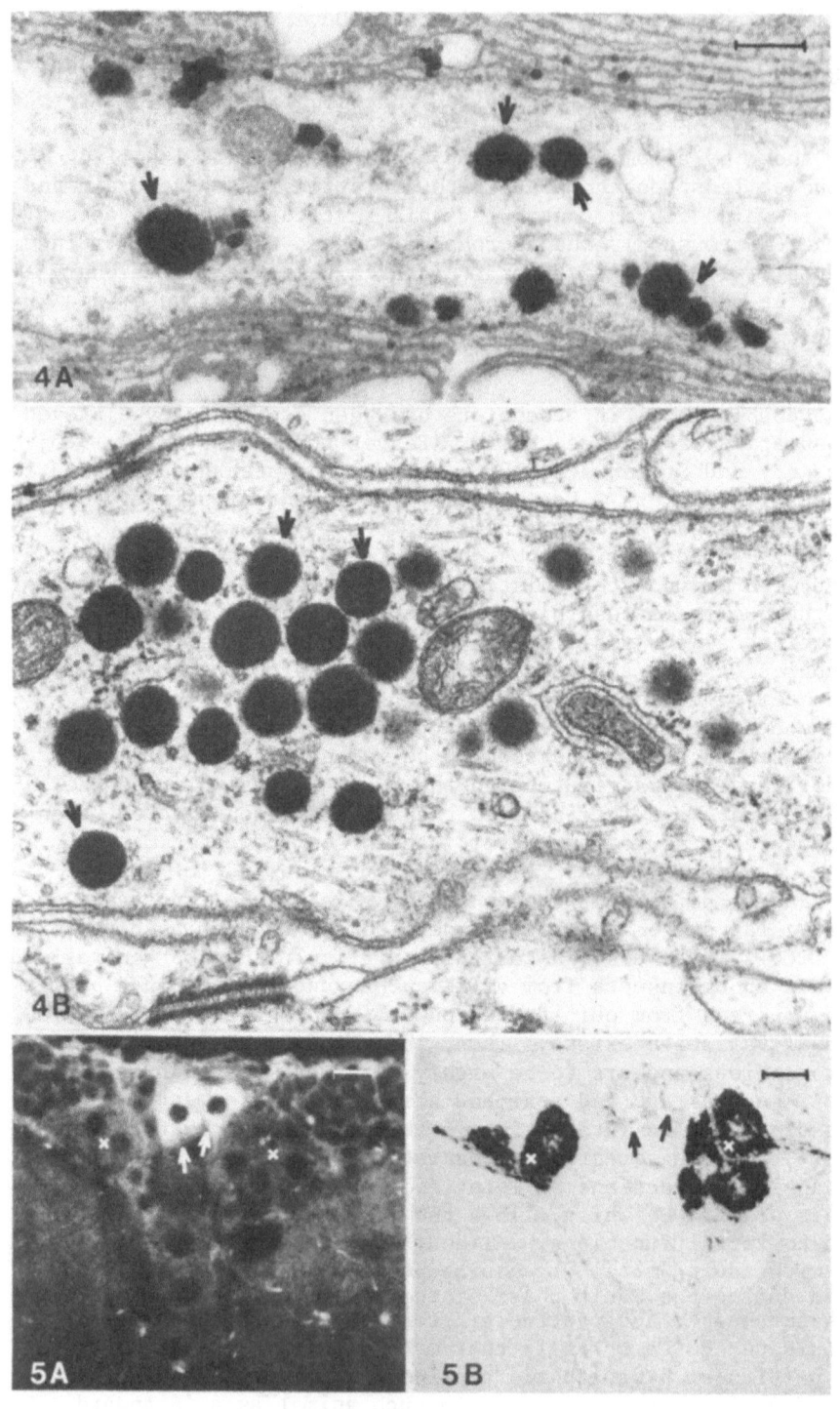

Fig. 4.   Longitudinal sections of median neurosecretory cell axons
          of *Calliphora* containing neurosecretory granules.
          (A) Section of tissue treated with α-endorphin antiserum
          in the PAP technique. Granules show covering of immuno-
          reaction product. (B) The same granules in a normal EM
          section. Bar = 0.25 μm.

Fig. 5.   Transverse section through the pars intercerebralis of
          Calliphora. (A) Certain of the median neurosecretory
          cells (MNCs) showing immunoreactivity towards a bovine
          insulin antiserum (indirect fluorescence technique)
          (arrows). (B) The same section destained and treated with
          an antiserum to ·α-endorphin (PAP technique). The α-
          endorphin immunoreactive cells may be distinguished from
          the insulin immunoreactive cells. Bar = 10 μm.
          Reproduced with the kind permission of Cell & Tissue
          Research, Duve and Thorpe (1983b) in press.

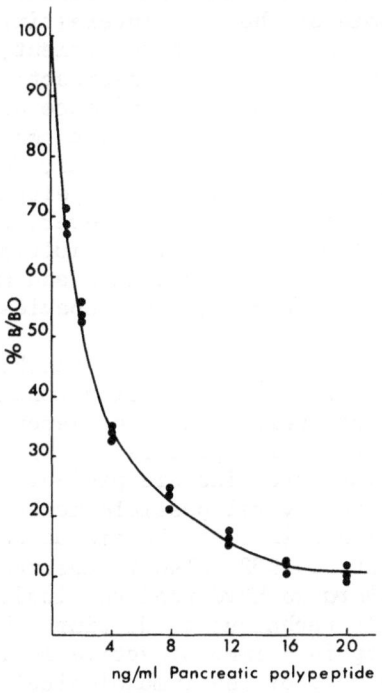

Fig. 6.   Standard curve for the radioimmunoassay of pancreatic
          polypeptide.

<u>Coexistence of peptides within single neurons.</u>   There are now several reports in the literature of the coexistence of 2 or more neuroactive substances within single neurons. The occurrence of biogenic monoamines and peptide hormones in certain endocrine cells is well documented (c.f. Chan-Palay et al., 1978; Hökfelt et al., 1977) but the coexistence of more than one type of peptide within endocrine cells or neurons is less well known (see Larsson, 1980b for review). It has been stressed by Larsson (1981b) that in order to verify simultaneous production and secretion of multiple peptides within cells it is necessary to demonstrate them at the cytochemical and biochemical levels and also to demonstrate their biosynthesis and secretion. This information has only been obtained in full for the ACTH/MSH/β-endorphin cells where only a single precursor molecule is involved (see for example Mains et al., 1977; Nilaver et al., 1979; Larsson, 1980c).

More recently, however, evidence has been presented from mammalian studies to show that peptides with dissimilar precursor molecules may also be produced in the same cell, e.g. ACTH-related peptides in gastrin cells (Larsson, 1981b and pituitary gastrins in corticotrophs and melanotrophs (Larsson & Rehfeld, 1981).

The foregoing comments are mammalian orientated and are included to illustrate some of the more interesting problems awaiting attention in insects where, at the moment, there are no published accounts on the coexistence of neuroactive substances. In *Calliphora* we have been accumulating evidence of neurons in both the central and autonomic nervous systems that contain at least two types of peptidergic immunoreactive material as, for example, gastrin/CCK-like material in association with either PP- or secretin-like immunoreactivities. At the moment, however, the criteria outlined earlier for the verification of the multiple coexistence of these peptides in *Calliphora* is incomplete and must await further experimental work on this and perhaps other species.

<u>Double staining.</u>   A frontal section of the brain of *Calliphora* through the median neurosecretory cell region contains several different types of secretory cells and a corresponding number of different antigens (see Duve & Thorpe refs.). The question arises as to whether these peptides/proteins are present in the same or different cells. There are several possible methods for the demon-stration of multiple antigens in a single tissue section (for reviews see Mason & Woolston, 1982, also Vandesande et al., 1983). In our studies on *Calliphora* we have used the indirect fluorescent method followed by the PAP technique to distinguish between the insulin – and the α-endorphin – immunoreactive median neurosecretory cells (Fig. 5). As with most of the immunological methods in insects this type of study is in its infancy and there is much scope for *Calliphora*.

RADIOIMMUNOASSAY

   Immunocytochemistry is a useful first guide to the location of
particular types of peptides within cells. Used alone, however,
it very rarely provides for complete peptide identification (see
Petrusz et al., 1980; Van Leeuwen, 1982). This is achieved only
by separating the peptides from tissue extracts, purifying them
and subjecting them to amino acid sequence analysis. This is
obviously a demanding and time-consuming procedure, made all the
more so when amounts of material are of necessity small and difficult
to collect, as in certain insect studies (Thorpe and Duve, 1984).
Perhaps the main problem in an exercise of this type is the need to
identify the peptides accurately and conveniently through all stages
of the purification procedure. For some peptides exclusive use is
made of bioassays (e.g., proctolin, Brown, 1967) and for others,
radioimmunoassay is the sole means of identification (e.g., pancreatic polypeptide-like substances) (Duve et al., 1981,1982). A combination of the two methods is probably the best way of ensuring
accurate identification. This has been the approach we have used
in our studies of the *Calliphora* insulin-like peptide (Duve et al.,
1979).

   As has been pointed out by Thorell (1983), radioimmunoassays
require a few relatively simple procedural steps such as pipetting,
centrifuging, decanting and counting but the results obtained can
be of extremely high sensitivity and specificity.

   Studies on insect vertebrate-type gastroenteropancreatic (GEP)
peptides have depended on homologous mammalian peptide RIAs (Duve
et al., 1979,1981,1982; LeRoith et al., 1981; Kramer, 1983) in association with mammalian and/or insect bioassays. The insulin RIA
is very straightforward and is actually available in kit form from
several manufacturers. We have made extensive use of labelled
^{125}I-insulin purchased from the Radiochemical Centre, Amersham,
Bucks., U.K. in conjunction with insulin binding reagent (a mixture
of guinea pig anti-insulin serum and rabbit anti-guinea pig-globulin
serum in buffer; Product RD12, Wellcome Reagents Ltd., Beckenham,
U.K. For the standard insulin solutions we have used either bovine
or porcine crystalline insulin in the range of 1-20 ng/ml. We have
found a short (1-4 h) equilibrium assay to be sufficient for routine
analysis of the freeze dried samples taken from the various chromatographic steps (see Duve et al., 1979). The most sensitive method
with which to separate the antibody bound from the free, labelled
insulin is filtration rather than the other alternative, i.e.,
centrifugation of the precipitate. With 0.45 μm acetate filters
(Oxoid, Basingstoke, U.K.) duplication within and between assays
is excellent.

The radioimmunoassay for pancreatic polypeptide is equally
simple and straightforward although it is not available in kit form.
The antiserum we have used exclusively for RIA is the antibovine PP
antiserum LOT 615-R110-146-10, very kindly donated by Dr. R. E.
Chance, Lilly Research Laboratories, Eli Lilly & Co., Indianapolis,
U.S.A. The components of the assay are, in order of pipetting,
100 µl of standard bovine PP (range: 0.5 - 20 ng/ml) or 100 µl of
appropriate dilutions of the freeze dried eluates from the chromatography columns, 100 µl of antiserum (1 : 100,000 in Clark-Lubb's
buffer) and 100 µl of ^{125}I-labelled bovine PP (the kind gift of
Dr. D. Baxter, Wellcome Foundation, Dartford, U.K.). An equilibrium assay of 24 h produces a standard curve with a high degree
of sensitivity in the 0-8 ng/ml range (Fig. 6), and, when necessary,
the assay time can be cut down to as little as 1 h or even less.
The separation of antibody bound from free tracer in this assay is
effected by adding 100 µl of activated charcoal 50 mg (Norit A
Grade)/ml buffer and centrifuging at 3,000 g for 30 min at 4^0C.
The supernatant contains the antibody-bound ^{125}I-labelled peptide
and the pellet of charcoal, the free ^{125}I-labelled peptide.

The key elements of a successful radioimmunoassay are a
suitable antiserum, accurate standards, a good effective separation
system and, especially important, a homogeneous preparation of
labelled antigen. The labelled antigen effectively determines the
population of antibodies within the antiserum that will form the
basis of the RIA. The other antibodies will not take part in the
assay and can be ignored. This of course is one of the major
differences between immunocytochemistry and radioimmunoassay and it
is important to stress that RIA data cannot be used as proof of
immunocytochemical specificity or vice versa. The whole question
of antigen labelling and its significance in RIA has recently been
reviewed by Hunter (1982) and it is not within the scope of this
paper to cover this ground again. It may be appropriate, however,
to discuss briefly some of the issues as they appertain to our ongoing studies of α-endorphin-like peptides in *Calliphora* (Duve et
al., unpublished results).

Firstly the purity of the α-endorphin is guaranteed by the
manufacturers and has been further assessed by us using HPLC with a
suitable analytical ODS column and a gradient of acetonitrile
+ 0.01% TFA/water. The single peak obtained mid-way through the
gradient is not definite proof of purity but it is a good indication.

The labelling procedure we have adopted is a modified lacto-
peroxidase iodination method with subsequent purification by means
of polyacrylamide gel electrophoresis and elution in ammonium bicarbonate buffer (Linde et al., 1980).

For the radioimmunoassay of the α-endorphin-like peptide of

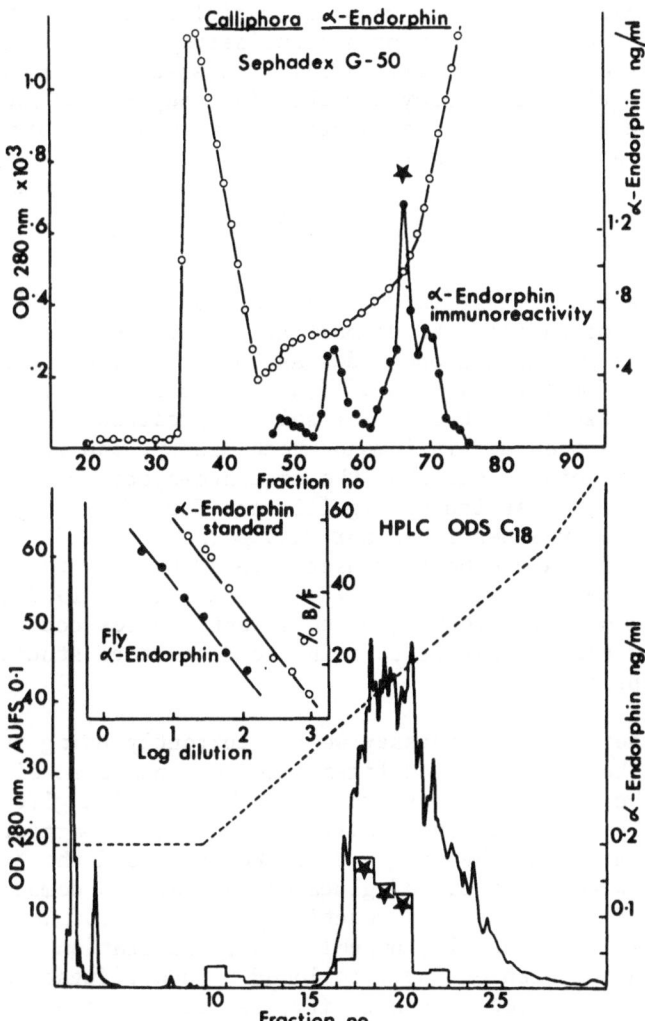

Fig. 7. Filtration on Sephadex G-50 (SF) of 0.2M HCl extract of
120 gm *Calliphora*. Column size 5 cm x 52 cm; 10 ml
fractions. Samples (0.2 ml) of eluate (1% formic acid)
were freeze-dried and immunoassayed for α-endorphin.
Protein absorbance at 280 nm-0-0- ; α-endorphin
immunoreactivity -●-●- . The peak fraction (66) was
freeze-dried and subjected to reverse phase HPLC (C18
column; size 3.9 mm i.d. x 30 cm; gradient 16-48%
acetonitrile/H_2O/0.0125% TFA; 30 min; flow rate 1.5 ml/min;
fraction size I.5 ml). Samples (0.2 ml) were evaporated
and immunoassayed for α-endorphin. Samples of the pooled
immunoreactive material were used to produce the dilution
curve (inset C). (See also Fig. 8).

Calliphora extracts we have used the radiolabelled ligand in conjunction with the same α-endorphin antiserum as used in immunocytochemistry (see Duve & Thorpe, 1983). The assay in this instance is
a 48 h dis-equilibrium assay and the final separation makes use of
polyethylene glycol (PEG) in association with horse serum, followed
by centrifugation and counting of the pellet (antibody bound faction).
The assay is extremely sensitive in the picogram/low nanogram range
and is therefore suitable for assessing the rather low amounts of
α-endorphin peptide present in *Calliphora* (Fig. 7).

For the RIA of gastrin/CCK-like peptides of *Calliphora* we have
used a range of antisera to demonstrate the COOH-terminal specific
nature of the insect neuropeptides as compared with the mammalian
counterparts (Dockray et al., 1981). The use of region specific
antisera are extremely useful for solving problems of specificity
in RIA just as they are in immunocytochemistry, and it has
recently been pointed out that the sequence-specific RIAs will
continue to be of particular significance for use in phylogenetic
studies of the type described for *Calliphora* (Rehfeld, 1983). The
specificity of RIA can be increased even further by the new development of residue-specific assays (Rehfeld & Morley, 1983), and the
introduction of these techniques to insect studies could be a
particularly useful adjunct to amino acid sequence studies of the
purified peptides.

The other radioimmunoassay we are currently using is for
glucagon. The radiolabelled ligand used was prepared either by Novo
Research Laboratories, Bagsvaerd, Denmark (the kind gift of Dr. Lise
Heding) or by Dr. D. Baxter, Wellcome Foundation, Dartford, U.K.,
to whom our grateful thanks are due. We have used the tracer in
conjunction with a variety of glucagon antisera to demonstrate
both C- and N-terminal cross reactivity of the *Calliphora* glucagon-
like molecule (Duve et al., unpublished). Separation is effected
by precipitation with ethanol, followed by centrifugation and
counting of the pellet. Sensitivity, as with α-endorphin, is
selected for in the picogram-low nanogram range.

CONCLUDING COMMENTS

In order to validate any RIA it is necessary to compare unknown
against standard antigens. The unknown substance which is inhibiting
the otherwise maximum binding of the standard antigen to the antibody
should dilute out linearly and in parallel to a dilution curve of
standards. We have made this an essential procedural step at each
point in our purification procedures (Fig. 8). This essential
precaution is necessary to eliminate the possibility of non-specific
interferences caused by a variety of factors such as pH, ionic
environment and a variety of chemicals that will tend to affect the
antibody-antigen reaction (see Yalow & Strauss, 1980). The pH and

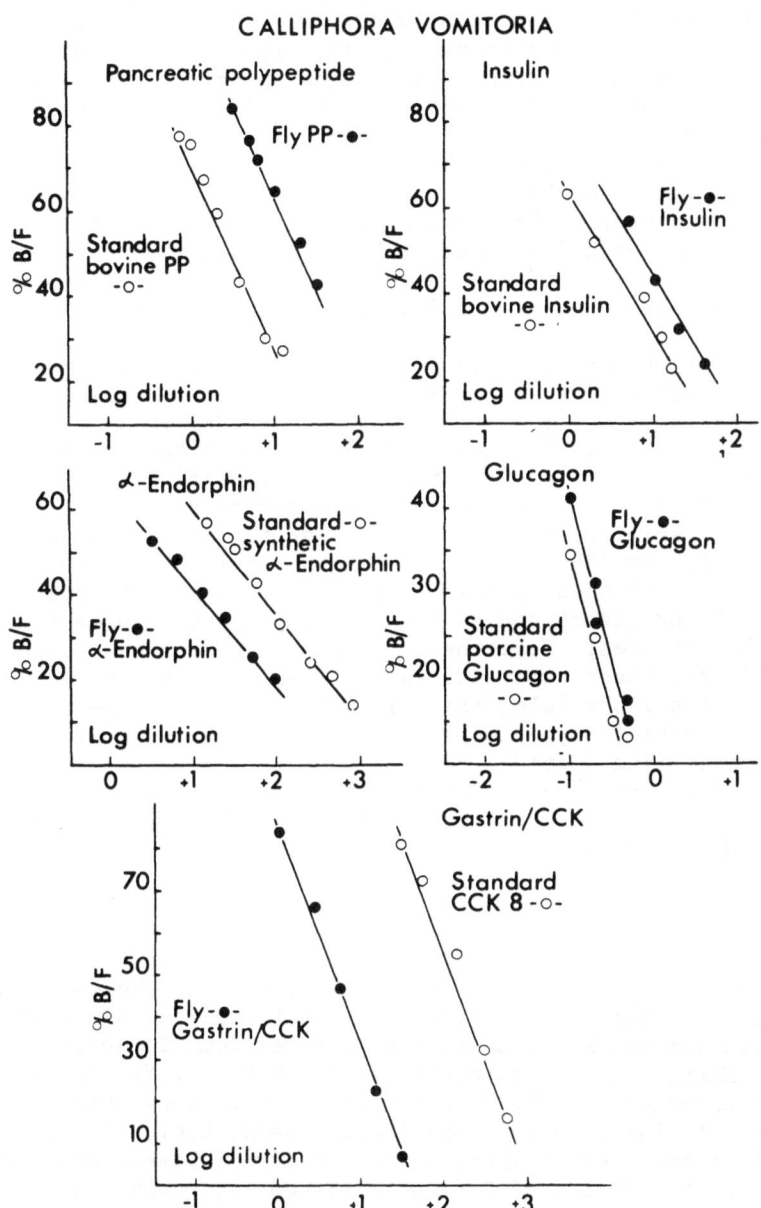

Fig. 8.   Dilution curves for five mammalian peptides compared
with the corresponding neuropeptides from *Calliphora*.
The parallelism over a threefold to a sevenfold concentration range shows identity of antigenic determinants.

ionic environment can easily be made identical in standard and
unknown RIA tubes but a potentially more difficult problem concerns
the possibility of contaminating proteins either from column eluates
or especially from haemolymph.   The large amounts of protein in
certain of the fractions from some of the *Calliphora* tissue extracts
we have prepared do not appear to give either false positives or
false negatives in the RIAs we have conducted (see Fig. 7).
Haemolymph on the other hand is difficult to work with on account
of interfering proteins which often give results substantially in
excess of the zero-binding tubes.   It is necessary to purify the
haemolymph therefore to make an accurate assessment of the peptides
present there (Kramer et al., 1980;   Duve & Thorpe, 1984).

     In conclusion, evidence from immunocytochemistry and radio-
immunoassay together with data from bioassay speaks strongly in
favour of the presence of vertebrate-type peptides in *Calliphora*.
As yet these techniques have not been fully exploited in relation
to the study of insect-specific peptides.   The main problem here
is in obtaining the peptides in pure form and in sufficient amount
to prepare antisera, standards and labelled tracers.   Immunocytochemical studies on AKH and proctolin have provided information on
the distribution of these peptides within the nervous tissues of
the cockroach and locust and as yet bioassay has been used to
quantify the peptides.   Nevertheless, it should soon be possible to
develop RIAs for these two peptides and when this is achieved it is
certain that their physiological role will be better understood.
As for the vertebrate-type insulin, glucagon-, PP-, gastrin/CCK-
and α-endorphin-like peptides of *Calliphora*, the two immunological
techniques discussed here are helping to create a much better
understanding of their structure and function than was considered
possible a decade ago.

ACKNOWLEDGMENTS

     We wish to express our sincere appreciation of the work and
continuing involvement in the project on GEP peptides of *Calliphora*
of our colleague Dr. N. R. Lazarus of the Wellcome Foundation,
Beckenham, Kent, U.K. We also wish to thank Mr. R. Neville of the
Wellcome Foundation, Dr. P. J. Lowry of the Pituitary Hormone
Laboratory, St. Bartholomew's Hospital, London, U.K., Professor G. J.
Dockray, Department of Physiology, University of Liverpool, and Mr.
Alan Scott of the School of Biological Sciences, Queen Mary College,
London University. We are grateful to Dr. Ursula Bassemir of EMBL,
Heidelberg, for collaboration in immunoelectronmicroscopy.

     We also wish to thank all those who have so generously
provided us with antisera (named in the text).

     Finally, grateful acknowledgment is made to the Science and

Engineering Research Council, U.K., for providing grant support
(Grant Nos. GR/B88136 and GR/C20789).

REFERENCES

Bishop C.A., O'Shea M. and Miller R.J. (1981)   Neuropeptide proctolin
    (H-Arg-Tyr-Leu-Pro-Thr-OH) Immunological detection and neuronal
    localisation in the insect nervous system. Proc.Natl.Acad.
    Sci. U.S.A. 78, 5899-5920.
Brandtzaeg P. (1982) Tissue preparation methods in immunocyto-
    chemistry. In: Techniques in Immunocytochemistry. Vol. 1.
    (Bullock G.R. and Petrusz P. eds.). Academic Press, London,
    pp. 1-75.
Brown B.E. (1967) Neuromuscular transmitter substance in insect
    visceral muscle. Science, N.Y. 155, 595-597.
Bullock G.R. and Petrusz P. (1982) Techniques in Immunocytochemistry.
    Vol. 1. Academic Press, London.
Burns J. (1982) The unlabelled antibody peroxidase anti-peroxidase
    method (PAP). In: Techniques in Immunocytochemistry. Vol. 1.
    Bullock G.R. and Petrusz P. eds). Academic Press, London,
    pp.91-105.
Chan-Palay V., Jonsson G. and Palay S.L. (1978) Serotonin and sub-
    stance P coexist in neurons of the rat's central nervous
    system. Proc.Natl.Acad.Sci. USA 75, 1852-1856.
Coons A.H. (1978) Fluorescent antibody methods. In: General Cyto-
    chemical Methods (Danielli J.F. ed.). Academic Press, New
    York, pp. 399-422.
De Mey J., Moeremans M., De Waele M., Geuens G. and De Brabander M.
    (1982) The IGS (immuno gold staining) method used with mono-
    clonal antibodies. In: Proteids in Biological Fluids (Peeters
    H. ed.). Pergamon Press, Oxford.
Dockray G.J., Duve H. and Thorpe A. (1981) Immunochemical characteri-
    sation of gastrin/cholecystokinin-like peptides in the brain of
    the blowfly, Calliphora vomitoria. Gen.Comp.Endocrinol. 45,
    491-496.
Duve H. and Thorpe A. (1979) Immunofluorescent localization of
    insulin-like material in the median neurosecretory cells of the
    blowfly, Calliphora vomitoria (Diptera). Cell Tissue Res. 200,
    187-191.
Duve H. and Thorpe A. (1980) Localisation of pancreatic polypeptide
    (PP)-like immunoreactive material in neurones of the brain of
    the blowfly, Calliphora erythrocephala (Diptera). Cell Tissue
    Res. 210, 101-109.
Duve H. and Thorpe A. (1981) Gastrin/cholestokinin (CCK)-like
    immunoreactive neurones in the brain of the blowfly, Calliphora
    erythrocephala (Diptera). Gen.Comp.Endocrinol. 43, 381-391.
Duve H. and Thorpe A. (1982) The distribution of pancreatic poly-
    peptide in the nervous system and gut of the blowfly,
    Calliphora vomitoria (Diptera). Cell Tissue Res. 227, 67-77.

Duve H. and Thorpe A. (1983) Immunocytochemical identification of
α-endorphin-like material in neurones of the brain and corpus
cardiacum of the blowfly, *Calliphora vomitoria* (Diptera). <u>Cell
Tissue Res.</u> (In press).

Duve H. and Thorpe A. (1984) Vertebrate-type brain/gut peptides in
neurones of the blowfly *Calliphora*. <u>In</u>: Symposium Proceedings
of the Ninth International Symposium on Comparative Endocrinology. (Lofts B. and Chan D.K.O. eds). Hong Kong University
Press (In press).

Duve H., Thorpe A. and Lazarus N.R. (1979) Isolation of material
displaying insulin-like immunological and biological activity
from the brain of the blowfly, *Calliphora vomitoria*. <u>Biochem.J.</u>
<u>184</u>, 221-227.

Duve H., Thorpe A. and Strausfeld N.J. (1983) Cobalt-immunocyto-
chemical identification of peptidergic neurons in *Calliphora*
innervating central and peripheral targets. <u>J.Neurocytol.</u>
(In press).

Duve H., Thorpe A., Lazarus N.R. and Lowry P.J. (1982) A neuropeptide
of the blowfly *Calliphora vomitoria* with an amino acid composition homologous with vertebrate pancreatic polypeptide.
Biochem.J. <u>201</u>, 429-432.

Duve H., Thorpe A., Neville R., Lazarus N.R. (1981) Isolation and
partial characterization of pancreatic polypeptide-like material
in the brain of the blowfly *Calliphora vomitoria*. <u>Biochem.J.</u>
<u>197</u>, 767-770.

Eckert M. and Ude J. (1983) Immunocytochemical techniques for the
identification of peptidergic neurons. <u>In</u>: Functional Neuroanatomy. Springer Series in Experimental Entomology (N.J.
Strausfeld, ed.). Springer, Heidelberg, Berlin, New York.
(In press).

Eisenbarth G.S. and Jackson R.A. (1982) Application of monoclonal
antibody techniques to endocrinology. <u>Endocr.Rev.</u> <u>3</u>, 26-39.

El-Salhy M., Falkmer S., Kramer K.J. and Speirs R.D. (1983) Immunohistochemical investigations of neuropeptides in the brain,
corpora cardiaca and corpora allata of an adult lepidopteran
insect, *Manduca sexta* (L.). <u>Cell Tissue Res.</u> (In press).

El-Salhy M., Abou-El-Ela R., Falkmer S., Grimelius L., Wilander E.
(1980) Immunohistochemical evidence of gastro-entero-pancreatic
neurohormonal peptides of vertebrate type in the nervous system
of the larva of a dipteran insect, the hoverfly, *Eristalis
aeneus*. <u>Regulatory Peptides</u> <u>1</u>, 187-204.

Engvall E. and Perlmann P. (1971) Enzyme linked immunosorbent assay
(ELISA). III. Quantitative assay of immunoglobulin G.
<u>Immunochemistry</u> <u>8</u>, 871.

Forsmann W.G., Pickel V., Reinecke M., Hock D. and Hetz J. (1981) Immunohistochemistry and immunocytochemistry of nervous tissue. <u>in</u>:
Techniques in Neuroanatomical Research (Heym Ch. and Forsmann
W.G., eds.). Springer Verlag, Berlin, Heidelberg, New York,
pp. 171-205.

Hansen B.L., Hansen G.N. and Scharrer B. (1982) Immunoreactive

material resembling vertebrate neuropeptides in the corpus cardiacum
    and corpus allatum of the insect *Leucophaea maderae*. Cell Tissue
    Res. 225, 319-329.
Heyderman E. (1980) The role of immunocytochemistry in tumour
    pathology: a review. Journal of the Royal Society of Medicine
    73, 655-658.
Hökfelt T., Elfovin L.G., Elde K., Schultzger M., Goldstein M. and
    Luft R. (1977) Occurrence of somatostatin-like immunoreactivity
    in some peripheral sympathetic noradrenergic neurons. Proc.
    Natl.Acad.Sci. USA 75, 2587-2591.
Hunter W.M. (1983) Assay design and the influence of radioligand
    quality. In: Radioimmunoassay Design and Quality Control
    (Thorell J.I. ed.). Pergamon Press, Paris, pp. 9-18.
Iwanaga T., Fujita T., Nishiitsutsuji-Uwo J. and Endo Y. (1981)
    Immunohistochemical demonstration of PP-, somatostatin-,
    enteroglucagon- and VIP-like immunoreactivities in the cockroach
    midgut. Biomed.Res. 2, 202-207.
Kramer K.J. (1983) Vertebrate hormones in insects. In: Comprehensive
    Insect Physiology, Biochemistry and Pharmacology. Vol. 7.
    Endocrinology 1. Chapter 10. Pergamon Press, New York. (In press).
Kramer K.J., Tager H.S. and Childs C.N. (1980) Insulin-like and
    glucagon-like peptides in insect hemolymph. Insect Biochem. 10,
    179-182.
Larsson L.-I. (1979) Simultaneous ultrastructural demonstration of
    multiple peptides in endocrine cells by a novel immunocyto-
    chemical method. Nature (London) 282, 743-746.
Larsson L.-I. (1980a) Problems and pitfalls in immunocytochemistry
    of gut peptides. In: Gastrointestinal Hormones (Jerzy Glass
    G.B. ed.). Raven Press, New York, pp. 53-70.
Larsson L.-I. (1980b) On the possible existence of multiple endocrine,
    paracrine and neurocrine messengers in secretory cell systems.
    Invest.Cell Pathol. 3, 743-85.
Larsson L.-I. (1980c) Corticotropin and α-melanotropin in brain
    nerves: Immunocytochemical evidence for axonal transport and
    processing. In: Neural Peptides and Neuronal Communication
    (Costa E. ed.). Raven Press, New York, pp. 101-107.
Larsson L.-I. (1981a) Peptide immunocytochemistry. Progr.Histochem.
    Cytochem. 13,1.
Larsson L.-I. (1981b) Multiple secretory messengers in endocrine
    cells and neurons: ACTH-related and opioid peptides in gut and
    brain. In: Cellular Basis of Chemical Messengers in the
    Digestive System (Grossman M.I., Brazier M.A.B. and Lechago J.
    eds.). UCLA Forum in Medical Sciences No. 23. Academic Press,
    London, pp. 151-157.
Larsson L.-I. (1981c) Immunocytochemistry of secretory peptides:
    Introduction of a new immunocytochemical technique. In:
    Cellular Basis of Chemical Messengers in the Digestive System.
    (Grossman M.I., Brazier M.A.B. and Lechago J. eds.). UCLA
    Forum in Medical Sciences No. 23. Academic Press, London,
    pp. 67-71.

Larsson L.-I. and Rehfeld J.F. (1977) Characterization of antral
    gastrin cells with region-specific antisera. J.Histochem.
    Cytochem. 25, 1317-1321.
Larsson L.I. and Rehfeld J.F. (1981) Pituitary gastrins occur in
    corticotrophs and melantrophs. Science 213, 768-770.
Larsson L.-I. and Schwartz T.W. (1977) Radioimmunocytochemistry -
    A novel immunocytochemical principle. J.Histochem.Cytochem.
    25, 1140-1148.
LeRoith D., Lesniak M.A. and Roth J. (1981) Insulin in insects and
    annelids. Diabetes 30, 70-76.
Levine R.B. and Truman J.W. (1982) Metamorphosis of the insect
    nervous system: changes in morphology and synaptic inter-
    actions of identified neurones. Nature (London) 299, 250-252.
Linde S., Hansen B. and Lernmark A (1980) Stable iodinated poly-
    peptide hormones prepared by polyacrylamide gel electrophoresis.
    Analytical Biochem. 107, 165-176.
Lutz H., Higgins J., Pederson N.C. and Theilen G.H. (1979) The demon-
    stration of antibody specificity by a new technique.
    J.Histochem. Cytochem. 27, 1216.
Mains R.E., Eipper B.A. and Ling N (1977) Common precursor to
    corticotropins and endorphins. Proc.Natl.Acad.Sci. USA 74,
    3014-3018.
Mason D.Y. and Woolston R.E. (1982) Double immunoenzymatic labelling.
    in: Techniques in Immunocytochemistry. Vol. 1. (Bullock G.R.
    and Petrusz P. eds.). Academic Press, London, pp. 135-153.
Nilaver G., Zimmerman E.A., Defendini R., Liotta A.S., Krieger D.T.
    and Brownstein M.H. (1979) Adrenocorticotropin and β-lipotropin
    in the hypothalamus. Localization in the same arcuate neurons
    by sequential immunocytochemical procedures. J.Cell Biol. 81,
    50-59.
O'Beirne A.J. and Cooper H.R. (1979) Heterogeneous enzyme immuno-
    assay. J.Histochem.Cytochem. 27, 1148.
O'Shea M. (1982) Peptide neurobiology. An identified neurone
    approach with special reference to proctolin. Trends in
    Neuroscience 15, 69-73.
Petrusz P., Ordronneau P. and Finley J.C.W. (1980) Criteria of
    reliability in light microscopic immunocytochemical staining.
    Histochem.J. 12, 333-348.
Petrusz P., Sar M. Ordronneau P., and DiMeo P. (1976) Specificity
    in immunocytochemical staining. J.Histochem.Cytochem. 24,
    1110-1115.
Pool C.W., Buijs R.M., Swaab D.F., Boer G.J. and van Leeuwen F.W.
    (1982) Specificity in immunocytochemistry. In: Neuroimmuno-
    cytochemistry (Cuello A.C., ed.). IBRO Handbook Series:
    Methods in Neurosciences. John Wiley & Sons, Chichester.
Reaves T.A. and Hayward J.N. (1979) Intracellular dye marked
    enkephalin neurons in the magnocellular preoptic nucleus of
    the goldfish hypothalamus. Proc.Natl.Acad.Sci. USA 76,
    6009-6011.
Reaves T.A. and Hayward J.N. (1980) Functional and morphological

studies of peptide-containing neuroendocrine cells in goldfish hypothalamus. J.Comp.Neurol. 193, 777-788.

Rehfeld J.F. (1983) The influence of antiserum properties on radio-immunoassay characteristics. In: Radioimmunoassay Design and Quality Control (Thorell J.I., ed.). Pergamon Press, Paris, pp. 19-27.

Rehfeld J.F. and Morley J.S. (1983) Residue-specific radioimmuno-analysis - a novel analytical tool. Analyt.Biochem. (In press).

Schooneveld H., Tesser G.I., Veenstra J.A. and Rombert-Privee H.J. (1983) Adipokinetic hormone and AKH-like peptide demonstrated in the corpora cardiaca and nervous system of *Locusta migratoria* by immunocytochemistry. Cell Tissue Res. 230, 67-76.

Sternberger L.A. (1979) Immunocytochemistry. 2nd edn. J. Wiley, New York.

Sternberger L.A., Hardy P.H., Cuculis J.J. and Meyer H.G. (1970) The unlabelled antibody enzyme method of immunohistochemistry. Preparation and properties of soluble antigen-antibody complex (horseradish peroxidase-anti horseradish peroxidase) and its use in identification of spirochetes. J.Histochm.Cytochem. 18, 315-333.

Thorell J.I. (1983) Radioimmunoassay Design and Quality Control. Pergamon Press, Paris.

Thorpe A. and Duve H. (1984) Insulin- and glucagon-like peptides in insects and molluscs. Molecular Physiology. (To be published, 1984).

Vacca L.L., Abrahams S.J. and Naftchi N.E. (1980) A modified peroxi-dase procedure for improved localization of tissue antigens: localization of substance P in rat spinal cord. J.Histochem. Cytochem. 28, 297-307.

Van der Kooy D. and Steinbusch H.W.M. (1980) Simultaneous immuno-histofluorescent identification and fluorescent retrogade tracing of neurons. J.Neurosc.Res. 5, 479-438.

Vandesande F. (1975) A critical review of immunocytochemical methods for light microscopy. J.Neurosci.Methods 1, 3-23.

Vandesande F., Dierickx K., De Mey J., Goossens N., Van Vossel-Daeninck J. and Van Vossel A. (1983) Immunocytochemical locali-zation of hypothalamic neuropeptides. In: Symposium Proceedings of the Ninth International Symposium on Comparative Endocrino-logy. (Lofts B. and Chan D.K.O. eds.). Hong Kong University Press. (In press).

Van Leeuwen F. (1982) Specific immunocytochemical localization of neuropeptides : A utopian goal? In: Technique in Immunocyto-chemistry. Vol. 1. (Bullock G.R. and Petrusz P. eds.). Academic Press, London, pp. 283-299.

Van Leeuwen F.W. and Swaab D.F. (1977) Specific immunoelectronmicro-scopic localization of vasopressin and oxytocin in the neuro-hypophysis of the rat. Cell Tissue Res. 177, 493-501.

Van Noorden S. and Polak J. (1981) Advances in immunocytochemistry. in: Gut Hormones (Bloom S.R. and Polak J.M., eds.). 2nd edn. Churchill Livingstone, Edinburgh, London, Melbourne and New York, pp. 80-89.

Yalow R.S. and Straus E. (1980) Problems and pitfalls in the radio-
    immunoassay of gastrointestinal hormones. In: Gastrointestinal
    Hormones (Jerzy Glass G.B., ed.). Raven Press, New York, pp.
    751-767.
Yui R., Fujita T. and Ito S. (1980) Insulin-, gastrin-, pancreatic
    polypeptide-like immunoreactive neurones in the brain of the
    silkworm, *Bombyx mori*. Biomed.Res. 1, 42-46.

THE ISOLATION OF INSECT NEUROPEPTIDES USING

REVERSE-PHASE HIGH PERFORMANCE LIQUID CHROMATOGRAPHY

Timothy K. Hayes and Larry L. Keeley

Texas Agricultural Experiment Station
Department of Entomology, Texas A&M University
College Station, Texas 77843

High performance liquid chromatography (HPLC) is an advanced technique for the preparation and analysis of small quantities of peptides and proteins. Some HPLC techniques follow the example of classical column chromatography and use the principles of ion exchange (Takahashi et al, 1981; Isobe et al, 1982) and gel filtration (Bennett et al, 1983) to separate compounds by charge or molecular weight. However, the hydrophobic properties of peptides have been more extensively exploited by researchers using HPLC to yield efficient separations of peptides and proteins(Smith and McWilliams, 1980; Hearn et al, 1982). The term "reversed phase" was used first by Howard and Martin (1950). Reversed phase distinguishes partition chromatography which uses a hydrophobic stationary phase with polar solvents from "normal" partition chromatography which uses a polar stationary phase and nonpolar organic solvents. The focus of this report is on applications of reversed phase technology to the isolation of small quantities of insect neuropeptides.

Like all chromatography, reversed phase high performance liquid chromatography (RP-HPLC) depends on a differential distribution of solutes between moving and stationary phases to achieve separation. The two phases must share a common interface and the movement of the mobile phase is responsible for the elution of samples from the column. The hydrophobic characteristics of individual peptides are responsible for their adsorption to the stationary phase. Elution from the column depends on increasing the polarity of the peptide or increasing the hydrophobicity of the mobile phase by adding organic modifiers so that the solubility of the peptide is shifted toward the solvent. The relative hydrophobicities of the components in a biological extract determine their individual retentions on the column. Generally, the more nonpolar a given peptide the stronger are its interactions with the bonded hydrophobic stationary phase and the longer are its retention times.

Research reports published in the late 1970's and early 1980's clearly show several advantageous properties of RP–HPLC for the separation of complex mixtures of peptides from biological sources. The high resolution of RP–HPLC is perhaps the most fundamental reason for using this method. Peptides elute from RP–HPLC in narrow peaks, undergo less dilution and remain at higher concentrations. Higher sample concetrations result in lower detection limits and increased peptide yields. Also, since peak widths remain narrow, minimal peak overlap is observed and the resolution of closely eluting peptides is optimized. Taken together, the properties of improved peptide resolution, detection sensitivity, recovery and separation speed make RP–HPLC an ideal method for insect neuropeptide purification.

PRELIMINARY PEPTIDE INFORMATION

Certain information can be obtained on insect neuropeptides before isolation and these observations are helpful if not necessary to the successful implementation of most purification procedures. Crucial information includes the peptidic nature of the factor, determination of critical functional groups, stability to physical and chemical treatments, and solubility in solvents useful for purification.

Before any of this information can be obtained the investigator must develope a useful bioassay system for the detection and quantitation of the neuropeptide. The bioassay should be optimized so that it is sensitive, specific, rapid and simple. Time spent properly developing the bioassay will pay off in the future because all fractions generated during the isolation, must be assayed to detect and quantify the peptide.

Inactivation by various proteases will provide evidence for the peptidic nature of the factor and a preliminary structural analysis. Inactivation of a neural peptide by a protease indicates an exposed and critical peptide bond in the peptide that corresponds to the structure–specificity requirement for that protease. For example, inactivation by chymotrypsin, indicates the presence of aromatic amino acids or by carboxypeptidase Y or aminopeptidase M to show unblocked termini. However, failure of a protease to inactivate a peptide is no guarantee that the specific structural element is absent. The protease–required element may be present yet concealed from proteolytic action by the folding of the peptide chain. Immobilized enzymes are prefered since they can be removed from the incubation solution by centrifugation. Subsequently, the incubation solution can be bioassayed directly for the neural peptide. Simultaneous–control experiments should be run to insure the separate stability of the peptide and the protease under the conditions used for the incubation. Finally, when a protease fails to inactivate the peptide, control experiments should be run to insure the absence of protease inhibitors in the crude extracts.

The molecular weight of a peptide is also an important characteristic to consider when planning a purification scheme. The method of choice for estimating the molecular weights of the smaller peptides in crude extracts still seems to be gel filtration on either polydextran, polyacrylamide or more recently vinyl supports. However, recent developments in gel filtration HPLC technology should be considered as a viable alternative. The major disadvantage of the modified silica supports used for HPLC gel filtration is that the lower limits for peptide molecular weight estimates was around 2,000 daltons. Several vertebrate neuropeptides and the 2 isolated insect neuropeptides are below this molecular weight limit. However, recent developments in mobile phase technology has lowered this size limit to around 500 daltons. Using solvent systems containing trifluoroacetic acid (Bennett et al, 1983) or triethylammonium phosphate (Rivier, 1980) the molecular weight of peptides as small as met-enkephlin have been estimated.

The physical stability of the peptide should be assessed before the initiation of isolation procedures. With this information the researcher will know the approximate stability of the peptide for long term storage at reduced temperatures. Also, answers will become available to such questions as: will the peptide tolerate lyophilization, storage in solution at 4°C, or boiling to destroy endogenous tissue proteases?

Finally, a good knowledge of the solubility of the peptide in various solvents is essential to the success of a planned purification. In preparing for reversed phase chromatography a wide range of solvents should be tested to determine their ability to solubilize the biological activity of the peptide. Polarity and pH are 2 important solvent characteristics which should be explored. Control experiments should be run to determine the influence of any potential solvent residues on the bioassay.

BASIC PRINCIPLES

A clear understanding of RP-HPLC requires a knowledge of the basic principles and definitions of chromatography. These principles include retention, selectivity, efficiency and resolution.

The degree of interaction between the sample peptide and the stationary phase is called retention. The rough quantitative measure of retention is retention volume (V_R) or, if the flow rate does not change, retention time (t_R). A sample which fails to interact with the column packing material elutes in the void volume (V_o). V_o represents the volume between the packed particles of the column, and the size of V_o is a function of the particular packing and column. Under isocratic (unchanging solvent

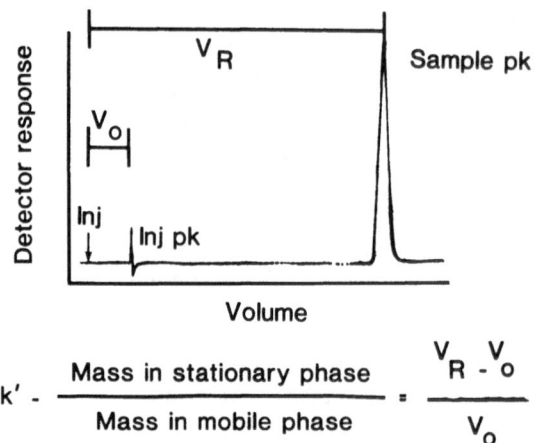

$$k' = \frac{\text{Mass in stationary phase}}{\text{Mass in mobile phase}} = \frac{V_R - V_o}{V_o}$$

Fig. 1. Calculation of a capacity factor (k').

composition) conditions a capacity factor (k') is an accurate measure of retention and the distribution of the solute between 2 phases. The term capacity factor is somewhat of a misnomer since k' does not indicate the total possible mass retained by a column but shows how the mass of a particular compound is distributed between 2 phases under a controlled set of conditions. The k' is thus more of an equilibrium constant and is calculated directly from chromatographic results (Fig. 1.).

RP–HPLC is an isolation method and as such the researcher is concerned with the retentions of more than one component. Each component of the mixture must be retained selectively by the column to achieve isolation. The selective retention of 2 compounds on a column is expressed as a selectivity factor which is the ratio of the k' for each compound. A successful separation is only possible when the selectivity factor does not equal 1.

$$\text{Selectivity factor} = \frac{k'_b}{k'_a} = \frac{V_{Rb} - V_o}{V_{Ra} - V_o}$$

However, a good selectivity factor does not guarantee a selective separation. The eluting components must come off the column within sufficiently small volumes so that the respective peaks do not overlap. A selectivity factor that differs from 1 means only that the points of maximum concentration are separated. Thus, overlaps can occur and selective separation must consider those factors which increase peak broadening and decrease column efficiency.

Several factors contribute to peak broadening in RP-HPLC systems. These factors are divided into three major categories: 1) eddy diffusion, 2) longitudinal diffusion and 3) mass transfer. The region of the HPLC where these problems influence peak broadening is between the injection valve and the collection port.

Eddy diffusion causes peak broadening to occur at several points in the chromatographic system. Any space that is unswept by solvent after sample injection or any channeling effect in the column will cause mixing or eddy diffusion. Typical locations for this problem occur in the system at tubing junctions, at fittings between the column to tubing, in the column bed and at the detector flow cell. Eddy diffusion can be minimized by evoking these precautions:

1) Be sure all tubing ends are polished, flat and fitted properly in the appropriate fitting.
2) Never connect a larger ID tubing downstream from a section of smaller ID tubing.
3) Use zero dead volume fittings where possible.
4) Detector design should be such that the flow cell volume is minimized, the path length maximized and all areas thoroughly swept by the solvent.

Additional eddy diffusion problems occur because of the column packing material. Anomalies in the packed bed result in channels with low flow resistance. These channels allow a fraction of the molecules that transverse the column to elute ahead of identical molecules which flow through the usual, more difficult paths. Because of particle alignment , the region of column packing adjacent to the column wall is less dense than in the center. This gives rise to the "wall effect" which allows molecules traveling next to the wall to elute before those which travel nearer the center of the column. The wall effect is easily controlled by using a column packed with a resin of small particle size (<10 u). If the size range is broad for the packing material particles, then the column will have dense areas of fine particles and diffuse areas of larger particles. This will result in an erratic elution pattern. Thus a narrow range of particle sizes is important to obtain a uniform column packing and elution pattern (Pharis, 1976).

Longitudinal diffusion results in band broadening due to random molecular motions similar to Brownian movement. This form of band broadening is a considerable problem in gas chromatography. However, since liquid diffusion coefficients are four orders of magnitude lower than gas diffusion coefficients, longitudinal diffusion usually has little influence on HPLC results. This factor may be ignored except when low flow rates are used (Pharis, 1976).

Mass transfer is necessary for any chromatographic procedure to separate individual compounds, and it is also a source of peak broadening. When a molecule is adsorbed on the stationary phase, other identical molecules in the mobile phase continue to move through the column and ahead of those that are adsorbed. If the mass transfer of molecules from one phase to another is slow compared the flow rate of the mobile phase, then significant band broadening will occur. The use of organic solvents of low viscosity (e.g. acetonitrile) will increase the rate of mass transfer and reduce band broadening. High solvent flow rates also increase band broadening. However, flow rates of 1-2 mls per min cause little peak broadening due to mass transfer in the standard 4.6 mm ID column (Pharis, 1976).

Another mass transfer problem develops when small pore sizes exist and the mobile phase or peptides fail to freely move through the support particles. Once in a particle, such molecules enter the equivalent of a second stagnant mobile phase and are retained as compared to those identical molecules outside the particle. The resulting dichotomy of the mobile phase results in peak broadening and loss of column efficiency. The new 300 angstrom columns eliminate this problem for most peptide separations by providing pores large enough for peptides and some proteins to move through unrestricted (Pearson et al, 1982).

A phase dichotomy can exist for the stationary phase as well. The most common example of this problem is the presence of residual silanol groups on reversed phase supports after reaction to attach the hydrocarbon to the silica (Regnier, 1980). The silanol groups compose a residual polar phase on the silica support. This polar phase interacts with compounds in an opposite manner than the reversed phase. Stronger solvents designed to remove compounds from the reversed phase intensify the polar interacts of the silanol groups. The result is peak broadening and the loss of sample on the column. An easy cure to this problem is to investigate the preparation of each column before purchase. Most columns made specifically for peptide and protein separations are advertised as having residual silanol groups "capped". Researchers who prepare and pack their own columns should use a monohalogenated organosilane for derivatization (Roumeliotis and Unger, 1978) followed by an end-capping reagent like trimethylchlorosilane (Evans et al, 1980).

Stationary phases are made nonhomogeneous by irreversible adsorption of sample components to the top of the column. Thus, peak broadening will occur with column use. These problems can be reduced by practicing proper sample preparation and using replaceable guard columns installed between the injector and the main reverse-phase column. Samples should undergo solid phase extraction, microfiltration or, at the least, centrifugation before injection. Guard columns contribute to the separation and efficiency observed in a chromatogram. When possible, these columns should contain the exact material used in the main column. Many microparticulate resins are now available as a replaceable cartridge at nominal cost.

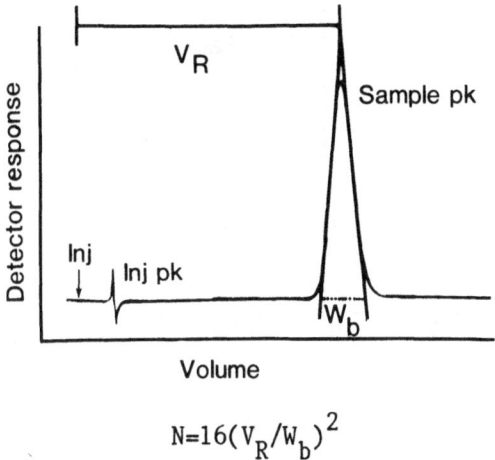

$$N=16(V_R/W_b)^2$$

Fig. 2. Calculation of HPLC system efficiency.
N = # of theoretical plates.

Since problems of inefficiency occur with time and can be introduced by many factors in a chromatographic system, it is important to routinely test the system's efficiency. System efficiency is tested by determining an elution volume and peak width for a standard sample under standard conditions. (Fig. 2.) Determine peak width by drawing an isosceles triangle using the peak slopes to construct tangential lines. Connect the lines at the base and measure the base width as shown. The value obtained from this equation is expressed as the "number of theoretical plates" (N). The term theoretical plates originally was coined to describe fractional distillation separations and is not directly applicable to chromatography. Nevertheless, the important fact to remember is the higher the number of plates the more efficient the system. The most common reason for plate loss is column damage. Examine all fittings, tubing and the detector for defects. If no defects are observed in these components then the alternative is to repack or replace the column.

The single most important property of the chromatographic system is its resolving power. The definition of peak width in the above discussion of efficiency will allow a mathematical expression of resolution to be obtained from a single chromatogram (Fig. 3.). The resolution equation uses the HPLC system's selectivity and peak widths to determine a resolution factor for a particular separation. A factor of 1.0 means that two peaks are completely separated with their bases just touching. A resolution factor above 1.0 means an even better separation (Pharis, 1978).

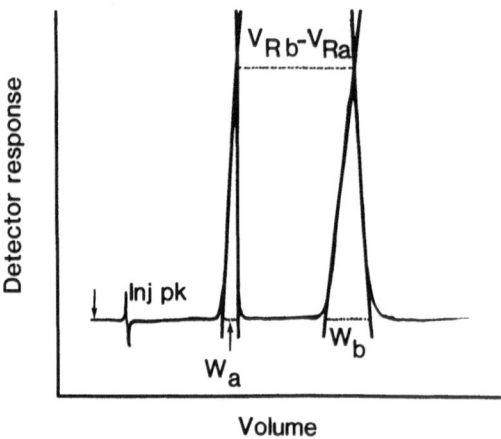

$$R = 2\left[\frac{V_{Ra} - V_{Rb}}{W_b + W_a}\right]$$

Fig. 3. Calculation of a resolution factor (R).

STATIONARY PHASE

The stationary phase used in RP-HPLC consists of a silica support, and a bonded hydrophobic layer. Totally porous microparticulate silica is the support most successfully used to prepare stationary phases. The best supports consist of a narrow size distribution of very small particles of 10 microns or less. Recent evidence has shown that microparticulate silica containing pores of around 300 angstroms is superior to the classical pore of only 60-100 angstrom for the separation of peptides containing more than about 30 amino acid residues (Pearson et al, 1982). Many insect neuropeptides currently under investigation are under this size and would logically chromatograph well on small and large pore RP modified silica. However, the investigator may not be completely certain that all peptide contaminants are smaller than this size limit. Therefore, use of the larger pore silica supports are still a logical choice for the isolation of small peptides under most conditions.

Useful bonded layers for peptide purification have largely included representatives of the alkyl class. Examples include most often the octadecyl (C18), followed by octyl (C8), and most recently butyl (C4) (Regnier and Gooding, 1980). Other bonded reversed phases include the phenyl and nitrile. The phenyl phase is reported to have additional selectivity for aromatic groups due to the ring-stacking phenomenon often observed in the interaction between 2 aromatic compounds (Supelco, 1982).

Recent advances in the bonding chemistry of reversed phases has favorably influenced the interaction of these stationary phases with peptides. The favorable results are primarily from a reduction of residual silanol groups in the final product. As discussed earlier these silanol groups act as a separate polar stationary phase and adversely influence the efficiency, resolution and recovery of RP peptide separations. Previously, most manufacturers of RP supports used trichloroalkylsilanes as reactants with silica to produce their products. For steric reasons only one and sometimes two chlorine moieties are capable of reacting with the silica. Thus, unreactive chlorosilanes were covalently attached to the support. When exposed to water these groups hydrolyze to form silanol groups. In addition, many of the original silanol groups were left unreacted on the silica. The problem of residual silanols on reversed phase supports is now resolved by replacing the trichloroalkylsilanes in the bonding reaction with monochlorodimethylalkylsilanes(Roumeliotis and Unger, 1978). Thus, the finished product is close to a pure reversed phase with no polar properties providing a homogeneous stationary phase for highly efficient peptide separations.

Care should be taken in the operation of RP-HPLC columns to prevent damage to the stationary phase. Operation of the column above the recommended pressure limit will crush the individual particles making smooth solvent flow difficult. Column operation should always be within the pH range where the silica support is stable (pH=2-7.5)(Hearn and Hancock, 1979). Since alkyl bonded silica with good surface coverage and end capping has little contact with the solvent, it is more resistant to pH damage than the naked support. However, pH limits should still be observed to insure long column life. When corrosive ion pairing reagents (discussed below) are used, the column and system should be flushed with pure water and organic solvent before overnight storage. Columns should be protected from damage by sample components by guard columns and proper sample preparation as discussed above.

We examined several commercially available RP-HPLC columns for their ability to separate small peptides. A standard peptide mixture was separated on each column in various isocratic mixtures of AcCN in 5 mM TFA. The standard sample mixture consisted of the vertebrate endogenous opioid peptides, met- and leu- enkephlins. These peptides are available commercially at a nominal cost and are good sample peptides to separate for the evaluation of different columns for peptide resolution or to detect damage to a single column after long term use. Solvent composition was altered so that the sample peptides eluted around 5 min after injection. The one exception was the Lichrosorb column which retained Leu-enkephlin for 22 min in 50% AcCN in 5 mM TFA.

Table 1. A comparison of several RP–HPLC columns for their
         efficiency and resolution of peptide separations.

Column	Particle size (microns)	Pore size (angstroms)	%AcCN in 5 mM TFA	HETP (mm)	R
Altex C18 ultrapore RPSC	5	300	20	0.036	4.43
Hi–Pore RP–318 C18	5	330	30	0.038	5.01
Vydac C4 214TP54	5	330	30	0.045	3.76
Micro- bondapak C18	10	55–105	30	0.057	4.81
Lichrosorb RP–8 C8	10	100	50	0.089	2.70

HETP=Height equivalent to a "theoretical plate".
HETP=L/N where L=column length and N=#of theoretical plates for
leu–enkephlin.
R=the resolution factor calculated for the separation of met– and
leu– enkephlins.

        All columns tested easily separated the test peptides (Table
1.). However, there was a considerable difference in the degree of
hydrophobic interaction between the peptides and some of the
individual columns.   For example, the Altex column (Beckman
Instruments, Fullerton, Calif.) only weakly bound the peptides,
and they eluted in 20% AcCN in 5 mM TFA. At the other extreme, a
much stronger solvent (50% AcCN in 5 mM TFA) was just able to
elute the peptides from the Lichrosorb column (Alltech Assc.,
Houston, Texas). The hydrophobic interaction of the other 3
columns was intermediate. The Altex column would appear to be
useful in the separation of very hydrophobic peptides and would
keep organic solvent consumption to a relative minimum.   The
Lichrosorb column appears useful for extremely hydrophilic peptide
mixtures which have little retention when chromatographed on the
other columns. Bio-Rad's Hi-Pore (Richmond, Calf.) gave the best
resolution but was followed closely by the Micro-bondapak (Water's
Assc., Milford, Mass.), Altex and Vydac (the Separations Group,

Hesparia, Calf.) columns. All are well suited for peptide separations.

MOBILE PHASE

Mobile phases for RP-HPLC consist of three or four components. These include water, a water miscible organic solvent, a buffer and sometimes an additional, ion-pairing reagent. Variation of these components will often change the retention, selectivity or efficiency of the chromatographic system. Solvents with a high composition of water allow for greater retention on the column. As the percentage of organic solvent is increased the solvent becomes "stronger" and elutes sample components. Buffers control the pH and contribute to the ionic strength of the solvent system. Ion-pairing reagents allow for selective elution based on charge and hydrophobicity (Regnier & Gooding, 1980).

Water purity is critical to the success of HPLC separations, particularly where solvent gradients are used to separate compounds of diverse polarity. Failure to use high purity water will result in elution of extra peaks which make analysis of results difficult or impossible (Sampson, 1977). Our HPLC water is first prepared by glass distillation. Trace organics are removed with a Norganic water purification apparatus (Millipore Corp., Bedford, Mass). The Norganic system also microfilters and degases the water.

Organic solvents increase the hydrophobicity of the mobile phase which allows the elution of adsorbed sample components. The most common organic solvents used in RP-HPLC are methanol, 1- or 2-propanol and acetonitrile (AcCN). AcCN is transparent in the UV spectrum to 190 nm and is the least viscous of the solvents listed above. AcCN has a good history for producing highly efficient peptide separations . Rubinstein (1979) has found that 1-propanol is useful for chromatography of larger peptides and proteins because it is the more hydrophobic of the water-miscible solvents used in RP-HPLC. Thus, less severe changes in concentration are required to elute larger proteins than with AcCN or MeOH. These mild solvent changes apparently limit peptide denaturation. Propanol is also less toxic and volatile than AcCN, but it has a higher viscosity and results in 2- to 3- fold higher operating pressures. Variation in the type of organic solvent usually plays a minor role in system selectivity for most separations (O'Hare & Nice, 1979).

On the other hand, pH and ionic strength play a large role in the selective separation of peptides on RP-HPLC systems. The retention of tyrosine-containing peptides increases with pH to a maximum of retention at around pH 8 (Krummen & Frei, 1977). Low pH causes acidic peptides to increase their retention on RP-HPLC. The increased retention of acid peptides at low pH is thought to be the result of the protonation of the carboxyl side chains and termini which increases hydrophobicity and interaction with the stationary phase (Hancock et al, 1978). At pH 7, increasing the ionic strength increases peptide retention at differing rates thus providing a selective route for changing HPLC conditions and achieving peptide separation (Krummen & Frei, 1977).

Retention times of ionic peptides can be selectively altered by the addition to the mobile phase of counter ions which appreciably alter peptide polarity (Hearn & Hancock, 1979). For example, hydrophilic ion–pairing agents like sodium phosphate can be added to increase the polarity of peptides thus reducing RP–HPLC retention. Conversely, hydrophobic ion–pairing agents like hexanesulfonate or dodecylsulfonate will decrease the polarity of selected peptides and increase retention times. Acidic conditions favor ion pair formation by amino groups and the ion suppression of carboxylic acid moities on peptides. The pH operational limits of RP–HPLC (2–7.5) favor the use of ion–pair reagents directed toward interaction with the basic amino acid residues. The hydrophilic and hydrophobic ion pair examples listed above all interact with the positive charges of arginine, lysine and N-terminal residues of a particular peptide. Since most peptides have different numbers and conformational arrangements of these positive charges, different degrees of influence can be expected on retention times for different peptides.

The use of perfluorinated carboxylic acids as ion pair reagents in the RP–HPLC of peptides deserves special consideration (Bennett et al, 1980). These reagents include trifluoroacetic acid (TFA), pentafluoropropanoic acid (PFPA), heptafluorobutyric acid (HFBA) and undecafluorocaproic acid (UFCA). The more hydrophobic acids (ie. HFBA and UFCA) increase retention times for all peptides over the more hydrophilic acids (ie. TFA and PFPA). The phenomenon of increased retention time was most notable for peptides with a large proportion of basic amino acid residues. In sample purifications, Bennett and coworkers (1980) showed retention order inversions and complete separation of peptides in HFBA solvent systems as opposed to TFA systems. Starrett and Stevens (1980) have separated diastereomers and other analogs of protolin using these same solvent systems. Thus, it is conceivable that a unique chromatographic species can be generated so that the isolation of any peptide is possible from a complex mixture. The perfluorinated carboxylic acids have the additional advantages for RP–HPLC of being completely volatile and usable at 210 nm in the UV spectrum.

Rivier and coworkers(1982) have shown the utility of triethylammonium phosphate (TEAP) as another useful ion pairing reagent for the isolation of peptides by RP–HPLC. Corticotropin releasing factor was completely lost when the first RP–HPLC isolation attempts were made using an ammonium acetate/acetonitrile solvent system. When TEAP was substituted for the ammonium acetate, the peptide was recovered in good yields. The authors believe the TEAP increases peptide yields by inactivating excess silanol groups left on RP–HPLC supports even after reaction with monochloroalkylsilanes and extensive end capping. TEAP solvent systems result in drastically different peptide elution patterns for peptide mixtures than TFA solvent systems.

HANDLING AND STORAGE OF PEPTIDES

Degradation and loss of small quantities of biological peptides is often a severe problem during isolation procedures thus it is important to discuss a few general observations which are helpful in preventing peptide loss before a discussion of a general approach to HPLC peptide purification. Adsorption of peptides from dilute solutions and formation of chemical artifacts are the two major reasons for peptide loss (Stein, 1976). Loss of peptides due to nonspecfic adsorption can be limited by using several precautions. Peptides should not be handled in overly dilute solutions. The researcher may consider intentionally contaminating the sample peptide with a defined standard peptide when very small amounts of purified sample are obtained. The contaminating standard should be easily separated from the sample. The contaminant will tend to "fill" nonspecific binding sites on tube and pipette surfaces leaving more of the active peptide in solution. Glass tubes and pipettes should be silanized. However, it is often better to use polypropylene or polyethylene tubes and pipettes. Polystyrene equipment should be avoided because it strongly adsorbs some peptides.

Perhaps the most common cause of peptide loss in early stages of a purification results from artifact formation. Artifacts are a chemical change in the peptide structure (Samanen & Chan, 1981). General causes of artifact formation include: 1) proteolysis 2) oxidation 3) exposure to conditions favorable to chemical hydrolysis, esterification, transpeptidation or disulfide exchange. The rate of artifact formation is minimized by keeping solutions at ice-cold temperatures.

Proteolytic artifacts (Stein, 1977) are a major problem in the initial extraction of peptides from neural tissues. Besides keeping samples cold, proteolysis is reduced by making the solution acidic. Small peptides are often extracted from larger proteases by using a high concentration of organic solvents (ie. methanol, acetonitrile or acetone) in the initial extractant. Excess protein (ie. BSA) can be added to the sample during extraction so that endogenous proteases have more protein substrate to digest and less time to degrade the sample peptide. Harsher treatments may selectively destroy proteolytic activity and leave the sample peptide intact. An example is heat denaturation of proteases. This treatment is not practical if the sample peptide is also heat labile; furthermore, some of the proteolytic enzymes may renature after cooling. When all else fails , specific proteolytic inhibitors should be included in the initial extraction.

Oxidation is another major cause of peptide degradation during purification particularly for peptides containing cystine,

methionine, histidine, tryptophan and tyrosine. Oxidative
reactions are controled by adding an antioxidant like thiodiglycol
(TDG) or butylated hydroxytoluene to the initial extractant and to
subsequent separated fractions. Storage under an inert atmosphere
such as nitrogen or argon is also helpful. Organic solvents (ie.
ethylether) which build up peroxides should be purified and tested
for these highly oxidative agents. Vigorous mixing of peptide
solutions should also be avoided (Samanen & Chan, 1981).

Other chemical reactions causing peptide degradation include
hydrolysis of amide side chains, esterification of carboxyl
groups, carboxyl exchange and disulfide exchange(Samanen & Chan,
1981). Extracting with solvents containing mineral acids promotes
partial deamination of the amide side chains of asparagine and
glutamine during concentration by evaporation or lyophilization.
Acids should be neutralized and the solution desalted prior to
concentration by evaporative methods. Transamination of aspartic
and glutamic acids, deamination, oxidation, or exchange of half
cystines between disulfide bonds may all occur in high pH
conditions. This too is limited by a return to neutral or slightly
acidic conditions.

Observation of these storage criteria will help reduce
peptide degradation.

1. DAILY USE STORAGE – Store peptide chilled in a slightly
   low pH solution (0.1% acetic acid) when possible.
2. OVERNIGHT STORAGE – Store solution under nitrogen in the
   refrigerator.
3. SHORT TERM STORAGE (a few days) – Degas solution and store
   under nitrogen in the refrigerator.
4. LONG TERM STORAGE (more than a week) – Freeze–dry and
   store under nitrogen or argon in a desiccator at -20°C or
   below (Samanen & Chan, 1981).

A vacuum centrifuge is most convenient to remove excess
solvent and concentrate sample peptide solutions. The centrifugal
force applied to the sample during this procedure prevents
bubbling and foaming in the vacuum. As a result the sample is
concentrated in the bottom of the tube so that recoveries are
increased and cross contamination decreased.

EXAMPLE SEPARATION

We have worked out a simple, yet flexible, plan for isolating
insect neuropeptides with reversed phase technology. The plan
consists of the following general steps:

1. The development of an optimal liquid extraction of the
   peptide from the best tissue source.

2. Analysis of the extract by reverse phase thin layer
   chromatography (RP–TLC) for comparison of the
   hydrophobicity of the sample peptide with a set of
   standard peptides.

3. Preparation of a crude peptide extract for HPLC by RP-solid phase extraction.*

4. Purification with different modes of RP-HPLC.*

*Conditions for these steps are developed with reference peptides instead of the sample peptide.

EXAMPLE PEPTIDE

The cytochromogenic hormone (CGH) from the corpora cardiaca (CC) of Blaberus discoidalis will be used as an example peptide for this discussion. CGH is involved in early adult development and the basal metabolic energy content of the primary biosynthetic tissue of the insect, the fat body. CGH specifically stimulates the production of mitochondrial cytochromes aa_3+b, which are 2 essential links in the electron transport chain (Keeley, 1978).

A simple in vivo assay is available to monitor the presence and relative quantity of CGH. The assay follows the hormones ability to stimulate fat body mitochondrial cytochrome synthesis by monitoring the incorporation of ^{14}C-aminolevulenic acid, a specific cytoheme precursor, into the cytohemes of cytochromes aa_3+b (Hayes & Keeley, 1981). The assay is quantitative between 0.01 and 0.08 CC equivalent injected.   The bioassay has been used to collect most of the preliminary information discussed above.   For example, CGH is inactivated by chymotrypsin, aminopeptidase M and carboxypeptidase Y. Thus, demonstrating the peptidic nature of the hormone and suggesting that it contains aromatic amino acid residue(s) and unblocked amino and carboxyl terminal ends. Sephadex G-25 chromatography suggests a CGH molecular weight of 1,400 daltons or less (Hayes & Keely, unpublished) . CGH is heat sensitive but can be stored at -20°C with an antioxidant like thiodiglycol for at least 5 weeks. CGH activity is recovered after lyophilization (Hayes & Keeley, 1981). CGH activity is not soluble in acidic aqueous solutions but is soluble in aqueous neutral and basic solutions. The hormone is also soluble in acidic organic solvents to as low as 20% acetonitrile (AcCN).

EXTRACTION OF THE PEPTIDE FROM THE TISSUE SOURCE

We developed a method for extracting tissue of B. discoidalis for CGH activity using information obtained from CGH solubility studies. Recall CGH was soluble in acidic organic solvents. CGH is localized in the corpora cardiaca (CC), thus, when CC or whole heads are homogenized in an ice-cold solvent system consisting of 90% spectral-analyzed acetone + 10% 0.1 N HCl + 0.01% thiodiglycol (TDG), CGH is completely solubilized. Repeated extractions with acid acetone failed to extract additional activity from the tissue and indicated that the first extraction was essentially 100 percent complete. The high concentration of acetone in the solvent

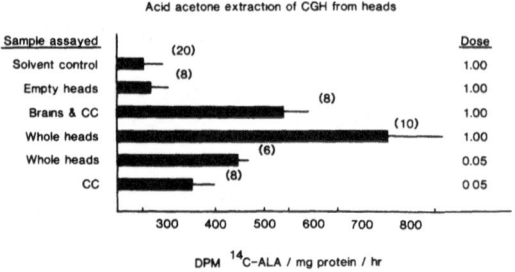

Fig. 4. The extraction of CGH from heads of adult male <u>B. discoidalis</u>. All samples were extracted in 90% spectral grade acetone: 10% 0.1 N HCl + 0.01% thiodiglycol. Acid was neutralized with NH$_4$OH. Acetone was removed and the remaining aqueous solution lyophilized in a vacuum centrifuge. The residue was resuspended in Ephrussi-Beadle Ringer for analysis by the bioassay. The large horizontal bars represent the total synthesis of cytohemes a+b + S.E.(thin bars) with the number of replicates in parentheses. Bioassay values of 350 dpm/mg mitchondrial protein/hr or less are a base-level response (< 0.01 CC pr, total dose). Values in excess of 500 dpm/mg mito-protein/hr are a maximal response (> 0.08 CC pr, total dose.

system precipitates most large proteins including proteases which could degrade the peptide we are attempting to isolate. The acid allows for the solubility of CGH in the acetone. TDG serves as an antioxidant.

The CGH-acid-acetone extract is treated in several different ways depending on the next use for the hormone. Samples destined for TLC analysis are applied directly to the thin layer plates in the acid acetone solution and dried under N$_2$. For those samples destined for bioassay or further purification, acetone is removed either under a stream of N$_2$ or in a in a vacuum centrifuge. Samples to be bioassayed are neutralized with a slight excess of NH$_4$OH, brought to near dryness and resuspended in Ephrussie-Beadle Ringer. The neutralization is necessary to insure CGH solubility in the increasingly polar solvent as the acetone is removed. Other samples are prepared for solid phase extraction by diluting with 0.1 N sodium phosphate (pH=7.4) after removal of acetone.

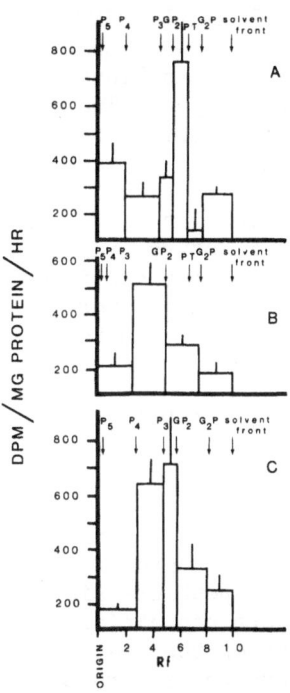

Fig. 5. Migration of CGH relative to reference peptides of differing hydrophobicities using three solvent systems and C18-reversed-phase, thin-layer chromatography on Whatman microslides. Reference peptide migration is indicated by the arrow position. Reference peptides identified by the following amino acid abreviations: phenylalanine (P), glycine (G) and tyrosine (T). Solvent systems: (A) 20 mM diaminoethane-TFA (pH 7.4):acetonitrile (6:4). (B) 0.1 N sodium perchlorate-5 mM sodium phosphate (pH 7.4):acetonitrile (7:3). (C) 5 mM TFA: acetonitrile (65:35). Samples were separated on the basis of the migration of the reference peptides, scraped from the TLC and extracted with acid acetone. Process of the acid acetone extracts and the bioassays were done as described in Fig. 4 except the total dose of CC was 2.

Control experiments confirmed that the extraction procedure isolated a unique factor from the brain+CC complex (Fig. 4.). No cytochromogenic response was found for extracts of empty heads from which the brain+CC complex has been removed or by the residue from dried solvents redissolved in Ringer solution. Whole head extracts gave a dose-dependent CGH response yet this was larger than the response of CC alone suggesting that more activity was recovered from whole heads.

ANALYTICAL REVERSE PHASE THIN LAYER CHROMATOGRAPHY

Progress in HPLC purification of specific insect peptides is hampered by the time required to bioassay the many fractions recovered from the chromatography. The CGH bioassay requires a minimum of 4 working days to obtain a result. Analytical RP-TLC has been used by our laboratory to overcome much of this handicap during the stage of methods development.

For example, CGH purification by RP-HPLC was first optimized by RP-TLC to determine the relative hydrophobicity of CGH or its ion pair complexes. First, solvent conditions are found to evenly separate a set of standard peptides increasing in hydrophobicity. The specific standards used in our studies are:

Glycylglycylphenylalanine (G_2P)
Phenylalanyltyrosine (PT)
Glycylphenylalanylphenylalanine (GP_2)
Triphenylalanine (P_3)
Tetraphenylalanine (P_4)
Pentaphenylalanine (P_5)

Then an extract of CGH is applied to an identical TLC plate and dried under nitrogen. The chromatogram is developed in the solvent system with a set of standard peptides simultaneously developed on an adjacent plate. The plate containing the standard peptides is visualized and the migration of the standard peptides are used as division points for fractions to be scraped from the CGH-TLC. Each fraction then represents an area of the TLC bounded by two standard peptides. Each fraction is extracted with the acid acetone solvent used above. and the acid acetone is neutralized and the resulting extract is dried in a vacuum centrifuge. The samples are resuspended in Ringer and tested with the bioassay.

Representative analytical TLC results are depicted in Fig. 5. These systems include 20 mM diaminoethane–trifluoroacetic acid (TFA) at pH 7.4: acetonitrile (6:4) (Fig. 5a); a denaturing system consisting of 0.1 N perchlorate– 5 mM sodium phosphate:acetonitrile at pH 7.4 (7:3) (Fig. 5b); and an ion-suppression system consisting of 5 mM TFA:acetonitrile (65:35) (Fig. 5c). The mobility of the CGH activity changes with the solvent systems. The first chromatogram shows CGH migrating between PT and GP_2 in a solvent system with a pH of 7.4 (Fig 5a). When a denaturant like perchlorate is added to the solvent, CGH migrates between GP_2 and phenylalanylphenylalanylphenylalanine (P_3) in a more hydrophobic position (Fig 5b). A larger increase in the hydrophobic interaction is fostered by the ion suppression system which causes some of the CGH activity to migrate behind P_3 (Fig 5c). The hydrophobic standards which bracket CGH in these test chromatograms predicts the elution of CGH for use in similar mobile-phase systems, for solid phase extraction and for HPLC purification. The following two criteria must be met to encourage success in predicting sample peptide elution with the hydrophobic

standards:
1. Additional RP-methods must use an aliphatic bonded solid phase similar to those used for the TLC.
2. The solvent must contain the same buffer, acid or ion-pair reagent concentrations and content as used for the TLC.

When the above conditions are met, hydrophobic standards which bracket the sample peptide may be used to predict mobile phase conditions favorable for the sample peptide elution (Von Arx E. & Faupel M., 1978). Thus, near optimal elution conditions for the sample peptide may be worked out using the bracketing hydrophobic reference peptides and spectrophotometric detection. Once elution conditions are defined by the reference peptides, neuropeptide extracts can be separated on the RP-system and the general position of the neuropeptide will be known. Thus, only limited fractions will require the bioassaying to locate the neuropeptide.

SOLID PHASE EXTRACTION

Reversed-phase, solid phase extraction is a low pressure-batch version of the more efficient RP-HPLC and offers several advantages over more traditional sample preparation methods such as gel filtration. For example, a neuropeptide can be extracted and concentrated from an aqueous solution by simply passing it through a low pressure cartridge (ie. a Water's Seppak) containing the same ODS-silica found in the more advanced HPLC columns (Brubaker P., et al, 1980). The cartridge can be washed to remove polar impurities and the neuropeptide can be eluted with a few mls of a stronger solvent (ie. more organic modifier). Strongly adsorbing hydrophobic impurities are left behind on the cartridge and thus do not damage expensive HPLC columns used in future purification steps. Higher peptide recovery should be expected because samples are not unduely diluted and sample handling is kept to a minimum.

The information obtained by the analytical RP-TLC (above) of CGH and the reference peptides was useful in determining optimal conditions for the solid phase extraction. The following steps were used to calibrate a typical Water's Seppak for a solid phase extraction of CGH.

1. Cure the Seppak with 10 mls of AcCN followed by 20 mls of HPLC grade water.
2. Load a separate cured Seppak with one standard peptide (50 ug) in 5 mM TFA.
3. Wash the Seppak with a 5% step gradient (5 mls/step) (20:80) to (50:50), AcCN: 5 mM TFA.
4. Check each step of the effluent for peptide elution with trinitrobenzenesulfonic acid as described by Habeeb(1966).

Table 2. Elution of reference peptides from the Seppak
in various concentrations of acetonitrile in 5 mM TFA

Peptide	20	25	30	35	40	45
	\multicolumn{6}{c}{% AcCN in 5 mM TFA}

Peptide	% AcCN in 5 mM TFA					
	20	25	30	35	40	45
GP_2	–	+	–	–	–	–
P_3	–	–	+	–	–	–
P_4	–	–	–	+	–	–

Elution of peptides located by orange color reaction
with trinitrobenzenesulphonic acid (Habeeb, 1966).
"+" = positive color reaction.
"–" = no color observed.

The results of this calibration (Table 2.) show that GP_2, the
standard peptide which precedes CGH activity in the 5mM TFA
solvent system, elutes with 25% AcCN. The standard which first
elutes behind all CGH activity elutes with 35% AcCN. Thus, CGH is
predicted to elute from a Seppak in a TFA solvent system between
20 and 35% AcCN.
       The TFA system was used because it has been reported to
increase peptide recovery and TFA is easily removed under vacuum.
This last advantage makes a solid phase extraction of CGH in TFA
an idea desalting method. In addition, TFA is nearly transparent
at 210 nm which allows the UV monitoring for peptide bonds (Dunlap
et al, 1978).

Elution of CGH from seppak in an acetonitrile in
5 mM trifluoroacetic acid solvent system

Solvent	Stimulation of cytochrome synthesis (DPM ^{14}C-ALA/mg protein/hour)
Water	141
20% AcCN in 5mM TFA	0
35% AcCN in 5mM TFA	638
80% AcCN in 5mM TFA	12

The utility of this solid phase extraction for CGH purification was confirmed by using acid acetone extracts of B. discoidalis (Table 3.). Acetone was driven off under a stream of prepurified nitrogen so that CGH would stick to the Seppak. Since, CGH solubility is adversely affected by acidic aqueous solutions, the pH of the solvent was brought to 7.4 with 0.1 N sodium phosphate (pH 7.6) and mixed gently for a few minutes. This is the method for solid phase extraction of CGH:

1. Cure Seppak with 10 mls AcCN followed by 20 mls of HPLC water.
2. The head extract is loaded and concentrated on the Seppak in the pH 7.4 phosphate solution. Each Seppak holds 50-100 head equivalents of the crude CGH extract.
3. Wash Seppak with 10 mls HPLC grade water to remove the phosphate and other salts.
4. Wash Seppak with 10 mls AcCN:5 mM TFA (2:8) to remove many polar impurities.
5. Elute CGH with 5 mls AcCN:5 mM TFA (4:6) leaving many hydrophobic impurities on the Seppak.
6. Neutralize with NH_4OH, protect with TDG, vacuum centrifuge and resuspend in Ringer for bioassay.

CGH elutes from the Seppak (Table 3.) in the solvent predicted by the standard peptide calibration. A bioassay titre shows no detectable loss of activity and a purification of 2,772 fold over the amount of hormone and material contained in the original heads. A purification without hormone loss is highly unlikely and the apparent recovery is probably the result of other factors. These other factors could include the purification of CGH away from other head components which may interfere with the bioassay.

Since, we detected no protein impurities eluting in the 20% AcCN in 5mM TFA, this step was omitted from the' extraction procedure for samples collected for HPLC purification.

PEPTIDE SEPARATION BY HPLC

Seppak-purified CGH was further separated from cockroach head impurities by RP-HPLC. Purification was attempted first using a Bio-Rad Hi-Pore C18 column designed especially for peptide and protein isolations. The general composition of the mobile phase consisted of various mixtures of 5 mM TFA (solvent A) and acetonitrile (solvent B).

Following our original plan, we first characterized this general RP-HPLC system using the reference hydrophobic peptides (GP_2, P_3, and P_4) associated with CGH retention. The reference peptides were chromatographed in the system using a step gradient of 10-50% B in A with steps of 10% for 5 min duration. The first reference peptide, GP_2, elutes in 30% B in A. CGH must follow GP_2 if our approach to developing a solvent system made the correct

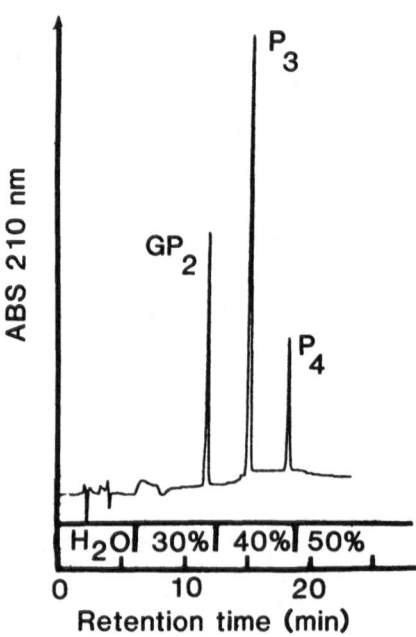

Fig. 6.   The separation of 3 CGH associated hydrophobic reference
peptides by RP–HPLC. The 3 peptides are glycylphenylalanylphenylalanine (GP$_2$), triphenylalanine (P$_3$) and tetraphenylalanine
(P$_4$). The downward hash marks of the upper horizontal axis
represent the time when step gradient changes first entered the
detector flow cell. The steps of the gradient are: 1) HPLC grade
water, 2)30% AcCN (B) in 5 mM TFA (A), 3) 40% B in A and 4) 50% B
in A. The bottom horizontal axis is the scale for retention time.
The flow rate was constant at 1 ml/min at 25°C.

assumptions. The special solubility properties of CGH must be
considered when deciding on correct HPLC conditions. CGH is not
soluble in an all–aqueous acidic medium. However, CGH is soluble
in as little as 20% B in A. Thus, CGH cannot be injected in 10% B
in A for fear that it may precipitate in the HPLC. Fig. 6. then
shows the standard peptide separation when injected in HPLC water
followed by a step gradient of 30%–50% B in A.

Seppaked–purified CGH residues are resuspended in 0.1 N
sodium phosphate (pH 7.4) and separated on the HPLC in the solvent
system described above. CGH extracts are injected on the HPLC

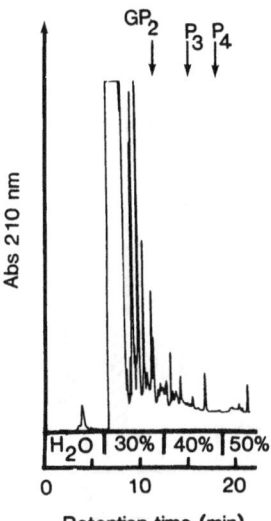

Fig. 7. The separation of CGH containing extracts of male B. discoidalis head extracts on RP-HPLC. Injection contained 1 head equivalent of the extract in 0.1 N sodium phosphate (pH 7.4). See the legend of Fig. 6. for futher details.

which is equilibrated with water (Fig. 7.) Many polar impurities elute ahead of the predicted elution position for CGH. When the separation is scaled up to accommodate 10–50 head equivalents, these impurities will elute in broad peaks, which will overlap with the area of predicted CGH elution. Thus, the solvent system should be adjusted to push these impurities off the column farther ahead of the predicted CGH elution time. This is accomplished by flushing with a mixture of 20% B in A directly after the pure water that was used as the mobile phase during the injection (Fig. 8.). The bioassay detects that CGH eluted between 16 and 17 min after the injection. This is the projected elution position for CGH as predicted by the reference peptides.

    CGH purification continues in our laboratory using different modes of RP-HPLC. The separation discussed above uses TFA to lower the pH of the solvent making it an ion–suppression system. Such a system promotes additional RP-retention for peptides based on the number and arrangement of carboxylic acids in the peptide structure. The separation currently underway in our laboratory is using a non-ion-suppresive system with diaminoethane trifluoroacetate at pH 7.4 as the solvent system. We hope this system will allow CGH to run ahead of many of the impurities left from the previous HPLC separation. Additional purification steps

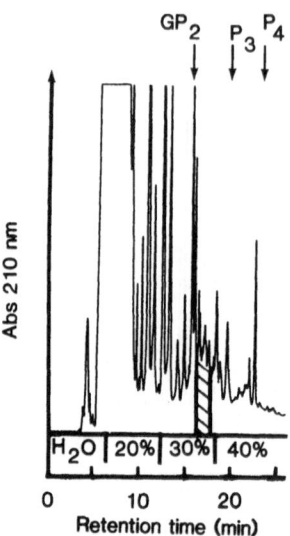

Fig. 8. The separation of CGH containing extracts of male B. discoidalis head extracts on RP-HPLC. Injection contained 10 head equivalents of the extract in 0.1 N sodium phosphate (pH 7.4). The step gradient differs from Fig. 7. by the inclusion of a 20% AcCN in 5 mM TFA step. The diagonal striped bar represents the elution position for CGH as detected by the bioassay. For additional information see Fig. 6.

will include hydrophobic ion-pairing with heptafluorobutyric acid to allow the basic properties of the peptides to contribute to the HPLC selectivity.

PREPARATIVE SCALE UP

A separation to isolate sufficient amounts of an insect neuropeptide for structural characterization will likely require the processing of gram quantities of tissue extracts. The standard 25 x 0.46 cm HPLC column can only handle mg amounts of a peptide mixture (Gesellchen, 1980). Thus, a scale-up of the solid phase extraction and prehaps the first HPLC separation should be anticipated when isolating neuropeptides from a very large number of insects.

An efficient and economical avenue for such a scale up operation has been shown by Gesellchen(1980) and coworkers at Eli Lilly and Company. These workers prepared their own RP-Silica packing by reacting Whatman 10-20 u LP-1 silica gel with octyldecyltrichlorosilane. Residual silanol groups were capped with trimethylchlorosilane. Recent studies suggest that replacement of the octyldecylchlorosilane with noctyldimethylchorosilane (Petrarch Systems, Levittown, Pa.) will result in a better product (Roumeliotis & Unger, 1978). Gesellchen (1979) and coworkers were able to isolate gram quantities of peptide on such columns with similar retention orders as are seen on comparable analytical HPLC columns. However, such separation took on the average 6 to 7 times longer to accomplish. Another economical alternative for a preparative RP-scale-up is by the use of E. M. Laboratories glass LOBAR RP-8 columns. Furthermore, several liquid chromatography companies offer semi-preparative HPLC columns made with the same chromatography medium found in their analytical columns, although, these are generally expensive.

CRITERIA FOR PURITY

Confirmation for homogeneity must precede definitive structural analysis of an isolated neuropeptide. The presence of a single UV peak for the neuropeptide in conjunction with biological activity following several modes of HPLC analysis is a preliminary qualitative measure of purity and will suffice if quantities of the peptide are severely restricted. The quantitative bioassay response should not change relative to the estimated protein content from one HPLC analysis to another. When sufficient peptide is available, quantitative amino acid analyses should be performed after each analytical HPLC. The molar ratio of constituent amino acids for the peptide should remain consistent and integral after each HPLC analysis. Finally, if the peptide is unblocked, end group analysis should show the same N- and C- terminal amino acid residues for each sample run.

ACKNOWLEDGEMENTS

The authors thank Mrs. Melanie Martin and Ms. Sheila Sowa for their excellent technical assistance with the research involved with this manuscript. These studies were funded by the following grants from NIH (TMP AI 15190), NSF (PCM 8103277) and the PHS (S07 RR07090-17) in addition to the Texas Agriculture Experiment Station.

LITERATURE CITED

Bennett H.P.J., Browne C.A. and Solomon S. (1980) The use of perfluorinated carboxylic acids in the reversed-phase HPLC of peptides. J of Liq. Chromatog. 3, 1353-1365.

Bennett H.P.J., Browne C.A. and Solomon S. (1983) N-acetyl-endorphin from the neurointermediary lobe of the rat pituitary: isolation, purification and characterization by HPLC. Anal. Biochem. 128, 121-129.

Brubaker P., Bennett H., Baird A. and Solomon S. (1980) Isolation of ACTH$_{1-39}$, ACTH$_{1-38}$ and CLIP from the calf anterior pituitary. Biomed. and Biophys. Res. Com. 96, 1441-1448.

Dunlap C.E., Gentleman S. and Lowney L.I. (1978) Use of trifluoroacetic acid in the separation of opiates and opiod peptides by RP-HPLC. J. of Chromatog. 160, 191-198.

Evans M.B., Dale A.D. and Little C.J. (1980) The preparation and evaluation of superior bonded phases for reversed-phase HPLC. Chromatographia 13, 5-10.

Habeeb A.F.S.A. (1966) Determination of free amino groups in proteins by trinitrobenzenesulfonic acid. Anal. Biochem. 14, 328-336.

Hancock W.S., Bishop C.A., Prestidge L., Harding D.R.K. and Hearn M.T.W. (1978) J. Chromatog. 153, 391-398.

Hayes T.K. and Keeley L.L. (1981) Cytochromogenic factor: a newly discovered neuroendocrine agent stimulating mitochondrial cytochrome synthesis in the insect fat body. Gen. Comp. Endocrinol. 45, 115-124.

Hearn M.T.W. and Hancock W.S. (1979) Ion pair partition reversed phase HPLC. Trends in Bio. Sci. 4, N58-N61.

Hearn T.W., Regnier F.E. and Wehr C.T. (1982) HPLC of peptides and proteins. Amer. Lab. 14, 18-39.

Howard G.A. and Martin A.J.P. (1950) The separation of C12-C18 fatty acids by reversed phase partition chromatography. Biochem. J. 46, 534-544.

Isobe T., Takayasu T., TakaI N. and Okuyama T. (1982) HPLC of peptides on a macroreticular cation-exchange resin: application to peptide mapping of Bence-Jones proteins. Anal. Biochem. 122, 417-425.

Keeley L.L. (1978) Development and endocrine regulation of mitochondrial cytochrome biosynthesis in the insect fat body. ^{14}C-aminolevulinic acid incorporation. Arch. Biochem. Biophys. 187, 87-95.

Krummen K. and Frei R.W. (1977) The separation of nonapeptides by reversed-phase HPLC. J. Chromatog. 132 27-36.

O'Hare M.J. and Nice E.C. (1979) Hydrophobic HPLC of hormonal polypeptides and proteins on alkylsilane-bonded silica. J. of Chromatog. 171, 209-226.

Pearson J.D., Lin N.T. and Regnier F.E. (1982) The importance of silica type for reverse-phase protein separations. Anal. Biochem. 124, 217-230.

Pharis N.A. (1976) Instrumental Liquid Chromatography. Elsevier, Amesterdam.

Regnier F.E. and Gooding K.M. (1980) HPLC of proteins. Anal. Biochem. 103, 1-25.

Rivier J.E. (1980) Evaluation of triethyl ammonium phosphate and formate/acetonitrile mixtures as eluents for high performance gel permeation chromatography. J. Chromatog. 202, 211-222.

Rivier J., Rivier C., Spiess J. and Vale W. (1982) HPLC purification of peptide hormones: Ovine hypothalamic amunine (CRF). Anal. Biochem. 127, 258-266.

Roumeliotis P. and Unger K.K. (1978) Structure and properties of n-alkyldimthylsilyl bonded silica for reversed phase packings. J. Chromatog. 149, 211-224.

Rubinstein M. (1979) Preparative high-performance liquid partition chromatography of proteins. Anal. Biochem. 98, 1-7.

Samanen J. and Chan D. (1981) Peptide pointers. Beckman Peptide News 2, 1.

Sampson R.L. (1977) High purity water for liquid chromatography. Amer. Lab. 9, 40-44.

Smith J.A. and McWilliams R.A. (1982) HPLC of peptides. Amer. Lab. 12, 23-30.

Starratt A.N. and Stevens M.E. (1980) Ion-pair HPLC of the insect neuropeptide proctolin and some analogs. J. of Chromatog. 194, 421-423.

Stien S. (1977) Application of fluorescent techniques to the study of peptides. in Peptides in Neurobiology (H. Gainor, ed.), Plenum Press, New York 9-38.

Supelco (1982) Separating proteins by reversed phase HPLC. The Supelco Reporter 1, 8-9.

Takahashi N., Isobe T., Kasai H., Seta K. and Okuyam T. (1981) An analytical and preparative method for peptide separation by HPLC on a macroreticular anion-exchange resin. Anal. Biochem. 115, 181-187.

Von Arx E. and Faupel M. (1978) Ubertragung von stark wasserhaltigen reversed-phase-hochdruckflussigkeits chromato-graphischen systemen auf die dunnschichtchromatographie und bioautographie von reversed-phase dunnschichtchromatogrammen. J. of Chromatography <u>154</u>, 68–72.

MICROSEQUENCING STRATEGIES FOR INSECT NEUROPEPTIDES

Renée M. Wagner and G. Mark Holman

VTERL, ARS, USDA
P.O. Drawer GE
College Station, Tx 77841

INTRODUCTION

The process of sequencing any protein or peptide consists of a minimum of four stages: purification of the native peptide, cleavage of the peptide into manageable fragments, methodical separation of these fragments, and subsequent sequencing of the purified fragments. Finally, the sequence information must be catalogued in such a way as to confirm the primary structure of the naturally-occurring peptide in question. Historically this process has been accomplished by ion-exchange or gel filtration column chromatography purification and chemical or enzymatic cleavage, followed by amino acid analysis and mapping of the resulting fragments.

Entomologists, however, are faced with additional limitations in this sequencing scheme. While the quantities of peptides needed for biological activity in insect systems are minute, the amounts available in a single insect are equally miniscule. For this reason, classical methodologies of peptide purification and sequencing are inadequate for elucidation of structures of nanomolar amounts of biological materials.

Recent advances in high-performance liquid chromatography have led to rapid isolation of insect neuropeptides (Hayes, 1983), while refinements in instrumentation have resulted in automated gas-phase systems capable of sequencing 0.1 picomole of purified peptide (Hewick et al., 1981). Nevertheless, practical modifications in N-terminal analysis and C-terminal degradation, together with the commercial availability of purified peptidases and advances in mass spectrometric methodology, have expanded the sequencing capabilities of the peptide chemist so that routine analysis of peptides of 15 or fewer residues is possible without the use of a costly automated Edman degradation system.

   In order to provide peptides of manageable size for sequencing, it is useful to cleave larger peptides into smaller fragments at specific sites. This may be accomplished either chemically or enzymatically. By changing the pH or by modifying some of the susceptible residues, more limited cleavage may be effected. Either singly or in combination, these cleavage methods produce peptide fragments, many of whose N- or C-terminal residues may be predicted. After separation of the fragments, each may be analyzed for complete amino acid content to ensure that the total of the amino acids present in the fragments is equivalent to that of the original peptide. Although selective cleavage of biologically active peptides appears to be the rational approach to sequence analysis, the amounts of peptide needed are usually greater than the quantities of insect peptides available by reasonable isolation schemes. Therefore, little mention will be made of such approaches here. (For a review of isolation, limited cleavage techniques, and separation of peptide fragments, see Konigsberg and Steinman, 1977).

AMINO ACID ANALYSIS

   The first step to complete sequence determination is an amino acid profile of the peptide sample. Once a peptide is isolated and purity is demonstrated on the basis of molecular weight and charge, biological activity should not increase, and repetitive amino acid analyses should yield a constant ratio of amino acids if further purification is attempted. Ninhydrin detection systems are now available for analyses of less than one nanomole (Andrews, 1982), while orthophthalaldehyde systems are capable of detection in the picomole range (Roth, 1971) for acid hydrolysates. However, tryptophan, asparagine, glutamine, and many derivatized amino acids are destroyed by acid cleavage. While base hydrolysis is useful for confirmation of acid-labile amino acids and derivatives, larger sample sizes are usually required. Addition of bis(I,I-trifluoroacetoxy)iodobenzene to an aliquot of a sample before hydrolysis is a useful method of quantitating asparagine and glutamine (Soby and Johnson, 1981). Rapid hydrolysis (15 minutes) of peptides with hydrochloric acid/propionic acid helps to minimize loss of acid-labile amino acids (Westall and Hesser, 1974), while use of mercaptoethanesulfonic acid (Penke et al., 1974) or methanesulfonic acid (Simpson et al., 1976) provides excellent recoveries of tryptophan. Of course, exhaustive enzymatic digestion is a useful alternative to chemical hydrolysis, providing that separation of amino acids from the enzyme is feasible and all amino acids are of the L-configuration. Once the amino acid composition of a peptide or of its cleavage fragments is established, one has a relative framework in which to choose a sequencing method.

While amino acid analysis is a necessary starting point for sequence determination, it is not sufficient in itself. At this point it is known only which amino acids may be present, not in what form or in what order they occur. As each sequencing method has limitations in terms of length of peptide, type of amino acid residues, isolation and identification of residue derivatives, sensitivity, and expense, one should never assume an absolute structure for a peptide based on amino acid analysis and a single sequencing method. For most samples, however, amino terminal degradation is a useful starting point. Even the lack of success of the first step in N-terminal sequencing provides information to the investigator, a suggestion that the amino terminus may be blocked. From this point alternative means of sequence analysis may be attempted.

N-TERMINAL ANALYSIS

Classical sequence determination involves a step-wise "Edman" degradation from the amino terminus of the peptide by reaction of the first amino acid with phenylisothiocyanate (PITC) and subsequent cleavage of the phenylthiohydantoin amino acid (PTH-amino acid) from the remainder of the peptide (Edman, 1950). The resulting PTH-amino acid is routinely identified by high-performance liquid chromatography (HPLC), thin-layer chromatography (TLC), or gas chromatography (GC). In some cases where separation of PTH-amino acids is a problem, PTH derivatives may be back-hydrolyzed to free amino acids and identified by conventional amino acid analysis techniques. Alternatively, a small portion of the peptide remaining after each degradation step may be hydrolyzed to confirm the identity of the amino acid derivatized in each cycle.

Classical Edman degradation techniques require milligram quantities of material for positive sequence analysis. A considerable percentage of basic PTH-amino acids (lysine, arginine, and to some extent, histidine) and hydrophobic phenylthiocarbamoyl-peptides may be lost during the wash procedures. Yields of the proline derivative are very low and require a second coupling with PITC. In addition, desulphurization of cysteine and methionine may occur due to oxidative conditions. Probably the most limiting factors in Edman degradation are, however, instability of PTH-amino acids and difficulty in separating and identifying them on a routine basis.

To a certain extent these limitations have been overcome by changes in buffers (N,N-dimethyl-N-allylamine) and by addition of reducing agents (ethanethiol), enabling manual sequencing of nanomolar quantities of small peptides (<10 residues) (Levy, 1977). Attachment of the peptide to a solid support (Laursen, 1971) or embedding it in a matrix (Hewick et al., 1981) also eliminates losses due to extraction. This latter development is the basis of the newest solid-phase/gas-phase sequenator. Purity of peptides, reagents, and solvents is the

limiting factor of sensitivity of this instrument, which has the capability of sequencing picomolar quantities of peptide (in rare circumstances, as little as 0.1 picomole).

In the past ten years, a number of investigators have attempted
to increase the sensitivity of the Edman reaction to produce derivatives which are formed more rapidly, are more stable, or have lower
limits of detection. The simplest modification has been the use of
radioactivity, either by use of cell-free translation systems for
incorporation of ^{14}C-amino acids into the peptides being studied
(e.g., Atger et al., 1979) or by use of ^{35}S-phenylisothiocyanate,
which has increased the sensitivity of this method to 0.1 picomole of
amino acid. Other limitations of the Edman procedure, however, are
not eliminated by these techniques.

For increased sensitivity of manual procedures, often the N-terminus of the remaining peptide is coupled with dansyl chloride
(dimethylaminonaphthalene 1-sulfonyl chloride) after each successive
cycle of Edman degradation (Gray and Hartley, 1963). The dansylated
peptide is then hydrolyzed and the fluorescent N-terminal amino acid
derivative is identified by HPLC or polyamide TLC. This method is
capable of determination of the primary structure of one nanomole of
peptide because of the low limit of detection of dansyl amino acids
(10 picomoles). Although a sensitive procedure, it is limited by
incomplete hydrolysis of the dansylated amino acid from the peptide
and requires amino acid analysis of the remaining peptide or simultaneous reaction of several aliquots of the sample. However, extraction
of unreacted reagents and PTH-amino acids is not necessary unless the
peptide under investigation is longer than five residues (Gray and
Smith, 1970). Thus, the dansyl-Edman procedure is extremely useful
for sequence determination of small peptides or cleavage products of
larger peptides.

Probably the most widely-used alternative to phenylisothiocyanate
in N-terminal sequencing is dabsyl isothiocyanate (4-N,N-dimethyl-
aminoazobenzene 4'-isothiocyanate; DABITC) (Chang and Creaser, 1976).
This reagent may be used equally efficiently in either manual solu-
tion- or solid-phase sequencing procedures, as well as in automated
ones (Hughes et al., 1979; Wilson et al., 1979). DABITC is not as
reactive as PITC, but because of the highly-colored dabsyl thiohy-
dantoin (DABTH) derivatives formed, it may be used to sequence between
two and ten nanomoles of a peptide containing six to twelve amino acid
residues (Chang, 1979). DABTH-amino acids may be separated by HPLC
(detection at 436 nm) or by polyamide TLC (visual detection) with as
little as five picomoles of derivative (Wilson, et al., 1979). While
more sensitive than classical Edman coupling, this method is of limited use for serine and threonine residues (Chang, 1979), although there
is more efficient coupling of DABITC to proline than of PITC (Chang et
al., 1978). The biggest drawback to the use of dabsyl isothiocyanate
is that reaction is not always complete. To overcome this difficulty

Chang (1979) has suggested a second treatment with PITC to achieve quantitative coupling and to prevent overlapping of residue identification.

In another attempt to avoid the limitations of the Edman degradation fluoresceinisothiocyanate (FITC) is used. As with DABITC, FITC is less reactive than phenylisothiocyanate, but the method is more sensitive. Although as little as five nanomoles are needed for sequencing ten residues, coupling is not always quantitative, and PITC is added to react with any residue not completely coupled to the fluorescent moiety (Muramoto et al., 1978). Like Edman degradation, this method gives low yields of serine and threonine derivatives, and fluoresceinthiohydantoin-leucine (FTH-leucine) and FTH-isoleucine are not resolved by either HPLC or TLC. To circumvent this latter problem, these FTH derivatives are often back-hydrolyzed, then converted to dansyl amino acids for identification (Muramoto and Tuzimura, 1977).

C-TERMINAL ANALYSIS

Sequencing of peptides from the carboxyl end is a more recent development than that of Edman degradation and associated techniques. Essentially, small peptides attached to glass beads or polystyrene are reacted with thiocyanate to yield amino acid thiohydantoins (Stark, 1968). The thiohydantoins are identified by either GC or TLC, or back-hydrolyzed to free amino acids. Yields decrease rapidly as each residue is removed from the peptide, and between ten and 100 nanomoles of peptide are needed to sequence six amino acid residues (Rangarajan and Darbre, 1976). Reaction of aspartic acid, glutamic acid, and asparagine is slow, and there is considerable loss of cysteine and proline. Kassell et al. (1977) have suggested conversion of the -COOH of the dicarboxylic amino acids and the -SH of cysteine to -CONHR groups (which are stable under subsequent reaction conditions). In addition, yields of proline hydantoin are increased by cleavage with anhydrous trifluoroacetic acid. The number of residues that may be sequenced has also been increased by the use of a thiocyanic acid reagent instead of thiocyanate salts (Meuth et al., 1982). Although automation has been reported for this method (Braunitzer and Pfletschinger, 1978), it is limited to small peptides with free amino termini available for coupling to a solid support.

ENZYMATIC ANALYSIS

Although chemical techniques for sequencing of peptides have been improved to a great extent in the past five years, they are still limited by the ability of the researcher to separate products from starting materials and by-products. Naturally, as more residues are sequenced by manual methods, the number of contaminants and the extent of salt formation increase.

Classical sequencing techniques utilize highly specific endopeptidases to provide smaller peptides which are hydrolyzed and analyzed
for amino acid content. By choosing a variety of enzymes with overlapping cleavage sites, one may determine the original structure of
the starting peptide.

In recent years, exopeptidase digestion of small peptides has
become a common method of determining N- and C-terminal amino acids.
Moreover, these enzymes may also be used to sequentially degrade peptides from one end or the other. Sequence determination may be performed by direct quantitative identification of the amino acid
released or by subtractive amino acid analysis of a portion of the
remaining peptide. Aliquots of amino acids and peptides formed by
enzymatic activity may be separated by TLC, ion-exchange chromatography, or HPLC prior to analysis.

The most commonly used exopeptidases include leucine aminopeptidase, aminopeptidase M, and carboxypeptidases A, B, C, and Y (CPA,
CPB, etc.), or combinations thereof. Use of aminopeptidase M and carboxypeptidase A is restricted because of their limited ability to
cleave prolyl bonds, and CPA releases lysine or arginine very slowly
if at all. However, prolyl residues may be cleaved prior to digestion
with proline-specific endopeptidase or with prolidase in conjunction
with one of the other peptidases. CPB cleaves only lysyl and arginyl
linkages, and thus may be used to compensate for the limitations of
CPA. Unless the N-terminus is pyroglutamate, which may be hydrolyzed
with pyroglutamate aminopeptidase, it is clear that enzymatic cleavage
is limited to peptides with free N- and C-termini and which contain
only L-amino acids.

In another ingenious scheme involving the use of dipeptide aminopeptidase (DAP), half of the sample is treated with one cycle of Edman
degradation to remove the N-terminal amino acid and both samples are
treated with DAP-I and DAP-IV. The enzymes hydrolyze peptides into
dipeptide fragments which may be converted to trimethylsilyl derivatives (Krutzch and Pisano, 1979) or pentafluoroacetyl dipeptide methyl
esters (Siefert et al., 1978) which are separated by gas chromatography and identified either by retention time or by mass spectrometry.
An obvious drawback to this method is that it relies heavily on probability and nearest neighbor analysis to provide a feasible sequence
for the sample, which necessitates the use of computer analysis for
larger peptides.

Although enzymatic digestion is simple in theory, there are many
complications of such procedures. Besides the limitations imposed by
the specificities of the enzymes themselves, the presence of the proteinaceous enzymes and the salts of the buffers may interfere with
separation of the components and may effectively complicate analysis.
It is preferable to select buffers which are volatile and compatible
with the chromatography system chosen. Aqueous combinations of

sequencing grade N-ethylmorpholine and acetic acid, for example, may be adjusted to give a range of pH values consistent with maximal enzymic activity. In the past few years the introduction of molecular-weight filter centrifugation and immobilization of proteins has greatly facilitated separation of peptidases from smaller peptides and amino acids. Immobilized enzymes are easily removed from the reaction mixture by low-speed centrifugation or by filtration. While immobilization slightly shifts the optimum pH and reduces activity a modest amount when compared to its soluble counterpart (Royer and Andrews, 1973; Liberatore et al., 1976), the convenience and simplicity of the method greatly outweigh decreases in activity.

Because of the ease of product separation, the need for little sophisticated equipment, and the relatively low cost, enzymatic digestion of unknown peptides is a viable alternative to standard chemical sequencing techniques.

While these N-terminal and C-terminal methodologies are useful for small peptides, they are limited only to those peptides with free amino and carboxyl termini. However, if the peptide to be sequenced is of sufficient length and contains an appropriate unique internal amino acid, a chemical or enzymatic cleavage could produce two smaller peptides, each of which could be sequenced from the internal cleavage site by appropriate means.

MASS SPECTROMETRIC ANALYSIS

In the past, mass spectrometry (MS) has been an alternative for analysis of phenylthiohydantoin amino acids, methylisothiocyanate amino acids, amino acid thiohydantoins, pentafluoroacetyl dipeptide methyl esters, and trimethylsilyl polyamino alcohols (Matsuo et al., 1981; Okada and Sakuno, 1978; Ziemer et al., 1979; Herlihy et al., 1980). However, only recently has mass spectrometry become a useful tool in sequence analysis of peptides. There are several obstacles which have prevented MS from becoming a routine procedure in peptide sequencing. Peptides as such are very involatile; temperatures that would be required for volatilization, even in direct probe work, lead to thermal decomposition of the sample. Secondly, some amino acids are not stable during electron bombardment and give rise to many signals rather than one peak which is characteristic of the amino acid fragment. In addition, peptides may cleave into multitudes of fragments yielding spectra which may become too complicated to interpret.

Early work in electron impact mass spectrometry (EIMS) of peptides involved chemical derivatization of unstable amino acids, peptide bonds, and the amino terminus. By acylating the terminal amino group a peptide is made more volatile. Methylation of all peptide bonds increases volatility and induces a preferred fragmentation at peptide bonds, producing a series of clear sequence ions from the

C-terminus of the sample (Laclercq and Desiderio, 1971). Normally
acetyl groups are used to block the amino terminus, but Larson (1982)
has recently introduced use of the ortho-phthalaldehyde group for this
purpose in order to increase sensitivity during purification of the
derivative. Conversion of arginine to ornithine (Shemyakin et al.,
1967) and conversion of methionine to α-amino butyric acid (Thomas et
al., 1968) eliminate some of the difficulties associated with these
residues. Later studies indicated that "short" permethylation schemes
prevent salt formation with several of the problem amino acids, such
that only arginine need be converted to ornithine before analysis
(Morris et al., 1973).

Because of the additional steps in derivatization (acylation,
permethylation) and purification necessary for meaningful EIMS, chemical ionization mass spectrometry (CIMS) has been applied to peptide
sequencing. Björkman (1982) has demonstrated that bombardment of the
sample with a variety of ionized reagent gases provides excellent
sequence information, even without prior derivatization. Advances in
field desorption mass spectrometry (FDMS) have also provided a means
of molecular ion determination of underivatized peptides, although
sequence data are limited without prior derivatization (Matsuo et al.,
1981).

Perhaps the most significant advance with respect to peptide
sequence determination is the development of fast-atom-bombardment
mass spectrometry (FAB-MS). In FAB-MS an ion beam is accelerated and
undergoes charge exchange on collision with an inert gas. Subsequently, these "fast" atoms are directed at the sample. In the process the
sample fragments into masses both indicative of sequence ions and
characteristic of individual amino acids. Molecular ions may be
determined for underivatized samples of up to 21 amino acid residues
(Williams et al., 1972), while up to fifteen residues may be sequenced
from both the carboxyl and amino termini by current techniques. Both
positive and negative ion spectra are possible, the $(M + H)^+$ formation facilitated in peptides containing arginine or lysine and the
$(M - H)^-$ in those with aspartic or glutamic acids (Williams et al.,
1982). In many cases the sample to be sequenced is dissolved in glycerol, as this solution provides a surface monolayer of sample which
is continuously regenerated during analysis.

Apart from the obvious expense to maintain and operate a mass
spectrometer, sample size is a limiting factor. While only 0.1 nanomole is required for molecular ion determination (Williams et al.,
1982), often as much as 5-10 nanomoles are required for complete
sequence analysis. Because identification of amino acid residues is
based on mass, leucine and isoleucine are indistinguishable, and
lysine and glutamine may be identified only after prior acetylation of
the lysine residue (Williams et al., 1982). However, only mass spectrometric techniques are feasible for peptides with blocked amino and/
or carboxyl termini, the most extreme case being a cyclic peptide. In

addition, only MS provides information for identification of unusual amino acids and derivatives missed by conventional sequence and amino acid analyses, such as pyroglutamate, γ-carboxyglutamate, phosphoserine, acetylated or glycosylated amino acids, etc. Therefore, while mass spectrometry might not be the primary method of choice for structure determination because of cost, it may be a very necessary adjunct to other sequencing techniques for unusual peptides.

X-RAY CRYSTALLOGRAPHY

X-ray crystallography is an excellent means of determining the secondary and tertiary structures of a peptide and has been utilized in a few cases to determine primary structure as well. One advantage of X-ray crystallography is the ability to sequence an intact peptide sample, rather than its cleavage products. In addition, X-ray crystallography provides information concerning spatial orientation of one residue with respect to others. This process involves identification of residues by the size, shape, or location (e.g., β-turn) of the side chain groups. This may lead, however, to ambiguous or incomplete amino acid assignment. Also, the necessity of growing a pure, single crystal of the peptide is a severe limitation of this method. Furthermore, the cost of X-ray crystallographic instruments and maintenance and the expertise needed to decipher cystallographic data prevent this technique from becoming a routine procedure for primary sequence determination.

SEQUENCY STRATEGY

Although the various sequencing strategies available provide a multitude of approaches to acquisition of structural information about peptides, sequencing of insect neuropeptides presents additional constraints on these methods. While the quantity of neuropeptide per insect is in the nanogram range, limits of detection for standard purification procedures are seldom this sensitive. Biological activity assays usually provide the only quantitative measure of a native peptide. Often much of the biological activity of neuropeptides is lost during purification because of lability under isolation conditions. If the peptide no longer has activity it is difficult to determine whether the primary structure of the inactive peptide represents that of the native peptide.

It is important, therefore, to acquire as much structural information as possible during isolation procedures. Very specific endopeptidases and non-specific exopeptidases may provide a wealth of information even prior to amino acid analysis. Does the peptide lose activity when incubated with an aminopeptidase or carboxypeptidase? . . . with prolidase? . . . with pyroglutamate aminopeptidase? Mass spectrometric methods are also useful after preliminary purification.

Mixtures of peptides may be distilled sequentially from the probe as a function of increasing volatility. Later the pure, biologically active peptide may be analyzed for other specifics of sequence.

Perhaps one of the more elegant sequencing designs is that of a combination of mass spectrometry following enzymatic digestion from each end of the peptide (Bradley and Williams, 1982; Self and Parente, 1983). Although sequence analysis by MS usually requires larger samples than by some other methods, the limit of detection for molecular ion determination is comparable to most other techniques. If exopeptidase digestion is carried out in a controlled fashion, a time course of "subtractive" mass spectrometry of the remaining peptide will indicate the amino acid residue cleaved at each step.

Sequencing of peptides has been viewed by some merely as physical characterization of a molecule that elicits a particular biological response. However, structural determination of insect neuropeptides to the smallest detail is crucial to study of binding and/or function, expression of genes encoding for their synthesis, and sequence similarity to peptides from other biological sources.

To date very few insect neuropeptides have been completely sequenced, although amino acid analysis by conventional means is available for several of them. Proctolin (Arg-Tyr-Leu-Pro-Thr), a myotropic peptide from the proctodeal muscles of the cockroach (Starratt and Brown, 1975), was sequenced by the dansyl-Edman procedure previously discussed. Such an approach was adequate for complete sequence determination, as proctolin contains only five L-amino acids, none of which are blocked or modified. On the other hand, the sequence determination of locust adipokinetic hormone (PCA-Leu-Asn-Phe-Thr-Pro-Asn-Trp-Gly-Thr-NH_2), a peptide eliciting the release of diglycerides from the fat body and their transport to the flight muscle, required a variety of sequencing methods. Enzymatic cleavage, dansyl-Edman degradation, and mass spectrometry were necessary to determine the presence of a pyroglutamate residue at the amino terminus and an amide group on the C-terminal tryptophan (Stone et al., 1976).

For initial sequence work, however, it is not necessary to have access to complex and expensive instrumentation for reasonably complete sequence determination. Many manual methods, including the dansyl-Edman procedure and the use of DABITC, have been developed and refined for sequencing nanomole quantities of purified peptides. Except for the gas-phase sequenator, automated methods offer little better sensitivity for the much greater expense. While standard methods of peptide sequencing may be adequate for many peptides, the presence of blocked amino acids or derivatives, which are very common in neuropeptides, may complicate sequence analysis. The failure of the first cycle of N- or C-terminal analysis is not sufficient for positive identification as a blocked terminus. There are always limitations inherent in end-group analysis, including possible chemical side reactions, and negative results are more believable if more than one sequencing method is attempted. (In some cases, rearrangements during Edman-type degradations may lead to blocked amino termini, especially with threonine and serine residues.) In addition, the presence of modified or D-amino acid residues may prevent hydrolysis by exopeptidase activity. Under these circumstances, mass spectrometry may be necessary in order to analyze samples which present difficulties for the more usual sequencing methods. Therefore, the researcher should not rely entirely on any one sequencing technique which is unable to provide total structural information.

With the rapid advance in HPLC purification and FAB-MS of peptides, it is likely that analysis of small quantities of unusual peptides will become more commonplace in the future. Interfaces are now being developed to transfer HPLC column eluant directly into the mass spectrometer with high sample enrichment and vaporization of low volatility HPLC solvents. Recent innovations in interfacing liquid chromatography and mass spectrometry hold a great deal of promise for future studies of insect neuropeptides.

REFERENCES

Andrews, R. P., 1982, High sensitivity amino acid analysis, Protein Chemistry Notes, 22:1.

Atger, M., Mercier, J. C., Haze, G., Fridlansky, F., and Milgrom, E., 1979, N-terminal sequences of uteroglobin and its precursor, Biochem. J., 177: 985.

Björkman, S., 1982, Electron impact and methane, isobutane and ammonia chemical ionization mass spectra of peptide amides related to melanostatin (Pro-Leu-Gly-NH$_2$), Biomed. Mass Spectrom., 9:315.

Bradley, C. V., and Williams, D. H., 1982, Peptide sequencing using the combination of Edman degradation, carboxypeptidase digestion, and fast atom bombardment mass spectrometry, Biochem. Biophys. Res. Commun., 104: 1223.

Braunitzer, G., and Pfletschinger, J., 1978, Kovalente, C-terminale fixierung von bromcyanpeptiden im flussigphasensequenator. Hoppe-Seyler's Z. Physiol. Chem., 359:1015.

Chang, J. Y., 1977, High-sensitivity sequence analysis of peptides and proteins by 4-N,N-dimethylaminoazobenzene 4'-isothiocyanate, Biochem J., 163:517.

Chang, J. Y., 1979, Manual solid-phase sequence analysis of polypeptides using 4-N,N-dimethylaminoazobenzene 4'-isothiocyanate, Biochim. Biophys. Acta. 578:188-195.

Chang, J. Y., and Creaser, E. H., 1976, A novel manual method for proteinsequence analysis, Biochem. J., 157:77.

Chang, J. Y., Brauer, D., and Wittman-Liebold, W., 1978, Microsequence analysis of peptides and proteins using 4-N,N-dimethylaminoazo-

benzene 4'-isothiocyanate/phenylisothiocyanate double coupling
method, FEBS Lett. 93:205.

Chang, J. Y., Lehmann, A., and Wittman-Liebold, B., 1980, Analysis of
dimethylaminobenzene-thiohydantoins of amino acid by high-pres-
sure liquid chromatography, Anal. Biochem., 102:380.

Edman, P., 1950, Method for the determination of the amino acid
sequence in peptides, Acta Chem. Scand., 4:283.

Gray, W. R., and Hartley, B. S., 1963, A fluorescent end-group reagent
for proteins and peptides, Biochem. J., 89:59P.

Gray, W. R., and Smith, J. F., 1970, Rapid sequence analysis of small
peptides, Anal. Biochem., 33:36.

Hayes, T. K., 1983, Application of reverse phase technology for the
isolation of insect peptides, in: "Proc. 1st Intl. Conf. Insect
Neurochem. Neurophysiol.," Plenum Press, New York.

Herlihy, W. C., Royal , N. J., Biemann, K., Putney, S. D., and
Schimmel, P. R., 1980, Mass spectra of partial protein hydrolysates as a multiple phase check for long polypeptides deduced
from DNA sequences: NH_2-terminal segment of alanine tRNA syn-
thetase, Proc. Natl. Acad. Sci. USA, 77:6531.

Hewick, R. M., Hunkapiller, M. W., Hood, L. E., and Dreyer, W. J.,
1981, A gas- liquid solid phase peptide and protein sequenator,
J. Biol. Chem., 256:7990.

Hughes, G. J., Winterhalter, K. H., Lutz, H., and Wilson, K. J., 1979,
Microsequence analysis III: Automatic solid-phase sequencing
using DABITC, FEBS Lett., 108:92.

Kassell, B., Krishnamurti, C., and Friedman, H. L., 1977, Studies on
solid phase carboxyl terminal sequencing, in: "Solid Phase Methods in Protein Sequence Analysis," Inserm Symposium No. 5, (A.
Previero and M.-A. Coletti-Previero, eds.), 39-48. Elsevier/
North Holland Biomedical Press, New York.

Konigsberg, W. H., and Steinman, H. M., 1977, Strategy and methods of
sequence analysis, in: "The Proteins," 3rd edition, (H. Neurath
and R. L. Hill, eds.), 3:1-110. Academic Press, New York.

Krutzch, H. C., and Pisano, J. J., 1979, Preparation of dipeptidyl
aminopeptidase IV for polypeptide sequencing, Biochim. Biophys.
Acta, 576:280.

Laclercq, P. A., and Desiderio, D. M., Jr., 1971, A laboratory procedure for the acetylation and permethylation of oligopeptides on
the microgram scale, Anal. Lett., 4:305.

Larson, B. R., 1982, IV. A model system for mass spectrometric
sequencing of ortho-phthalaldehyde peptide derivatives, Life
Sciences, 30:1003.

Laursen, R. A., 1971, Solid-phase Edman degradation: An automatic
peptide sequencer, Eur. J. Biochem., 20:89.

Levy, W. P., 1977, Manual Edman sequencing techniques for proteins and
peptides at the nanomole level, Methods Enzymol., 79:27.

Liberatore, F. A., McIsaac, J. E., Jr., and Royer, G. P., 1976, Immobilized carboxypeptidase Y. Applications in protein chemistry,
FEBS Lett., 68: 45.

Matsuo, T., Matsuda, H., Katakuse, I., Shimonishi, Y., Maruyama, Y.,
    Higuchi, T., and Kubota, E., 1981, Field desorption-collision
    activation mass spectrometry with accumulated linked-scan tech-
    nique for peptide structure, Anal. Chem., 53:416.
Meuth, J. L., Harris, D. E., Dwulet, F. E., Crowl-Powers, M. L., and
    Gurd, F. R. N., 1982, Stepwise sequence determination from the
    carboxyl terminus of peptides, Biochemistry, 21:3750.
Morris, H. R., Dickinson, R. J., and Williams, D. H., 1973, Studies
    towards the complete sequence determination of proteins by mass
    spectrometry: derivatisation of methionine, cysteine and argi-
    nine containing peptides, Biochem. Biophys. Res. Commun., 51:
    247.
Muramoto, K., and Tuzimura, K., 1977, Indirect identification of fluo-
    resceinthiohydantoin-amino acids in sequence analysis of pep-
    tides, Agric. Biol. Chem., 41:2469.
Muramoto, K., Kawauchi, H., and Tuzimura, K., 1978, Sequence determi-
    nation of peptides by the combined use of fluoresceinisothiocya-
    nate and phenylisothiocyanate, Agric. Biol. Chem., 42:1559.
Okada, K., and Sakuno, A., 1978, Identification of amino acid thiohy-
    dantoin derivatives by chemical ionization mass spectrometry,
    Org. Mass Spectrom. 13:535.
Penke, B., Ferenczi, R., and Kovacs, K., 1974, A new acid hydrolysis
    method for determining tryptophan in peptides and proteins, Anal.
    Biochem., 60:45.
Rangarajan, M., and Darbre, A., 1976, Studies on sequencing of pep-
    tides from the carboxyl terminus by using the thiocyanate method,
    Biochem. J., 157:307.
Roth, M., 1971, Fluorescence reaction for amino acids, Anal. Chem.,
    43:880.
Royer, G. P., and Andrews, J. P., 1973, Immobilized derivatives of
    leucine aminopeptidase and aminopeptidase M, J. Biol. Chem.,
    248:1807.
Self, R., and Parente, A., 1983, The combined use of enzymatic hydro-
    lysis and fast atom bombardment mass spectrometry for peptide
    sequencing, Biomed. Mass Spectrom., 10:78.
Shemyakin, M. M., Ovchinnikow, Y. A., Vinogradova, E. I., Feigina, M.
    Y., Kiryushkin, A. A., Aldanova, N. A., Alakhov, Y. B., Lipkin,
    V. M., and Rosinov, B. V., 1967, Mass spectrometric determination
    of the amino acid sequence in arginine-containing peptides,
    Exper., 23:428.
Shimonishi, Y., Hong, Y-M., Takao, T., Aimoto, S., Matsuda, H., and
    Izumi, Y., 1981, A new method for carboxyl-terminal sequence
    analysis of a peptide using carboxypeptidases and field-desorp-
    tion mass spectrometry, Proc. Japan. Acad., 57B:304.
Siefert, W. E., McKee, R. I., Beckner, C. F., and Caprioli, R. M.,
    1978, Characterization of mixtures of dipeptides by gas chroma-
    tography/mass spectrometry, Anal. Biochem., 88:149.
Simpson, R. J., Neuberger, M. R., and Liu, T. Y., 1976, Complete amino
    acid analysis of proteins from a single hydrolysate, J. Biol.
    Chem., 251: 1936.

Sobey, L. M., and Johnson, P., 1981, Determination of asparagine and glutamine in polypeptides using Bis(I,I-trifluoroacetoxy)iodobenzene, Anal. Biochem. 113:149.

Stark, G. R., 1968, Sequential degradation of peptides from their carboxyl termini with ammonium thiocyanate and acetic anhydride, Biochemistry, 7: 1796.

Starratt, A. N., and Brown, B. E., 1975, Structure of the pentapeptide proctolin, a proposed neurotransmitter in insects, Life Sciences, 17:1253.

Stone, J. V., Mordue, W., Batley, K. E., and Morris, H. R., 1976, Structure of locust adipokinetic hormone, a neurohormone that regulates lipid utilisation during flight, Nature, 263:207.

Thomas, D. W., Das, B. C., Gero, S. D., and Lederer, E., 1968, Mass spectrometry of permethylated peptide derivatives: extension of the technique to peptides containing arginine or methionine, Biochem. Biophys. Res. Commun., 32:519.

Westall, F., and Hesser, H., 1974, Fifteen-minute hydrolysis of peptides, Anal. Biochem., 61:610.

Williams, D. H., Bradley, C. V., Santikarn, S., and Bojesen, G., 1982, Fast-atom-bombardment mass spectrometry. A new technique for the determination of molecular weights and amino acid sequences of peptides, Biochem. J., 201:105.

Wilson, K. J., Hunziker, P., and Hughes, G. J., 1979, Microsequencing analysis IV: Automatic liquid-phase sequencing using DABITC, FEBS Lett., 108:98.

Ziemer, J. N., Perone, S. P., Caprioli, R. M., and Seifert, W. R., 1979, Computerized pattern recognition applied to gas chromatography/mass spectrometry identification of pentafluoropropionyl dipeptide methyl esters, Anal. Chem., 51:1732.

THE INVESTIGATION OF INSECT NEUROPEPTIDE

STRUCTURE-ACTIVITY RELATIONSHIPS

Alvin N. Starratt

Research Centre, Agriculture Canada
University Sub P.O., London, Ontario
Canada N6A 5B7

Biologists and chemists have long been intrigued by the connection between the structure of a chemical and its biological activity. Practical considerations, such as the desire to develop more potent or selective drugs and insecticides, have provided much of the impetus for the investigation of such relationships. Consequently, many studies concerning structure-activity or structure-function relationships have been described in the literature. This brief report will attempt to provide an introduction to this field emphasizing its application to the insect neuropeptide area.

It is supposed that biological activity results from the recognition of a specific molecule by a receptor or enzyme. The idea that bioactive substances have to bind or interact in a specific way with the biological object in order to be effective can be traced to publications of Ehrlich early in this century (see ARIENS, 1977; FERSHT, 1982). However, activity is usually not dependent only on receptor binding or activation of the biological response once binding has been achieved. When the point of application of a chemical differs from its site of action, other factors including metabolism, elimination, transport and penetration into tissues also contribute to the observed bioactivity and complicate the development of an understanding of structure-activity relationships. Further complication occurs when bioactive substances bind in different ways to the receptor or bind to different receptors or when the biological response measured is a complex result of several processes (MARTIN, 1981).

265

An empirical approach has been utilized in the majority of investigations to date. Generally, substances structurally related to a bioactive agent are prepared and tested to discover the effect of changes in molecular architecture on activity. As well as identifying essential parts of the molecule, this approach has frequently led to compounds with more desirable biological properties or even to the discovery of compounds with quite different effects. Observations of a relationship between certain physical properties of a molecule and biological activity led to more quantitative attempts to correlate structure and bioactivity (HANSCH and FUJITA, 1964; FREE and WILSON, 1964). The most popular approach to the study of quantitative structure-activity relationships (QSAR) is by the Hansch multiple parameter method which assumes that the physicochemical factors governing transport and drug-receptor interaction can be separated into electronic, hydrophobic and steric components (HANSCH, 1969; FUJITA, 1972). Recently, MARTIN (1981) has discussed the development of QSAR and pointed out its value and limitations from the standpoint of a medicinal chemist. Especially noteworthy are the number of correct predictions of biological activity from QSAR studies. However, despite the success of this technique in the design of many classes of compounds, there have been very few QSAR studies of biologically active peptides (FAUCH-ÈRE, 1982). Computers with their capacity to handle large amounts of data have and will, no doubt, in the future play an increasingly important role in such studies (HANSCH, 1972; MARTIN, 1981). Computer graphics, which can be used to create and manipulate three-dimensional molecular structures, have recently been employed in the design of peptide analogues and other substances (KRÖHN, 1982; TUTE, 1983).

Although structurally related naturally occurring peptides as well as substances obtained by chemical modification (e.g. acetylation, methylation) or enzymatic cleavage of the peptide being investigated may be used in the study of structure-activity relationships, most analogues are obtained by synthesis. Many excellent methods are available for the synthesis of peptides (see BODANSZKY et al., 1976; JONES, 1979; BODANSZKY, 1979; MEIENHOFER, 1979a,b; RICH and SINGH, 1979; BARANY and MERRIFIELD, 1980; LUKAS et al., 1981) and the necessary equipment, chemicals, and most of the required amino acids are readily obtainable commercially.Care must be taken to ensure that the preparations of synthetic peptides are pure before they are bioassayed. Over the years, countercurrent distribution, high-voltage electrophoresis and ion-exchange, partition and gel permeation chromatography have been and continue to be very useful separation techniques. Recently, high-performance liquid chromatography (HPLC) has become the method of choice for assessing the purity of peptides as well as for the efficient separation of peptide mixtures (KRUMMEN, 1980; VOELTER, 1981).

In recent years, a large number of neuropeptides have been identified within the mammalian central nervous system. Extensive studies, involving the synthesis and biological testing of many analogues, have led to a partial understanding of the relation between structure and activity for a number of these, e.g. the enkephalins (GORIN et al., 1980; MORLEY, 1980; PAJUSZ, 1982). In an excellent report on the design of peptide hormone analogues, STEWART (1982) has listed several reasons for their synthesis including the need to understand the nature of the peptide-receptor interaction, the need to know which structural features of the molecule are responsible for a particular response when the peptide has several activities, the need for analogues which will inhibit the action of the endogenous peptide and the desire for analogues with enhanced potency and extended lifetime in vivo. The many structure-function studies in the mammalian peptide area provide an excellent background and source of ideas for anyone planning similar studies of insect neuropeptides.

Despite the discovery and recognition of the importance of many peptides in the insect nervous system (FRONTALI and GAINER, 1977; RAABE, 1982), only two, proctolin (STARRATT and BROWN, 1975) and locust adipokinetic hormone (STONE et al., 1976), have been identified to date. Both of these have been the subject of structure-activity investigations. It is informative to examine these studies quite closely since they illustrate approaches which will be generally applicable to the study of other insect neuropeptides.

Proctolin, H-Arg-Tyr-Leu-Pro-Thr-OH, first isolated from the cockroach Periplaneta americana (BROWN and STARRATT, 1975), is widely distributed among insects (BROWN, 1977; HOLMAN and COOK, 1979a,b) and is highly active on a number of invertebrate muscle preparations (see STARRATT and STEELE, 1983; WATSON et al., 1983). Most tests of the relative activities of proctolin and its synthetic analogues have been made on the hindgut (proctodeum) of the cockroach P. americana. Preliminary findings based on 4 analogues indicated that small structural changes

resulted in substantial losses of activity (STARRATT and
BROWN, 1975).

Later, a larger number of pentapeptide analogues of
proctolin as well as several di-, tri- and tetrapeptides
with partial amino acid sequences of proctolin were synthesized and assayed (STARRATT and BROWN, 1979). Relative activities were ascertained by determining the
concentration of each analogue required to produce a response equivalent to that of a standard amount of proctolin. Results are presented in Table 1.

Each amino acid was successively replaced by alanine
to determine if a single amino acid is essential for the
proctolin response. Of the 5 analogues, only [Ala^4]-
proctolin, with about 15% of the activity of proctolin,

Table 1. Relative myotropic activities of proctolin and
various proctolin analogues on P. americana hindgut (from
STARRATT and BROWN, 1979)

Peptide	Relative Activity
H-Arg-Tyr-Leu-Pro-Thr-OH(Proctolin)	100.00
H-Ala-Tyr-Leu-Pro-Thr-OH	0.05
H-Arg-Ala-Leu-Pro-Thr-OH	0.06
H-Arg-Tyr-Ala-Pro-Thr-OH	0.08
H-Arg-Tyr-Leu-Ala-Thr-OH	15.70
H-Arg-Tyr-Leu-Pro-Ala-OH	0.14
H-D-Arg-Tyr-Leu-Pro-Thr-OH	3.50
H-Arg-D-Tyr-Leu-Pro-Thr-OH	10.97*
H-Arg-Tyr-D-Leu-Pro-Thr-OH	0.06
H-Arg-Tyr-Leu-D-Pro-Thr-OH	0.03
H-Arg-Tyr-Leu-Pro-D-Thr-OH	1.80
H-Arg-Phe-Leu-Pro-Thr-OH	15.40
H-Arg-Phe(p-OMe)-Leu-Pro-Thr-OH	278.80
H-Orn-Tyr-Leu-Pro-Thr-OH	2.83
H-Arg-Tyr-Leu-Pro-Ser-OH	5.44
H-Arg-Tyr-Leu-Pro-Thr-NH$_2$	0.04
H-Arg-Tyr-Leu-Pro-OH	0.15
H-Tyr-Leu-Pro-Thr-OH	< 0.0001
H-Arg-Tyr-Leu-OH	0.0005
H-Tyr-Leu-Pro-OH	0.0009
H-Leu-Pro-Thr-OH	0.06
H-Arg-Tyr-OH	< 0.0001
H-Tyr-Leu-OH	< 0.0001

* Activity attributable in large part to proctolin, an
impurity in this peptide (HPLC evidence).

showed appreciable activity. This indicates that the side chains of all the amino acids are required for maintaining a favorable peptide conformation, binding to the receptor or subsequently eliciting the biological response.

Results were obtained which showed that the terminal amino acids were important for full activity. Substituting serine for threonine or ornithine for arginine resulted in a large loss of potency while shortening the peptide chain from either end yielded substances with little or no significant activity. Proctolin amide had only slight activity indicating the importance of the free carboxyl.

Successive replacement of the amino acids of proctolin with the respective D-amino acids also yielded analogues with low activity. The most active, [D-Tyr^2]-proctolin, was subsequently shown chromatographically to contain sufficient proctolin, arising from partial racemization of tyrosine during the synthesis, to account for most of the observed activity. This illustrates the importance of knowing the purity of the compounds being tested. In the proctolin work, HPLC has proven to be very useful for determining homogeneity and for purifying small quantities of the synthetic peptides (STARRATT and STEVENS, 1980). The low activity of the analogues containing the D-amino acids supported the earlier supposition that only L-amino acids constitute the natural neuropeptide.

In discussing the antagonism of proctolin by tyramine, BROWN (1975) speculated that the phenolic group of tyramine has considerable affinity for the same receptor site as the phenolic group of proctolin. However, it was later found that a free phenolic group was not a prerequisite for high myotropic activity. [Phe(p-OMe)^2]-Proctolin, in which the phenolic group of the tyrosine moiety of proctolin was methylated, was nearly three times as active as proctolin. [Phe^2]-Proctolin which lacks a phenolic group was 15% as active. With regard to tyramine as an antagonist of proctolin, it should be noted that PIEK (1982) has recently reported variation in the activity of different samples of tyramine and suggested that the antagonistic action may be due to an impurity.

Since a specific antagonist would be useful in studying the function of proctolin, the analogues were tested for antagonistic activity. None inhibited neural-

ly evoked contractions of the hindgut at concentrations
below the level at which they caused contractions themselves.

SULLIVAN and NEWCOMB (1982) also used the hindgut of
P. americana for a comparison of the activities of proctolin and some analogues. The forces of the contractions
induced by the test substances were measured. Using data
obtained at several concentrations, dose-response relationships were compared and apparent equilibrium dissociation constants (Kd_{app}) obtained for the most active
compounds. The results (Table 2) complement and extend
the earlier findings of STARRATT and BROWN (1979).No analogues with activity greater than proctolin were discovered and none acted as proctolin antagonists.

Replacement of arginine, the basic N-terminal amino
acid of proctolin, with the neutral amino acid glycine or
the acidic amino acid glutamic acid resulted in total
loss of activity up to 10^{-5}M. This result was similar to
that observed when arginine was replaced by alanine
(STARRATT and BROWN, 1979). Substitution with lysine,
however, gave a fully active agonist (capable of causing
maximum gut contraction) of slightly less affinity
(higher Kd_{app}). A similar affinity was shown for the
peptide resulting from the addition of arginine to the
N-terminal end of the peptide chain and a slightly lower
affinity was exhibited when glycine was added. Although

Table 2. Activities of proctolin and various proctolin
analogues on P. americana hindgut (from SULLIVAN and
NEWCOMB, 1982)

Peptide	Kd_{app}*
(a) proctolin	
H-Arg-Tyr-Leu-Pro-Thr-OH	2×10^{-8}M
(b) N-terminal alterations	
H-Lys-Tyr-Leu-Pro-Thr-OH	1×10^{-7}M
H-Arg-Arg-Tyr-Leu-Pro-Thr-OH	1×10^{-7}M
H-Gly-Arg-Tyr-Leu-Pro-Thr-OH	3×10^{-7}M
H-Gly-Tyr-Leu-Pro-Thr-OH	inactive up to 10^{-5}M
H-Glu-Tyr-Leu-Pro-Thr-OH	inactive up to 10^{-5}M
(c) 2-position alterations	
H-Arg-Trp-Leu-Pro-Thr-OH	5×10^{-8}M
H-Arg-Phe-Leu-Pro-Thr-OH	1×10^{-7}M
H-Arg-His-Leu-Pro-Thr-OH	$>1 \times 10^{-6}$M

* Apparent equilibrium dissociation constant.

a basic side chain is important for productive binding, this latter result indicates that it does not have to be at the N-terminus of the peptide chain.

As was observed when alanine was substituted for the tyrosyl residue of proctolin, substitution of glycine resulted in an inactive analogue. Active compounds were obtained when tyrosine was replaced by other aromatic amino acids in agreement with results of STARRATT and BROWN (1979).

[Gly^4]-Proctolin and [Leu^4]-proctolin, analogues having glycine and leucine, respectively, in place of proline, were proctolin agonists but showed much lower affinities than proctolin for the receptor. Together with the earlier observation for [Ala^4]-proctolin, these results indicate that proline in the 4-position is not essential for productive binding.

[Ala^5]-Proctolin exhibited weak activity at 10^{-6} to 10^{-5}M and H-Arg-Tyr-Leu-Pro-Thr-Gly-OH, having an additional glycine residue at the C-terminal end, caused a small contracture of the hindgut at 10^{-6}M. In comparison, H-Arg-Tyr-Leu-Pro-OH, which lacks the C-terminal amino acid, was without effect below 5 x 10^{-6}M. Activity was not observed for several peptides of shorter chain-length having sequences found in proctolin. The retro-isomer (see GOODMAN and CHOREV, 1979) H-Thr-Pro-Leu-Tyr-Arg-OH was also inactive.

In addition to studying proctolin analogues, SULLI-VAN and NEWCOMB (1982) tested a number of unrelated neuropeptides including α-MSH, β-lipotropin 60-65, met-enkephalin, leu-enkephalin, vasoactive intestinal peptide and elodoisin related peptide. The only one to reproduc-ibly cause contracture of the hindgut at micromolar concentrations was the crustacean erythrophore or red pigment concentrating hormone, an octapeptide with no structural similarity to proctolin, which produced a slight increase in tonus and enhanced spontaneity at concentrations exceeding 10^{-6}M.

Results of these two studies indicate that the activity of proctolin cannot be attributed to the presence of a single amino acid or to any sequence of amino acids shorter than that found in the natural peptide. Most modifications led to marked losses in activity. The most active analogues were those in which other aromatic amino acids replaced tyrosine ([Phe(p-OMe)^2]-proctolin, [Phe^2]-proctolin, [Trp^2]-proctolin) or in which the amino

terminus was altered ([Lys[1]]-proctolin, Arg-proctolin).
The number of analogues studied is too small to draw
conclusions about the relative influence of lipophilic,
steric and electronic factors which could affect the
interaction of proctolin with its receptors.

It is well recognized that other peptides with myotropic activity on the hindgut occur in the cockroach
(FRONTALI and GAINER, 1977). BROWN (1965) reported the
presence of two active peptides in extracts of the corpora cardiaca of P. americana and HOLMAN and COOK (1972,
1979b) have presented evidence for a hindgut-stimulating
neurohormone in extracts of Leucophaea maderae. The
apparent rather precise structural requirements for high
activity on the proctolin receptors of the cockroach
hindgut invites the speculation that these unidentified
myotropic peptides bear some structural resemblance to
proctolin if they act on the same receptors.

PIEK et al. (1979) reported the activity of proctolin and a number of other peptides on the extensor tibia
of Locusta migratoria. Small myogenic rhythmic contractions are induced by 10^{-10} to 10^{-9}M proctolin and 10^{-9}M
Glu-proctolin. The latter result was similar to results
of SULLIVAN and NEWCOMB (1982) for single amino acid
additions to the N-terminal of the proctolin chain.
H-Arg(NO_2)-Tyr-Leu-Pro-Thr-OH, an intermediate
in the synthesis of proctolin, was about 1000 fold less
active than proctolin. The hexapeptide β-lipotropin
60-65, H-Arg-Tyr-Gly-Gly-Phe-Met-OH, which has two amino
acids in common with proctolin, and 8 pentapeptides with
no structural relationship to proctolin were inactive.
Their most interesting finding was that the potent   ·
bradykinin-potentiating peptide BPP_{5a}, <Glu-Lys-Trp-Ala-
Pro-OH, was active at concentrations from 10^{-9} to 10^{-8}M.
Although this pentapeptide has only one amino acid in
common with proctolin, both peptides contain the sequence
of a basic amino acid (arginine or lysine), an aromatic
amino acid (tyrosine or tryptophan), an aliphatic amino
acid (leucine or alanine) and proline. Since H-Val-Lys-
Trp-Ala-Ala-OH with three amino acids identical to those
in BPP_{5a} was inactive, it was proposed that proline is
necessary for proctolin-like activity on the extensor
tibia preparation. This contrasts with the results on
the cockroach hindgut which show that peptides in which
alanine, glycine or leucine replace proline are still
active. As well as being highly active on the locust extensor tibia, BPP_{5a} was also active on locust neurones
(WALKER et al., 1981). In both cases, it was about 10
times less potent than proctolin. However, somewhat sur-

prisingly in view of these results, it was without effect at concentrations up to 10^{-4}M on the cockroach hindgut (SULLIVAN and NEWCOMB, 1982). These differences in the responses of the cockroach hindgut and the locust extensor tibia raise the possibility of two classes of receptors with slightly differing structural specificities. While the entire amino acid sequence of proctolin is required for full activity on the hindgut, these results suggest that peptides lacking the C-terminal amino acid may produce good activity on the extensor tibia. If so, H-Arg-Tyr-Leu-Pro-OH may be much more active on the extensor tibia than it is on the hindgut. Also, it would be interesting to see if <Glu-Lys-Trp-Ala-Pro-Thr-OH would retain the activity of BPP_{5a} on the extensor tibia and show high activity on the cockroach hindgut.

Adipokinetic hormone, a peptide which regulates lipid utilization during flight, was extracted from the locusts Schistocerca gregaria and L. migratoria and identified as <Glu-Leu-Asn-Phe-Thr-Pro-Asn-Trp-Gly-Thr-NH_2 by STONE et al. (1976). The fat body and the flight muscle represent the two major sites of action in the locust although a number of other biological responses for this substance have also been discovered (MORDUE and STONE, 1981). The structure of this hormone closely resembles the crustacean red pigment concentrating hormone, <Glu-Leu-Asn-Phe-Ser-Pro-Gly-Trp-NH_2 (FERNLUND, 1974). Adipokinetic hormone is five times more active than the crustacean hormone in the locust system but less active on prawn erythrophores indicating that each is much better at eliciting a response in its own system (MORDUE and STONE, 1977). A larger difference in activities was observed in tests with two butterflies (DALLMANN et al., 1981). Two other peptides from insect corpora cardiaca, compound II from S. gregaria and L. migratoria (CARLSEN et al., 1979) and neurohormone D from P. americana (BAUMANN and GERSCH, 1982) also appear to be structurally related to locust adipokinetic hormone (GREENBERG and PRICE, 1983).

HERMAN et al. (1977) tested several synthetic analogues of the red pigment concentrating hormone for ability to elevate lipid levels in adult S. gregaria. <Glu-Leu-Asn-Tyr-Ser-Pro-Gly-Trp-NH_2 with tyrosine in place of phenylalanine at position 4 was as effective as the natural hormone while analogues of shorter chain length were inactive. In order to obtain some understanding of adipokinetic hormone-receptor interactions, STONE et al. (1978) tested 16 peptides structurally related to adipokinetic hormone. Relative activities of the active ana-

logues are shown in Table 3. The results indicate that a
minimum of the first 8 amino acid residues from the amino
terminal are necessary to produce the adipokinetic response. Removal of more than two amino acids from the C-
terminus or the first 4 to 6 amino acids from the N-terminus yielded peptides with no significant activity. Although all the active compounds are uncharged, it cannot
be concluded that this is a prerequisite for activity.
Substitution of the L-enantiomer of pyroglutamic acid
with the D-form causes a large reduction in the activity
(95% lower in the decapeptides and 60% in the octapeptides). This indicates that the L-enantiomer is required
for maximum biological activity. Interchange of the two
C-terminal amino acids or deletion of the terminal threonine also reduces the activity to less than that of the
octapeptides. Substitution of glycine for the asparagine
residue at position 7 reduces the activity by 70%. This
result indicates that the asparagine side chain is not
essential for activity although it obviously contributes.
The activity of the octapeptide resulting from deletion
of the two C-terminal amino acids was equal to that of
the red pigment concentrating hormone which differs by
having glycine instead of asparagine at position 7 and
serine instead of threonine at position 5.

As indicated before, activities determined for a
series of analogues of a biologically active compound may
reflect the structural requirements of the receptor, differences in rate of transport or differences in degradation rates. In considering these factors, STONE et al.

Table 3. Relative activities of adipokinetic hormone and
some structurally related peptides (from STONE et al.,
1978)

Peptide	Relative Activity
$<$Glu-Leu-Asn-Phe-Thr-Pro-Asn-Trp-Gly-Thr-NH$_2$*	100
$<$D-Glu-Leu-Asn-Phe-Thr-Pro-Asn-Trp-Gly-Thr-NH$_2$	5
$<$Glu-Leu-Asn-Phe-Thr-Pro-Asn-Trp-Thr-Gly-NH$_2$	3
$<$Glu-Leu-Asn-Phe-Thr-Pro-Gly-Trp-Gly-Thr-NH$_2$	30
$<$Glu-Leu-Asn-Phe-Thr-Pro-Asn-Trp-Gly-NH$_2$	7
$<$Glu-Leu-Asn-Phe-Ser-Pro-Gly-Trp-NH$_2$**	20
$<$D-Glu-Leu-Asn-Phe-Thr-Pro-Asn-Trp-NH$_2$	8
$<$Glu-Leu-Asn-Phe-Thr-Pro-Asn-Trp-NH$_2$	20

*   Adipokinetic hormone.
** Red pigment concentrating hormone.

(1978) investigated the time courses for lipid mobilisation to obtain a measure of the speed of response and an estimate of the half-life in the haemolymph of the various peptides compared with the natural hormone. They found that the analogues and adipokinetic hormone are transported to their site of action and degraded over similar time periods indicating that the relative activities reflect the different abilities of the analogues to activate the receptors. Therefore, the observed structure-activity correlations are probably indicative of the structural requirements of the locust fat body adipokinetic hormone receptors.

Recently, HARDY and SHEPPARD (1983) have reported the synthesis of two further analogues of locust adipokinetic hormone, [4,5-dehydroLeu^2]- and [3,4-dehydro-Pro^6]-adipokinetic hormone. The former, which has 4,5-dehydroleucine in place of leucine in the 2-position, is as active as adipokinetic hormone while the latter, which has 3,4-dehydroproline in place of proline in the 6-position, has only about 20% of the activity. These results indicate that neither receptor binding or resistance to enzymic attack are enhanced by the introduction of the unsaturation. However, the prolyl residue appears to be an important contributor to the hormone agonist activity since a slight change, the introduction of a double bond, resulted in a significant loss of activity.

GÄDE (1979) observed the accumulation of cyclic AMP in the fat body of adult L. migratoria after injection of synthetic adipokinetic hormone. He also tested the ability of four analogues which were previously evaluated for adipokinetic hormone activity by STONE et al. (1978) to raise fat body cyclic AMP levels. The octapeptide <Glu-Leu-Asn-Phe-Thr-Pro-Asn-Trp-NH$_2$, lacking the two C-terminal amino acids of adipokinetic hormone, and the decapeptide with glycine substituted for the asparagine at position 7 were substantially less effective than adipokinetic hormone. A much greater reduction in activity occurred when the two C-terminal amino acids of adipokinetic hormone were reversed. Little effect was shown for a high concentration of a heptapeptide, <Glu-Leu-Asn-Phe-Thr-Pro-Trp-NH$_2$. These results are consistent with the results obtained by STONE et al. (1978) and support the proposal (GÄDE and HOLWERDA, 1976; SPENCER and CANDY, (1976) that the action of adipokinetic hormone is mediated by cyclic AMP.

In summary, approaches utilized in the very limited studies of insect neuropeptide structure-activity relationships to date include: (1) alteration of the termin-

al amino acids, (2) extension of peptide chain, (3) trun-
cation of peptide chain, (4) substitution of an individual amino acid with amino acids of the same class (e.g.
replacement of tyrosine with other aromatic amino acids),
(5) substitution of amino acids of peptide chain with an
aliphatic amino acid (e.g. glycine, alanine), (6) substitution of L-amino acid moieties with the respective D-
amino acids, (7) substitution of amino acid moieties with
unnatural amino acids, (8) change of more than one amino
acid in the peptide, (9) modification of an amino acid
residue (e.g. methylation), and (10) tests of peptides
from other natural sources or obtained commercially.

Published work concerning mammalian or other invertebrate neuropeptides provide additional strategies which
may be useful. Frequently, analogues containing unnatural amino acids have been very informative, especially in
assessing the importance of electronic, hydrophobic and
steric factors in determining bioactivity. For example,
such substances have been used for the preparation of
peptides with greater hydrophobicity (LEDUC et al., 1983)
and with reduced conformational flexibility (HRUBY,
1982). Unnatural amino acids have also been employed to
produce peptide analogues with enhanced resistance to
enzymatic degradation (HAZUM et al., 1981; SHIMOHIGASHI
et al., 1982). Incorporation of unnatural amino acids at
selected positions in the peptide chain can lead to analogues with either greatly enhanced agonist (NESTOR et
al., 1982) or antagonist (ERCHEGYI et al., 1981) activity.

A large number of possible conformational states
differing only slightly in energy are available to most
peptides. Because molecular geometry is so important in
determining peptide-receptor affinity and the activation
of the biological response, many conformational studies
have been made on peptides found in the mammalian central
nervous system (HRUBY et al., 1983; MARSHALL, 1982;
STEWART, 1982). Most have been based on physical measurements of peptides in solution. Although peptides probably act on a particular receptor via a single conformation, the conformation of the peptide in combination with
the receptor may not be the one in greatest abundance in
solution. There is also evidence that peptides can induce conformational changes in the receptor upon effective combination with it (STEWART et al., 1976). Because
of the problems associated with attempts to extrapolate
results from solution studies, other methods for determining information about the receptor-bound conformations
of biologically active peptides are necessary.

Structure-activity studies, particularly using analogues having restricted conformational possibilities, have been very fruitful (HRUBY, 1982; HRUBY et al., 1983).

No direct evidence concerning the receptor bound conformations of the two known insect neuropeptides is available. Both are probably very flexible molecules able to adopt a wide variety of conformations. STONE et al. (1978) have reported that application of the second-ary structure predictive model of CHOU and FASMAN (1974) to adipokinetic hormone indicated a highly favored con-formation in which residues 5-8 form a β-bend (LEWIS et al., 1971). Some support for this conformation is obtained from the fact that a minimum of the first 8 res-idues are required for activity. In a theoretical study, BETIN'SH and NIKIFOROVICH (1979) calculated the energies of various conformations of proctolin and found that a quasicyclic structure with the tyrosyl side chain point-ing outward was favored.

Although only two insect neuropeptides have been identified to date, the recognition that peptides may serve as hormones, transmitters or modulators in the insect nervous system coupled with recent technical achievements such as HPLC and improved sequencing methods (HUNKAPILLER and HOOD, 1978, 1980) should soon lead to the structural elucidation of others. The desire to de-velop new and more selective insect control strategies also contributes to the increased attention given such compounds and will provide incentive for the investiga-tion of the relationships between structure and activity once the amino acid sequences are known. Agonists and antagonists, useful for the elucidation of the mode of action of the natural biologically active peptides, may result from the latter investigations. It is clear that studies of insect neuropeptide structure-activity rela-tionships, especially using synthetic analogues, will be actively pursued in the future.

REFERENCES

ARIENS E.J. (1977) Excursions in the field of SAR. A consideration of the past, the present and the future. In Biological Activity and Chemical Structure (Ed. by BUISMAN J.A.K.), pp. 1-35. Elsevier, Amsterdam.

BAJUSZ S. (1982) Structure and function relationships among synthetic enkephalin analogues. In Hormonal-ly Active Brain Peptides, Structure and Function

(Ed. by MCKERNS K.W. and PANTIĆ V.), pp. 1-24.
Plenum Press, New York.

BARANY G. and MERRIFIELD R.B. (1980) Solid-phase peptide
synthesis. In The Peptides, Analysis, Synthesis,
Biology (Ed. by GROSS E. and MEIENHOFER J.) 2,
1-284. Academic Press, New York.

BAUMANN E. and GERSCH M. (1982) Purification and identification of neurohormone D, a heart accelerating
peptide from corpora cardiaca of the cockroach
Periplaneta americana. Insect Biochem. 12, 7-14.

BETIN'SH Y.R. and NIKIFOROVICH G.V. (1979) Theoretical
conformational analysis of the proctoline molecule.
Sov. J. Bioorg. Chem. 5, 1175-1177 (Bioorg. Khim.
5, 1581-1583).

BODANSZKY M. (1979) Active esters in peptide synthesis.
In The Peptides, Analysis, Synthesis, Biology (Ed.
by GROSS E. and MEIENHOFER J.) 1, 105-196. Academic
Press, New York.

BODANSZKY M., KLAUSNER Y.S. and ONDETTI M.A. (1976)
Peptide synthesis, 2nd edition. John Wiley & Sons,
New York.

BROWN B.E. (1965) Pharmacologically active constituents
of the cockroach corpus cardiacum: resolution
and some characteristics. Gen. comp. Endocr. 5,
387-401.

BROWN B.E. (1975) Proctolin: a peptide transmitter candidate in insects. Life Sci. 17, 1241-1252.

BROWN B.E. (1977) Occurrence of proctolin in six orders
of insects. J. Insect Physiol. 23, 861-864.

BROWN B.E. and STARRATT A.N.(1975) Isolation of proctolin, a myotropic peptide, from Periplaneta americana. J. Insect Physiol. 21, 1879-1881.

CARLSEN J., HERMAN W.S., CHRISTENSEN M. and JOSEFSSON
L. (1979) Characterization of a second peptide with
adipokinetic and red pigment-concentrating activity
from the locust corpora cardiaca. Insect Biochem.
9, 497-501.

CHOU P.Y. and FASMAN G.D. (1974) Prediction of protein
conformation. Biochemistry 13, 222-245.

DALLMANN S.H., HERMAN W.S., CARLSEN J. and JOSEFSSON L.
(1981) Adipokinetic activity of shrimp and locust
peptide hormones in butterflies. Gen. comp. Endocr.
43, 256-258.

ERCHEGYI J., COY D.H., NEKOLA M.V., PEDROZA E., COY E.J.,
MEZO I. and SCHALLY A.V. (1981) LH-RH antagonists:
further analogs with ring-substituted aromatic
residues. Peptides 2, 251-253.

FAUCHÈRE J.-L. (1982) A quantitative structure-activity
relationship study of the inhibitory action of a
series of enkephalin-like peptides in the guinea

pig ileum and mouse vas deferens bioassays. J. med.Chem. 25, 1428-1431.

FERNLUND P. (1974) Structure of the red-pigment-concentrating hormone of the shrimp, Pandalus borealis. Biochim. biophys. Acta 371, 304-311.

FERSHT A.R. (1982) Chemical basis of biological specificity. Pure Appl. Chem. 54, 1819-1824.

FREE S.M. and WILSON J.W. (1964) A mathematical contribution to structure-activity studies. J. med. Chem. 7, 395-399.

FRONTALI N. and GAINER H. (1977) Peptides in invertebrate nervous systems. In Peptides in Neurobiology (Ed. by GAINER H.), pp. 259-294. Plenum Press, New York.

FUJITA T. (1972) The extrathermodynamic structure-activity correlations. Background of the Hansch approach. In Biological Correlations - The Hansch Approach (Ed. by GOULD R.F.), Adv. in Chem. Ser. 114, 1-19. American Chemical Society, Washington.

GÄDE G. (1979) Studies on the influence of synthetic adipokinetic hormone and some analogs on cyclic AMP levels in different arthropod systems. Gen. comp. Endocr. 37, 122-130.

GÄDE G. and HOLWERDA D.A. (1976) Involvement of adenosine 3':5'-cyclic monophosphate in lipid mobilization in Locusta migratoria. Insect Biochem. 6, 535-540.

GOODMAN M. and CHOREV M. (1979) On the concept of linear modified retro-peptide structures. Acc. Chem. Res. 12, 1-7.

GORIN F.A., BALASUBRAMANIAN T.M., CICERO T.J., SCHWIETZER J. and MARSHALL G.R. (1980) Novel analogues of enkephalin: identification of functional groups required for biological activity. J. med. Chem. 23, 1113-1122.

GREENBERG M.J. and PRICE D.A. (1983) Invertebrate neuropeptides: native and naturalized. Annu. Rev. Physiol. 45, 271-288.

HANSCH C. (1969) A quantitative approach to biochemical structure-activity relationships. Acc. Chem. Res. 2, 232-239.

HANSCH C. (1972) A computerized approach to quantitative biochemical structure-activity relationships. In Biological Correlations - The Hansch Approach (Ed. by GOULD R.F.), Adv. in Chem. Ser. 114, 20-40. American Chemical Society, Washington.

HANSCH C. and FUJITA T. (1964) ρ-σ-π Analysis. A method for the correlation of biological activity and chemical structure. J. Am. chem. Soc. 86, 1616-1626.

HARDY P.M. and SHEPPARD P.W. (1983) Synthesis of [4,5-dehydroLeu[2]]- and [3,4-dehydroPro[6]]-locust adipo-kinetic hormone. J. chem. Soc., Perkin Trans. 1, 723-729.

HAZUM E., FRIDKIN M., BARAM T. and KOCH Y. (1981) Synthesis, biological activity and resistance to enzymic degradation of luteinizing hormone-releasing hormone analogues modified at position 7. FEBS Lett. 123, 300-302.

HERMAN W.S., CARLSEN J.B., CHRISTENSEN M. and JOSEFSSON L. (1977) Evidence for an adipokinetic function of the RPCH activity present in the desert locust neuroendocrine system. Biol. Bull. mar. biol. Lab., Woods Hole, 153, 527-539.

HOLMAN G.M. and COOK B.J. (1972) Isolation, partial purification and characterization of a peptide which stimulates the hindgut of the cockroach, Leucophaea maderae (Fabr.). Biol. Bull. mar. biol. Lab., Woods Hole 142, 446-460.

HOLMAN G.M. and COOK B.J. (1979a) The analytical determination of proctolin by HPLC and its pharmacological action in the stable fly. Comp. Biochem. Physiol. 62C, 231-235.

HOLMAN G.M. and COOK B.J. (1979b) Evidence for proctolin and a second myotropic peptide in the cockroach, Leucophaea maderae, determined by bioassay and HPLC analysis. Insect Biochem. 9, 149-154.

HRUBY V.J. (1982) Conformational restrictions of biologically active peptides via amino acid side chain groups. Life Sci. 31, 189-199.

HRUBY V.J., MOSBERG H.I., SAWYER T.K., KNITTEL J.J., ROCKWAY T.W., ORMBERG J., DARMAN P., CHAN W.Y. and HADLEY M.E. (1983) Conformational and dynamic considerations in the design of peptide hormone analogs. Biopolymers 22, 517-530.

HUNKAPILLER M.W. and HOOD L.E. (1978) Direct microsequence analysis of polypeptides using an improved sequenator, a nonprotein carrier (Polybrene), and high pressure liquid chromatography. Biochemistry, 17, 2124-2133.

HUNKAPILLER M.W. and HOOD L.E. (1980) New protein sequenator with increased sensitivity. Science, Wash. 207, 523-525.

JONES J.H. (1979) The formation of peptide bonds: a general survey. In The Peptides, Analysis, Synthesis, Biology (Ed. by GROSS E. and MEIENHOFER J.) 1, 65-104. Academic Press, New York.

KRÖHN A. (1982) Molecular graphics in the design of peptide analogues. Biochem. Soc. Trans. 10, 309-310.

KRUMMEN K. (1980) HPLC in the analysis and separation of pharmaceutically important peptides. J. Liq. Chromat. 3, 1243-1254.

LEDUC R., BERNIER M. and ESCHER E. (1983) Angiotensin-II analogues. I: Synthesis and incorporation of the halogenated amino acids 3-(4'-iodophenyl)alanine, 3-(3',5'-dibromo-4'-chlorophenyl)alanine,3-(3',4', 5'-tribromophenyl)alanine, and 3-(2',3',4',5',6'-pentabromophenyl)alanine. Helv. chim. Acta 66,960-970.

LEWIS P.N., MOMANY F.A. and SCHERAGA H.A. (1971) Folding of polypeptide chains in proteins: a proposed mechanism for folding. Proc. natn. Acad. Sci. U.S.A. 68, 2293-2297.

LUKAS T.J., PRYSTOWSKY M.B. and ERICKSON B.W. (1981) Solid-phase peptide synthesis under continuous-flow conditions. Proc. natn. Acad. Sci. U.S.A. 78, 2791-2795.

MARSHALL G.R. (1982) Conformational studies of peptide hormones as a basis for analog design. In The Chemical Regulation of Biological Mechanisms (Ed. by CREIGHTON A.M. and TURNER S.),pp. 279-292. The Royal Society of Chemistry, London.

MARTIN Y.C. (1981) A practitioner's perspective of the role of quantitative structure-activity analysis in medicinal chemistry. J. med. Chem. 24, 229-237.

MEIENHOFER J. (1979a) The azide method in peptide synthesis. In The Peptides, Analysis, Synthesis, Biology (Ed. by GROSS E. and MEIENHOFER J.) 1, 197-239. Academic Press, New York.

MEIENHOFER J. (1979b) The mixed carbonic anhydride method of peptide synthesis. In The Peptides, Analysis, Synthesis, Biology (Ed. by GROSS E. and MEIENHOFER J.) 1, 263-314. Academic Press, New York.

MORDUE W. and STONE J.V. (1977) Relative potencies of locust adipokinetic hormone and prawn red pigment-concentrating hormone in insect and crustacean systems. Gen. comp. Endocr. 33,103-108.

MORDUE W. and STONE J.V. (1981) Structure and function of insect peptide hormones. Insect Biochem. 11,353-360.

MORLEY J.S. (1980) Structure-activity relationships of enkephalin-like peptides. Annu. Rev. Pharmacol. Toxicol. 20,81-110.

NESTOR J.J., HO T.L., SIMPSON R.A., HORNER B.L., JONES G.H., MCRAE G.I. and VICKERY B.H. (1982) Synthesis and biological activity of some very hydrophobic superagonist analogues of luteinizing hormone-releasing hormone. J. med. chem. 25, 795-801.

PIEK T. (1982) Solitary wasp venoms and toxins as tools

for the study of neuromuscular transmission in
insects. In Neuropharmacology of Insects (Ed. by
EVERED D., O'CONNOR M. and WHELAN J.), Ciba Fdn.
Symp. 88, 275-290. Pitman, London.

PIEK T., VISSER B.J. and MANTEL P. (1979) Effect of
proctolin, BPP_{5a} and related peptides on rhythmic
contractions in Locusta migratoria. Comp.
Biochem. Physiol. 62C, 151-154.

RAABE M. (1982) Insect neurohormones. Plenum Press,
New York.

RICH D.H. and SINGH J. (1979) The carbodiimide method.
In The Peptides, Analysis, Synthesis, Biology (Ed.
by GROSS E. and MEIENHOFER J.) 1, 241-261.
Academic Press, New York.

SHIMOHIGASHI Y., CHEN H.-C. and STAMMER C.H. (1982) The
enzyme stability of dehydro-enkephalins.
Peptides 3, 985-987.

SPENCER I.M. and CANDY D.J. (1976) Hormonal control of
diacyl glycerol mobilization from fat body of the
desert locust, Schistocerca gregaria. Insect
Biochem. 6, 289-296.

STARRATT A.N. and BROWN B.E. (1975) Structure of the
pentapeptide proctolin, a proposed neurotransmitter
in insects. Life Sci. 17, 1253-1256.

STARRATT A.N. and BROWN B.E. (1979) Analogs of the insect
myotropic peptide proctolin: synthesis and
structure-activity studies. Biochem. biophys.
Res.Commun. 90, 1125-1130.

STARRATT A.N. and STEELE R.W. (1983) In vivo inactivation
of the insect neuropeptide proctolin in Periplaneta
americana. Insect. Biochem. In press.

STARRATT A.N. and STEVENS M.E. (1980) Ion-pair high-
performance liquid chromatography of the insect
neuropeptide proctolin and some analogs. J.
Chromat. 194, 421-423.

STEWART J.M. (1982) The design of peptide hormone analogs. Trends Pharmacol. Sci. 3, 300-303.

STEWART J.M., FREER R.J., REZENDE L., PENA C. and
MATSUEDA G.R. (1976) A pharmacological study of the
angiotensin receptor and tachyphylaxis in smooth
muscle. Gen. Pharmacol. 7, 177-183.

STONE J.V., MORDUE W., BATLEY K.E. and MORRIS H.R. (1976)
Structure of locust adipokinetic hormone, a neurohormone that regulates lipid utilisation during
flight. Nature, Lond. 263, 207-211.

STONE J.V., MORDUE W., BROOMFIELD C.E. and HARDY P.M.
(1978) Structure-activity relationships for the
lipid-mobilising action of locust adipokinetic
hormone. Synthesis and activity of a series of
hormone analogues. Eur. J. Biochem. 89, 195-202.

SULLIVAN R.E. and NEWCOMB R.W. (1982) Structure function
     analysis of an arthropod peptide hormone: proctolin
     and synthetic analogues compared on the cockroach
     hindgut receptor. Peptides 3, 337-344.
TUTE M. (1983) QSAR and drug design. Chem. Ind.
     (London), 10-13.
VOELTER W. (1981) High performance liquid chromatography
     in peptide research. In High Performance Liquid
     Chromatography in Protein and Peptide Chemistry
     (Ed. by LOTTSPEICH F., HENSCHEN A. and HUPE K.-P.),
     pp. 1-53. Walter de Gruyter, Berlin.
WALKER R.J., JAMES V.A. and ROBERTS C.J. (1981) The
     action of FMRF-amide and proctolin on Helix,
     Hirudo, Limulus and Periplaneta neurones. In Adv.
     Physiol. Sci., Proc. Int. Congr., 28th (Ed. by
     PETHES G. and FRENYÓ V.L.) 20, 411-416. Acad.
     Kiado, Budapest.
WATSON W.H., AUGUSTINE G.J., BENSON J.A. and SULLIVAN
     R.E. (1983) Proctolin and an endogenous proctolin-
     like peptide enhance the contractility of the
     Limulus heart. J. exp. Biol. 103, 55-73.

SULLIVAN R.E. and NEWCOMB R.W. (1982) Structure function analysis of an arthropod peptide hormone: proctolin and synthetic analogues compared for the activity of hindgut receptor. Peptides 12, 337-344.

NEUROSECRETION IN INSECTS:

STRATEGIES FOR CELLULAR ANALYSIS

Grant M. Carrow

The Biological Laboratories
Harvard University
Cambridge, MA

INTRODUCTION

The intercellular communicative functions of neurons are generally initiated by secretory processes and completed by the binding of secreted molecules to target cell receptors. Neurosecretory cells are distinguished from nonendocrine neurons by characteristics unique to the secretion of neurohormones. Thus, neurosecretory cells have a large number of terminals that end at extracellular spaces rather than synaptic clefts. Furthermore, all neurohormones have to date turned out to be polypeptides, packaged in large electron-dense vesicles (reviewed by Rowell, 1976; Maddrell and Nordmann, 1979).

Neurosecretory cell somata and neurohemal organs have been located and studied throughout insect central and peripheral nervous systems by the histochemical, ultrastructural, and physiological detection of neurosecretory material (reviewed by Rowell, 1976; Raabe, 1982). Classically, standard histological techniques employing chrome-hematoxylin, paraldehyde-fuchsin, or other specific stains have been used to label neurosecretory material. More recently, neurosecretory cells or neurohemal organs have been labeled with invertebrate or vertebrate neuropeptide antisera (Remy and Girardie, 1980; Hansen et al., 1982; El-Salhy et al., 1983; Schooneveld et al., 1983). At the ultrastructural level, neurosecretory cells are distinguished by electron-dense vesicles with diameters of 100 - 300 nm located in somata, axons, and terminals. Extraction of hormone activity from neural tissue has assisted in the assignment of endocrine function to particular groups of cells. Finally, induced release of neurohormones has been used as a probe for identifying neurohemal sites as well as for

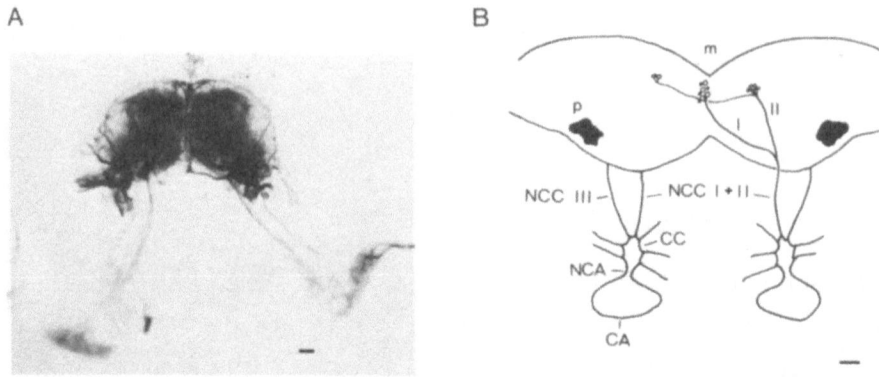

Fig. 1.   The brain-retrocerebral complex of <u>Manduca</u> <u>sexta</u>.
          (A) Light micrograph of a living, intact complex isolated
          from a wandering (premetamorphic) larva; darkfield
          illumination. Connective tissue sheath is visibly
          separated from cortex. (B) Diagram of the complex of a
          pupa showing locations of neurosecretory cell bodies and
          axon tracts. Only those cells whose axons exit from the
          right hemisphere are shown; their counterparts are
          omitted for clarity. CA, corpus allatum; CC, corpus
          cardiacum; m, midline of brain; NCA, nervus corporis
          allati; NCC, nervus corporis cardiaci; p, pigment
          granules in optic lobe; I, axon tract I; II, axon tract
          II. Bars equal 100 μ.

studying mechanisms of neurosecretion.

     These approaches have led to the identification of a few simple
systems of potentially accessible and identifiable neurosecretory
cells. It is likely that characterization of the endocrine function
of these neurons will contribute to the elucidation of the cellular
basis of neurosecretion. A particularly promising subject for
cellular analysis is the neuroendocrine system in the insect brain.
In the moth, <u>Manduca</u> <u>sexta</u>, the structure of the cerebral
neurosecretory cells has been described (Nijhout, 1975; Buys and
Gibbs, 1981; Carrow, 1983). Moreover, the cell bodies and
neurohemal organs for prothoracicotropin -- a neuropeptide regulator
of growth and postembryonic development -- have been identified
(Gibbs and Riddiford, 1977; Agui et al., 1979, 1980; Carrow et al.,
1981; Carrow, 1983). Thus, the cerebral neuroendocrine system of
this species is sufficiently well defined to be accessible to
concommitant physiological, anatomical, and biochemical analysis.

     The cerebral neuroendocrine system of <u>Manduca</u>, illustrated in
Fig. 1, will serve here as a focus for discussion of strategies for
the cellular analysis of neurosecretion. The discussion will

encompass the identification of neurohemal storage and release
sites, detection and measurement of neurohormone release in vitro,
and determination of mechanisms of excitation-secretion coupling.

INTRACELLULAR STAINING OF NEUROSECRETORY TERMINALS

Identification and characterization of neurosecretory cells and
their neurohemal organs is a prerequisite for studying
neurosecretion at the cellular level. Cephalic neurosecretory
endings have been found in the retrocerebral complex (corpora
cardiaca and corpora allata), the aorta, and the brain itself (see
Raabe, 1982 for review). Retrograde cobalt filling was used to
locate the cells that project to the retrocerebral complex by
filling the nervi coporis cardiaci in Orthoptera (Mason, 1973; Pipa,
1978; Koontz and Edwards, 1980) and Lepidoptera (Nijhout, 1975; Buys
and Gibbs, 1981). Details of the morphology of dendritic fields and
axon tracts were revealed by silver intensification of the
precipitated cobalt sulfide. Although anterograde filling of the
nervi corporis cardiaci can be used similarly to stain neurohemal
areas with cobalt, any nonendocrine axons that share these nerves
may be filled simultaneously (Nijhout, 1975).

Previous studies of cephalic neurohemal areas have suffered
from the inability to ascribe the sites to particular cells in the
brain. Recently, identification and characterization of individual
neurosecretory cells has been achieved by intracellular marking
(Carrow et al., 1982; Taghert and Truman, 1982; Zaretsky and Loher,
1982; Carrow, 1983). As shown in Fig. 2, intracellular injection of
horeseradish peroxidase into the cerebral neurosecretory somata of
pupal Manduca has revealed at least five classes of monopolar
neurons distinguishable by their soma locations, dendritic
architecture, axonal pathways, and neurohemal projections. Cells of
two classes have terminal ramifications in the corpora cardiaca
while the remainder have terminals in the corpora allata. The
terminals of any single cell ramify throughout either organ or on
the surface of the corpus allatum. Furthermore, varicosities were
often apparent, suggesting neurohemal storage or release sites;
these may be equivalent to the dilations observed in scanning
electron micrographs of these organs in the beetle, Xyleborus
ferrugineus (Chu and Norris, 1979). Thus, there is now direct
morphological evidence that the corpora cardiaca and corpora allata
are distinct neurohemal organs for cerebral neurosecretory cells.

Access to these cells for dye injection with glass
micropipettes was made practicable by a combination of several
factors. The cell bodies are relatively large (15-30 μ diameter)
and are observable in the living brain due to their reflective
opalescence under spot fibre-optic illumination. Removal of the
connective tissue sheath promoted access to the somata for

microelectrodes without compromising the integrity of the
blood-brain barrier. Further, stiff electrode glass was employed
for penetrating the remaining perineurium.

    Cobalt, lucifer yellow, and horseradish peroxidase were each
injected to mark these cells; all the dyes resulted in equivalent
neurosecretory cell profiles but peroxidase gave the best results
for neurohemal organs. When injected into the somata with pressure,
horseradish peroxidase moved more rapidly and effectively to the
distal terminal areas than to the more proximal neurites and
dendrites. This may reflect active protein transport along the axon
and therefore may provide a tool for studying that process. An
additional advantage of marking cells with horseradish peroxidase is
that the chromophore that results from the reaction of the enzyme
with substrate is electron opaque. Thus, the ultrastructure of
individually identified cells can be examined. This approach could
help identify storage and release sites not apparent with light
microscopy. In addition, variations in the ultrastructure of an
identified cell under different physiological conditions could be
studied since the homologous cell can be marked in conspecifics.

RELEASE OF NEUROHORMONE IN VITRO

Assay

    Although evidence for neurosecretory cell terminals and
neurohormone storage is necessary for defining a neurohemal area, it
is insufficient in the absence of direct evidence for physiological
release of a neurohormone from the organ. Because of the small
amounts of neurohormone involved, measurement of release requires a
sensitive assay and an economical means of collecting
neurosecretions. Both of these requirements are met by studying
neurosecretion in vitro.

Fig. 2.   Architecture of the cerebral neurosecretory cells of
           pupal M. sexta and details of terminal projections as
           revealed by horseradish peroxidase injection. (A) Camera
           lucida drawings of cells whose somata lie in the left
           hemisphere. Five classes were distinguished on the basis
           of soma size and location, axon pathway, and neurohemal
           projection. (B) Light micrograph of a group IIa cell --
           a putative prothoracicotrope -- with soma and dendritic
           field in right protocerebrum, decussating axon, and
           projection to the corpus allatum. (C) Detail of terminal
           ramifications and varicosities in the corpus cardiacum of
           a group Ib cell. (D) Group Ia cell terminals and
           varicosities on the surface of the corpus allatum.
           Abbreviations as in Fig. 1. Bars equal 100 μ .

A

Group	Structure	Projection
Ia		CA
Ib		CC
IIa		CA
IIb		CC or CA

B

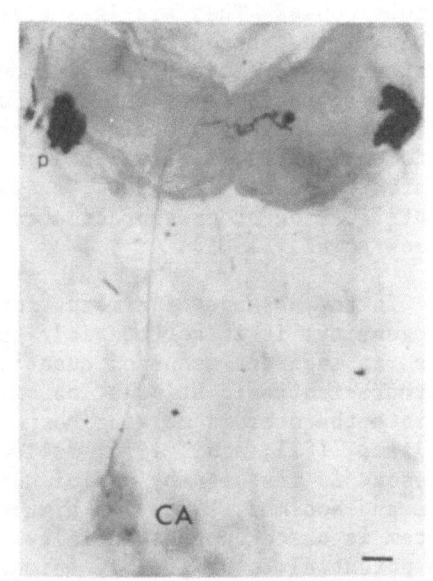

C

D

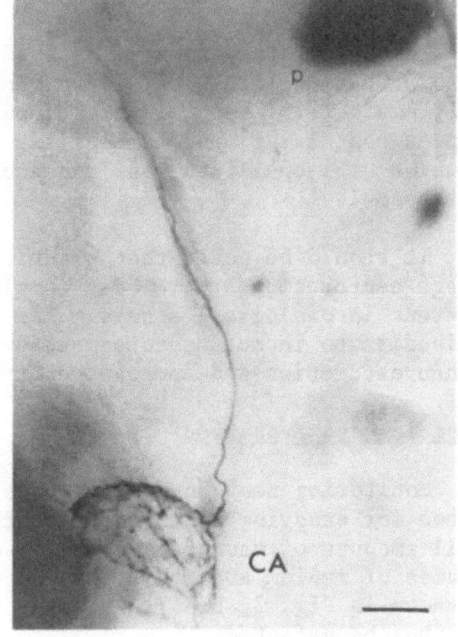

In vitro biological assays are efficient means for measuring neurohormones. They are less wasteful of material than are assays involving the introduction of neurohormone into the hemolymph where substances are subject to dilution and enzymatic degradation. In vitro preparations for the assay of insect neurohormones include isolated Malpighian tubules for the detection of diuretic hormone (Maddrell, 1980), isolated hindgut for measurement of proctolin (Starratt and Steele, 1980), and isolated prothoracic glands for the quantification of prothoracicotropin (Bollenbacher et al., 1979; Carrow et al., 1981).

In the absence of a means for purifying prothoracicotropin to homogeneity, it is not possible to measure the hormone directly. Thus, an indirect means of quantification was developed based on prothoracicotropic stimulation of the prothoracic glands in vitro to secrete the steroid molting hormone, ecdysone (Kambysellis and Williams, 1971; Agui, 1975; Bollenbacher et al., 1979); secreted ecdysone is measured by radioimmunoassay (Borst and O'Connor, 1974; Reum and Koolman, 1979). A highly sensitive and efficient assay system is based upon essentially inactive larval glands which have the potential to be greatly stimulated (Carrow et al., 1981). In addition, the use of larval prothoracic glands circumvents the problem of possible stimulation by juvenile hormone (Gilbert and Schneiderman, 1959; Williams, 1959). The response is log-linear over a 100-fold range of prothoracicotropin concentration and as little as a few percent of the total neurohormone extractable from a single brain can be detected.

Initially, tissues were assayed for prothoracicotropic activity by coincubation with prothoracic glands (Kambysellis and Williams, 1971; Agui, 1975) but this procedure was cumbersome. Instead, samples containing neurohormone are routinely stored at $-20 \,^{\circ}C$, in solution or lyophilized, and subsequently assayed along with many other samples.

It should be noted that quantitative measurements of most insect neurohormones have been complicated by the high variance inherent in biological assays. The development of monospecific antibodies to insect neurohormones would greatly facilitate studies of neurosecretion and neurosecretory cells.

Detection and Evocation

Monitoring neurohormone release in vitro provides a direct method for studying neurosecretion as well as a means for collecting small amounts of neuropeptide. Tissues may be incubated in small volumes of medium and there is no loss due to extraction procedures. As shown in Fig. 3, basal levels of release from individual organs can thus be detected.

In vitro neurosecretion may be used as a probe for identifying
neurohemal organs for specific neurohormones. Prothoracicotropin in
Manduca is extractable from the brain and the retrocerebral complex;
in the latter, activity is found primarily in the corpora allata
(Gibbs and Riddiford, 1977; Agui et al., 1980; Carrow et al., 1981).
In order to localize prothoracicotropin release sites, a wax/mineral
oil barrier was placed across the nervi corporis cardiaci in vitro.
In this manner, the pool bathing the corpora cardiaca and corpora
allata was separate from that bathing the brain. All released
neurohormone was recovered from the medium bathing the retrocerebral
complex (Carrow et al., 1981). In subsequent experiments,
summarized in Fig. 3, intact or reduced preparations were incubated,
without barriers, and the bathing media assayed for prothoracicotropic activity. The activity released spontaneously from isolated
corpora allata could account for the activity released from intact
brain-retrocerebral complexes indicating that the corpus allatum is
the primary neurohemal organ for prothoracicotropin.

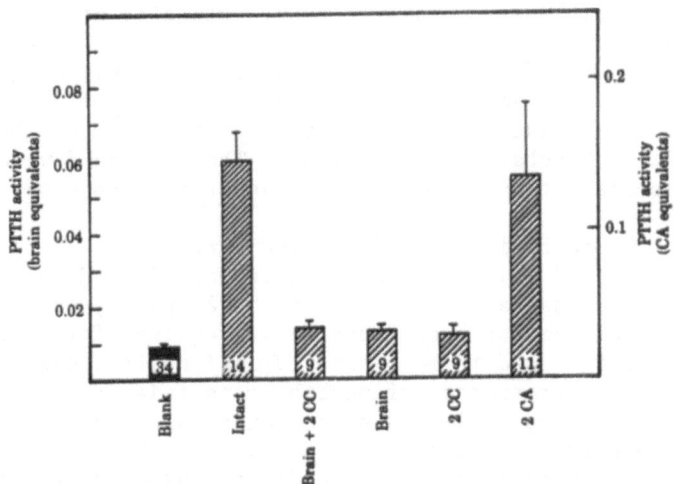

Fig. 3. Spontaneous release of prothoracicotropin (PTTH) from
         day-5 fifth instar brains and retrocerebral organs of
         M. sexta during 4-hr incubation periods. The means ± SEM
         are plotted in terms of the amount of PTTH extractable
         from a single brain (left ordinate) or corpus allatum
         (right ordinate); see text for explanation of assay. The
         number in each bar denotes the number of trials for that
         organ type. Blank, background due to basal level of
         ecdysone production by prothoracic glands; intact, intact
         brain-retrocerebral complex; brain + 2 CC, intact but
         lacking the pair of CA; brain, isolated brain; 2 CC,
         isolated pair of CC; 2 CA, isolated pair of CA. (From
         Carrow et al., 1981)

Neurohormone release can also be physiologically evoked in vitro. Incubation of the mesothoracic ganglionic mass of the bug Rhodnius in high-potassium medium led to the calcium-dependent release of diuretic hormone from abdominal nerves (Maddrell and Gee, 1974). As shown in Fig. 4, calcium-dependent release of prothoracicotropin was similarly evoked from the isolated corpora cardiaca and corpora allata of Manduca thus demonstrating that both organs are neurohemal sites for at least one cerebral neurohormone (Carrow et al., 1981). High potassium medium (greater than 50 mM) depolarizes insect neurons, including neurosecretory cells (Hoyle, 1953; Jego et al., 1970; Pichon et al., 1972; Carrow, 1983). Since the basis of the release mechanism in neurons is excitation-secretion coupling (Douglas, 1978), it is assumed that the release of substances from nerve terminals in vitro evoked by high potassium media is due to depolarization of the cells (Iversen et al., 1980).

Neurosecretory cell products may also be released by electrical stimulation. For example, both radioactively labelled proteins and gonadotropic activity were recovered from perfusates of brain-corpora cardiaca after electrical stimulation of the pars intercerebralis of Locusta (Girardie and Girardie, 1977). Similarly, immunoprecipitable radiolabelled proteins and diuretic hormone activity were recovered from the corpora cardiaca after electrical stimulation of the nervi corporis cardiaci (Orchard et al., 1981). The calcium-dependence of release was demonstrated only for the immunoprecipitable proteins.

Although the mechanism underlying the correlation of electrical activity and chemical release remains to be established for any neuron, the two phenomena are believed to be linked by calcium mobilization. Calcium has been found to be required for secretion in vertebrate neurons and in most non-neuronal cells (Douglas, 1978). The calcium dependence of neurohormone release established in Rhodnius and Manduca indicates a similar requirement for calcium in insect neurosecretion. Thus, the possibility of artifact in experiments on neurosecretion can be addressed by demonstrating the calcium-dependence of release and the reversibility of calcium block, as shown in Fig. 4.

The calcium-dependence of the release process can also be used as a tool for the preparation of tissues for study in vitro. Figure 4 shows that neurohemal tissue can be maintained in calcium-free medium to block neurosecretion prior to an experiment without subsequent loss of its ability to release neurohormone (Carrow et al., 1981).

Applications

In vitro neurosecretion can be used as a probe for locating and exploring neurohemal sites, measuring releasable stores of neurohormone, studying excitation-secretion coupling, and obtaining

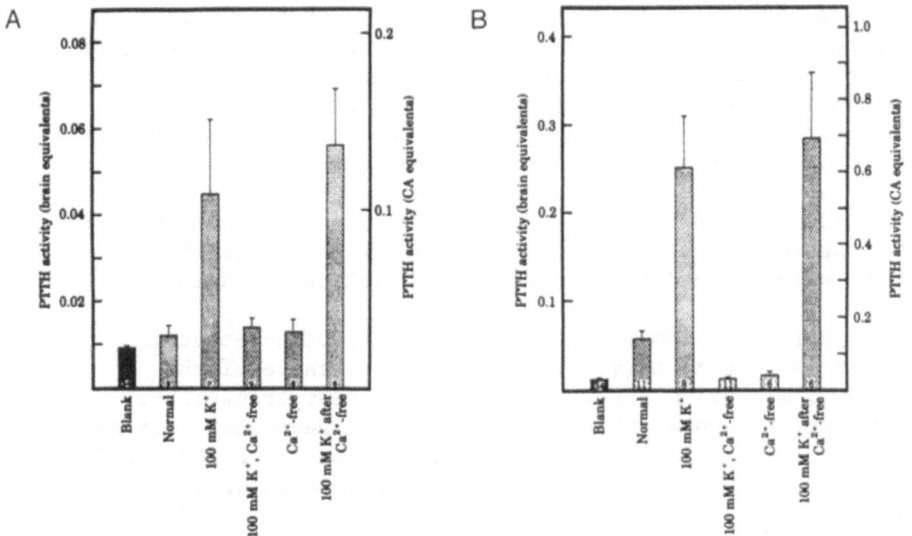

Fig. 4. Prothoracicotropin release from isolated pairs of CC (A) and CA (B) evoked by high potassium and calcium-dependence of release. Incubations were for 4 hr, except as noted below, in Grace's media that were isosmotic (350 mosmol/liter). The two bars at the right of each graph represent PTTH released from pairs of neurohemal organs preincubated for 4 hr in Ca^{2+}-free, normal-K^+ medium followed by incubation for 2 hr in high-K^+, normal-Ca^{2+} medium. All other variables were as in Fig. 3. (From Carrow et al., 1981)

the circulating forms of neurohormones.

In conjunction with intracellular staining and tissue extraction, the stimulation of neurosecretion in vitro is a powerful technique for identifying and characterizing neurohemal organs for identified neurohormones. The findings reviewed here of calcium-dependent release of prothoracicotropin from both the corpora cardiaca and corpora allata of Manduca helped establish that both are distinct neurohemal organs for cerebral neurosecretory cells. Neuropeptide release has been demonstrated by ionic or electrical extracellular stimulation; it should now be feasible to evoke neurosecretion by intracellular stimulation of identified neurons.

Differences in releasable stores of neurohormone have been correlated with the developmental or physiological stages of insects. In Rhodnius there is more releasable diuretic hormone after a bloodmeal than prior to it, suggesting that the neurohormone is mobilized as it is needed (Berlind, 1981). Whether mobilizaton reflects new synthesis, increased transport or enhanced secretion mechanisms is a question that could be addressed by the use of specific inhibitors of those processes. Prothoracicotropin is available for release in Manduca prior to and after the last larval molt but not during the molt itself (unpublished observations); this may indicate depletion of stores after release in vivo. One caveat for interpretation of data on releasable stores of neurohormone is that release in response to prolonged depolarization may be more limited by calcium inactivation than by depletion (Nordmann, 1976).

The opportunity now exists to study excitation-secretion coupling in individual neurosecretory cells; because of the difficulty of measuring the small amounts of neurotransmitter released from individual nonendocrine neurons, neurosecretory cells may serve as models in which to study excitation-secretion coupling in neurons in general. The dependence on extracellular calcium of neurohormone release in insects has been discussed. In addition, the induction of release of the diuretic hormone of Rhodnius by calcium ionophores in the presence of extracellular calcium further supports the thesis that release is dependent upon transmembrane calcium flux rather than the mobilization of intracellular stores (Berlind, 1981). The mechanisms governing coupling are otherwise unknown.

Some neurohormones are processed prior to release so that stored and circulating forms differ. In vitro release provides a means for obtaining relatively clean preparations of the circulating forms of neurohormones for biochemical analysis (Aston and White, 1974; Aston, 1979). The biological activity of prothoracicotropin extracted from brains or corpora allata of Manduca was indistinguishable from that released from the neurohemal organ (Carrow et al., 1981). Thus, although their specific acitivities and the degree of homogeneity between the forms could not be determined without purification, the different forms appeared to have a common origin.

CONCLUSIONS

Analysis of the process of neurosecretion encompasses the localization and characterization of neurohemal organs as well as the assessment of the dynamics and mechanisms of neurohormone storage and release. Intracellular injection of light and electron opaque dyes and evaluation of neurohormone release processes in vitro provide complementary means for investigating these aspects of neurosecretion in insects. A framework has been established in a few neuroendocrine systems for the analysis of neurosecretion at the level of individually identified neurosecretory cells.

ACKNOWLEDGEMENTS

The research was supported by grants from the National Science Foundation, the National Institutes of Health, and the Rockefeller Foundation.

REFERENCES

Agui, N., 1975, Activation of prothoracic glands by brains in vitro, J. Insect Physiol., 21:903.
Agui, N., Bollenbacher, W.E., Granger, N.A., and Gilbert, L.I., 1980, Corpus allatum is release site for insect prothoracicotropic hormone, Nature, 285:669.
Agui, N., Granger, N.A., Gilbert, L.I., and Bollenbacher, W.E., 1979, Cellular localization of the insect prothoracicotropic hormone: In vitro assay of a single neurosecretory cell, Proc. Natl. Acad. Sci. USA, 76:5694.
Aston, R.J., 1979, Studies on the diuretic hormone of Rhodnius prolixus. Some observations on the purification and nature of the hormone and the dynamics of its release in vitro, Insect Biochem., 9:163.
Aston, R.J. and White, A.F., 1974, Isolation and purification of the diuretic hormone from Rhodnius prolixus, J. Insect Physiol., 20:1673.
Berlind, A., 1981, Mobilization of a peptide neurohormone for release during a physiological secretion cycle, Gen. Comp. Endocrinol., 44:444.
Bollenbacher, W.E., Agui, N., Granger, N.A., and Gilbert, L.I., 1979, In vitro activation of insect prothoracic glands by the prothoracicotropic hormone, Proc. Natl. Acad. Sci. USA, 76:5148.
Borst, D.W. and O'Connor, J.D., 1974, Trace analysis of ecdysones by gas-liquid chromatography, radioimmunoassay and bioassay, Steroids, 24:637.
Buys, C.M. and Gibbs, D., 1981, The anatomy of neurons projecting to the corpus cardiacum from the larval brain of the tobacco hornworm, Manduca sexta (L.), Cell Tissue Res., 215:505.
Carrow, G.M., 1983, Physiology and architecture of insect cerebral neurosecretory cells, Ph.D. thesis, Harvard University.
Carrow, G.M., Calabrese, R.L., and Williams, C.M., 1981, Spontaneous and evoked release of prothoracicotropin from multiple neurohemal organs of the tobacco hornworm, Proc. Natl. Acad. Sci. USA, 78:5866.
Carrow, G.M., Calabrese, R.L., and Williams, C.M., 1982, Protocerebral neuroendocrine system of the tobacco hornworm: Morphology and physiology of identifiable neurosecretory cells, Soc. Neurosci. Abstr., 8:532.
Chu, H.M. and Norris, D.M., 1979, Comparative morphology and ultrastructure of the corpora allata in newly emerged and

sexually mature female <u>Xyleborus ferrugineus</u> (Fabr.)
(Coleoptera: Scolytidae), <u>Int</u>. <u>J</u>. <u>Insect Morphol</u>. <u>&
Embryol</u>., 8:359.

Douglas, W.W., 1978, Stimulus-secretion coupling: Variations on
the theme of calcium-activated exocytosis involving
cellular and extracellular sources of calcium, <u>Ciba Fdn</u>.
<u>Symp</u>., 54:61.

El-Salhy, M., Falkmer, S., Kramer, K.J., and Speirs, R.D., 1983,
Immunohistochemical investigations of neuropeptides in the
brain, corpora cardiaca, and corpora allata of an adult
lepidopteran insect, <u>Manduca sexta</u> (L), <u>Cell Tissue Res</u>.,
232:295.

Gibbs, D. and Riddiford, L.M., 1977, Prothoracicotropic hormone in
<u>Manduca sexta</u>: Localization by a larval assay, <u>J</u>. <u>exp</u>.
<u>Biol</u>., 66:255.

Gilbert, L.I. and Schneiderman, H.A., 1959, Prothoracic gland
stimulation by juvenile hormone extracts of insects,
<u>Nature</u>, 184:171.

Girardie, J. et Girardie, A., 1977, Liberation provoquee <u>in vitro</u>
du produit de neurosecretion des cellules protocerebrales
medianes chez le criquet migrateur. <u>J</u>. <u>Physiol</u>. <u>(Paris)</u>
73:707.

Hansen, B.L., Hansen, G.N., and Scharrer, B., 1982, Immunoreactive
material resembling vertebrate neuropeptides in the corpus
cardiacum and corpus allatum of the insect <u>Leucophaea maderae</u>,
<u>Cell Tissue Res</u>. 225:319.

Hoyle, G., 1953, Potassium ions and insect nerve muscle, <u>J</u>. <u>exp</u>.
<u>Biol</u>., 30:121.

Iversen, L.L., Lee, C.M., Gilbert, R.F., Hunt, S., and Emson, P.C.,
1980, Regulation of neuropeptide release, <u>Proc</u>. <u>R</u>. <u>Soc</u>.
<u>Lond</u>. <u>B</u>, 210:91.

Jego, P., Callec, J.J., Pichon, Y., et Boistel, J., 1970, Etude
electrophysiologique de corps cellulaires excitables
du VI[e] ganglion abdominal de <u>Periplaneta americana</u>.
Aspects electriques et ioniques. <u>C.R</u>. <u>Seances Soc</u>. <u>Biol</u>.
<u>Filiales</u>, 164:893.

Kambysellis, M.P. and Williams, C.M., 1971, <u>In vitro</u> development
of insect tissues. II. The role of ecdysone in the
spermatogenesis of silkworms, <u>Biol</u>. <u>Bull</u>., 141:541.

Koontz, M. and Edwards, J.S., 1980, The projection of neuroendocrine
fibers (NCC I and II) in the brains of three orthopteroid
insects, <u>J</u>. <u>Morph</u>., 165:285.

Maddrell, S.H.P., 1980, Bioassay of diuretic hormone in Rhodnius,
<u>in</u>: "Neurohormonal Techniques in Insects", T.A. Miller, ed.,
Springer-Verlag, NY.

Maddrell, S.H.P. and Gee, J.D., 1974, Potassium-induced release of
the diuretic hormone of <u>Rhodnius prolixus</u> and <u>Glossina
austeni</u>: Ca dependence, time course and localization of
neurohaemal areas, <u>J</u>. <u>exp</u>. <u>Biol</u>., 61:155.

Maddrell, S.H.P. and Nordmann, J.J., 1979, "Neurosecretion",
    Wiley and Sons, NY.
Mason, C.A., 1973, New features of the brain-retrocerebral
    neuroendocrine complex of the locust Schistocerca vaga
    (Scudder), Z. Zellforsch., 141:19.
Nijhout, H.F., 1975, Axonal pathways in the brain-retrocerebral
    complex of Manduca sexta (L.) (Lepidoptera: Sphingidae),
    Int. J. Insect Morphol. & Embryol., 4:529.
Nordmann, J.J., 1976, Evidence for calcium inactivation during
    hormone release in the rat neurohypophysis, J. exp. Biol.,
    65:669.
Orchard, I., Friedel, T., and Loughton, B.G., 1981, Release of a
    neurosecretory protein from the corpora cardiaca of
    Locusta migratoria induced by high potassium saline and
    compound action potentials, J. Insect Physiol., 5:297.
Pichon, Y., Sattelle, D.B., and Lane, N.J., 1972, Conduction
    processes in the nerve cord of the moth Manduca sexta in
    relation to its ultrastructure and haemolymph ionic
    composition, J. exp. Biol, 56:717.
Pipa, R. L., 1978, Locations and central projections of neurons
    associated with the retrocerebral neuroendocrine complex of
    the cockroach Periplaneta americana (L.). Cell Tissue Res.,
    193:443.
Raabe, M., 1982, "Insect Neurohormones", Plenum, NY.
Remy, C. and Girardie, J., 1980, Anatomical organization of two
    vasopressin-neurophysin-like neurosecretory cells throughout
    the central nervous system of the migratory locust, Gen.
    Comp. Endocrinol., 40:27.
Reum, L. and Koolman, J., 1979, Analysis of ecdysteroids by
    radioimmunoassay: Comparison of three different antisera,
    Insect Biochem., 9:135.
Rowell, H.F., 1976, The cells of the insect neurosecretory system:
    Constancy, variability, and the concept of the unique
    identifiable neuron, Adv. Insect Physiol., 12:63.
Schooneveld, H., Tesser, G.I., Veenstra, J.A., and Romberg-Privee,
    H.M., 1983, Adipokinetic hormone and AKH-like peptide
    demonstrated in the corpora cardiaca and nervous system of
    Locusta migratoria by immunocytochemistry, Cell Tissue Res.,
    230:67.
Starratt, A.N. and Steele, R.W., 1980, Proctolin: Bioassay,
    isolation, and structure, in: "Neurohormonal Techniques
    in Insects", T.A. Miller, ed., Springer-Verlag, NY.
Taghert, P.H. and Truman, J.W., 1982, Identification of the
    bursicon-containing neurones in abdominal ganglia of the
    tobacco hornworm, Manduca sexta, J. exp. Biol., 98:385.
Williams, C.M., 1959, The juvenile hormone. I. Endocrine activity
    of the corpora allata of the adult cecropia silkworm,
    Biol. Bull., 116:323.
Zaretsky, M.D. and Loher, W.J., 1982, Structure of individual
    neurosecretory cells of the brain in crickets, Soc. Neurosci.
    Abstr., 8:532.

THE USE OF A CLONED cDNA PROBE AND PEPTIDE RADIOIMMUNOASSAY TO
EXAMINE TRANSCRIPTIONAL AND TRANSLATIONAL REGULATION OF ENKEPHALIN
EXPRESSION IN EUKARYOTIC CELLS

Lee E. Eiden

Laboratory of Cell Biology
National Institute of Mental Health
Bethesda, Maryland 20205

The enkephalin pentapeptides leu- and met-enkephalin were first
purified from pig brain, and characterized, by Hughes et al. (1975)
on the basis of their ability to antagonize the electrically-induced
contraction of guinea-pig ileum strips in vitro. Since then,
enkephalins and their receptors have been localized throughout the
central and peripheral nervous system, and the enteric nervous
system, of mammals (Hughes et al., 1977; Finley et al. 1981; Pert and
Snyder, 1973) as well as the nervous systems of several insects (Remy
and Dubois, 1981; Stefano et al., 1982). In mammals, the enkephalins
are recognized to possess physiological properties of both neuromodulators and neurotransmitters (Frederickson, 1977). In general,
enkephalins exhibit an inhibitory function in the brain and
peripheral nervous system (Hughes, 1975; Frederickson, 1977; Stefano
et al., 1980; Eiden and Ruth, 1982). Enkephalins may modulate the
actions of other neurotransmitters by blockade of hormone-linked
calcium channels (Bixby and Spitzer, 1983).

In 1978, Schultzberg et al. visualized enkephalin immunoreactivity in the rat adrenal medulla. Subsequently, Viveros and co-
workers established that large amounts of enkephalin are present in
the bovine adrenal medulla and appeared to be co-stored with
catecholamines, since both could be released from the perfused gland
in vitro by acetylcholine, the natural secretagogue for the adrenal
medulla in vivo (Viveros et al., 1979; Viveros et al., 1980).

Udenfriend, and co-workers, using secretory granules of the
bovine adrenal medulla as starting material, purified and sequenced
several enkephalin-containing peptides ranging in size from 500 to

299

18,000 daltons (see Udenfriend and Kilpatrick, 1983). From this
information, and the assumption that the enkephalin precursor is a
single polypeptide chain, these workers deduced that the enkephalin
precursor weighs about 30 kilodaltons, and contains four copies of
met-enkephalin, one copy of leu-enkephalin, and one copy of met-
enkephalin $arg^6 gly^7 leu^8$, all flanked by pairs of basic amino acids,
as well as one copy of met-enkephalin $arg^6 phe^7$, the latter comprising
the C-terminus of the proenkephalin molecule and preceded by a pair
of basic amino acids.

    The entire coding sequence of the enkephalin precursor, called
preproenkephalin or proenkephalin A, was deduced by sequencing cloned
DNA complementary to the enkephalin messenger RNA from bovine adrenal
medulla (Gubler et al., 1982; Noda et al., 1982) and human pheochromocytoma (Comb et al., 1982). Elucidation of the complete primary
sequence of preproenkephalin has afforded a more complete
understanding of enkephalin biosynthesis; more importantly, the
availability of enkephalin cDNA allows the detection and quantitation
of the messenger RNA ($mRNA^{enk}$) coding for preproenkephalin. Thus, an
experimental avenue has been opened, for the examination of how the
expression of this neuropeptide is regulated at the level of mRNA
synthesis and translation in all enkaryotic neuroendocrine cells.
The remainder of this report will describe experiments carried out to
determine the cellular loci at which enkephalin expression is
regulated upon induction of enkephalin biosynthesis by various
pharmalogical agents, with an emphasis on recombinant DNA/DNA-RNA
hybridization methods employed. These methods are generally
applicable to the study of the regulation of enkephalin expression in
both vertebrate and invertebrate neuroendocrine cells.

    As a cell system in which to examine enkephalin biosynthesis and
its regulation, we have chosen primary cultures of bovine chromaffin
cells. Chromaffin cells may be obtained from the bovine adrenal
gland by cell dispersion with 0.1% collagenase followed by differ-
erential centrifugation to obtain a population of cells consisting
mainly of chromaffin cells (Eiden et al., 1983). When these cells
are plated in a medium containing 5% fetal calf serum and cytosine
arabinofuranoside to inhibit overgrowth by fibroblasts and other
dividing cells, they may be maintained for up to eight days in
culture with minimum loss of viability. Chromaffin cells in culture
behave similarly to cells of the intact bovine adrenal medulla: they
store both catecholamines and enkephalin peptides and release them in
a calcium-dependent manner in response to acetylcholine as well as
nicotine and carbachol, as shown in Table I. Thus enkephalins and
catecholamines are appropriately released from chromaffin cells by
stimulation of nicotinic receptors on their surfaces, as they are in
vivo following release of acetylcholine from the splanchnic nerve
innervating the adrenal gland (Viveros et al., 1979; Viveros et al.,
1980; Livett et al., 1981; Livett et al., 1982; Eiden et al., 1982;
Eiden et al., 1983).

Table 1.   Release of Enkephalin and Catecholamines from
            Cultured Chromaffin Cells by Acetylcholine, Nicotine
            and Carbachol.

ACh	CA (% release)	Met-Enk (% release)
5×10^{-7}M	0.6	0.7
2×10^{-6}M	0.2	0.0
1×10^{-5}M	0.8	0.6
5×10^{-5}M	3.7	3.7
2×10^{-4}M	10.9	5.9
Nicotine		
5×10^{-7}M	0.1	0.7
2×10^{-6}M	6.3	2.9
1×10^{-5}M	8.2	3.8
5×10^{-5}M	7.7	3.8
2×10^{-4}M	8.7	4.5
Carbachol		
1×10^{-5}M	0.6	0.8
4×10^{-5}M	6.0	2.5
2×10^{-4}M	9.3	3.7
1×10^{-3}M	11.4	3.8
5×10^{-3}M	2.5	1.7

Values are the mean of 2-3 separate experiments. % Release
is corrected for basal release during a 15 min period in
the absence of secretagogue. For details, see Eiden et al.,
1983.

In order to examine the regulation of enkephalin expression,
i.e., biosynthesis, in these cells, two pharmacological agents were
chosen as potential inducers of enkephalin biosynthesis. The first,
forskolin, is a potent stimulator of adenylate cyclase (Seamon et
al., 1981). This agent was chosen since cyclic AMP appears to be
involved in activation of synthesis of pituitary hormones (Labrie et
al., 1975; Dannies et al., 1976). Reserpine is known to increase
protein synthesis in mucous-secreting cells of the intestine, and
enkephalin synthesis in cultured chromaffin cells (Wilson et al.,
1980; Misch and Kim, 1982) although its mechanism of action is
unknown. Table 2 shows the time course of induction of enkephalin
immunoreactivity, measured by radioimmunoassay (Giraud et al., 1981),
after exposure of cultured chromaffin cells to reserpine or forskolin
over a 72-hour period. The induction of enkephalin caused by
forskolin and reserpine are qualitatively dissimilar: reserpine
elicits a rise in cellular enkephalin levels which is maximal after a
24-hour exposure to the drug, while forskolin causes a much slower
induction of enkephalin, detectable only after a 72-hour exposure.

Table 2.   Time Course of Increased Cellular Enkephalin
           Immunoreactivity Following Exposure to Forskolin and
           Reserpine.

Time	Forskolin,50 uM	Reserpine, 10 uM
0-4 hr	86 $\pm$ 4.6%	87 $\pm$ 5.8%
24 hr	105 $\pm$ 17	175 $\pm$ 20
48 hr	108 $\pm$ 17	210 $\pm$ 28
72 hr	272 $\pm$ 47	193 $\pm$ 16

Drugs were added to cells on day three of culture,
and cells harvested for peptide radioimmunoassay at
the times indicated. Values represent the mean $\pm$ S.E.M.
of 6-8 individual determinations (wells) and are expressed
as percent of immunoreactive enkephalin levels in
untreated cultures harvested at corresponding times.

To further examine the cellular loci at which enkephalin biosynthesis is regulated by forskolin and reserpine, it is necessary to be
able to quantitate changes in enkephalin mRNA (mRNA^{enk}) elicited by
these agents. Thus, transcriptional control is suggested if there is
an approximately equal increase in the cellular levels of peptide and
its corresponding mRNA, while an increase in cellular peptide content
without a concommitant change in the amount of cellular messenger RNA
implies regulation of peptide expression at the level of mRNA
translation, protein precursor processing, or both.

Experiments to measure changes in enkephalin mRNA after
induction of enkephalin peptide by forskolin or reserpine were
carried out using the human enkephalin cDNA clone pHPE9 generated
from a human pheochromocytoma cDNA "library" by Comb et al.
(1982b). (A bovine enkephalin clone kindly provided by Dr. Peter
Seeburg was also employed in some experiments.) The cloning strategy
employed by Comb et al. is outlined in Figure 1. Briefly, an
oligodeoxynucleotide pentadecamer probe complementary to the
enkephalin pentapeptide portion of the putative enkephalin messenger
RNA was constructed based on the knowledge of the enkephalin peptide
sequence (tyr-gly-gly-phe-met) and the genetic code (Comb et al.,
1982a). This probe was labeled with α -^{32}P-ATP and polynucleotide
kinase and used to screen nitrocellulose filters which contained
immobilized DNA from individual colonies of the pheochromocytoma
clone bank or "library". The library was generated by making
complementary DNA (cDNA) copies of all mRNA species from
pheochromocytoma in the size class of the putative proenkephalin
message (about 1.4 kilobases) and inserting these into the Pst I
restriction endonuclease site of the plasmid vector pBR322 using the
homopolymer tailing method (for a general review of cloning
strategies for proteins of neurobiological interest, see Eiden,
1982). Positive clones (containing a cDNA copy of all or part of the

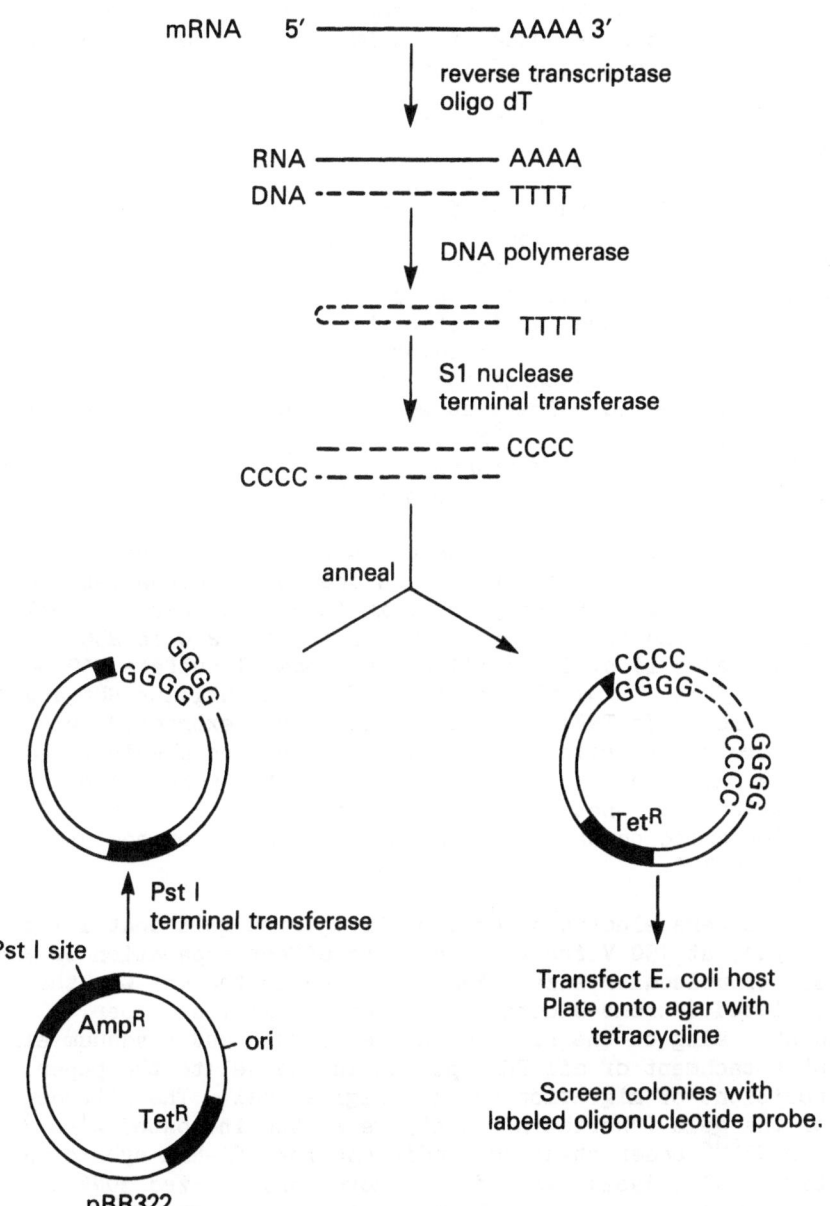

Fig. 1. Strategy for Cloning and Isolation of Human Proenkephalin
         cDNA

         See text for additional details.

enkephalin mRNA) were identified by hybridization to the labeled
pentadecamer probe and subsequently sub-cloned into the coliphage
vector M13mp7 and the sequence of the DNA inserts determined by the
chain-termination technique of Sanger.

In order to employ the cDNA clone to quantify enkephalin mRNA in
cultured chromaffin cells, the plasmid was grown in the E. coli host
MC1061 and harvested without amplification from 500 ml overnight
cultures. Plasmid was purified from a cleared lysate by SDS-alkali
extraction/cesium chloride purification as described by Maniatis et
al. (1982). This procedure yielded plasmid preparations which could
be treated with PstI, which cut the pHPE9 plasmid at two sites within
the insert, allowing the purification by gel electrophoresis of a 400
base pair fragment within the reading frame of the enkephalin
messenger RNA. Deoxyguanosine (G) and deoxyadenosine (A) residues
within the 400 base pair fragment were replaced with ^{32}P-labeled G
and A by incubation with α -^{32}P-labeled G and A triphosphates, DNase
I and DNA polymerase ("nick translation"). This probe was used for
quantifica- tion of mRNA^{enk} as shown in Figure 2.

Chromaffin cells in monolayer culture exposed to drugs for
various periods of time were washed in ice-cold phosphate-buffered
saline (PBS) to remove RNase present in fetal calf serum or released
from the cells, and incubated for 90 minutes at 42^{0}C in 250 ul per
well (10^{6}cells/well) of SETpK (1% sodium dodecyl sulfate, 10 mM Tris
pH 7.6, 5 mM EDTA, 70 ug/ml proteinase K) to solubilize RNA and DNA
(total nucleic acids-TNA). Incubates were then extracted twice with
phenol-chloroform to remove protein, and TNAs were precipitated from
the aqueous phase in 65% ethanol at -20^{0}C. TNAs were collected by
centrifugation and denatured for 5 min at 65^{0}C in 2.2 M formalde-
hyde/50% formamide containing 20 mM MOPS, 5 mM sodium acetate and 1
mM EDTA at pH 7.0.

Samples were electrophoresed on 1% agarose gels containing 2.2 M
formaldehyde, at 150 V for 2-3 hours, to effect separation of RNAs on
the basis of molecular size. RNA contained in the gel was then
electro-eluted ("trans-blotted") onto nitrocellulose paper.
Subsequent baking of the nitrocellulose at 80^{0}C under vacuum allows
covalent attachment of all RNA species in the gel to the paper, at
their positions of migration on the original gel. The nitrocellulose
blots of the gels ("Northern blots") were then incubated with ^{32}P-
labeled cDNA^{enk} under standard conditions for DNA-RNA hybridization
(Maniatis et al., 1982), washed to remove unhybridized cDNA and
exposed to autoradiographic film for 2-12 hours. The intensity of
autoradiographic exposure at the position of mRNA^{enk} (about 14S) was
quantitated by scanning densitometry using an LKB soft laser scanning
densitometer. Since cDNA-RNA hybridization is carried out in vast
cDNA excess (the concentration of cDNA in the hybridization reaction
is 10 ng/ml, while each Northern blot contains 1-20 picogram per lane
of mRNA^{enk}) the intensity of the autoradiographic signal

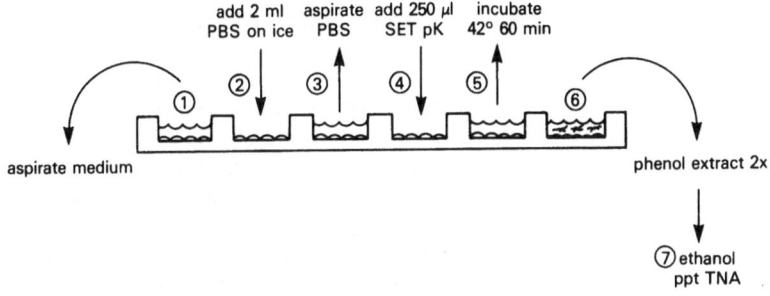

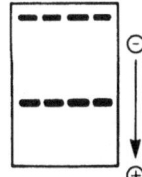

⑩ top part of gel
   a) trans-blot unto nitrocellulose paper.
   b) hybridize with enkephalin cDNA probe.
   c) quantify by autoradiography/soft laser scanning.

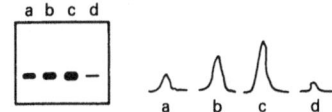

⑪ bottom part of gel
   a) Stain with ethidium bromide
   b) photograph
   c) quantify RNA by soft laser scanning

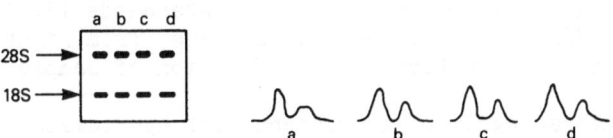

Fig. 2.   Measurement of Enkephalin mRNA in Cultured Chromaffin
               Cells.

               See text for abbreviations and additional details.

for each sample (each lane of the gel blot) represents essentially
complete hybridization of the labeled cDNA to mRNA^{enk} and is
therefore directly proportional to the amount of mRNA^{enk} in the
sample.

Duplicate gels were run for each experiment and these were
stained with ethidium bromide (EtBr) to visualize RNA applied to the
gel. Since ribosomal RNA represents about 85% of the total RNA in a
eukaryotic cell, the major bands visualized by EtBr staining are the
28 and 18 S ribosomal subunit RNAs. Negatives of photographs of the
stained gels were then scanned densitometrically, to quantitate the
amount of RNA applied in each lane, allowing correction for recovery
in each sample. A negative of a typical EtBr-stained gel is shown in
Figure 3.

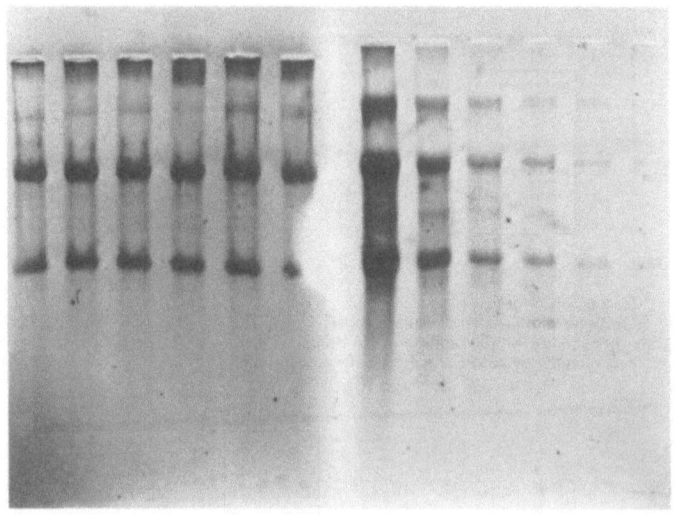

Fig. 3.   Ethidium Bromide-Stained RNA Gel.

Six lanes on the left represent samples of RNA extracted
from 200,000 chromaffin cells per sample. Six lanes on the
right are bovine adrenal RNA standards (2, 1, 0.5, 0.25,
0.125 and 0.0625 ug RNA per lane, from right to left) used
as a calibration for quantitation of sample RNA.

An autoradiogram of a Northern blot hybridized with the lableled
enkephalin cDNA is shown in Figure 4. Cells had been treated for 72
hours with 50 uM forskolin (F), 10 uM reserpine (R) or no drug (C)
prior to harvesting and RNA extraction. Densitometric scanning
reveals that exposure to forskolin results in an about 400% increase
in mRNA^{enk} relative to the untreated cells, while exposure to
reserpine results in a 40-70 % decrease in mRNA^{enk}. Thus the
increase in enkephalin peptide (Table 2) elicited by forskolin is

paralleled by an increase in the amount of $mRNA^{enk}$, consistent with
an action of forskolin to induce enkephalin peptide levels via
transcriptional regulation. The increase in cellular levels of
enkephalin caused by reserpine, however, is not paralleled by an
increase in $mRNA^{enk}$; instead $mRNA^{enk}$ concentrations are significantly
lower in reserpine-treated cells. One possiblity for this fall in
$mRNA^{enk}$ levels is that reserpine causes an increased utilization of
$mRNA^{enk}$, without altering enkephalin gene transcription.
Alternatively, $mRNA^{enk}$ may rise initially (due to increased
transcription) but decrease by 72 hours (due to increased $mRNA^{enk}$
utilization or decreased message stability).

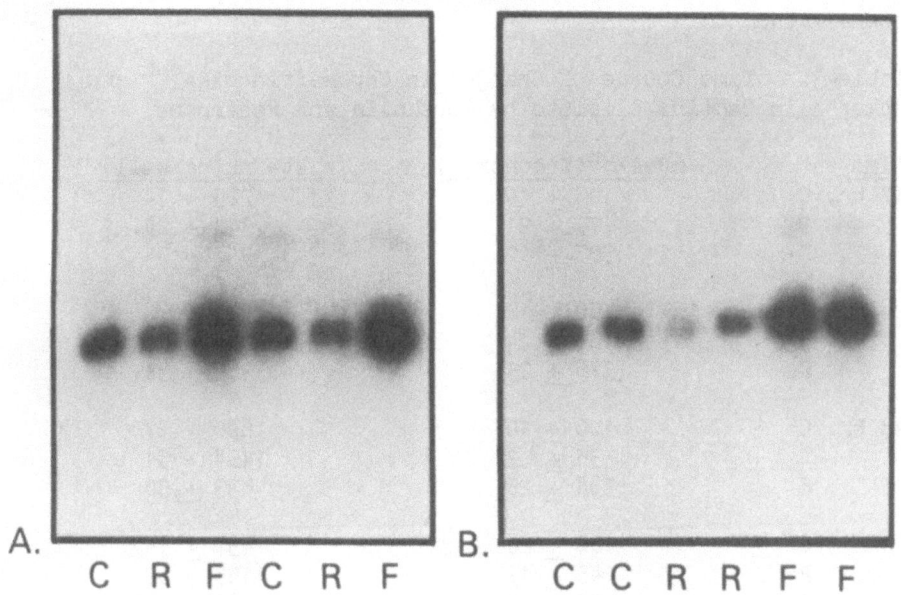

Fig. 4   Autoradiographs of Chromaffin Cell RNA Northern Blots
          Hybridized with ^{32}P-labeled Enkephalin cDNA.

          C - untreated cells, F - 50 uM forskolin-treated cells,
          R - 10 uM reserpine-treated cells. Chromaffin cells in
          culture were exposed to drugs for 72 hours, harvested and
          processed for hybridization as shown schematically in Fig.
          2. Each lane represents RNA extracted from 300,000 cells.
          A. and B. represent two separate experiments, each performed
          in duplicate.

     To test this possibility, enkephalin peptide and mRNA levels
were measured 17, 24, 48, and 72 hours after exposure of cultured
chromaffin cells to 50 uM forskolin or 10 uM reserpine. As shown in
Table 3, exposure to forskolin results in a 3-5 fold increase in the
cellular concentration of $mRNA^{enk}$ relative to untreated cells. This
induction is apparent at the earliest time-point measured (17 hours),

whereas a significant increase in enkephalin peptide is not measured
until 48 hours following addition of forskolin. The delay presumably
represents the time necessary for mRNA translation, preproenkephalin
packaging into secretory vesicles, and proteolytic conversion of
proenkephalin to the mature pentapeptide hormone. The regulation of
enkephalin biosynthesis by reserpine, however, follows a completely
different pattern: peptide levels are already significantly elevated
by 24 hours after exposure to the drug, and continued elevation of
cellular enkephalin peptide is accompanied by a fall in cellular
mRNA^{enk}. This suggests that in fact reserpine effects an elevation
in enkephalin peptide levels by an increased utilization of mRNA^{enk}
which is not compensated by transcription of the enkephalin
structural gene.

Table 3.   Time Course of Changes in Chromaffin mRNA^{enk} and
Enkephalin Peptide Elicited by Forskolin and Reserpine

Time		mRNA^{enk}(% control)	Met-Enk(pg/well)
17 hr.	C	100 ± 19	229 ± 14
	R	79 ± 21	344 ± 24
	F	507 ± 10	247 ± 34
24 hr.	C	100 ± 12	313 ± 8
	R	88 ± 14	544 ± 26
	F	376 ± 23	385 ± 34
48 hr.	C	100 ± 10	600 ± 27
	R	35 ± 3	1454 ± 51
	F	384 ± 25	833 ± 30
72 hr.	C	100 ± 20	437 ± 14
	R	45 ± 11	1155 ± 52
	F	521 ± 62	982 ± 127

Values represent the mean ± S.E.M. of 3 individual
determinations.

    Several questions remain to be answered. It is not clear, for
example, why increased translation of mRNA^{enk} upon exposure to
reserpine, if indeed this occurs, should lead to degradation of the
enkephalin message. Perhaps some portion of the enkephalin mRNA
exists in the cell in a non-translated "storage" form, analogous to
stored mRNA in frog and sea urchin embryos (Lodish, 1976) and it is
this RNA pool which is made translationally active, and RNase-
sensitive, by reserpine. It also remains to be determined by in
vitro transcription of the enkephalin gene if the increase in mRNA^{enk}
elicited by forskolin is indeed solely a consequence of increased
gene transcription, or if a cAMP-dependent mechanism for mRNA
stabilization may exist in chromaffin cells, and account for an
accumulation of mRNA^{enk} following exposure to forskolin. Finally,

the question of physiological regulation of enkephalin biosynthesis, during development and in response to secretagogues such as acetylcholine must be addressed.

The preliminary studies described here serve merely to indicate that neuropeptide biosynthesis may be regulated at multiple steps. It remains to be seen what portion of this regulational repertoire is actually relevant to the physiological expression of neuropeptides in neuroendocrine cells.

Acknowledgements

These studies were carried out in collaboration with Urs Affolter (NIMH), Adair Hotchkiss (NIADDK, NIH), Edward Herbert (Dept. of Chemistry, University of Oregon) and Peter Seeburg (Genentech). The author wishes to thank them as well as Anna Iacangelo and Chang-Mei Hsu (NIMH) for invaluable assistance in preparation of cDNA and peptide radioimmunoassay. The preparation of the manuscript by Betty Moulis is gratefully acknowledged.

References

Bixby, J.L., and Spitzer, N.C., 1983, Enkephalin reduces quantal content at the frog neuromuscular junction, Nature, 301:431.

Comb, M., Herbert, E., and Crea, R., 1982a, Partial characterization of the mRNA that codes for enkephalins in bovine adrenal medulla and human pheochromocytoma, Proc. Natl. Acad. Sci. USA, 79-360.

Comb, M., Seeburg, P.H., Adelman, L., Eiden, L., and Herbert, E., 1982b, Primary structure of the human met- and leu- enkephalin precursor and its mRNA, Nature, 295:663.

Dannies, P.S., Gantrik, K.M., and Tashjian, A.H. Jr., 1976, A possible role of cyclic AMP in mediating the effects of thyrotropin-releasing hormone on prolactin release and on prolactin and growth hormone synthesis in pituitary cells in culture, Endocrinology, 98:1147.

Eiden, L.E., 1982, Recombinant DNA methods in neuroendocrinology: new answers to old questions, Peptides, 3:217.

Eiden, L.E., Eskay, R.L., Scott, J., Pollard, H., and Hotchkiss, A.J., 1983, Primary cultures of bovine chromaffin cells synthesize and secrete vasoactive intestinal polypeptide (VIP), Life Sci., 33:687.

Eiden, L.E., Giraud, P., Hotchkiss, A., and Brownstein, M.J., 1982, Enkephalins and VIP in human pheochromocytomas and bovine adrenal chromaffin cells, in; "Regulatory Peptides: From Molecular Biology to Function, E. Costa and M. Trabucchi, eds,. Raven Press, New York.

Eiden, L.E., and Ruth, J.A., 1982, Enkephalins modulate the responsiveness of rat atria in vitro to norepinephrine, Peptides, 3:475.

Finley, J.C.W., Maderdrut, J.L., and Petrusz, P., 1981, The

immunocytochemical localization of enkephalin in the central
nervous system of the rat, J. Comp. Neurol., 198:541.

Frederickson, R.C.A., 1977, Enkephalin pentapeptides - a review of
current evidence for a physiological role in vertebrate
neurotransmission, Life Sci., 21:23.

Giraud, P., and Eiden, L.E., 1981, Cell-free translation of human
pheochromocytoma messenger RNA yields protein(s) containing
methionine-enkephalin, Biochem. Biophys. Res. Commun., 99:969.

Giraud, P., Eiden, L.E., Audigier, Y., Gillioz, P., Conte-Devolx,
B., Boudouresque, F., Eskay, R., and Oliver, C., 1981,
Enkephalins, ACTH, α-MSH, and β-endorphin in human
pheochromocytomas, Neuropeptides, 1:237.

Gubler, U., Seeburg, P., Hoffman, B.J., Gage, L.P., and Udenfriend,
S., 1982, Molecular cloning establishes proenkephalin as the
precursor of enkephalin-containing peptides, Nature, 295:206.

Hughes, J., 1975, Isolation of an endogenous compound from the brain
with pharmacological properties similar to morphine, Brain Res.,
88:295.

Hughes, J., Kosterlitz, H.W., and Smith, T.W., 1977, The
distribution of methionine-enkephalin and leucine-enkephalin in
the brain and peripheral tissues, Br. J. Pharmacol., 61:639.

Hughes, J., Smith, T.W., Kosterlitz, H.W., Fothergill, L.A., Morgan,
B.A., and Morris, H.R., 1975, Identification of two related
pentapeptides from the brain with potent opiate agonist activity,
Nature, 258:577.

Labrie, F., Borgeat, P., Lemay, A., Lemaire, S., Barden, N., Drouin,
J., Lemaire, I., Jolicoeur, P., and Belanger, A., 1975, Role of
cyclic AMP in the action of hypothalamic regulatory hormones,
Adv. Cyclic Nucleotide Res., 5:787.

Livett, B.G., Day, R., Elde, R.P., and Howe, P.R.C., 1982, Co-
storage of enkephalins and adrenaline in the bovine adrenal
medulla, Neuroscience, 7:1323.

Livett, B.G., Dean, D.M., Whelan, L.G., Udenfriend, S., and Rossier,
J., 1981, Co-release of enkephalin and catecholamines from
cultured adrenal chromaffin cells, Nature, 289:317.

Lodish, H.F., 1976, Translational control of protein synthesis,
Annu. Rev. Biochem., 45:39.

Maniatis, T., Fritsch, E.F., and Sambrook, J., 1982, Molecular
cloning. A laboratory manual, Cold Spring Harbor Laboratory,
Cold Spring Harbor.

Misch, D.W., and Kim, W.-K., 1982, Action of reserpine on mucous
secretion in hamster intestine: development of an animal model
for autonomic control of exocrine secretion, J. Cell Biol.,
95:403a.

Noda, M., Furutani, Y., Takahashi, H., Toyosata, M., Hirose, T.,
Inayama, S., Nakanishi, S., and Numa, S., 1982, Cloning and
sequence analysis of cDNA for bovine adrenal preproenkephalin,
Nature, 295:202.

Pert, C.B., and Snyder, S.H., 1973, Opiate receptor: Its
demonstration in nervous tissue, Science, 179:1011.

Remy, C., and Dubois, M.P., 1981, Immunohistochemical evidence of methionine enkephalin-like material in the brain of the migratory locust, Cell Tissue Res., 218:271.

Schultzberg, M., Hökfelt, T., Lundberg, J., Terenius, L., Elfrim, L. - G., and Elde, R., 1978, Enkephalin-like immunoreactivity in nerve terminals in sympathetic ganglia and adrenal medulla and in adrenomedullary gland cells, Acta Physiol. Scand., 103:475.

Seamon, K.D., Padgett, W., and Daly, J.W., 1981, Forskolin: unique diterpene activator of adenylate cyclase in membranes and in intact cells, Proc. Natl. Acad. Sci. USA, 78:3363.

Stefano, G.B., Scharrer, B., and Assanah, P., 1982, Opioid binding sites in the midgut of the insect Leucophaea maderae (Blattaria), Life Sci., 31:1397.

Stefano, G.B., Vadasz, I., and Hiripi, L., 1980, Methionine enkephalin inhibits the bursting activity of the Br-type neuron in Helix pomatia L., Experientia, 36:666.

Udenfriend, S., and Kilpatrick, D.L., 1983, Biochemistry of the enkephalins and enkephalin-containing peptides, Arch. Biochem. Biophys., 221:309.

Viveros, O.H., Diliberto, E.J. Jr., Hazum, E., and Chang, K.-J., 1979, Opiate-like materials in the adrenal medulla: evidence for storage and secretion with catecholamines, Mol. Pharmacol., 11:1101.

Viveros, O.H., Diliberto, E.J. Jr., Hazum, E., and Chang, K.-J., 1980, Enkephalins as possible adrenomedullary hormones: storage, secretion, and regulation of synthesis, in: Neural Peptides and Neuronal Communication, E. Costa and M. Trabucchi, eds., Raven Press, New York.

Wilson, S.P., Chang, K.-J., and Viveros, O.H., 1980, Synthesis of enkephalins by adrenal medullary chromaffin cells: reserpine increases incorporation of radiolabeled amino acids, Proc. Natl. Acad. Sci. USA, 77:4364.

ACTIONS OF SYNTHETIC ADIPOKINETIC HORMONE AND RELATED PEPTIDE ANALOGS ON DISPERSED LOCUST FAT BODY CELL PREPARATIONS

Carol Asher, Pnina Moshitzky, J. Ramachandran* and
S. W. Applebaum

Faculty of Agriculture, The Hebrew University
Rehovot 76100, Israel

The insect fat body is functionally analogous to the vertebrate liver and cytologically similar to adipose tissue. This duality entails a high capacity for protein synthesis and for storage of energy reserves. In adult locusts triglycerides serve as the major storage reserve, and are mobilized as diglycerides. The majority of haemolymph proteins originate in the fat body and in mature female adults vitellogenin is produced therein. Although apparently similar to comparable vertebrate systems, the biosynthetic activities in the multifunctional insect fat body are regulated by hormones different from those described in vertebrates. One of these, adipokinetic hormone (AKH-I, Fig. 1), produced in the corpora cardiaca is responsible for mobilization of diglycerides from the locust fat body (Stone et al., 1976; Carlisle et al., 1979)(Fig. 1).

(AKH-I)*   <Glu-Leu-Asn-Phe-Thr-Pro-Asn-Trp-Gly-Thr-NH$_2$

(II)*   <Glu-Leu-Asn-Phe-Ser-Thr-Gly-Trp-NH$_2$

(III)*   <Glu-Leu-Asn-Tyr-Ser-Thr-Gly-Trp-NH$_2$

(IV)**   <Glu-Leu-Asn-(3,5-diiodo)Tyr-Ser-Thr-Gly-Trp-NH$_2$

(V)**   <Glu-Leu-Asn-(^{3}H)Tyr-Ser-Thr-Gly-Trp-NH$_2$

Fig. 1. Amino acid sequence of adipokinetic hormone and synthetic analogues
 *Yamashiro, Applebaum and Li (1983)
 **Muramoto, Ramachandran, Moshitzky and Applebaum (1983)

---

*Permanent address: Hormone Research Laboratory and Dept. of Biochemistry and Biophysics, Univ. of California, San Francisco, CA 94143

It has been characterized as a blocked decapeptide and synthesized
(Stone et al., 1976,1978). AKH-I affects cAMP levels and inhibits
protein synthesis of excised locust fat bodies incubated in vitro
(Carlisle and Loughton, 1979; Orchard et al., 1982). We have recently synthesized several AKH analogues (Fig. 1) for the specific
purpose of studying structure-function relationships and the interaction of adipokinetic hormones with their target tissues. Peptide
II (Fig. 1) is structurally similar to the red-pigment-concentrating-
hormone of crustaceans and has the same amino composition as that
of locust AKH-II (Carlson et al., 1979). Peptide III is a tyrosine
analogue of II and its iodination yields IV, the 3,5-diiododerivative
of III. Peptide IV was catalytically tritiated to yield a [^{3}H] tyrosine analogue (peptide V) with specific radioactivity of 57.2
Ci/mmol, indicating that essentially complete exchange had occurred.

Synthetic AKH-I as well as the synthetic analogues II and III
stimulated lipid mobilization in vivo at comparable rates (approximately 3 fold increase over the basal rate). The iodinated peptide
was inactive.

A major difficulty in assessing adipokinetic activity in vitro
resides in the inherent variability of the excised fat body response
even when the donor locusts are developmentally staged. This often
renders the statistical significance of response questionable and
precludes the possibility of ascertaining minor but potentially important differences in structure-function relationships. In order
to overcome this problem we have devised a procedure for the preparation of dispersed locust fat body cells which are functionally
similar in synthetic activity and response to whole adult locust fat
bodies, and with greatly reduced variability. This procedure involves
collagenase treatment of minced fat body in locust physiological medium containing 1% bovine serum albumin (BSA). After dispersion the
cells are filtered through cheescloth, washed, and resuspended in the
same medium. AKH-I (40 nM) stimulated lipid release into the medium
at a rate comparable to that obtained from equivalent in vitro incubations of whole fat bodies. No addition of haemolymph or other insect components was necessary for the adipokinetic response. Cyclic-
AMP (cAMP) levels in the isolated cell suspension, measured by radioimmunoassay, were stimulated 6-fold by AKH-I.

The incorporation of [^{3}H] leucine into proteins synthesized by
isolated fat body cells was inhibited by AKH-I and by several of the
synthetic analogues up to 70% in a dose-dependent manner and was
linear for at least a six-hour period. In all cases, the variability
of results obtained with isolated fat body cells was much less than
that obtained with comparable whole fat bodies, whereas the activities
were within the same range. This system is currently being used to
evaluate other characteristics of AKH-receptor interactions.

Acknowledgments

This work was supported by a grant from the United States-Israel Binational Agricultural Research and Development Fund (BARD I-47-79). J. R. was a Lady Davis Visiting Professor of the Hebrew University of Jerusalem.

References

CARLISE J. A., and LOUGHTON B. G. (1979) Nature (London) 282, 420-421.

CARLSEN J., HERMAN W. S., CHRISTENSEN M., and JOSEFSSON L. (1979) Insect Biochem. 9, 497-501.

MURAMOTO K., RAMACHANDRAN J., MOSHITZKY P., and APPLEBAUM S. W. (1983) Int. J. Peptide Protein Res. (in press).

STONE J. V., MORDUE W., BATLEY K. E., and MORRIS H. R. (1976) Nature (London) 263, 207-211.

STONE J. V., MORDUE W., BROOMFIELD C. E., and HARDY P. M. (1978) Eur. J. Biochem. 89, 195-202.

YAMASHIRO D., APPLEBAUM S. W., BIRK Y., and LI C. H. (1981) Int. J. Peptide Protein Res. 17, 546-548.

YAMASHIRO D., APPLEBAUM S. W., and LI C. H. (1983) Int. J. Peptide Protein Res. (in press).

CHOLINERGIC NEURONES IN NEURONAL CULTURES

FROM PERIPLANETA AMERICANA

David J. Beadle, George Lees and Roger P. Botham*

School of Biological Sciences, Thames Polytechnic
London SE18, U.K. and *Wellcome Research Laboratories
Berkhamsted, U.K.

An insect tissue culture system has been developed from central nervous tissue of embryonic cockroaches that facilitates neurophysiological studies. The dissociated cells, of which 90% are neuronal, can be maintained in vitro for several weeks. During this period the cells differentiate to produce a typical neuronal ultrastructure with axonal processes aggregating to form fibre bundles that form a complex network on the floor of the culture vessel. Within these fibres ultrastructural profiles characteristic of synapses can be identified (Beadle et al., 1982). This in vitro preparation has been used to investigate choline accumulation and putative acetylcholine receptors. The accumulation of choline via a high affinity system and its association with acetylcholine synthesis is considered to be a presynaptic marker of cholinergic systems in vertebrates (Simon et al., 1975) and acetylcholine receptors are a reliable postsynaptic marker (Sattelle, 1980).

The cultures were prepared from the brains of 23-26 day old embryos of Periplaneta americana. They were initiated in a medium consisting of 5 parts Schneider's Drosophila medium and 4 parts Eagle's basal medium and after seven days were transferred to a medium consisting of equal parts Leibovitz's L-15 medium and Yunker's modified Grace's medium containing 10 µg/ml ecdysone. The cells were grown in air at 29°C in 50 mm Falcon petri dishes (Beadle et al., 1982). The cultures contained approximately 50,000 neurones and were used after fourteen days growth in vitro when there was 10 µg of protein.

For all biochemical experiments a single culture was used as an assay unit. Choline uptake was investigated by incubating cultures in a solution containing 170 mM NaCl, 4 mM KCl, 2mM

317

$MgCl_2$, 2mM $CaCl_2$, 0.2mM NaH_2PO_4, 1.8mM Na_2HPO_4, 35mM sucrose and various concentrations of methyl-^{3}H-choline (78 Ci/mmol) for periods up to 60 min. Choline uptake was also measured in the absence of sodium ions and in the presence of various inhibitors. Putative acetylcholine receptors were studied using the nicotinic ligand, α-bungarotoxin (αBTX), from the venom of <u>Bungurus multi-cinctus</u>. Cultures were incubated in 50mM NaCl, 260mM sucrose, 0.1% BSA, 10mM Tris HCl buffer at pH 7.2 and ^{125}I-αBTX (200 Ci/mmol, Amersham International). Non-specific binding was determined by co-incubation in excess cold toxin and the pharmacological profile by preincubation with various cholinergic ligands. After incubation the cultures were washed several times and digested overnight in 0.05M NaOH. The digested cells were washed into a scintillation vial and counted in a Corumatic Scintillation Counter. The fate of the accumulated choline was determined by high-voltage paper chromatography. The response of the neurones to acetylcholine and nicotine was measured by conventional intracellular recording techniques using glass microelectrodes filled with 1M potassium acetate and tip impedence of 80-150 MΩ. Agonists were applied by pressure ejection (N_2, 10 psi) and antagonists by bath application. Full details of these methods can be found in Lees et al.,(in press), and Bermudez et al.,(submitted).

The cultured neurones rapidly accumulated ^{3}H-choline and the uptake was linear for 20 min. reaching a maximum at around 60 mins. The uptake was almost completely abolished by the absence of sodium ions. The kinetics of sodium-dependent uptake was studied across a choline concentration of 0.1 - 2 μM and 2 - 75 μM. The uptake increased with increasing choline concentrations and eventually became saturable. Double reciprocal analysis of the data revealed two linear components, a high affinity system with a K_m of 0.57 $\pm$ 0.28 μM and a V_{max} of 2.99 $\pm$ 0.69 pmol/10 min/culture, and a low affinity system with a K_m of 5.35 $\pm$ 1.82 μM and V_{max} of 11.85 $\pm$ 0.86 pmol/10 min/culture. By fitting this data to the Michaelis-Menten equation at various concentrations of choline the relative preponderance of the two systems can be determined. At 0.5 μM choline the high affinity system predominates and this concentration was used to further characterise this mechanism.

The high affinity mechanism was dependent on sodium since its replacement by lithium reduced choline uptake by 87%. The low affinity system was found to be less dependent on sodium ions. Increasing concentrations of potassium also significantly reduced choline uptake. The accumulation of ^{3}H-choline appeared to be unaffected by the metabolic inhibitors, sodium azide and 2,4-dinitrophenol, although iodoacetamide, a glycolysis inhibitor, reduced uptake by 56%. Hemicholinium-3 was a potent inhibitor of high affinity choline accumulation with as little as 1 μM reducing uptake by 80% and 100 μM by more than 95%. In order to determine the ability of the neurones to convert accumulated choline to acetylcholine, the identity of the labelled compounds extracted from neurones after 10 min. exposure to 0.5 µM ^{3}H-choline was determined by paper chromatography. At 0.5 µM ^{3}H-choline 34% of total accumulated radioactivity was present as ^{3}H-acetylcholine and in the absence of sodium the acetylation of choline was reduced to 10.6%. These results indicate that these cultures contain a high proportion of neurones that possess a high affinity choline uptake mechanism that is associated with acetylcholine synthesis (Bermudez et al., submitted).

At nM concentrations ^{125}I-αBTX bound specifically to the cultured neurones and the binding increased with time reaching equilibrium after 100 min. Under equilibrium conditions the specific binding component was saturable and at 10 nM, a dose sufficient to saturate the receptor population, non-specific binding accounted for approximately 10% of total binding. Scatchard analysis of the binding data revealed an apparent dissociation constant of 3.51 x 10^{-9}M and a maximal number of binding sites of 41.7 fmol./culture. The Hill coefficient of 0.929 indicated the essentially non-cooperative nature of the binding. The pharmacological specificity of the toxin binding component was determined by preincubation with a range of cholinergic compounds and was found to be essentially nicotinic. The most potent competitors for the αBTX binding sites after the unlabelled toxin were d-tubocurarine and nicotine with I.C.$_{50}$s of 3 x 10^{-7} and 4 x 10^{-6}M respectively and one of the least potent was muscarine (I.C.$_{50}$ = 2 x 10^{-4}M). The results of these experiments indicate that the cultured neurones possess a specific ^{125}I-αBTX binding component with many of the properties expected of an acetylcholine receptor (Beadle et al., in press; Lees et al., in press).

When 5 µM acetylcholine was applied to the cells by pressure ejection 18 of 19 cells tested were depolarised. The depolarisation was accompanied by a decrease in membrane resistance. 4 out of 5 cells tested also responded in a qualitatively similar manner to pressure ejected nicotine at 0.1 - 1 µM. The ability of bath applied αBTX at 25 nM to inhibit responses to acetylcholine and nicotine was tested on a number of cells. The results showed that whereas 18 out of 19 cells were depolarised by acetylcholine this was reduced to 3 out of 8 after the cells had been exposed to αBTX for periods in excess of one hour. A statistical analysis of this data indicated that the observed antagonism was significant at a probability level of 99%. In a few cases cells responded to acetylcholine and nicotine even after they had been exposed to saturating doses of the toxin for periods up to three hours. This suggested that while the majority of the neurones tested possessed nicotinic receptors that were sensitive to αBTX there was a subpopulation of neurones that were insensitive to the toxin.

The results of this investigation show that cultures produced

from the CNS of embryonic cockroaches contain neurones that both
possess a high affinity choline uptake mechanism that is associated
with acetylcholine synthesis and physiologically functional
acetylcholine receptors.   These cultures provide a unique preparation for more detailed studies of the cholinergic system in insects.
The results also provide additional information to support the
view that acetylcholine is a major neurotransmitter in the insect
CNS.

REFERENCES

Beadle, D. J., Hicks, D. and Middleton, C., 1982, J. Neurocytol.,
        11:611-626.
Beadle, D. J., Botham, R. P. and Lees, G., in press, J. Physiol.
        (Lond.).
Bermudez, I., Lees, G., Middleton, C., Botham, R. P. and Beadle,
        D. J., submitted, J. Neurochem.
Lees, G., Beadle, D. J. and Botham, R. P., in press, Brain Res.
Sattelle, D. B., 1980, Adv. Insect Physiol., 15:215-315.
Simon, J. R., Atweh, S. and Kuhar, M. J., 1975, J. Neurochem.,
        26:909-922.

ROLE OF THE FRONTAL GANGLION IN LEPIDOPTEROUS INSECTS

Robert A. Bell

USDA-ARS, Otis Methods Development Center, Otis
ANGB, MA 02542

Most physiological studies on the frontal ganglion have been carried out on certain Orthopterous insects primarily locusts and cockroaches (Roussel, 1972). Little attention has been given to the role of the frontal ganglion in the Lepidopterous insects although Bounhiol (1938) reported that it was essential for metamorphosis in the silkworm, Bombyx mori. However, previous studies in which the presence of neurosecretory cells was demonstrated in the frontal ganglion of the tobacco hornworm, Manduca sexta (Bell et al., 1974) prompted the experiments presented here.

Frontal ganglionectomy and other surgical operations on the stomatogastric nervous system were carried out on larvae within 12 hours following the molt to the final (5th) instar and just prior to initiating of feeding. All larvae were reared on artificial diet in a daily photoperiod of 15 hours at $25^{\circ}C$ to prevent occurrence of the pupal diapause (Bell et al., 1975; Bell and Joachim, 1976). Following CO_2 anaesthesia, the frontal ganglion and associated nerves or connectives were exposed by excision of the triangular part of the head (frons area) located between the main epicranial suture and the clypeus. The underlying frontal ganglion could then be removed with minimal damage to the mandibular and pharngeal muscles required for normal feeding and swallowing. Various operations included removal of the frontal ganglion, transection of the recurrent nerve, transection of one or both frontal connectives and a sham-operated control in which the cuticle was excised and the underlying ganglion was disturbed but otherwise left intact.

All larvae (12/treatment) were observed for growth, completion of larval feeding and pupation as well as adult differentiation, eclosion and wing expansion.

Results showed that all sham operated (control) larvae developed into normal adults (Tables 1,2). Moreover, the moths arising from the sham-operated larvae were found to be fully capable of feeding and reproduction. Surprisingly, most of the larvae deprived of frontal ganglia or that had their recurrent nerves severed successfully completed obligatory feeding although growth and rate of development was somewhat less than that of the sham-operated group (Table 1). It should also be noted that all larvae in a starved control group died within 7 days after the test was initiated thereby showing that final instar larvae must undergo a period of obligatory feeding to reach the pupal stage. Larvae without intact recurrent nerves or frontal ganglia attained pupation in 14.5 and 17 days, respectively as compared to 13.3 days for the sham-operated group. Furthermore, larvae deprived of the frontal ganglia or without

Table 1. Effects of surgical alterations of the stomatogastric nervous system on growth and development of Manduca sexta

Treatment*	Initial larval wt.(g)	Pupal wt. (g)	Wt. increase (% of control)	% pupated
Sham operated control	1.44	4.05	100.0	100.0
Frontal connectives severed	1.34	3.22	85.4	66.7
Recurrent nerve severed	1.49	3.52	84.0	66.7
Frontal ganglion removed	1.50	3.27	77.5	58.3
Starved control	1.38	(all died within 7 days)		

* n = 12 insects per treatment

intact recurrent nerves not only successfully pupated but also developed to pharate adults. Those insects without frontal ganglia generally died as pharate adults within their pupal enclosures; however, one individual successfully eclosed but failed to complete the process of wing expansion (Table 2). In contrast the group of insects with severed recurrent nerves showed a somewhat mixed response. Nearly half of this group was able to undergo adult eclosion and normal wing expansion whereas the remainder died as pharate adults. Finally the group of insects which had their frontal connectives severed showed an altogether different response from that of the other treated insects. All larvae with severed frontal connectives that attained pupation, except one, underwent metamorphosis, adult eclosion and wing expansion but invariably showed abnormally distended or ruptured abdomens. The ruptured or distended abdomens were caused by overinflation due to excessive intake of air the regulation of

Table 2. Effects of surgical alteration of the stomatogastric nervous system on eclosion and wing expansion of Manduca sexta

Treatment*	% completing adult eclosion	wing expansion	Remarks
Sham operated control	100.0	100.0	All adults normal.
Frontal connectives severed	66.7	58.5	All adults with abnormally inflated or ruptured abdomens.
Recurrent nerve severed	33.3	25.5	4 died as pharate adults; 1 eclosed without expanding wings; 3 were normal.
Frontal ganglion removed	8.5	0.0	All died as pharate adults except one.

* n = 12 insects/treatment

which appears to involve the frontal connectives.

It is evident from these studies that in Manduca sexta the frontal ganglion is not absolutely essential for feeding, larval growth, pupation or metamorphosis. However, it seems necessary to conclude that the frontal ganglion and the recurrent nerve appear to be involved in mediating critical events that occur later on in pharate adult life. It is suspected that the frontal ganglion and the associated recurrent nerve may play some role in the process of adult eclosion perhaps by regulating the event of air swallowing to facilitate escape of the pharate moth from the pupal case. Air swallowing is evidently involved in the subsequent expansion of the wings of the adult following eclosion. It is further suggested from the experiments previously described that the frontal connectives are involved in conveying nervous stimuli back to the frontal ganglion to terminate air swallowing since overinflation occurs in those insects with severed frontal connectives. Further experiments will certainly be necessary to unravel the various physiological events that are regulated by the frontal ganglion and other components of the stomatogastric nervous system in lepidopterous insects.

References

ROUSELL J. P. (1972) Physiologie du ganglion frontal des insectes. Ann. Biol. 11, 235-255.
BELL R. A., BORG T. K., and ITTYCHERIAH P. I. (1974) Neurosecretory cells in the frontal ganglion of the tobacco hornworm, Manduca

sexta. _J. Insect Physiol_. 20, 669-678.
BELL R. A., RASUL C. G., and JOACHIM F. G. (1975) Photoperiodic
    induction of the pupal diapause in the tobacco hornworm, _Manduca
    sexta_. _J. Insect Physiol_. 21, 1471-1480.
BELL R. A. and JOACHIM F. G. (1976) Techniques for rearing labora-
    tory colonies of tobacco hornworms and pink bollworms. _Ann. Ent.
    Soc. Am_. 69, 365-373.

CHEMICAL AND BIOLOGICAL PROPERTIES OF THE PROTHORACICOTROPINS

OF MANDUCA SEXTA

Walter E. Bollenbacher

Department of Biology
Wilson Hall 046A
University of North Carolina at Chapel Hill
Chapel Hill, North Carolina 27514

Although the primary effect of the prothoracicotropic hormone
(PTTH) is that of activating the prothoracic glands (PG) to synthesize ecdysone, in doing so it initiates a cascade of endocrine events
that lead to molting and metamorphosis in insects (see Granger and
Bollenbacher, 1981). Of the several hormones involved in insect
postembryonic development, PTTH was the first to be discovered, and
yet its endocrinology, particularly its chemistry, is the least
understood.

Studies investigating the chemical nature of PTTH have actually
spanned about two decades, and their results have suggested that
PTTH could be a sterol, a mucopolysaccharide and/or a protein (see
Bollenbacher and Bowen, 1983). More recent studies, however, have
firmly established that PTTH is proteinaceous and that in Bombyx
there is probably one hormone having an apparent molecular weight
(M_r) of 4.4 kd (Suzuki et al., 1982), but, recently, attempts to
purify PTTH from the brain of Manduca sexta (Bollenbacher and Gilbert,
1981; Kingan, 1981; Gilbert et al., 1981; Bollenbacher and Bowen,
1983), have yielded results which are in marked contrast to those on
the PTTH present in adult heads of Bombyx. From these studies, it
appears that there is at least one large PTTH having a M_r of $\sim$25-30
kd, and possibly a second neurohormone having a M_r of $\sim$7 kd. The
existence of more than one moiety with PTTH activity is not a new
discovery, since early experiments attempting to purify the PTTH
activity in brains of Bombyx pupae and Periplaneta nymphs suggested
this possibility (see Bollenbacher and Granger, 1984). The potential
physiological significance of multiple moieties with PTTH activity
underlines the importance of the purification of these factors, and

325

of demonstrating that they function in vivo as neurohormones, and
this is the problem that this laboratory is addressing.

Initial attempts to purify PTTH from Manduca utilized brains
from both day 3 last instar larvae and day 1 pupae. Although a big
factor and a small factor were found in both sources, the pupal brain
possessed approximately three times more activity than the larval
brains, and thus only pupa brains were used in further purification
studies. For the purpose of discussion, these two factors will be
considered to be neurohormones, but it must be cautioned that their
identification as such awaits further characterization.

The extraction protocol used with the pupal brains was less
complex and harsh than those developed for Bombyx, and was developed
so that the final product of the procedure could be used directly in
the in vitro PG assay for the hormone (Bollenbacher et al., 1979)
with minimal changes in buffer molarity, ionic strength, pH etc.
The method yielding the best results in this respect entailed extraction in 25% Grace's medium or distilled water at 4°C. Sonication did
not inhanse the efficiency of extraction. The centrifuged extract
was then heated for 2 min at 100°C, centrifuged again, the supernatant
concentrated and diafiltrated with distilled H_2O and lyophilized.
The PTTH activity present in the lyophilate was quantified by the
in vitro assay (Bollenbacher et al., 1979, 1983). This procedure
yielded approximately a ten fold purification of PTTH with near quantitative recovery of activity. Gel filtration of the lyophilized
material yielded two peaks of PTTH activity with mean M_rs of ∿28 kd
and ∿7 kd. These moieties were termed "big" and "small" PTTH, respectively, and their apparent M_rs were confirmed by other gel filtration methods. Guanidine·HCl (Gn·HCl) chromatography of the two
activities was attempted to acertain more precisely their M_r, but in
the presence of this denaturant big PTTH activity was irreversibly
lost, presumably due to the susceptability of this larger peptide to
degradation by harsh salt denaturants. Small PTTH activity was re-
naturable after Gn·HCl treatment, and its M_r was the same as that
obtained by conventional gel filtration methods. These different
stabilities of the two putative PTTHs indicated that the big moiety
was probably not an aggregate of the small, nor was it comprised of
subunits of the small form. Thus, these findings provided additional
evidence of the existence of a big and small form of PTTH.

After gel filtration big PTTH was purified by DEAE ion exchange
chromatography, followed by isoelectric focussing. This procedure
yielded two activities having pIs of 5.35 and 5.15, with the majority
of the recoverable activity in the pH 5.35 fraction. SDS gel electrophoresis of the IEF purified material (both pIs) yielded a single
band with an M_r of ∿29 kd, in close agreement to the M_r determined
previously by gel filtration. The fact that there were two factors
from the IEF that had PTTH activity may have been an artifact of the

method itself, or perhaps, may have resulted from slight changes in
the charge of the peptide produced by the purification procedure.
Purification of the small form of PTTH is in progress, and preliminary analyses have revealed that it does not bind to DEAE, as does
big PTTH, despite the fact it appears to have acidic pIs (5.2 and
3.9) (Bollenbacher and Granger, 1984).

The big and small prothoracicotropic factors appear to affect
different, developmentally specific responses from the PG. Each
moiety was shown to activate the PG in vitro in a dose-dependent
manner similar to crude PTTH (Bollenbacher and Granger, 1984), but
the amount of each form of PTTH necessary to activate the PG varied
depending on the developmental stage of the glands. For example, the
dose responses of activation and ED_{50}s obtained for big PTTH with day
3 larval (ED_{50}=0.075) and day 0 pupal PG (ED_{50}=0.08) were the same.
However, small PTTH activates day 3 larval glands (ED_{50}=0.16) more
effectively than it does pupal glands (ED_{50}=3.8), and this difference
proved of practical significance during the purification of big and
small PTTH, since it could be used to verify which of the two moieties
was present.

In further support for the two form-two function hypothesis, it
appears that the PTTH(s) released into the hemolymph during the IV
larval instar HCP and first HCP of the V larval instar are different
in their chemical and biological properties. The hormone released
into the hemolymph during the IV larval instar HCP appears to be big
PTTH, as judged by its M_r, pI, and ability to activate both larval
and pupal PG at comparable ED_{50}. By some of the same criteria, the
PTTH activity found in the hemolymph during the first HCP of the V
larval instar has preliminarily been identified as small PTTH.

If it can be shown unequivocally that there is more than one
PTTH, the PTTH axis (at least in Manduca) would be a unique system,
in which several functionally related neurohormones would be released
at selective times during development to evoke a common response from
the same target gland, but a very different type of development. A
conclusive demonstration of the two hormone-two function hypothesis,
however, must await the development of methods such as a PTTH radioimmunoassay with which the relationship of these two moieties can
be probed more critically.

ACKNOWLEDGMENTS

The author wishes to thank Ms. J. Hightower for the preparation
of this manuscript. This research was supported by grants from NIH,
NS18791 and AM31642, to W. E. Bollenbacher.

REFERENCES

Bollenbacher W. E., Agui N., Granger N. A. and Gilbert L. I. (1979) In vitro activation of insect prothoracic glands by the pro-thoracic hormone. Proc. Nat. Acad. Sci. 76, 5148–5152.

Bollenbacher W. E. and Bowen M. F. (1983) The prothoracicotropic hormone. In Endocrinology of Insects (Edited by Laufer H. and Downer R. G. H.), in press. A. R. Liss, New York.

Bollenbacher W. E., O'Brien M., Katahira E. J. and Gilbert L. I. (1983) A kinetic analysis of the action of the insect prothor-acicotropic hormone. Mol. Cell. Endocr. in press.

Bollenbacher W. E. and Gilbert L. I. (1981) Neuroendocrine control of postembryonic development in insects, the prothoracicotropic hormone. In Neurosecretion: Molecules, Cells, Systems (Edited by Farner D. S. and Lederis K.), pp. 361–370. Plenum, New York.

Bollenbacher W. E. and Granger N. A. (1984) Endocrinology of the prothoracicotropic hormone. In Comprehensive Insect Physiology, Biochemistry and Pharmacology (Edited by Kerkut G. A. and Gilbert L. I.) Vol. 7, in press. Pergamon Press, New York.

Gilbert L. I., Bollenbacher W. E., Agui N., Granger N. A., Sedlak B. J., Gibbs D. and Buys C. M. (1981) The prothoracicotropes: Source of the prothoracicotropic hormone. Amer. Zool. 21, 641–653.

Granger N. A. and Bollenbacher W. E. (1981) Hormonal control of insect metamorphosis. In Metamorphosis: A Problem in Develop-mental Biology (Edited by Gilbert L. I. and Frieden E.), 2nd Edition, pp. 105–137. Plenum Press, New York.

Kingan T. G. (1981) Purification of the prothoracicotropic hormone from the tobacco hornworm Manduca sexta. Life Sci. 28, 2585–2594.

Suzuki A., Nagasawa H., Kataoka H., Hori Y., Isogai A., Tamura S., Guo F., Zhong X., Ishizaki H., Fujishita M. and Mizoguchi A. (1982) Isolation and characterization of prothoracicotropic hormone from silkworm, Bombyx mori. Agr. Biol. Chem. 46, 1107–1109.

NEUROCHEMICAL ANALYSIS OF CHOLINERGIC ELEMENTS

IN INSECT SYNAPTOSOMES

Heinz Breer

Department of Zoophysiology
University Osnabrück
Osnabrück, West Germany

The proper interest of comparative neurochemistry
is to know all the variations which nature has invented
to fulfil a certain molecular function in the nervous
system. For the group of insects neurochemical approaches are highly recommendable also for immediately
practical reasons; in order to control harmful insects
while sparing their beneficial relatives and other
creatures, a more detailed knowledge of molecular mechanisms in neuronal function of insects might be helpful to develop more selective and effective insecticides. The availability of isolated nerve endings from
insects would allow neurochemical research, which might
provide informations bridging the gap between fine
structure as seen at the electronmicroscope level and
the functional activity of neuronal interconnections
usually defined by physiological techniques. Since all
attempts to isolate synaptosomes from insects using
standard techniques have failed, a microscale flotation
procedure was designed specifically for use with insects
(Breer, 1981a). In a Ficoll solution a marked separation
of free mitochondria and intact, pinched off and resealed nerve terminals was achieved. Ultrastructural analysis revealed that insect synaptosomes are obviously
well sealed structures containing spherical vesicles
and frequently glycogen granules. Isolated nerve endings
accumulated exogenous choline via a high affinity transport system and were capable to convert most of it to
acetylcholine. The mechanism of choline transport was
studied on membrane vesicles derived from synaptosomes,
which retain only the limiting membrane and lack the

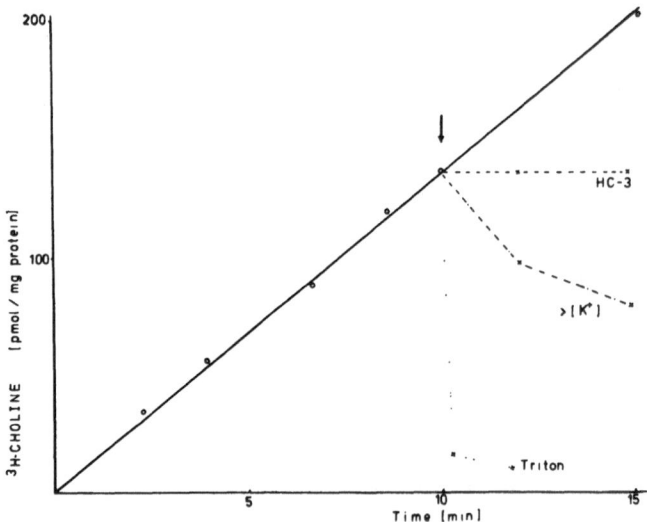

Choline uptake of labelled choline by synaptosomes from locusts. The addition of hemi-
cholinium-3 blocked the transport, detergent
(0,1 %) released all the accumulated material, increased potassium concentration caused
a depolarization-dependent release.

complex energetics, metabolism and compartimentation of
intact nerve endings. Such vesicles retained the capability to accumulate choline using the artificially
created ion gradients as sole energy source. In experiments varying the saline compositions it could be established that the presence of external sodium as well as
a sodium gradient (out > in) was essential for the choline uptake (Breer, 1983). Furthermore the sodium electrochemical potential was obviously the main driving
force for the carrier-mediated choline translocation.
The results are suggestive of a sodium-linked cotrans-
port of choline across the membrane. The effects of
ionophores (monensin, valinomycin) as well as neurotoxins (veratridine, tetrodotoxin) demonstrated that the
membrane potential may contribute to the driving energy.

    Exposing insect synaptosomes, preloaded with tri-
tiated choline, to depolarizing conditions (veratridine,
high potassium) induced a release of labelled acetylcholine. The evoked release of acetylcholine from insect
nerve terminals was demonstrated to be essential dependent on extracellular calcium, and the importance of
Ca^{2+} for the release process was emphasized by experi-

ments using calcium ionophores, which induced release without membrane depolarization. The extent of transmitter release was significantly reduced in the presence of muscarinic agonist, like oxotremorine, but was even enhanced by atropine, suggesting a feedback regulation via muscarinic presynaptic mechanisms.

There are several physiological and pharmacological evidences for the existence of nicotinic cholinergic synapses in the CNS of insects and the binding of α-bungarotoxin, a specific ligand for nicotinic receptors to locust ganglionic tissue fulfils the criteria of saturability and pharmacological specificity as required for a transmitter receptor (Breer, 1981b). Furthermore some of the monoclonal antibodies raised against the acetylcholine receptor of Torpedo marmorata cross-reacted with the toxin binding sites from insects. These results point to considerable similarities in the molecular structure of α-toxin binding components in locust ganglionic tissue and Torpedo electroplaques and suggest the equivalence of toxin binding site and true nicotinic receptor. Solubilized in locust CNS the binding component, sedimented in a linear sucrose density gradient as a single species with a sedimentation coefficient of 9 S. The toxin binding sites were further purified by affinity chromatography on α-bungarotoxin coupled to Sepharose 4 B. The adsorbed material, eluted with d-tubocurarine, contained the specific binding sites. Analysis of the eluted protein on microgelelectrophoresis under denaturing conditions gave only one major band corresponding to an apparent molecular weight of about 65,000. The availibility of synaptosomes, membrane vesicles and receptor macromolecules from insect nervous tissue, will allow neurochemical approaches on cholinergic transmission in insects at various levels.

References

Breer, H., 1981a, Characterization of synaptosomes from the CNS of insects. Neurochem. Int., 3:155.
Breer, H., 1981b, Properties of putative nicotinic and muscarinic cholinergic receptors in theCNS of Locusta migratoria. Neurochem. Int., 3:43.
Breer, H., 1983, Choline transport by synaptosomal membrane vesicles isolated from insects nervous tissue. FEBS-Letters, 153:345.

ENDOGENOUS N-ACETYL-5-HYDROXYTRYPTAMINE (NA-5-HT)

IN INSECT CEREBRAL GANGLIA

G.L. Brookhart and L.L. Murdock

Department of Entomology
Purdue University
West Lafayette, IN 47907

Two major enzymes involved in the metabolism of aromatic biogenic amines in mammalian nervous tissues are monoamine oxidase (MAO) and catechol-O-methyltransferase (COMT). Present evidence would favour the view that these enzymes, if present at all in insect central nervous structures, are quantitatively minor. Evidence that an alternative pathway, N-acetylation, might be a major route of metabolism of aromatic biogenic amines in insect nervous tissues was first presented by DEWHURST, CROKER, IKEDA, and McCAMAN (1972). They found high levels of N-acetyltransferase activity utilizing acetyl-coenzyme A and several aromatic amine substrates in Drosophila melanogaster nervous tissue. The hypothesis that N-acetylation may be an important route of metabolism of aromatic biogenic amines in nervous tissues was supported by enzyme and radiotracer studies by EVANS and FOX (1975a,b) working with Apis mellifera central nervous tissues; VAUGHN and NEUHOFF (1976) and MIR and VAUGHN (1981a,b) with Schistocerca gregaria; HAYASHI, MURDOCK, and FLOREY (1977) with Locusta migratoria; MAXWELL, MOORE, and HILDEBRAND (1980) with Manduca sexta; and EVANS, SODERLUND, and ALDRICH (1980) with Ostrinia nubialis. We thought it desireable to test that hypothesis further by asking if we could detect the products of the N-acetyltransferase pathway (MURDOCK and OMAR, 1981): just as the products of MAO and COMT are chemically detectable in mammalian brain, we expected to be able to detect N-acetyldopamine (NADA), N-acetyl-5-hydroxytryptamine (NA-5-HT), and N-acetyloctopamine (NAOA) in insect CNS tissues. NADA and NA-5-HT are easily amenable to quantitative analysis by liquid chromatography with electrochemical detection. With these methods we demonstrated the existence of endogenous NADA in CNS structures of a variety of insect species (Murdock and Omar, 1981). Its concentration in the

brain of P. americana adults was about the same as for dopamine, c.
10 nmol/g wet wt.. Injection of L-DOPA, a precursor of dopamine,
into P. americana adults increased the level of NADA, and injection
of the aromatic amine depleter reserpine, decreased it. Although
our observations of endogenous NADA supported the operation of the
N-acetyltransferase pathway in vivo, our failure to detect NA-5-HT
(OMAR, MURDOCK and HOLLINGWORTH (1982) did not: we could easily
show that the P. americana cerebral ganglion was capable of
accumulating NA-5-HT when incubated with 5-hydroxytryptophan
(5-HTP), a probable precursor of 5-HT and NA-5-HT, but we were
unable to find evidence for endogenous NA-5-HT. This raised a
question as to the generality of the N-acetyltransferase pathway:
Why should NADA be synthesized and accumulate and NA-5-HT not?

We have recently addressed this problem using improved sample
preparation and higher efficiency separation techniques and have
obtained prima facie evidence for the existence of endogenous
NA-5-HT in the cerebral ganglia of adult insects of several orders.

NA-5-HT was extracted from freshly dissected insect nervous
tissue using 50 or 100 ul of 0.1 M perchloric acid. The homogenate
was centrifuged at 12,800 x g for 2 min and the supernatant
injected directly into the liquid chromatograph. NA-5-HT, NADA,
and 5-HT were separated using a Brownlee MPLC system RP-18 column
(25 cm x2.1 mm i.d.; 5 u particle size). Mobile phase was composed
of 60 mM citric acid/40 mM Na$_2$HPO$_4$/1 mM EDTA containing 25%
methanol. Flow rate was 0.3 ml/min. Under these conditions 5-HT
eluted at 8.6 min, NADA at 11.2 min, and NA-5-HT at 16.5 min.

Cerebral ganglion from P. americana contained a substance
which coeluted with authentic NA-5-HT with the above mobile phase.
Extraction p-values (BOWMAN and BEROZA, 1965) for this substance
and authentic NA-5-HT were identical in the four binary solvent
systems tested: water/ethyl acetate; water/benzyl alcohol;
water/hexane; water/toluene. Hydrodynamic voltammograms for the
substance and authentic NA-5-HT were also identical. This evidence
supports the conclusion that NA-5-HT occurs naturally in the
cerebral ganglion of P. americana. Further support for this were
observations (1) that incubation of cerebral ganglion halves with
5-HTP caused a marked enhancement of the NA-5-HT peak; (2) that
reserpinization, which depletes 5-HT and other aromatic biogenic
amines from nervous tissues, caused the NA-5-HT peak from cerebral
ganglion to disappear. Evidently the NA-5-HT pool (c.1.3 nmol/g
wet weight) in P. americana cerebral ganglion, though small
relative to NADA, turns over rapidly and may owe its existence to
the availability of neural stores of 5-HT.

NA-5-HT was detected in cerebral ganglion from every insect
species assayed, including adults of Apis mellifera (Hymenoptera),

Phormia regina (Diptera), Manduca sexta (Lepidoptera), Oncopeltus fasciatus (Hemiptera), and Plathemis lydia (Odonata). Concentrations ranged from 0.5-4 nmol/g wet wt., and in most species were lower than those for NADA (range 0.4-13 nmol/g wet wt.)

These observations support the concept that the N-acetyl-transferase pathway is an important route of metabolism of aromatic biogenic amines in the CNS of insects. The functional role of the N-acetylated amines is not known.

REFERENCES

BOWMAN M.C. and BEROZA M. (1965) Extraction p-values of pesticides and related compounds in six binary solvent systems. JAOAC. 48, 943.

DEWHURST S., CROKER S.G., IKEDA K., and McCAMAN R.E. (1972) Metabolism of biogenic amines in Drosophila nervous tissue. Comp. Biochem. Physiol. 43B, 975-981.

EVANS P.H. and FOX P.M. (1975a) Enzymatic N-acetylation of indolealkylamines by brain homogenates of the honeybee, Apis mellifera. J. Insect Physiol. 21, 343-353.

EVANS P.H. and FOX P.M. (1975b) Comparasons of various biogenic amines as substrates for acetyltransferase from Apis mellifera (L.) CNS. Comp. Biochem. Physiol. 51C, 139-141.

EVANS P.H., SODERLUND D.M. and ALDRICH J.R. (1980). In vitro N-acetylation of biogenic amines by tissues of the European corn borer, Ostrinia nubialis Huebner. Insect Biochem 10, 375-380.

HAYASHI S., MURDOCK L.L. and FLOREY E. (1977) Octopamine metab-olism in invertebrates (Locusta, Astacus, Helix): Evidence for N-acetylation in arthropod tissues. Comp. Biochem. Physiol. 58C, 183-191.

MAXWELL G.D., MOORE M.M., and HILDEBRAND J.G. (1980). Metabolism of tyramine in the central nervous system of the moth Manduca sexta. Insect Biochem. 10, 657-665.

MIR A.K. and VAUGHN P.F.T. (1981a) The conversion of N-acetyl-tyramine to N-acetyldopamine by Schistocerca gregaria thoracic ganglia. Insect Biochem. 11, 571-577.

MIR A.K. and VAUGHN P.F.T. (1981b) Biosynthesis of N-acetyl-dopamine and N-acetyloctopamine by Scistocerca gregaria nervous tissue. J. Neurochem. 36, 441-446.

MURDOCK L.L. and OMAR D. (1981) N-Acetyldopamine in insect nervous tissue. Insect Biochem. 11, 161-166.

OMAR, D., MURDOCK L.L. and HOLLINGWORTH R.M. (1982) Actions of pharmacological agents on 5-hydroxytryptaime and dopamine in the cockroach nervous system (Periplaneta americana L.) Comp. Biochem. Physiol. 73C, 423-429.

VAUGHN P.F.T. and NEUHOFF V. (1976) The metabolism of tyrosine, tyramine and L-3,4-dihydroxyphenylalanine by cerebral and thoracic ganglia of the locust, Schistocerca gregaria. Brain Res. 117, 175-180.

NONLINEAR MASERLIKE RADIATION IN BIOLOGICAL SYSTEMS

Philip S. Callahan

Insect Attractants, Behavior, and Basic Biology Research
Laboratory, Agricultural Research Service, U.S.D.A.
Gainesville, Florida 32604 USA

This work was initiated in the belief that nonlinear maserlike radiation from insect scents (semiochemicals) may eventually
be utilized for control of pest species if some method of ampli-
fying the weak far infrared coherent emissions can be developed.

CALLAHAN (1957) first pointed out that the attraction of
moths to light was not to the visible radiation source itself,
but rather to the airspace around the source. He postulated that
the attraction was due to secondary scatter radiation from molecular or plant surfaces. He later termed his hypothetical frequencies maserlike and maintained that the effect involved Raman
scatter phenomenon (CALLAHAN, 1967).

CALLAHAN (1981) later pointed out that the system was extremely complex and probably involved, depending on the scent,
different types of scatter such as Rayleigh center, Rayleigh wing
Brillouin or Mandel'shtam-Brillouin, besides normal Stoke and
anti-Stoke Raman wavelengths. He evoked a dielectric waveguide
-- Feshback model to explain his coherent-resonant maserlike
lines. Feshback generalized the Bohr-Feshback atomic elastic
scatter model to include inelastic scatter phenomenon. It seems
logical to generalize such a dielectric waveguide-Feshback inelastic scatter model for the coherent frequencies obtained by
vibrating a light reflective (electrode-like) antenna, needle or
reed in flowing molecules of vapor (CALLAHAN, 1976).

Insects, upon entering a flowing plume of attractant scent
vibrate their antenna at specific frequencies depending on the
species (CALLAHAN, 1975).

This paper reports inelastic scatter due to surface-enhanced emission from semiochemicals blown across a waxy coated aluminum punch-hole antenna.

Cabbage looper, Trichoplusia ni (Hübner), pheromone and human breath are stimulated to emit maserlike radiation when the molecules collide with the curled edges of small holes punched in a 50-μm thick piece of weakly paramagnetic aluminum. The coherent emissions are due to the surface-enhanced Raman effect. Surface-enhanced Raman effects have been only recently discovered by physicists (VAN DWYNE, 1979), although this author has been observing them for well over a dozen years. This paper and previous ones are presented as irrefutable proof that room-temperature (ambient) coherent frequencies are available, from free floating scents, for resonant tuning by insect sensilla (spine) dielectric open resonators when such resonators are thrust into the flowing semiochemicals.

Figure 1 is a spectrum taken at 0.5 cm^{-1} resolution (16 scans) of human breath blown at 2 kph through the holes of the punch-hole scatter antenna. The CO_2 10.6 line is the well known IR line that physicists were first able to lase following the invention of the laser in 1953. This is the first report of the generation of coherent 10.6 μm IR at room temperature. The coherent 10.6 μm radiation has been shown to excite probobsis probing behavior in Aedes aegypti (L.) (CALLAHAN and GOLDMAN, 1970). The 26.6-μm line is the strongest line obtained from human breath and skin scent emissions, and peaks at a temperature of 37°C (98.6°F, human temperature). Both the CO_2 line and the "breath" 26-μm lines are pumped to maximum emission by near infrared in the 1-μm region. The human skin is highly reflective of 1 μm environmental (sky) radiation.

RAMAN (1928), in his inaugural address before the South Indian Science Association, speculated that a good portion of scatter radiation might be coherent.

F. A. POPP (1977) has demonstrated coherent visible radiation from wheat seeds. He states, "...that biological systems generally have the capacity to store coherent photons which come from the external world."

Since efficient communication systems require coherent signals, there is little doubt that nature utilizes such room-temperature coherent radiation, particularly in the IR region, in systems mediated by pheromones, hormones, peptides, and enzymes.

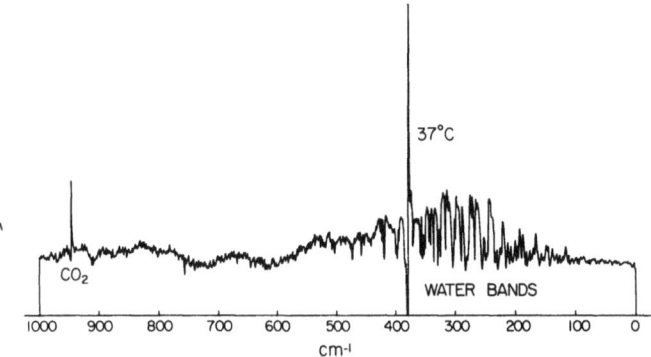

FIG. 1. Maserlike (surface enhanced Raman scatter) radiation from human breath at 37°C (98.6°F).

REFERENCES

CALLAHAN P. S. (1957) Oviposition response of the imago of the corn earworm, Heliothis zea (Boddie), to various wavelengths of light. Ann. Entomol. Soc. Am. 50, 444-452.

CALLAHAN P. S. (1967) Insect molecular bioelectronics. Misc. Publ. Entomol. Soc. Am. 5, 315-347.

CALLAHAN P. S. (1981) Nonlinear IR radiation in a biological system. Applied Optics 20, 3827.

CALLAHAN P. S. (1976) Insect antenna vibrating frequency modulator and resonating maserlike IR emitter. U. S. Patent No. 3,997,785.

CALLAHAN P. S. (1975) Insect antennae with special reference to the mechanism of scent detection and the evolution of the sensilla. J. Insect Morph. & Embryol. 4, 381-430.

CALLAHAN P. S. and LEONARD GOLDMAN. (1970) Response of Aedes aegypti to 10.6 micron radiation. First Quarterly Report, Insect Attractants, Behavior, and Basic Biology Research Laboratory, Gainesville, Florida.

POPP F. A. (1977) Photon storage in biological systems. Electromagnetic Bio-Information, (Ed. by F. A. POPP, G. BECKER, H. L. KÖNIG, and W. PESCHKO) pp. 123-149. München-Wien-Baltimore.

RAMAN C. (1928) A New Radiation. South Indian Science Association, Bangalore, India.

VON DWYNE R. P. (1979) Chemical and Biological Applications of Lasers. (Ed. by C. B. MOORE). Academic Press, New York.

A PROTEIN SYNTHESIS STIMULATING HORMONE IN THE LOCUST

James A. Carlisle and Barry G. Loughton

Department of Biology, York University
Downsview, Ontario, Canada M3J 1P3

In locusts, a massive release of neurosecretory material from the corpus cardiacum (CC) occurs immediately after feeding (FRIEDEL AND LOUGHTON, 1980). But there has been no direct demonstration of a physiological role for this released material. Though extracts of the glandular lobes of the CC inhibit protein synthesis, extracts of the whole CC do not (CARLISLE AND LOUGHTON, 1980). We report here that homogenates of the brain, homogenates of the storage lobes of the CC and haemolymph of animals which have been fed all contain factor(s) which stimulate protein synthesis. Both the neural and haemolymph factor(s) exhibit the same apparent molecular weight when subjected to gel filtration and possess similar stimulatory activity on protein synthesis. It would appear they are the same molecule.

When we determined the rate of incorporation of injected ^{3}H-leucine by D10 adult virgin male locusts, the incorporation of the label declined after feeding but when we adjusted this rate for the haemolymph pool of leucine, which increases by almost twenty-fold, the results indicated that protein synthetic rate in the fed animals increased by about 221%.

D10 virgin adult male locusts which had fasted for 24 hours were injected with 2uCi. ^{3}H-leucine in 25 ul of haemolymph collected from mature locusts (1) fasted for 24 hours or (2) 1 hour after feeding. After one hour haemolymph samples were taken and incorporation of radioactive label into haemolymph protein was determined. Haemolymph from fasting animals did not induce an increase in incorporation rate of ^{3}H-leucine over saline-injected controls whereas haemolymph from fed animals stimulated incorporation(fig.1). Water extracts of storage lobes also showed incorporation-stimulating activity (fig.1).

In order to purify the active factor, brains and storage lobes were homogenized separately. The supernatants were partitioned by anion-exchange chromatography on a DEAE Sephadex A-25 column. Fractions were tested for activity by injection into fasting locusts. The active fractions were pooled and further purified by separation on a Biogel P-2 column calibrated with peptides of known molecular weight. The fractions were then tested for their ability to stimulate incorporation of radioactive label **in vivo**. Haemolymph collected from mature animals within one hour of the commencement of feeding was separated in a similar manner. The stimulatory activity from brain, CC storage lobe and haemolymph resided in a similar fraction from each tissue after both ion exchange chromatography and gel filtration. This fraction exhibited an apparent molecular weight of 400-650 (fig.1).

A time-course was determined by injecting unfed locusts with 2 uCi. ^{3}H-leucine along with 25 ul of haemolymph taken from mature locusts at various times after feeding. Haemolymph from fed animals stimulated incorporataion of radioactive label within five minutes after the commencement of feeding (fig. 3).

Fat bodies from D10 adult virgin males were removed and incubated in Schneider's modified **Drosophila** medium (Grand Island Biological Co.) with 2 uCi ^{3}H-leucine for one hour at 36°C. Aliquots of the incubation medium were assayed for the incorporation of radioactive label into protein. When fat bodies from animals which had been fed one hour prior to dissection were compared to similar tissues from those remaining unfed for 24 hours the fat bodies from fed animals incorporated 71% more label and the difference between the tissues remained for an additional hour (fig.2).In other **in vitro** experiments fat bodies from D10 virgin male locusts which were unfed for 24 hours were removed from the animal and then split longitudinally and incubated in the manner described above. One half was incubated with the substance under test and the other half incubated separately with an equal volume of saline as a control. Haemolymph from fed animals was shown to act directly in stimulating incorporation by the fat body. Storage lobe homogenates and the partially purified small molecular weight fraction from the storage lobe also stimulated the incorporation of radioactive label into released protein. As the external pools of leucine were the same for all tissues in the **in vitro** experiments, it was concluded that increase in incorporation of label indicated an increase in protein synthesis by the tissues.

Unfortunately, the semi-purified factors from the brain, storage lobe and haemolymph proved to be unstable after freezing and thawing, at 4C and at room temperature, so the enzyme digestions necessary to determine whether they are proteinaceous or not could not be carried out. Although the identity and nature of the factors have not been determined their similar action, UV absorption, separatory characteristics and source argue strongly that they are a single small peptide. Our experiments represent the first evidence of a neurosecretory factor in insects which appears to be released into the haemolymph and acts directly to stimulate haemolymph protein synthesis.

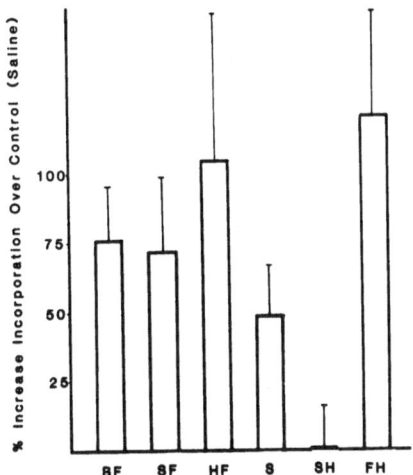

Figure 1:
The change in incorporation of ^{3}H-leucine into haemolymph proteins of locusts (D10 virgin adult males) after injection of: BF-brain extract (0.05 brain equivalents/insect; SF-CC storage lobe extract 0.03 storage lobe equivalents/insect); HF-extract of haemolymph from fed locusts (20 ul haemolymph equivalents/insect); S-CC storage lobe homogenate (0.1 lobe/insect); SH-haemolymph from fed locusts (50 ul/insect).

Figure 2:
The change in incorporation of ^{3}H-leucine into released protein by fat bodies of adult male locusts (D10) incubated <u>in vitro</u> (see text). Fat bodies from fed locusts incubated for 1h (F$_1$) or 2h (F$_2$) compared to fat bodies from fasting locusts (incorporation corrected for the protein content of the fat bodies). In experiments designated H, S and SF fat bodies were dissected out

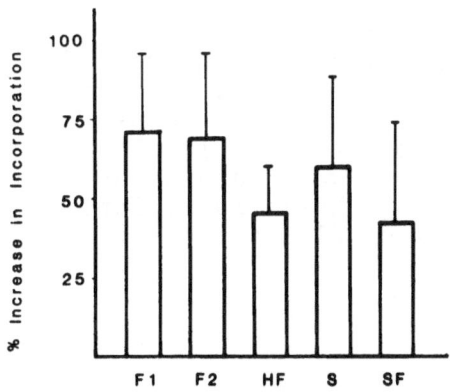

and divided longitudinally. The histograms show the increase in incorporation of ½ fat bodies treated with the test substance over the corresponding ½ fat body control. HF- 40 ul of haemolymph from a locust fed 1h earlier. The control ½ fat body received 40 ul of haemolymph from fasting locusts. S- Salibe homogenate of 0.1 CC storage lobes. SF- 4-600 MW fraction from CC storage lobe extracts (0.02 storage lobe equivalents). Incorporation into released protein is expressed as a function of fat body protein. The results represent the results of at least 6 incubations. The vertical lines show standard error of of the means.

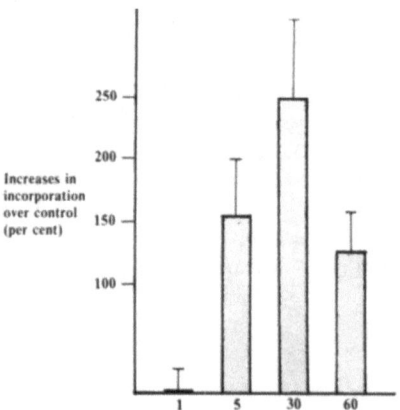

Time after commencement of feeding (minutes)

Figure 3:
Haemolymph was withdrawn from locusts at various times after feeding and 25 ul of the plasma fraction was injected into unfed locusts along with 2 uCi ^{3}H-leucine. Incorporation of label into haemolymph proteins after one hour was measured as an indication of relative protein synthetic rate. Control animals were injected with 25 ul of plasma from an unfed animal.

REFERENCES

CARLISLE J. and LOUGHTON B. G. (1979) Adipokinetic hormone inhibits protein synthesis in **Locusta. Nature 282,** 420-421
FRIEDEL T. and LOUGHTON B. G. (1980) Induced changes in neurosecretory activity of adult female **Schistocerca gregaria** in relation to feeding. **J. Insect Physiol., 26,** 33-37

PROCTOLIN AND A SECOND MYOTROPIC PEPTIDE FROM THE HINDGUT OF THE

COCKROACH LEUCOPHAEA MADERAE

Benjamin J. Cook and G. Mark Holman

VTERL, ARS, USDA
P.O. Drawer GE
College Station, TX 77841

When proctolin was discovered it was proposed that this peptide was the excitatory transmitter for visceral muscles of the hindgut (Brown, 1975). This hypothesis was supported by the demonstrated release (after neural stimulation) of a substance into perfusates that caused a potentiation of neurally evoked contractions similar to that caused by proctolin. However, in several release experiments we obtained a substance that caused a contractile response different from that of proctolin and the present study was initiated to isolate the active principle and determine its chemical nature.

Hindguts isolated from the central nervous system of the cockroach were dissected (Holman and Cook, 1970) and prepared for recording as previously described (Cook and Holman, 1978). Groups of 25 hindguts removed from adult cockroaches of both sexes were placed into a 125 ml Erlenmyer flask containing 100 ml methanol-water-acetic acid (90:9:1). After a 3-min homogenization with a Polytron™ the homogenizing vessel was rinsed with an additional 100 ml of the methanol solution. Following centrifugation (30 min, 13,800 g, 4°C), the supernatants were decanted into a 250-ml round-bottom flask and the solvent was removed (to near dryness) by rotary vacuum evaporation at 40°C. Additional steps in purification and the final separation of the active substances by high-performance liquid chromatography (HPLC) on a μ-Bondapak phenyl column have been described elsewhere (Holman and Cook, 1983).

All hindgut-stimulating substances were eluted from the Sep-pak by the 25% acetonitrile fraction. The 0.1% TFA solution comprised of the sample and the 10 ml 0.1% TFA rinse was not active, although it contained about 95% of the dry residue. Two active fractions were isolated from the HPLC separation. Proctolin was eluted between 36

and 40 min. The retention time of authentic proctolin in this HPLC
system was was 38 min. A second hindgut-stimulating material was
eluted between 56 and 60 min. Both the proctolin and the new peptide
retained full biological activity after being held at 105°C for 45
min. The biological activity of proctolin was completely abolished
after 2 hr incubation with aminopeptidase M, and was diminished by 80%
after 4 hr incubation with carboxypeptidase Y. The reverse situation
occurred with the new peptide all biological activity was destroyed
after 4 hr incubation with carboxypeptidase Y, whereas similar treatment with aminopeptidase M for 2 hr reduced the biological activity by
about 90%. Incubation with the heat denatured enzymes had no effect
on the biological activity of either peptide.

The response of the isolated hindgut to increasing amounts of
proctolin and the second hindgut-stimulating peptide is shown in Fig.
1. In terms of hindgut equivalents, the stimulating effect of the new
peptide (Figs. $1B_1$ and B_2) is twice that of proctolin. The difference
in the isolated hindgut response to the two peptides is best demonstrated by Figs. $1A_4$ and B_4. When one hindgut equivalent of protolin
was applied, the hindgut responded with an immediate large contraction
followed by an increase in tonus along with increases in frequency and
amplitude. Within 1 min after application of the proctolin the tonus
began to decay although frequency and amplitude of contraction was
maintained for some time. The initial tonic response was not present
when one hindgut equivalent of the new peptide was applied (Fig. $1B_4$).
Instead, tonus, frequency, and amplitude of contraction gradually
increased over 1-4 min.

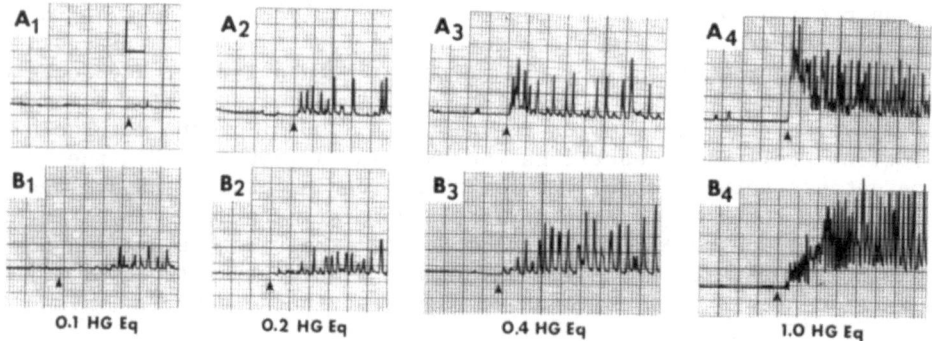

Fig. 1.   Response of a single hindgut preparation to both proctolin
A_{1-4} and the new peptide B_{1-4}. A_1 and B_1 show the response
of the hindgut to 0.1 hindgut equivalents (arrow) of proctolin and the new peptide. The space between myographs indicates a rinse in fresh saline solution. Time base 1 min and
vertical calibration = 2 mm of tissue movement.

The discovery of a second myotropic peptide in the hindgut of the cockroach complicates any simple theory of neural regulation in that visceral organ. Already monosodium glutamate (Holman and Cook, 1970) and proctolin (Brown, 1975) have been proposed as transmitters in the control mechanism. A dichotomy of action for the two candidates does exist in the hindgut of Locusta migratoria. Dunbar and Piek (1982) have found that the venom of Philanthus triangulum can selectively inhibit glutamate contractions without having any effect on those of proctolin. However, a measure of uncertainty still remains about the proposed transmitter function of proctolin because the muscle fiber membrane has shown a greater sensitivity to the peptide than the myoneural junction (Cook and Holman, 1979) and evidence for extra junctional receptors have also been discovered (Cook and Holman, 1980). Consequently, any speculation on the precise physiological function for the new peptide would be premature in spite of its hormone-like response at low tissue equivalent levels.

REFERENCES

Brown, B. E., 1975, A peptide transmitter candidate in insects, Life Sci., 17:1241.

Cook, B. J., and Holman, G. M., 1978, Comparative pharmacological properties of muscle function in the foregut and the hindgut of the cockroach Leucophaea maderae, Comp. Biochem. Physiol., 61C:291.

Cook, B. J., and Holman, G. M., 1979, The action of proctolin and L-glutamic acid on the visceral muscles of the cockroach Leucophaea maderae, Comp. Biochem. Physiol., 64C:21.

Cook, B. J., and Holman, G. M., 1980, Activation of potassium depolarized visceral muscles by proctolin and caffeine in the cockroach Leucophaea maderae, Comp. Biochem. Physiol. 67C:115.

Dunbar, S. J., and Piek, T., 1982, The action of the venom of Philanthus Triangulum F. on an insect visceral muscle, Comp. Biochem. Physiol, 73C:79.

Holman, G. M., and Cook, B. J., 1970, Pharmacological properties of excitatory neuromuscular transmission in the hindgut of the cockroach, Leucophaea maderae, J. Insect Physiol., 16:1891.

Holman, G. M., and Cook, B. J., 1983, Isolation and partial characterization of a second myotropic peptide from the hindgut of the cockroach Leucophaea maderae, Comp. Biochem. Physiol., (In press).

ESTIMATION OF OCTOPAMINE AND ITS ROLE IN THE RELEASE OF A
HYPERTREHALOSEMIC FACTOR IN THE AMERICAN COCKROACH, PERIPLANETA
AMERICANA

R.G.H. Downer, B.A. Bailey, J.W.D. Gole, R.J. Martin
and G.L. Orr

Department of Biology
University of Waterloo
Waterloo, Ontario, Canada N2L 3G1

The role of octopamine as a mediator of several important
physiological processes in insects is well established (Orchard,
1982). The nature of octopamine interaction with target tissues
has also been studied and, at least in some instances, octopamine
appears to bind with a specific membrane receptor to effect an
increase in adenylate cyclase activity and consequent elevation
of cyclic AMP levels in the target cell (Nathanson and Greengard,
1973; Harmer and Horn, 1977; Nathanson and Hunnicutt, 1981; Gole
et al. 1983).

The present report summarises two studies designed to
further elucidate the nature and physiological significance of
octopamine in insects. The first uses an analytical procedure
for the simultaneous estimation of octopamine and other monoamines in insect tissues to determine levels for the brain,
specific nerve cord ganglia and haemolymph of the American
cockroach, Periplaneta americana. The second study examines the
possible role of octopamine in regulating the release of a
hypertrehalosemic factor from cockroach corpus cardiacum.

ESTIMATION OF OCTOPAMINE

Most estimations of octopamine levels in insect tissues have
used radioenzymatic assays in which radiolabelled synephrine,
formed by enzymatic methylation of octopamine, is isolated and
counted (Molinoff et al., 1969; Evans, 1980). The procedure has
been criticised for lack of specificity and, indeed, Goosey and
Candy (1980) have demonstrated that additional purification steps
are required in order to avoid erroneously high estimates of

haemolymph octopamine. High performance liquid chromatography (HPLC) with electrochemical detection (ElCD) of eluted amines has been employed extensively for analysis of catecholamines in biological tissues (Shoup, 1982) but the technique has not been widely used for estimation of octopamine or other monohydroxy analogues. Recently an analytical procedure has been developed which allows simultaneous estimation of catecholamines, phenolamines and 5-hydroxytryptamine in insect tissues using HPLC/ElCD (Martin et al., 1983). The procedure employs a dual coulometric detection system with the first detector set at a relatively low potential in order to effect electro-oxidation of catecholamines while the second detector is set at the higher potential required to achieve oxidation of octopamine and other phenolamines (Bailey et al., 1982).

HPLC/ElCD has been used to determine the levels of octopamine and other monoamines present in the nervous system of P. americana. The levels compare favourably with those obtained by Evans (1978) using the radioenzymatic assay. Octopamine levels in the nerve cord ganglia are about 2 times those of dopamine whereas, in the brain, the levels of dopamine are double those of octopamine. The ratio of 5-hydroxytryptamine to octopamine or dopamine varies in different nerve cord ganglia. Relatively low amounts of tyramine and norepinephrine were detected in whole brain.

The effects of excitation and flight on haemolymph octopamine levels were also studied and demonstrate a rapid elevation of octopamine levels within 60 sec of initially disturbing a resting insect. The rapid increase in octopamine resembles the release of catecholamines that occurs during sympathetic stimulation in vertebrates and lends further credence to the proposal that octopamine serves a `sympathomimetic role in insects. A marked depletion of octopamine from cockroach nerve cord was observed in response to poisoning with the formamidine insecticide, chlordimeform (CDM). CDM and its more active demethylated derivative, demethylchlordimeform (DCDM) elicit a characteristic sequence of poisoning symptoms in which a period of hyperexcitability is succeeded by paralysis and ultimately death. The maximal depletion of nerve cord octopamine occurs at the end of the hyperexcited state thus providing additional support for the suggestion that octopamine is implicated in the expression of or response to excitation.

REGULATION OF RELEASE OF HYPERTREHALOSEMIC FACTOR FROM CORPUS CARDIACUM

Analysis of monoamine levels in cockroach corpus cardiacum demonstrates the presence of octopamine, dopamine and

5-hydroxytryptamine. The incubation of corpora cardiaca in physiological saline containing any of these monoamines results in elevated levels of cyclic AMP within the gland; furthermore, additivity studies in which maximal response-eliciting concentrations of the amines are included together in the incubation medium indicate that separate, independent receptor sites exist for each of the compounds. Electrical stimulation of the nervi corporis cardiaca II (NCCII) also elevates cyclic AMP levels. This observation, together with reports that the release of adipokinetic hormone from locust corpus cardiacum is under the control of aminergic neurons of NCCII (Rademakers, 1977; Orchard and Loughton, 1981; Orchard et al., 1982), suggests that the three amines identified in cockroach corpus cardiacum may be involved in the modulation of corpus cardiacum function. The possibility that the amines promote release of a hyper-trehalosemic factor from the corpus cardiacum was tested by bioassay of the medium following incubation of glands in the presence of the various amines. The results indicate that octopamine effects the greatest release of a hypertrehalosemic factor from cockroach corpus cardiacum although dopamine and 5-hydroxytryptamine also demonstrate a positive response, possibly as a result of interaction with the octopamine binding site. A peptidergic or proteinaceous nature for the active factor is indicated by deactivation of the factor following incubation in pronase, thereby suggesting that the released hypertrehalosemic factor may be trehalagon. Thus, it is proposed that the release of trehalagon, or some other hypertrehalosemic factor, from the cockroach corpus cardiacum is under the regulation of aminergic (probably octopaminergic) neurons contained within NCCII.

ACKNOWLEDGEMENTS

Supported by an operating grant from Natural Sciences and Engineering Research Council of Canada.

REFERENCES

Bailey, B.A., Martin, R.J., and Downer, R.G.H., 1982, Simultaneous determination of dopamine, norepinephrine, tyramine and octopamine by reverse phase liquid chromatography with electrochemical detection. J. Liquid Chromatogr., 5: 2435.

Evans, P.D., 1978, Octopamine distribution in the insect nervous system, J. Neurochem., 30: 1009.

Evans, P.D., 1980, Biogenic amines in the insect nervous system, Adv. Insect Physiol., 15: 317.

Gole, J.W.D., Orr, G.L., and Downer, R.G.H., 1983, Interaction of formamidines with octopamine-sensitive adenylate cyclase receptors in the nerve cord of <u>Periplaneta americana</u> L., <u>Life Sciences</u>, 32: 2939.

Goosey, M.W., and Candy, D.J., 1980, The d-octopamine content of the haemolymph of the locust, <u>Schistocerca americana gregaria</u> and its elevation during flight, <u>Insect Biochem.</u>, 10: 393.

Harmar, A.J., and Horn, A.S., 1977, Octopamine-sensitive adenylate cyclase in cockroach brain: effects of agonists, antagonists and guanylyl nucleotides, <u>Mol. Pharmacol.</u>, 13: 512.

Martin, R.J., Bailey, B.A., and Downer, R.G.H., 1983, Rapid estimation of catecholamines, octopamine and 5-hydroxytryptamine from biological tissues using high performance liquid chromatography with coulometric detection, <u>J. Chromatogr.</u> (in press).

Molinoff, P.B., Landsberg, L., and Axelrod, J., 1969, An enzymatic assay for octopamine and other β-hydroxylated phenylethylamines, <u>J. Pharmac. exp. Therap.</u>, 170: 253.

Nathanson, J.A., and Greengard, P., 1973, Octopamine-sensitive adenylate cyclase: evidence for a biological role of octopamine in nervous tissue, <u>Science</u>, 180: 308.

Nathanson, J.A. and Hunnicutt, E.J., 1981, N-demethylchlordimeform: a potent partial agonist of octopamine-sensitive adenylate cyclase, <u>Molec. Pharmacol.</u> 20: 68.

Orchard, I., 1982, Octopamine in insects: neurotransmitter, neurohormone and neuromodulator, <u>Can. J. Zool.</u>, 60: 659.

Orchard, I., and Loughton, B.G., 1981, Is octopamine a transmitter mediating hormone release in insects? <u>J. Neurobiol.</u>, 12: 143.

Orchard, I., Loughton, B.G., Gole, J.W.D., and Downer, R.G.H., 1982b, Synaptic transmission elevates adenosine 3',5'-monophosphate (cyclic AMP) in locust neurosecretory cells, <u>Brain Res.</u>, 258: 152.

Rademakers, L.H.P.M., 1977, Effects of isolation and transplantation of the corpus cardiacum on hormone release from its glandular cells after flight in <u>Locusta migratoria</u>, <u>Cell Tiss. Res.</u>, 184: 213.

Shoup, R.E., 1982, "Recent reports on liquid chromatography/electrochemistry", BAS Press, West Lafayette, Indiana, U.S.A.

MORPHOLOGICAL BASIS FOR A NEW TYPE OF PHOTONEUROENDOCRINE

MECHANISM IN THE CRICKET MELANOGRYLLUS DESERTUS

Semahat Geldiay and Sabire Karaçali

Department of Biology
Ege University
İzmir, Turkey

Both in invertebrates and vertebrates, interrelations between the endocrine and nervous systems mediated by light sensitive cells of the photoneuroendocrine mechanism have been demostrated. It has been observed experimentally in some investigated insects that known photoreceptors, eyes, and ocelli are not directly involved in the regulation of the endocrine system (De Wilde et al.,1959; Geldiay, 1969; Truman,1972). There is evidence that in this system the most sensitive region underlies the dorsal and medial regions of the head. Photoreceptors were postulated within the brain, in the neuroendocrine mechanisms for the photoperiodic control of diapause (Williams and Adkisson,1964; Geldiay,1970), of ecdysis (Truman 1972), and of polymorphism (Less, 1964,1981). In this study, cells ressembling photoreceptors structurally are described in the corpus cardiacum. It has not been proposed previously that nonvisual photoreceptor cells could be found at any site within the head other than the brain itself.

The postulated photoreceptor cells structures occur in the posterior end of the corpus cardiacum. These cells are much larger than the intrinsic secretory cells and are few in number. They lie close to the surface. Intrinsic secretory cells and axons bearing different types of granules are associated with the rhabdome cells (Fig.1). The rhabdome structure of these newly recognized cells is composed of microvilli arranged in a hexagonal array to form rhabdomers (Fig.2,3). Rhabdomeric structures with different orientations form the rhabdome. Large mitochondria are conspicuous in these cells (Fig.3). Neurosecretory axons lie in close association with the rhabdome-bearing cells and have synaptic connections with them (Fig.2). Direct close contacts between the membranes of

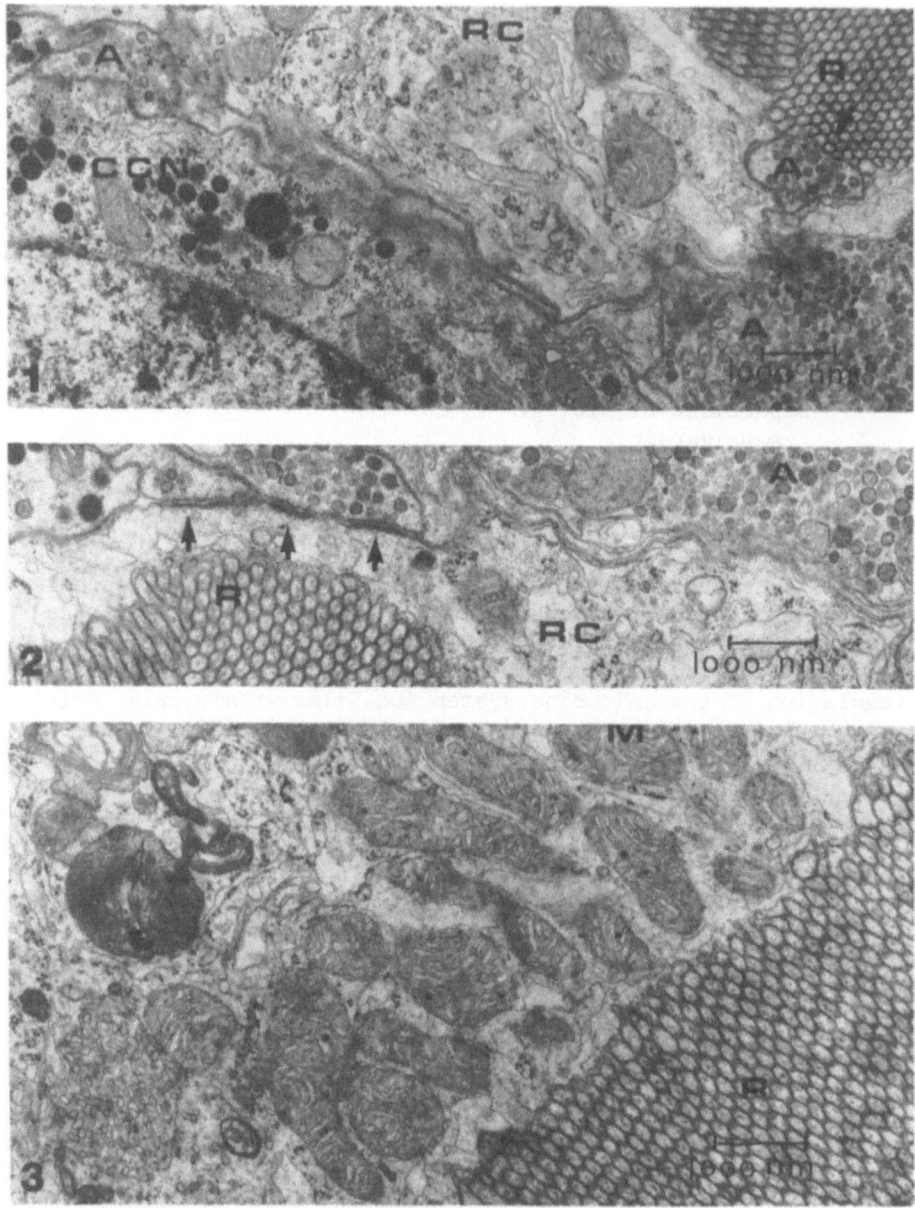

Fig. 1 and 2. small parts of the corpus cardiacum.
Fig. 3. The rhabdome-bearing cell.
        Corpus cardiacum neurosecretory cell (CCN);rhabdome-bearing
        cell (RC); rhabdome (R); neurosecretory axon (A);
        mitochondrion (M); close contacts between the neurosecretory
        axons and rhabdome-bearing cell. (→).

neurosecretory axons and rhabdomeric microvilli are also observed (Fig.1).

The arrangement of microvilli that contributes to the rhabdome structure in the corpus cardiacum cells and the appearance of all the other cellular components such as mitochondria, endoplasmic reticulum and multivesicular bodies are closely similar to retinular cells (White,1967; Varela and Porter,1969; Chu and Norris,1976).

REFERENCES

Chu,H., and Norris,D.M.,1976, Ultrastructure of the compound eye of the haploid male beetle, *Xyleborus ferrugineus*. Cell tissue Res. 168: 315-324.

De Wilde,J.,Duintjer,C.S.,and Mook,L.,1959, Physiology of diapause in the adult Colorado beetle (*Leptinotarsa decemlineata* Say). I. The photoperiod as a controlling factor. J.Insect.Physiol. 3:75-85.

Geldiay,S.,1969, The location of photoperiodic receptors and the activity of the neurosecretory cells of *Anacridium aegyptium* L.under different photoperiods. Sci.Rep.Fac.Sci.Ege Univ.89.

Geldiay,S.,1970, Photoperiodic control of neurosecretory cells in the brain of aegyptian grasshopper, *Anacridium aegyptium* L. Gen.Comp. Endocrinol. 14: 35-42.

Lees,A.D.,1964, The location of the photoperiodic receptors in the *Megoura viciae* Buckton. J.Exp.Biol.41: 119-133.

Lees,A.D.,1981, Action spectra for the photoperiodic control of polymorphism in the aphid *Megoura viciae*. J.Insect Physiol. 27;761-771.

Truman,J.W.,1972, Physiology of insect rhythms. I.Circadian organization of the endocrine events underlying the moulting cycle of larval tobacco hornworms. J.Exp.Biol.57:805-820.

Varela,F.G., and Porter,K.P.,1969, Fine structure of the visual system of the honeybee *(Apis mellifera)* I. The retina, J. Ultrastruct.Res. 29: 236-259.

White,R.H.,1967, The effect of light and light deprivation upon the ultrastructure of the larval mosquito eye. II.The rhabdom. J.Exp. Zool. 166: 405-426.

Williams C.M., and Adkisson,P.L.,1964, Physiology of insect diapause XIV. An endocrine mechanism for the photoperiodic control of pupal diapause in the oak silkworm, *Antheraea pernyi*. Biol Bull. (Woods Hole, Mass.) 127: 511-525.

THE INFLUENCE OF TEMPERATURE ON PHOTOPERIODIC CONTROL OF DIAPAUSE INDUCTION IN DIFFERENT STRAINS OF THE EUROPEAN CORN BORER[1]

Dale B. Gelman and Dora K. Hayes

Insect Reproduction and Livestock Insect Laboratories
ARS, USDA
Beltsville, MD USA

The European Corn Borer (ECB), Ostrinia nubilalis, undergoes a facultative diapause, i.e., it enters diapause only in response to short days and decreasing temperatures (Beck and Hanec, 1960). Diapause is believed to be caused in part by an inhibition of PTTH release and hence the lack of sufficient ecdysone production and/or release by the prothoracic glands to cause a molt (Williams, 1946; Cloutier et al., 1962). Gelman and Woods (in press) reported that ecdysteroid levels were much lower in diapause-bound than in non-diapause-bound last instars.

Since photoperiod is so important in programming ECBs for diapause or development, there must be physiological mechanisms by which photoperiodic signals are received (receptor system), recorded and summed (clock), and a response initiated (effector system). These systems are believed to be located in the head (probably the brain), but little is known about the operation of any one of them. Temperature can modify the effects of photoperiod on diapause induction by altering critical daylength (Beck and Hanec, 1960). Exposing ECBs to a thermoperiod (alternating high and low temperatures) can reduce the incidence of diapause if the thermophase (higher temperature) coincides with the scotophase (period of darkness) (Beck, 1962a). Thermoperiod can even substitute for appropriate photoperiodic regimens in inducing diapause if ECBs are kept in constant darkness (Beck, 1962b). We found that the role of temperature cannot be underestimated, especially when comparing different strains of insects. Thus, ECBs from Ankeny, Iowa when subjected to 10 or 12 hr of light and a temperature of

---

[1]Mention of a commercial or proprietary product in this paper does not constitute an endorsement of this product by the USDA.

30°C (optimum conditions for diapause induction in the Northern Wisconsin strain; Beck, 1962a) exhibited considerable pupation. Since diapause-bound and diapausing ECBs cannot be distinguished from nondiapause-bound larvae, we undertook to determine conditions that would be optimal for inducing diapause in our strain.

ECBs were reared and staged as described in Gelman and Hayes (1982). Five different temperatures (22°, 24°, 26°, 28° and 30°C) were combined with 11 different photoperiodic regimens (LD 7:17 = 7 hours light, 17 hours of dark, LD 8:16, LD 9:15, LD 10:14, LD 11:13, LD 12:12, LD 13:11, LD 14:10, LD 15:9, LD 16:8 and LD 17:7) to produce fifty-five different experimental combinations. Each combination was tested on a group of 40-50 insects and percent diapause was recorded.

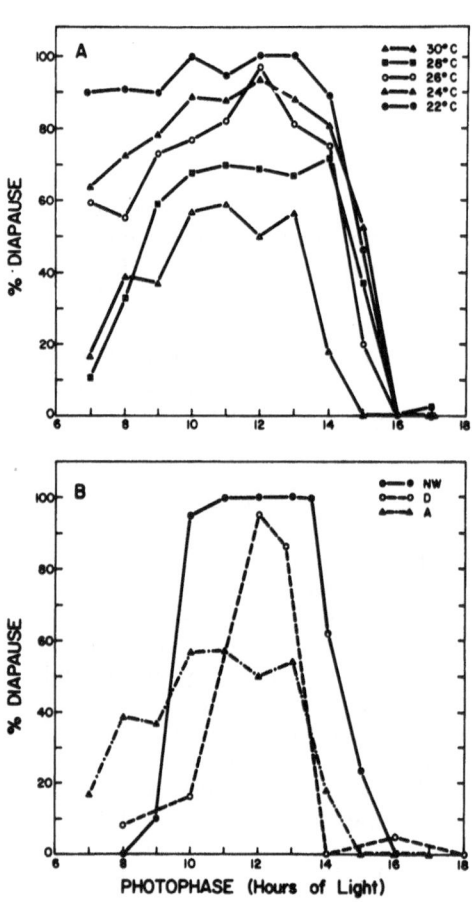

Fig. 1a shows that diapause induction was high for 10-12 hr photophases when temperatures were kept at 22-24°C. At 26°C, LD 12:12 was the only photoperiodic regimen that induced diapause in more than 90 percent of the animals. At 28 and 30°C, the ability of even LD 12:12 to induce diapause fell to between 50 and 60%. However, we did observe that as in other species, lower temperatures allowed for shorter critical daylengths to induce higher percentages of diapause (Beck, 1980). The results for the Ankeny (A) borer were different from results for the Northern Wisconsin (NW) (Beck, 1962a) and Delaware (D) (Skopik and Bowen, 1976) strains (Fig. 1b). At 30°C, NW ECBs exhibited 100% diapause for photophases of 11-13½ hr and D ECBs exhibited more than 90% diapause at LD 12:12. Surprisingly, while environmental conditions in the natural habitat are more similar for the NW and A strains than for the NW and D strains, in terms of temperature effects on diapause induction, it is

Fig. 1 Effect of photoperiod on diapause induction in A, the Ankeny strain of ECB at 22-30°C; B, the Northern Wisconsin (NW), Delaware (D), and Ankeny (A) strains at 30°C.

the D strain that is more similar to the NW strain.

According to Beck (1980), the more Northern the population, the less the effect of relatively high temperatures on diapause induction and the fewer the number of short day cycles needed for diapause induction. Since a minimum of 11-12 short day cycles are needed for the NW strain (latitude, 46°N) it may be that the Ankeny strain (latitude, 42°N) needs 13 or 14 short day cycles. At higher temperatures (e.g., 28 and 30°C) development may be proceeding too quickly to allow for the accumulation of a sufficient number of short day cycles to induce diapause. However, temperature also may interact in other ways at the level of the biological clock or at the level of the effector system.

REFERENCES

BECK S. D. (1962a) Photoperiodic induction of diapause in an insect. Biol. Bull. 122, 1-12.

BECK S. D. (1962b) Temperature effects on insects: relation to periodism. Proc. N. cent. Brch. Ent. Soc. Am. 17, 18-19.

BECK S. D. and HANEC W. (1960) Diapause in the European corn borer, Ostrinia nubilalis. Biol. Bull. 4, 304-318.

BECK S. D. (1980) Insect Photoperiodism. Academic Press, New York.

CLOUTIER E. J., BECK S. D., MCLEOD D. G. R., and SILHACEK D. L. (1962) Neural transplants and insect diapause. Nature, Lond., 195, 1222-1224.

GELMAN D. B. and HAYES D. K. (1982) Methods and markers for synchronizing the maturation of fifth stage diapause-bound and nondiapause-bound larvae, pupae and pharate adults in the European corn borer, Ostrinia nubilalis (Hübner). Comp. Biochem. Physiol. 73A, 81-87.

GELMAN D. B. and WOODS C. W. Haemolymph ecdysteroid titers of diapause- and nondiapause-bound fifth instars and pupae of the European corn borer, Ostrinia nubilalis (Hübner). Comp. Biochem. Physiol. (in press).

SKOPIK S. D. and BOWEN M. F. (1976) Insect photoperiodism: an hour glss measures photoperiodic time in Ostrinia nubilalis. J. Comp. Physiol. 111, 249-259.

WILLIAMS C. M. (1946) Physiology of insect diapause: the role of the brain on the production and termination of pupal dormancy in the giant silkworm, Platysamia cecropia. Biol. Bull. 90, 234-243.

PICROTOXIN ACTIVATES ADULT DEVELOPMENT AND

NEUROHAEMAL NERVE SPIKES IN MANDUCA SEXTA

Daniel Gibbs

Department of Biological Sciences
De Paul University
Chicago, IL 60614

Pupae developing from Manduca larvae reared under short days (LD 12:12) at 26°C normally enter pupal diapause. Spontaneous development is then infrequent until the termination of diapause after 1-3 months at 26°C. Adult development occurs in response to the secretion of alpha-ecdysone by the prothoracic glands (PTG) which are in turn activated by one or more forms of prothoracicotropic hormone (PTTH), a peptide neurohormone produced by neurosecretory neurons in the pupal brain and released primarily from neurohaemal endings in the corpus allatum (CA) (Carrow et al., 1981; Gibbs and Riddiford, 1977; Gilbert et al., 1981). It is a reasonable hypothesis that pupal diapause in these insects occurs because of neural inhibition mediated through synaptic inputs to the PTTH neurons. Dye-filling procedures demonstrate substantial dendritic arborizations in putative PTTH cells in both larval and pupal brains (Carrow et al., 1982; Buys and Gibbs, 1981).

Since gamma-amino butyric acid (GABA) is a common inhibitory transmitter in many animals and possibly in Manduca (Maxwell et al., 1978), I tested the effect of two GABA antagonists, picrotoxin and bicuculline, on the initiation of adult development. Pupae from larvae reared under short days (LD 12:12) at 26°C were injected one day after pupation with a drug solution or diluent alone in a volume of 0.10 ml. Animals were scored for development 7 days after injection using a table of normal developmental stages. Picrotoxin at a dose of 500 μg/animal typically induces development in 70-90% of the pupae injected at day 1 or day 2. Diluent controls (9.5% ethanol) typically showed 0-10% spontaneous development. The response on day 1 is dose-dependent, with a threshold near 100 μg/animal. Using 500 μg/animal as a test dose, sensitivity gradually declines to control levels (<10%) by day 12 after pupation, then

361

rises to 45% by day 20. By day 42, 62% of the picrotoxin-injected animals initiated development within 7 days after injection, compared to 26% of the controls.

To test whether picrotoxin might be stimulating the PTG directly, the brain, corpora cardiaca (CC) and CA were removed from day 1 pupae and the head resealed with Tackiwax. In 15 animals so treated and injected within a few hours with 500 µg picrotoxin, no development was observed. 85% of 13 sham-operated controls developed after a similar injection, suggesting a direct effect of picrotoxin on the brain.

For a more direct test, day 1 pupal brains were isolated in vitro with tracheal trunks floated intact to provide oxygen. Spike trains were recorded from the fused nervus corpus cardiacum (NCC) I+II, which is the nerve through which the putative PTTH neurons project to the CC and CA. The nerve was cut at or proximal to the CC and sucked into a suction electrode containing an AgCl pellet connected to a sensitive (1mV/picoampere) current/voltage converter of the type used for patch clamp studies (Neher et al., 1978). This allowed detection of 10-60 pA spikes over a period of a few hours. A CMOS analog switch allowed rapid switching of the single electrode between the recording amplifier and stimulator. Antidromic stimulation typically resulted in transient stimulation of a large number of units, presumably from neurosecretory axons, with a return to prestimulus levels by 2-3 minutes. In contrast, addition of 10^{-4}M picrotoxin to the bath resulted in activation of only a few units which often remained active for 30 minutes or more. Both long-day and short-day reared brains showed this response. The units thus activated were usually "narrow" (2-5 msec monophasic or biphasic) while spontaneous or stimulated activity usually contained both narrow and "broad" units (20-40 msec). In some preparations, a second application of picrotoxin was used to elicit a second sustained bout of activity.

Bicuculline methiodide (Sigma) injected at an equivalent dose (8.3×10^{-7}moles/animal) on day 1 did not stimulate development, nor did it elicit sustained spiking activity at 10^{-4}M in vitro. In fact, it appeared to suppress the spontaneous activity present. Mann and Enna (1980) report finding no measurable bicuculline-sensitive GABA receptor binding in Periplaneta brains or in those of crabs, tarantulas, squid or planaria, in contrast to all vertebrates tested.

It has previously been suggested that GABA plays a role in the control of development and neurosecretory activity in Manduca (Jungreis and Omilianowski, 1980; Jungreis, 1982). The present results support this possibility, but we do not know the identity of the neurons which are stimulated, nor can we rule out nonsynaptic effects of picrotoxin.

REFERENCES

Buys, C. M., and Gibbs, D., 1981, The anatomy of neurons pro-
     jecting to the corpus cardiacum from the larval brain of
     the tobacco hornworm, Manduca sexta (L.), Cell Tissue
     Res., 215:505.
Carrow, G. M., Calabrese, R. L., and Williams, C. M., 1981,
     Spontaneous and evoked release of prothoracicotropin
     from multiple neurohaemal organs of the tobacco hornworm,
     Proc. Natl. Acad. Sci. USA, 78:5866.
Carrow, G. M., Calabrese, R. L., and Williams, C. M., 1982,
     Protocerebral neuroendocrine system of the tobacco horn-
     worm: morphology and physiology of identifiable neuro-
     secretory cells, Soc. Neurosci. Abstr., 8:532.
Gibbs, D., and Riddiford, L. M., 1977, Prothoracicotropic
     hormone in Manduca sexta: localization by a larval
     assay, J. Exp. Biol., 66:255.
Gilbert, L. I., Bollenbacher, W. E., Agui, N., Granger, N. A.,
     Sedlak, B. J., Gibbs, D., and Buys, C. M., 1981, The
     prothoracicotropes: source of the prothoracicotropic
     hormone, Am. Zool., 21:641.
Jungreis, A. M., 1982, Evidence supporting a proposed associ-
     ation between hemolymph gamma-amino butyric acid and
     neurosecretion in lepidopterous insects, Comp. Biochem.
     Physiol., 71C:169.
Jungreis, A. M., and Omilianowski, D. R., 1980, Gamma-amino
     butyric acid and glutamic acid in Manduca sexta: pro-
     posed roles in insect development, Comp. Biochem.
     Physiol., 67C:173.
Mann, E., and Enna, S. J., 1980, Phylogenetic distribution of
     bicuculline-sensitive gamma-aminobutyric acid (GABA)
     receptor binding, Brain Res., 184:367.
Maxwell, G. D., Tait, J. P., and Hildebrand, J. G., 1978, Re-
     gional synthesis of neurotransmitter candidates in the
     CNS of the moth Manduca sexta, Comp. Biochem. Physiol.
     61C:109.
Neher, E., Sakmann, B., and Steinbach, J. H., 1978, The extra-
     cellular patch clamp: a method for resolving currents
     through individual open channels in biological membranes,
     Pflügers Arch., 375:219.

REFERENCES

STIMULATION OF JH III SYNTHESIS IN VITRO BY AN ALLATOTROPIC FACTOR

FROM THE BRAIN OF THE TOBACCO HORNWORM, MANDUCA SEXTA

Noelle A. Granger[+], Laura J. Mitchell[+] And Walter E.
Bollenbacher[++]

Departments of Anatomy[+] And Biology[++]
University of North Carolina at Chapel Hill
Chapel Hill, North Carolina 27514

The regulation of juvenile hormone (JH) synthesis by the corpora
allata (CA) is recognized as a complex process involving both nervous
and neurosecretory mechanisms (Gilbert et al., 1980). There is evidence to suggest that these mechanisms are highly integrated and may
involve separate and distinct controls of the synthesis of each of
the JH homologs (deKort and Granger, 1981; Granger et al., 1981;
Granger et al., 1982). On the basis of numerous in vivo and several
in vitro investigations (see Gilbert et al, 1980; deKort and Granger,
1981), neurohormonal control has been proposed to involve both activating and inhibiting factors, termed allatotropins and allatohibins/
allatostatins, respectively. The existence of these putative neurohormones has never been conclusively demonstrated, in part because
of the indirect approaches previously used to investigate them.
However, with the recent implementation of in vitro systems, prospects
are good that the physiological occurrence and the nature of these
regulatory moieties will soon be established.

One such in vitro system, developed to investigate the regulation of the CA of the tobacco hornworm, Manduca sexta, during larval-
pupal development, utilizes radioimmunoassays to quantify the synthesis of JH I and JH III, two of the three JH homologs known to be
synthesized by the CA during this period (Granger et al., 1982). In
initial in vitro studies of the activity of the CA, it was observed
that synthesis of JH III by glands left attached to the brain was
higher than that by CA alone, a result suggesting some type of tropic
control of the CA by the brain (Granger et al., 1982). Subsequent
investigations of this phenomenon have demonstrated the presence of
a cerebral neuropeptide that specifically stimulates JH III synthesis.

The effect of this factor on the CA was first probed with co-incubations of brain-corpora cardiaca (BRCC) and CA, each taken from different developmental stages. It was found that the incubation of CA from day 7 of the last larval instar (V/7) with BRCC from day 0 pupae (P/0) resulted in a significant three-fold increase in JH III synthesis, while the level of JH I synthesis did not change (Granger et al., 1981). To demonstrate that the stimulation of JH III synthesis by the pupal brain in vitro was due to a hormone-like factor, an attempt was then made to activate the CA in a dose-dependent manner. This required the development of a suitable assay with which to measure activation, and the first parameter of the assay to be defined was the optimal time for activation. This was determined by generating a time course of activation of V/7 CA by 0.5 equivalents (BE) of a P/0 brain extract. The extract was prepared by homogenization of brains in Grace's medium and centrifugation of the homogenate to obtain a post-mitochondrial supernatant, which was then used as the extract. The degree of activation at each time point during a 10 hr incubation was ascertained by the determination of an activation ratio (A_r), i.e. the value obtained when the amount of JH synthesized by the activated glands is divided by the amount synthesized by the control glands. The A_rs revealed that the maximum activation obtained was again three-fold, and that this occurred within two hr in vitro. Thus, this incubation time was used in subsequent studies.

To demonstrate a dose response of activation of JH III synthesis, synthesis by a pair of right glands incubated in the crude extract was compared to synthesis by a pair of left glands from the same two animals, for each concentration of extract assayed. This paired gland approach was used to minimize inherent variability in basal gland activity and was possible because right and left CA from the same animal synthesize comparable amounts of the JH homologs. Thus, a basal rate of JH synthesis can be determined precisely with one pair of glands and the degree of activation with the contralateral pair. Activation of V/7 CA by a P/0 brain extract was found to be dose-dependent for JH III, but JH I synthesis was not affected. The CA were activated by a fairly narrow dynamic range of extract concentration, from an A_r of 1 to the maximum of 3 within a dosage range of 0.1 to 0.25 BE.

Chemical characterization of the JH III allatotropic factor (ATF) first required an assessment of its stability. The JH III ATF was found to be stable with repeated freezing and thawing for at least one month, while treatment at 100°C for only 15 sec destroyed its activity. The proteinaceous nature of the ATF was suggested by its precipitation with ammonium sulfate (40-65%) and its sensitivity to proteolysis by trypsin digestion, which destroyed greater than 95% of the activity under defined assay conditions. Gel filtration of the crude extract yielded an apparent molecular weight of 35 kd,

and a determination of an isoelectric point for the activity has
revealed a pI of 5.5.

Although the information generated on the ATF thus far suggests
that it acts as a neurohormone, the possibility did exist that this
effect was non-specific. This possibility was addressed by assessing
with a dose response protocol activation of the CA by extracts of different tissues. Homogenates of brain, CC, subesophageal plus prothoracic ganglia, and muscle were tested for their ability to activate the CA. Of these, only neural tissue exhibited ATF activity.
Activity was detected in the brain, CC, and abdominal ganglia, with
the CC and abdominal ganglia having 1.4% and 23% of the activity in
the brain, respectively. The activity in the brain was discretely
localized to the lateral regions of the cerebral hemispheres, suggesting that the lateral neurosecretory cells might be the source of
the ATF.

Investigations are currently in progress to define more precisely the ATF activity, including its purification and the localization
of its site of synthesis in the brain. In addition, the ATF activity
in the brain is being titered during larval and pupal development.
In summary, the preponderance of evidence thus far obtained suggests
the existance of a cerebral peptide capable of specifically stimulating an increased rate of JH III synthesis by the CA. With further
substantiation, this will represent the first direct demonstration
of the existence of an allatotropin in an insect.

ACKNOWLEDGEMENTS

The authors are indebted to Janet Hightower for her assistance
in the preparation of the manuscript. This work was supported by
grants NS18791 and AM31642 from the National Institutes of Health
to W. E. Bollenbacher and N. A. Granger, respectively.

REFERENCES

DEKORT C. A. D. and GRANGER N. A. (1981) Regulation of the juvenile
    hormone titer. A. Rev. Ent. 26, 1-28.
GILBERT L. I., BOLLENBACHER W. E., and GRANGER N. A. (1980) Insect
    endocrinology: Regulation of endocrine glands, hormone titer and
    hormone metabolism. A. Rev. Physiol. 42, 493-510.
GRANGER N. A., BOLLENBACHER W. E., and GILBERT L. I. (1981) An in
    vitro approach for investigating the regulation of the corpora
    allata during larval-pupal metamorphosis. In Insect Endocrinol-
    ogy and Nutrition (Ed. by BHASKARAN G., FRIEDMAN S., and
    RODRIGUEZ J. G.), pp. 83-105, Plenum Press, New York.
GRANGER N. A., NIEMIEC S. M., GILBERT L. I., and BOLLENBACHER W. E.
    (1982) Juvenile hormone synthesis in vitro by larval and pupal
    corpora allata of Manduca sexta. Mol. Cell. Endocrinol. 28,
    587-604.

ANALYSIS OF GUSTATORY ACTIVITY USING COMPUTER TECHNIQUES

Frank E. Hanson

Department of Biological Sciences
University of Maryland Baltimore County
5401 Wilkens Avenue
Catonsville, Maryland 21228

Chemical cues play an important role in feeding, sexual, and oviposition behavior of insects. The chemoreceptors represent the crucial link between the animal and the environment. Thus the characterization of these receptors is important to our understanding of the physiological basis of behavior.

Proper characterization of chemoreceptors usually requires large quantities of data. One reason for this is that the variance can be large both within and across animals. Secondly, it is important to average across analogous receptor cells of many animals to insure that the "species receptor" is described. Studies which intensively analyse responses of a few receptor cells run the risk of focusing too narrowly on the peculiarities of atypical sensilla.

To solve the problem of handling large amounts of data required for proper characterization of receptor responses, our laboratory has enlisted computer-aided analysis techniques. Chemoreceptor responses are recorded using electrophysiological methods. Analog data are obtained using a high impedance amplifier designed to stabilize baseline fluctuations upon stimulus onset, yet preserve action potential shape as much as possible. These data are recorded on an FM tape recorder, and later played back into a laboratory computer programmed to digitize action potentials but not baseline. The digitized spikes are stored on disk and can be recalled for display or analysis. An automated sorting routine then separates spikes according to shape using a "goodness of fit" criterion set by the operator. The method is a modification of the template matching algorithm of Gerstein and

369

Clark (1964). It employs a least-squares comparison of the sample points of the digitized action potential with those of the template. (The template itself is a composite of several very similar spikes automatically selected from the raw data using criteria set by the operator.) If the least-squares value is less than criterion, the spike is accepted into that class; if not, it is reconsidered on the next pass through the data using another template. Data classified are then stored, displayed, graphed, averaged, and analysed by the computer.

The above algorithm provides the best resolution for different action potential shapes. Algorithms using other parameters of the spikes have been used as discriminators, such as peak-to-peak height or standard deviation, but in our hands these did not separate similar spikes as well as did template matching.

Using the above techniques, characterization of the salt receptor of the tarsal type D sensilla of the blowfly Phormia regina was attempted. A semi-intact preparation was developed that produced reliable results for 8 hours or more. Response to the alkali chlorides followed the typical insect chemoreceptor kinetics having a phasic onset which adapts rapidly to a slowly decaying tonic phase. Averaging ca. 30 such adaptation curves from several animals provided smooth curves to which equations could be fit. A single exponential decay function fit poorly (R^2 = .4 to .7) when applied to the entire curve. However, fitting separate decreasing exponential functions to the phasic (defined as prior to 0.1 sec.) and tonic (subsequent to 0.1 sec.) respectively led to average R^2 values of ca. 0.8 and 0.95 respectively. For example, the averaged responses of 30 sensilla to 1.0 M NaCl has a phasic portion that can be described by

$$y = 149\ e^{-14x} \qquad (R^2 = .88)$$

and a tonic portion described by

$$y = 41.9\ e^{-.19x} \qquad (R^2 = .98)$$

This suggests that two different mechanisms are responsible for the phasic and tonic curves. In Aplysia, the phasic and tonic portions appear to have different controlling mechanisms (Lewis and Wilson, 1982) and perhaps the same is true for the blowfly.

Recovery from adaptation was slow relative to many other insect receptors. For example, recovery of the receptor to a 5-sec. stimulus of NaCl was complete after 30 sec. for a 0.5 M solution and after 10 min. following a 4.0 M stimulus.

An important characteristic of a receptor is its dose-response curve. Repeated stimuli of six concentrations of 5 alkali chloride salts were applied to a total of several hundred sensilla. Responses were averaged across 8-10 sensilla from each of 6-8 animals at each concentration. Curves thus obtained can be constructed with considerable confidence of being reliable estimates of the average species' response.

The response hierarchy was KCl $\geq$ NaCl $\geq$ RbCl $>$ CsCl $>$ LiCl, which is similar to that found by Gillary (1966) for the labellar receptor of this species. The shape of the curves fell into two classes: For KCl, NaCl, and RbCl, the responses increase proportionally to the logarithm of concentration, but do not reach a plateau even at saturated solutions. Thus R_{max} of the receptor is not demonstrably achieved. The other two salts tested, CsCl and LiCl, elicit lowered responses that increase with the logarithm of concentration, peak at 4.0 M, and then decrease at higher concentrations.

These dose-response curves thus appear to be unlike those of receptor models developed from enzyme-substrate reactions having an S-shaped curve with peak and plateau at R_{max}. This suggests that the salt cell does not have the typical receptor site specific for the most effective substrate, e.g., monovalent cations. Perhaps an alternative mechanism, such as that hypothesized by den Otter (1972a,b) in which salts disturb the binding by calcium of distributed membrane phospholipids, would be a more appropriate model. The dose-response curves obtained in the present study fit predictions of den Otter better than those of the classical model.

Acknowledgements

The techniques were developed with support by the Whitehall Foundation, USDA/CRGO, USDA/FS, and NSF.

REFERENCES

Gerstein and Clark (1964) Stimultaneous studies of firing patterns in several neurons. Sci. 143: 1325-1327.
Gillary H.L. (1966) Stimulation of the salt receptor of the blowfly. III. The alkali halides. J. Gen. Physiol. 50: 359-368.
Den Otter C.J. (1972a) Interactions between ions and receptor membrane in insect taste cells. J. Insect Physiol. 18: 389-402.
Den Otter C.J. (1972b) Mechanism of stimulation of insect taste cells by organic substances. J. Insect Physiol. 18: 615-625.

COMPARATIVE PHARMACOLOGY OF INSECT AND TICK NERVE MUSCLE SYNAPSE

Richard J. Hart, David J. Beadle* and Roger G. Wilson

Wellcome Research Laboratories, Berkhamsted, U.K.
and *Thames Polytechnic, London, U.K.

There is considerable evidence that L-glutamic acid is the chemical transmitter at excitatory neuromuscular junctions in mandibulate arthropods, but it is based on studies of relatively few species. There is no report in the literature of glutamate as a neurotransmitter in arachnids, although octopamine (Grega 1978) and 4-aminobutyric acid (Brenner 1972) do affect the contraction of spider muscle. A cattle tick, *Amblyomma variegatum,* was used in the present investigation of neuromuscular pharmacology as it is representative of an economically important group. However, as little basic information is available, some initial studies were made of the physiology of peripheral excitation in the tick, to enable a useful comparison with insects to be made. The insect used for comparison was the desert locust, *Schistocerca gregaria.* Flexor muscles from the tibia of the fourth leg pair in the tick and retractor unguis and extensor tibia muscles from the metathoracic leg of the locust were examined. Cuticle was dissected away to allow a saline solution to perfuse the muscle fibres *in situ*. Tick saline composition was based on an analysis of the haemolymph and differed slightly from the representative range of insect salines quoted by Huddart (1970). Ionic composition was, sodium 203mM; potassium 6mM; calcium 10mM; magnesium 1mM; bicarbonate 3mM; phosphate 1mM. Locust saline was, sodium 151mM; potassium 11mM; calcium 2mM; magnesium 2mM; bicarbonate 4mM; phosphate 7mM. Recordings of evoked synaptic events following stimulation of the pedal nerve in the tick and crural nerve in locust were made using intracellular glass pipette electrodes. A miniature strain gauge was used to measure twitch contractions of both tick and locust muscle.

Records from resting tick muscle showed the presence of post-synaptic miniature potentials. The amplitude and frequency were low (80- 600 mv at about 1 Hz). The peaks in any one record showed varying rise times. The high gain used to record the miniature potentials frequency revealed a low amplitude rhythmic contraction of the muscle. Stimulation of the pedal nerve caused a small twitch contraction which had a fusion frequency of close to 10 Hz and a tetanus twitch ratio of 9:1. Intracellular recording showed distinct electrical responses. One was a depolarisation with a slow rise time (> 5 ms), peak typically 10-12 mv. A second response showed a fast rise time (<5 ms) with an amplitude of approximately 60 mv. The majority of fibres gave both responses but the events could often be separated by careful adjustment of the stimulus pulse width and amplitude. These events could be further distinguished by repetitive stimulation, the 'slow' events summated and the 'fast' showed antifacillitation.

Perfusion of the tick muscle with saline containing putative receptor agonists at mM concentration appeared to have no significant, measurable effect on evoked events with acetylcholine, D,L-octopamine, 4-aminobutyrate, histamine and 5-hydroxytryptamine. There did seem to be some effect on the spontaneous rhythm with octopamine but this was not confirmed. Perfusion of saline containing L-glutamate caused a block of the twitch contraction, a slight depolarisation of the muscle cell membrane and abolished the evoked potentials. Contractions and evoked potentials were restored to normal amplitude after perfusion with fresh saline.

A post-synaptic site of action for L-glutamate was confirmed by iontophoresis. Ejection of small amounts of glutamate from a micro-pipette caused transient depolarisation over circumscribed areas of the muscle cell membrane (maximum response was 8 mv/nC). Measurement of membrane conductivity showed that glutamate caused a reversible increase.

The effect of a range of glutamate receptor agonists was compared on tick and locust muscle to classify the response. The dose response curve for glutamate on both insect and tick muscle was found to be almost identical. The EC_{50} for twitch contraction block was 5.5×10^{-4}M and EC_{50} for excitatory post-synaptic potential block was 5×10^{-4}M. There was a difference, however, in the activity of other agonists. From the range of compounds : L-quisqualic acid, L-α-kainic acid, L-aspartic acid, N-methyl,D-aspartic acid and D-homocysteic acid, tick muscle was affected by all except N-methyl,D-aspartic acid and D-homocysteic acid. The activity, with equipotent molar ratios (EPMR) being quisqualate (0.02)>kainate (0.03)>aspartate = glutamate (1). The fast twitch fibres of locust muscles, as found by previous workers (Clements and May 1974, Daoud and Usherwood 1975), did not appear to be markedly affected by kainate, aspartate, N-methyl,D-aspartate or

D-homocysteic acid. Quisqualate was found to be much more potent than glutamate with an EPMR of 0.02. The glutamate antagonists, L-glutamic acid diethyl ester and L-glutamic acid dimethyl ester at mM concentrations had no measurable effect on evoked events in either tick or locust muscles.

The tick somatic muscles examined in this investigation would appear to resemble several insect muscles in possessing multiple excitatory innervation. The peripheral neurotransmitter is probably glutamate or a very closely related compound. The transmitter receptors in the muscle cell would appear to have a wider range of acceptance than is typical of insect, or mammals. Glutamate receptors, classified in mammalian pharmacology, are glutamate, aspartate or kainate preferring, with quisqualate, N-methyl,D-aspartate and kainate as type agonists. The major difference between the tick and insect appears to be aspartate sensitivity. A very selective response to aspartate has been demonstrated in the 'slow' excitation of dipteran muscles (Irving and Miller 1980) but the tick muscle receptors for both 'slow' and 'fast' innervation appear to be sensitive to aspartate.

REFERENCES

BRENNER H.R. (1972) Evidence for peripheral inhibition in arachnoid muscle. *J. Comp. Phys.* 80, 227-231.
CLEMENTS A.N. and MAY T.E. (1974) Pharmacological studies on a locust neuromuscular preparation. *J. exp. Biol.* 61, 421-442.
DAOUD A. and USHERWOOD P.N.R. (1975) Action of kainic acid on a glutaminergic synapse. *Comp. Biochem. Physiol.* 52C, 51-53.
GREGA D.S. (1978) The effects of monoamines on tarantula skeletal muscle. *Biochem. Physiol.* 61C, 337-340.
HUDDART H. (1971) Contraction of insect muscle, In : *Experiments in Physiology and Biochemistry,* Vol. 4, p 219-288, (Kerkut G.A. ed.) Academic Press, London.
IRVING S.N. and MILLER T.A. (1980) Aspartate and glutamate as possible transmitters at the 'slow' and 'fast' neuromuscular junctions of the body wall muscle of *Musca* larvae. *J. comp. Physiol.* 135, 299-314.
USHERWOOD P.N.R. and MACHILI P. (1968) Pharmacological properties of excitatory neuromuscular synapses in the locust. *J. exp. Biol.* 49, 341-361.

FUNCTIONAL ORGANIZATION OF CHEMICAL-SENSORY PATHWAYS IN LARVAL AND

ADULT *MANDUCA SEXTA* (Lepidoptera)

J.G. Hildebrand, K.S. Kent, I.D. Harrow, S.M. Camazine,
R.A. Montague, P. Quartararo and M. Imperato

Department of Biological Sciences
Columbia University
New York, NY 10027

INTRODUCTION

Because insects are experimentally favorable for neurobiological studies of the mechanisms of olfaction and gustation, and owing
to the agricultural and medical importance of numerous insect species
whose harmful or beneficial behaviors are at least partly controlled
through chemical signals, the chemical senses of insects are the
focus of growing interest and research. The structure and physiology
of chemical-sensory receptor cells and the sensory organs and sensilla they innervate have been studied extensively in a variety of
insect species, but only recently have the corresponding chemical-
sensory pathways in the central nervous system (CNS) been successfully explored. Toward the goals of understanding the mechanisms of
neural development, integration, and plasticity in those pathways
and of elucidating the roles of such mechanisms in important insect
behaviors, we study the olfactory and gustatory systems of the experimentally favorable sphinx moth *Manduca sexta*. In the anatomical
studies that guide our other work, we have explored the central projections of sensory axons from sensilla of the antenna and other
sensory organs of the head in larvae and adults by means of intracellular staining methods in order to begin to characterize the
organization of the CNS pathways and integrative neuropil regions
into which those sensory afferents send chemical-sensory information.

ADULT ANTENNAL PATHWAY

In our earlier work, through studies employing intracellular
recording and staining of neurons, histological and ultrastructural
observations, neurochemical and histochemical methods, and develop-

mental perturbations, we and our coworkers have described the structure and development of the antennae (Sanes and Hildebrand, 1976a-
c; Sanes et al., 1976; Schweitzer et al., 1976), traced the growth
and CNS projections of antennal sensory axons in the CNS of meta-
morphosing and adult *Manduca* (Sanes and Hildebrand, 1975; Camazine
and Hildebrand, 1979), characterized the neuropil centers -- including the antennal lobes (ALs) -- to which the sensory axons project
and their target neurons and synapses there (Hildebrand et al.,
1979; Matsumoto and Hildebrand, 1981; Tolbert and Hildebrand, 1981),
studied the development of synapses in the ALs (Tolbert et al.,
1983), probed the development of sexually dimorphic elements in the
AL (Schneiderman et al., 1982), and begun to explore the synaptic
mechanisms in the ALs (Harrow and Hildebrand, 1982).

About 260,000 sensory neurons, largely olfactory, are associated with ca. 100,000 sensilla in the antennal flagellum and send
axons through the antennal nerve (AN) into the ipsilateral AL of
the deutocerebrum. In the AL neuropil, these axons terminate within
glomeruli and form synapses upon dendrites of the relatively small
groups of interneurons in the AL. The antennae and ALs are sexually
dimorphic. Male ALs have a macroglomerular complex (MGC), which receives synaptic inputs from the axons of male-specific pheromone-
sensitive sensory neurons, in addition to ca. 57 "ordinary" spher-
oidal glomeruli present in both sexes and receiving inputs from
neurons that respond to plant odors. Of 11 distinct types of AL
neurons so far recognized morphologically, 4 are multiglomerular,
local amacrine interneurons (2 male-specific, arborizing in the
MGC; 1 apparently female-specific) and 7 are uniglomerular output
neurons (3 male-specific, arborizing in the MGC) whose axons project
in characteristic patterns to the calyces of the mushroom bodies,
the lateral protocerebrum, and the mid-rostral protocerebrum. A
"mechanosensory and motor" center in the deutocerebrum, distinct
from the AL, receives inputs from the axons of mechanosensory neurons in the scape and pedicel segments at the base of the antenna
and also encompasses the somata and dendrites of antennal motor
neurons.

POSSIBLE "ACCESSORY" OLFACTORY PATHWAY FROM THE LABIAL PALPS

Each labial palp has within its distalmost segment a flask-
like pit open to the environment and containing numerous peg-like
sensilla. The sensory neurons of this "labial-pit organ" (LPO) send
their axons via the first labial nerve to the subesophageal ganglion
(SEG) and, unlike other labial-palp inputs (which terminate in the
SEG), through the SEG to form bilateral projections to a single,
identified ordinary glomerulus in the posteroventral region of each
AL. Both morphological and preliminary physiological findings suggest that this input to the AL may be olfactory.

LARVAL ANTENNAL AND OTHER SENSORY PATHWAYS

Mechanosensory axons from tactile hairs on the larval antenna and neighboring head capsule enter the brain via the AN and ramify before they enter the circumesophageal connective (CEC) to project to the SEG. Axons from the antennal chemosensory sensilla project through the AN primarily to a nodular ball of neuropil in the deuto-cerebrum --the larval antennal center (LAC)-- medial to the point of entry of the AN (Kent and Hildebrand, 1982). Certain axons from the labrum (via the labral nerves) and maxillae (via the maxillary nerves to the SEG and thence via the CECs), which are probably from chemosensory neurons, project to neuropil regions adjacent to the LAC and possibly in the tritocerebrum and deutocerebrum. Mechano-sensory fibers in the labial and mandibular nerves, as well as mechanosensory and probably gustatory axons in the maxillary nerves, project to characteristic regions of SEG neuropil. Certain labial, mandibular, and tegumentary tactile fibers also project through the SEG and connectives to the first thoracic ganglion.

(This research has been supported by research grants from NIH and NSF.)

REFERENCES

Camazine S.M. and Hildebrand J.G. (1979) Central projections of antennal sensory neurons in mature and developing *Manduca sexta*. *Soc. Neurosci. Abstr.* 5:155.

Harrow I.D. and Hildebrand J.G. (1982) Synaptic interactions in the olfactory lobe of the moth *Manduca sexta*. *Soc. Neurosci. Abstr.* 8:528.

Hildebrand J.G., Hall L.M., and Osmond B.C. (1979) Distribution of binding sites for ^{125}I-labeled α-Bungarotoxin in normal and deafferented antennal lobes of *Manduca sexta*. *Proc. Nat. Acad. Sci. USA* 76:499-503.

Kent K.S. and Hildebrand J.G. (1982) Central projections of afferent axons in the antennal nerve of larval *Manduca sexta*. *Soc. Neurosci. Abstr.* 8:688.

Matsumoto S.G. and Hildebrand J.G. (1981) Olfactory mechanisms in the moth *Manduca sexta*: Response characteristics and morphology of central neurons in the antennal lobes. *Proc. Roy. Soc. (Lond.)* B213:249-277.

Sanes J.R. and Hildebrand J.G. (1975) Nerves in the antennae of pupal *Manduca sexta* (Lepidoptera:Sphingidae). *Wilhelm Roux' Arch.* 178:71-78.

Sanes J.R. and Hildebrand J.G. (1976a) Structure and development of antennae in a moth, *Manduca sexta*. *Devel. Biol.* 51:282-299.

Sanes J.R. and Hildebrand J.G. (1976b) Origin and morphogenesis of sensory neurons in an insect antenna. *Devel. Biol.* 51:300-319.

Sanes J.R. and Hildebrand J.G. (1976c) Acetylcholine and its meta-bolic enzymes in developing antennae of the moth, *Manduca sexta*. *Devel. Biol.* 52:105-120.

Sanes J.R., Hildebrand J.G., and Prescott D.J. (1976) Differentia-
    tion of insect sensory neurons in the absence of their normal
    synaptic targets. *Devel. Biol.* 52:121-127.
Schneiderman A.M., Matsumoto S.G., and Hildebrand J.G. (1982) Trans-
    sexually grafted antennae influence development of sexually
    dimorphic neurones in moth brain. *Nature* 298:844-846.
Schweitzer E.S., Sanes J.R., and Hildebrand J.G. (1976) Ontogeny
    of electroantennogram responses in the moth, *Manduca sexta*.
    J. Insect Physiol. 22:955-960.
Tolbert L.P. and Hildebrand J.G. (1981) Organization and synaptic
    ultrastructure of glomeruli in the antennal lobes of the moth
    Manduca sexta: A study using thin sections and freeze-fracture.
    Proc. Roy. Soc. (Lond.) B213:279-301.
Tolbert L.P., Matsumoto S.G., and Hildebrand J.G. (1983) Develop-
    ment of synapses in the antennal lobes of the moth *Manduca
    sexta* during metamorphosis. *J. Neurosci.* 3:1158-1175.

ISOLATION OF MYOTROPIC FACTORS FROM THE HEAD OF THE

COCKROACH LEUCOPHAEA MADERAE

G. Mark Holman, Benjamin J. Cook, and Renée M. Wagner

VTERL, ARS, USDA
P.O. Drawer GE
College Station, TX 77841

For a number of years it has been known that the corpus cardiacum (cc) of insects contains substances that stimulate contraction of the alimentary tract and heart muscles (Cameron, 1953). Davey (1962) demonstrated that exposure of the isolated cockroach hindgut to cc extracts produced increases in tonus, amplitude, frequency, and coordination of contractions. Based upon paper chromatographic separations, Brown (1965) demonstrated that two factors were responsible for the hindgut stimulating activity in cc extracts and presented evidence that the two stimulators were peptides. We have recently developed a purification method and a high-performance liquid-chromatography system, both based upon reverse-phase separations, that has allowed us to demonstrate that at least six myotropic stimulators of cockroach hindgut are present in cockroach head extracts.

Hindguts isolated from the central nervous system of the cockroach were dissected (Holman and Cook, 1970) and prepared for recording as previously described (Cook and Holman, 1978). Groups of 30 heads were removed from adult cockroaches and elsewhere processed through homogenization and purification procedures described elsewhere (Holman and Cook, 1983). HPLC separation of the myotropic components in the sample was carried out on a Waters μ-Bondapak phenyl column with a linear solvent program. The initial solvent was 0.1% TFA. Eight min after injection the linear gradient against 25% acetonitrile (made to 0.1% TFA) was begun; final concentration (25% acetonitrile) was achieved after 1 hr. Fractions (containing 3 ml) were arbitrarily collected every 2 min throughout the run. In order to locate the active fractions, a volume of solvent equivalent to 2 head equivalents was removed from each of the tubes for bioassay.

Table 1. Relative Biological Activity of the Five
          Myotropic Factors on the Isolated Hindgut
          of the Cockroach

Fraction	Elution Time (min)	Threshold Concentration head equivalent/ml saline
22	42–44	.015
28	54–56	.009
30	58–60	.028
32	62–64	.034
34	66–68	.083

Analyses of each of the active fractions for thermal stability,
enzymatic degradation, and threshold activity were performed as previously described elsewhere (Holman and Cook, 1983).

Five active fractions were isolated by HPLC fractionation of
Sep-pak purified head extracts (Table 1).

Fractions 22, 28, and 32 retained full biological activity after
incubation for 45 min at 105°C. Fractions 30 and 34 were heat
labile.

The myotropic peptides present in Fractions 30, 32, and 34 were
completely inactivated by a 4-hr incubation with either carboxypeptidase Y (CPY) or aminopeptidase M (APM), as was the pentapeptide proctolin. The myotropic activity of Fraction 22 was abolished completely
by APM but was reduced by about 20% by CPM. The biological activity
of Fraction 28 was unaffected by APM but reduced to zero by CPY.

Based upon retention time, the myotropic peptide in Fraction 22
appeared to be proctolin. However, bioassay showed that each head
equivalent of Fraction 22 contained about 11 activity units, which
would indicate that each head contained about 1.5 ng proctolin (in our
bioassay system one activity unit is equivalent to 0.15 ng proctolin).
Bishop et al. (1981) showed by radioimmunoassay that the brain and
subesophageal ganglion of the closely related cockroach, Periplaneta
americana, contain a total of about 0.3 ng proctolin (about 2 activity
units). Digestion of Fraction 22 with CPY suggests that proctolin and
another myotropic peptide are present. One peptide, accounting for
80% of the biological activity in the sample, was not degraded by CPY,
but a second material responsible for about 20% (2 activity units) of
the activity was destroyed. Since 2 activity units equal about 0.3 ng
proctolin (which was degraded by CPY), our results would seem to
correlate well with the results of Bishop et al. (1981). Recently,

Kingan and Titmus (1983) demonstrated by radioimmunoassay that the brain of Leucophaea maderae contains about 0.13 ng (0.2 picomole) proctolin, but they did not assay the subesophageal ganglion. However, the value obtained for the brain was similar to the results of Bishop et al. (1981), and suggests that our comparison of Leucophaea to Periplaneta is valid. The CPY refractive substance was verified to be a peptide by the total loss of biological activity following incubation with APM. The biological activity of the other four active fractions was completely destroyed by incubation with CPY, verifying that the myotropic factors in those fractions are indeed peptides. The activity in three fractions (30, 32, 34) was also destroyed by incubation with APM. However, the myotropic factor in Fraction 28 was not destroyed, even after 24 hr incubation. This would suggest that the N-terminus was blocked, perhaps by pyroglutamic acid (pGlu). At least one insect neurohormone, the locust adipokinetic hormone, has an N-terminal pGlu. However, the myotropic peptide in Fraction 28 was not the locust adipokinetic hormone, as that hormone has no myotropic activity in our bioassay system and its HPLC retention time was substantially longer than Fraction 28 (Holman and Wagner, unpublished).

We do not intend to infer that the cc are the source of the five myotropic peptides; this must await a distribution study. Finally, we do not suggest that the specific and primary function of the five peptides is the stimulation of hindgut contractile activity.

REFERENCES

Bishop, C. A., O'Shea, M., and Miller, R. J., 1981, Neuropeptide proctolin (H-Arg-Tyr-Leu-Pro-Thr-OH): Immunological detection and neuronal localization in insect central nervous system, Proc. Natl. Acad. Sci. USA, 79:5899.

Brown, B. E., 1965, Pharmacologically active constituents of the cockroach corpus cardiacum: Resolution and some characteristics, Gen. Comp. Endocr., 5:387.

Cameron, M. L., 1953, Secretion of an orthodiphenol in the corpus cardiacum of the insect, Nature, Lond., 172:349.

Cook, B. J., and Holman G. M., 1978, Comparative pharmacological properties of muscle function in the foregut and the hindgut of the cockroach, Leucophaea maderae, Comp. Biochem. Physiol., 61C:291.

Davey, K. G., 1962, The mode of action of the corpus cardiacum on the hindgut in Periplaneta americana, J. Exp. Biol., 39:319.

Holman, G. M., and Cook B. J., 1970, Pharmacological properties of excitatory neuromuscular transmission in the hindgut of the cockroach, Leucophaea maderae., J. Insect Physiol., 16:1891.

Holman, G. M. and Cook, B. J., 1983, Isolation and partial characterization of a second myotropic peptide from the hindgut of the cockroach Leucophaea maderae, Comp. Biochem. Physiol., (In press).

Kingan T., and Titmus M., 1983, Radioimmunologic detection of proctolin in arthropods, Comp. Biochem. Physiol., 74C:75.

PROTHORACICOTROPIC HORMONE - LIKE EFFECTS OF BIOGENIC AMINES IN LEPIDOPTEROUS LARVAE

Mamdouh Idriss, Shebl Sherby, Mahmoud Morshedy, and
Nabil A. Mansour

Plant Protection Department, Faculty of Agriculture
University of Alexandria, Alexandria, Egypt

The neurosecretory system of insects appears to contain several factors of different biochemical and physiological activities. The biogenic amines are important neurosecretory products of the central nervous system, in which dopamine the most abundant while epinephrine and norepinephrine (WELSH, 1972), and octopamine (DYMOND and EVANS, 1979) occur in smaller amounts. Beside the above mentioned biogenic amines, L-dihydroxyphenylalanine (L-DOPA) have been detected in the ventral nerve cord of the boll weevil, *Anthanomus grandis* (NESTLER *et al.*, 1981).

The present study was undertaken to test the function of certain biogenic amines and several other nonhydroxylated phenethylamine derivatives on the molting process of the decapitated larvae of the Eri-silkworm *Philosamia ricini* and the Egyptian cotton leafworm *Spodoptera littoralis*. Biochemical studies were undertaken to identify the biogenic amines interactions with the specific receptive site(s) on the prothoracic glands.

Fifth *Ph. ricini* and sixth *S. littoralis* larval instars were used for bioassay studies, both before the critical period of the action of ecdysone. Larvae were ligated at a level between the head and thorax to obtain a decapitated larva or between the thorax and abdomen to obtain an isolated abdomen. The decapitated *Ph. ricini* larvae pupated in 35 days (3 days for normal), while the isolated abdomens did not show any sign of morphological changes for more than 50 days. On the other hand the decapitated *S. littoralis* larvae pupated after 7 days (2 days for normal).

Prothoracic glands (PTG) of *S. littoralis* larvae were dissected and collected in ice cold Tris-HCl buffer, (50 mM Tris-HCl / 15 mM

385

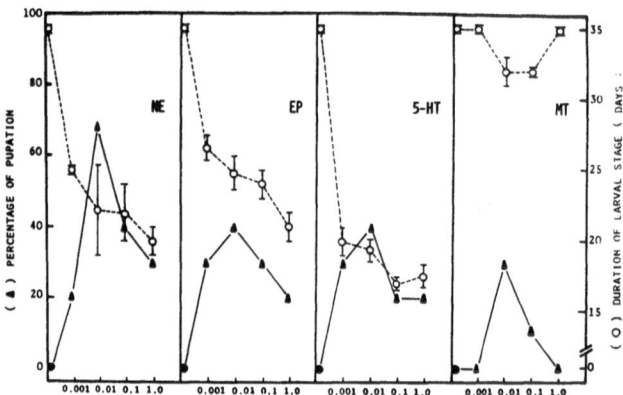

Fig. 1. The effect of Norepinepherine (NE), Epinephrine (EP), Serotonin (5-HT) and Melatonin (MT) on the percentage of pupation and duration of larval stage of decapitated *Ph. ricini* (Drug concentration μg / larva).

Mg Cl$_2$, pH 7.4) using OPM-6 microscope. The isolated glands (100 glands / ml buffer represent about 0.7 ± 0.1 mg protein / ml) were homogenized in glass hand homogenizer and centrifuged at 3000 X g for 15 min at 4°C. The supernatant was used as a crude membrane protein for β-adrenergic receptors binding assay. Membrane prepration was incubated with (-)-[³H]dihydroalprenolol, ([³H]DHA, sp. act. 44.9 Ci/mmol) in a total volume of 150 μl of the same buffer for 30 min with shaking at 37°C. In competition experiments, the stock solutions of unlabelled compounds of biogenic amines were used in the incubation mixture, which has been terminated by diluting 125 μl of the incubated aliquots into 2 ml of ice cold buffer followed by rapid vacuum filtration on Whatman GF/C glass fiber filter (WILLIAMS *et al.*,1976). Specific, nonspecific and biogenic amines blockade of [³H]DHA binding to its receptors were counted in a Packard liquid scintillation spectrometer at an efficiency of 40% .

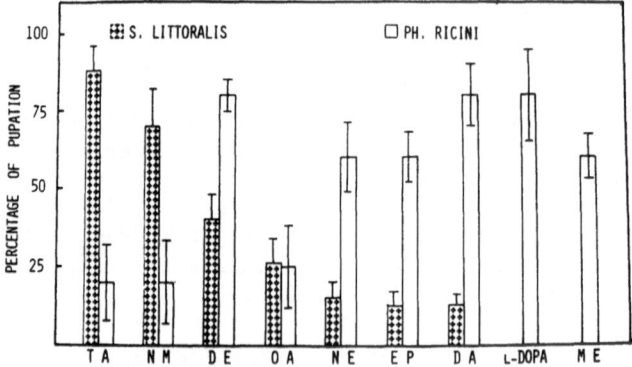

Fig. 2. The effect of biogenic and catecholamines on the percentage of pupation of decapitated *S. littoralis* and *Ph. ricini* larvae.

Certain biogenic amines, dopamine (DA), L-DOPA, norepinephrine (NE), epinephrine (EP), deoxyepinephrine (DE), metanephrine (ME), normetanephrine (NM), tyramine (TA), octopamine (OA), serotonin (5-HT), and melatonin (MT), have been tested on apolysis of the decapitated larvae. The apparent effects (Fig. 1) of NE, EP, 5-HT and MT on apolysis of decapitated *Ph. ricini* larvae, has initiated a clearly defined concept for the function of catecholamines and tryptophan derivatives on the induction of insect development. This also reveals the absence of a typical dose-response effect, whereas there is a direct correlation between the dosage and the time required for development except for MT. The relation between the chemical structure of biogenic amines and the molting induction was conflicting between the two lepidopterous species, by using the optimum dosage (Fig. 2).

The usefulness of EP and NE as sympathomimetic modulators of biochemical and physiological activities, have served as prototypes to guide the classical development of more significantly active amine derivatives. The main structure of β-chloro-β-phenethylamine were designed on conformationally difined bioisosteres of NE, in which pharmacophoric groups are oriented in proper special arrangement for probable optimal interactions on the receptive site(s). It was found that the absence of hydroxyl group(s) from the basic structure abolishes the PTTH - like effect activity.

The early suggction that some catecholamines act as a synaptic transmitters (FRONTALI and NORBERG, 1966) has encouraged us to study their interactions with β-adrenergic receptors of the PTG preparation in comparison with standard frog erythrocytes preparation. There is remarkable specific [^{3}H]DHA binding to the PTG receptors, which comprise 35 p mole/mg protein at 1.0 μM. *DL*-normetanephrine, DE and *DL*-EP inhibited [^{3}H]DHA binding to its receptors in both preparations which might affect the mechanism of biogenic amines interactions.

REFERENCES

DYMOND G.R. and EVANS P.D. (1979) Biogenic amines in the nervous system of the cockroach, *Periplaneta americana:* Association of octopamine with mushroom bodies and dorsal unpaired median (DUM) neurons. *Insect Biochem.* **9**, 535-545.

FRONTALI N. and NORBERG K.A. (1966) Catecholamines in neurones of cockroach brain. *Acta physiol Scand.* **66**, 243-244.

NESTLER C., BROWN C.S. and WHEELER A.P. (1981) Biogenic amines in the nervous system of the boll weevil, *Anthanomous grandis.* *Comp. Biochem. Physiol.* **69C**, 53-60.

WELSH J.H. (1972) Catecholamines in the invertebrates. *Handbk exp. Pharmac.* **33**, 79-109.

WILLIAMS L.T., JARETT L. and LEFKOWITZ R.J. (1976) Adipocyte β-Adrenergic Receptors: Identification and subcellular localization by (-).[^{3}H] dihydroalprenolol binding. *J. Biol. Chem.* **251**, 3096-3104.

PRECOCIOUS PUPATION IN CHELONUS PARASITIZED TRICHOPLUSIA NI:

THE ENDOCRINE BASIS OF THIS ANTI-JUVENILE HORMONE EFFECT

Davy Jones and Shalini Sreekrishna

Department of Entomology
University of Kentucky
Lexington, Kentucky 40506

Metamorphosis in insects is initiated when the juvenile hormone (JH) titer declines prior to ecdysone release. The mechanism
by which this decline is achieved has been the subject of intensive
study. Elucidation of the mechanism of JH suppression has important
implications for applied concerns, beyond increases in our basic
understanding of insect metamorphosis.

During an investigation of the ecology of host-parasite interaction it was observed that the wasp Chelonus insularis caused host
Trichoplusia ni to precociously spin a cocoon (Jones 1984). Reports
in the literature indicated that a number of wasps in this genus
caused similar behavioral effects in their hosts. More detailed study
using Chelonus sp. near curvimaculatus as the parasite of T. ni
demonstrated that this behavior was part of a true precocious pupa-
tions behavior (Jones et al. 1981). The endocrine basis for this apparent anti-JH effect has now been investigated and is reported here.

The host and parasites were reared as described by Jones et al.
(1981). Stung hosts prematurely cease feeding, begin wandering behavior and spin a cocoon in what would normally be the penultimate
instar. Some stung larvae precociously spin a cocoon but no parasite
emerges from them. Dissections reveal no obvious parasite during the
instar which shows this precocious behavior. Such 'apparently parasitized' larvae were used as experimental animals since no parasite
emerges to kill them during an experiment.

When a JH analog (Ro 13-5223, (ethyl(2-phenoxy-phenoxy)-ethyl)
carbamate) was applied to larvae apparently parasitized by C. near
curvimaculatus (Table 1) or C. insularis precocious initiation of
metamorphosis was prevented, and many larvae molted to another larval

instar. This larval-larval molt indicates that precocious metamorphosis is initiated by a JH titer decline, rather than by absent or
nonfunctional JH receptors.

A decline in JH biosynthesis is one mode of regulation that may
cause a JH titer decline. Allatectomy of penultimate instar $\underline{T}$. $\underline{ni}$
results in precocious pupation, indicating that suppression of cor-
pora allata (CA) activity can result in precocious metamorphosis.
When CA from normal, wandering larvae are implanted into abdomens
of ligated unparasitized wandering larvae, their JH production induces a level of JH esterase (JHE) activity significantly higher than
that occurring in sham-operated controls (5.49 vs. 1.22 nmoles JH III
hydrolyzed/min-ml hemolymph, respectively, $p < 0.05$). However, bioassay of CA from wandering larvae apparently parasitized by $\underline{C}$. near
$\underline{curvimaculatus}$ indicated no significant increase in JHE activity
(1.44 nmoles/min-ml). Little or no JHE activity occurs in the hemolymph of wandering larvae apparently parasitized by either parasite, in
contrast to normal wandering larvae, suggesting that their CA are
not active. Prelimenary tests in which CA from normal wandering
larvae are implanted into apparently parasitized abdomens indicates
that JHE activity would be induced in unligated apparently parasitized larvae if active CA were present. All these data suggest that
suppression of CA activity is a component to the JH decline in
apparently parasitized larvae.

It remains to be tested whether decreased JH biosynthesis is the
only premature biochemical event occurring in host larvae. JHE appears during the late feeding stage of normal larvae to aid in clear-
ance of hemolymph JH. This enzyme appears only in last instar larvae
which are initiating metamorphosis. However, JHE activity appears
in the hemolymph of feeding apparently parasitized larvae even though
they are in what would normally be the penultimate instar. When a
JHE inhibitor (EPPAT, O-ethyl-S-phenyl phosphoramidothiolate) was applied to normal feeding last instar larvae, metamorphosis was blocked
and the larvae molted to another larval instar (Table 1). Similar
treatment to feeding apparently parasitized larvae also caused molting
to another larval instar. These data demonstrate that extra larval
molts would occur in normal larvae were it not for the action of JHE.
These data also indicate that the JHE in apparently parasitized larvae functions in a manner similar to that in normal larvae.

Isoelectric focusing data demonstrate that JHE is not the only
protein unique to the last instar which appears in penultimate instar
apparently parasitized larvae. Also, unique 4th instar proteins
disappear in these larvae.

The data presented indicate that the anti-JH effect in precocious initiation of metamorphosis is due to the normal 5th instar developmental pattern being prematurely expressed in the 4th instar. Although a way of inducing anti-JH effects in insect pests has long

been sought, none have been found which are practical for Lepidoptera. The factor(s) produced by the parasites examined here are very potent in that they can induce precocious initiation of metamorphosis in several hosts from three families of Lepidoptera, including Noctuidae. Whether the applied goal is development of new third generation insecticides or of factors genetically engineered into plants for insect resistance (Barton and Brill 1983) the study of interactions like that of Chelonus and T. ni may provide unsuspected leads as to how novel manipulations of pest development may be achieved.

Table 1.   Effect of various chemical treatments on metamorphosis in last instar T. ni.

Treatment (n)	% apolysis to a larva
Normal last instar	
JHA application (15)	67
EPPAT (29)	45
Ethanol (11)	0
Apparently parasitized	
JHA (11)	73
EPPAT (34)	16
Ethanol (10)	0

100 nmoles of the JH analog (JHA) Ro 13-5223 were applied to larvae one day prior to cocoon spinning. 100 nmoles of EPPAT was applied to normal and feeding 'last instar' apparently apparasitized larvae 3-4 times a day until wandering or apolysis.

The investigation reported here is in connection with a project of the Kentucky Agricultural Experiment Station.

Barton, K. A. and Brill, W. J., 1983, Prospects in plant genetic engineering, Science 219: 671-76.

Jones, D., 1984, Predators and parasites of temporary row crop pests: agents of irreplaceable mortality or scavengers acting prior to other mortality factors? Entomophaga (accepted)

Jones, D., Jones, G., and Hammock, B. D., 1981, Developmental and behavioral responses of larval Trichoplusia ni to parasitization by an imported braconid parasite Chelonus sp., Physiol. Entomol. 6: 387-94.

REGULATION OF JUVENILE HORMONE ESTERASE ACTIVITY

IN LARVAE OF THE CABBAGE LOOPER, TRICHOPLUSIA NI

Grace Jones and B. D. Hammock

Department of Entomology, University of Kentucky
Lexington, Kentucky and Departments of Entomology
and Environmental Toxicology, University of California
Davis, California

Using a Trichoplusia ni model system, the question of the timing
and mechanisms of regulation of larval-pupal ecdysis were investigated. We report here results indicating that disturbances of just
several hours in timing of appearance and disappearance of a single
developmental hormone, juvenile hormone (JH), can totally disrupt
metamorphosis. Also, modulation of the enzyme which degrades JH appears to play a key role in regulation of the hormone titer. Furthermore, absence of JH is classically considered to be necessary for the
larval-pupal transformation. In contrast our work with the family
Noctuidae has shown that the brief presence of JH during the prepupal
period immediatelyfollowed by its rapid disappearance is crucial for
pupation. The appearance and disappearance of JH must occur during
a period of less than 12 hours for normal development to proceed.

The endocrine events in last instar larvae of our test insect
demonstrate the importance of precise timing and proper regulation
in successful metamorphosis. Toward the end of the feeding stage of
the last larval stadium (day 2) the JH titer must decline, at least
in part due to degradation to allow release of the larval steroid
hormone ecdysone (Sparks and Hammock 1980) and to prevent a larval-
larval molt (D. Jones, this conference). The combinations of low JH
and high ecdysone commits the larva to initiate the metamorphic trans-
ition from immature to adult, but it is not until a second period of
juvenile hormone in the prepupa that the ecdysis-inducing surge of
ecdysone appears. Although the prepupal JH can prevent expression of
certain adult characters, there has been no evidence published that
this burst of JH is required for sucessful molting.

The importance of the appearance of prepupal JH in T. ni is demonstrated by the removal of the corpora allata (allatectomy) and the

subsequent prevention of successful larval-pupal ecdysis which normally occurs late the following day. Application of exogenous JH
or JH analog (JHA) to replace the missing JH induces the completion
of ecdysis. Allatectomy at progressively earlier time points results
in decreasing numbers of larvae successfully completing the larval-
pupal molt (Table 1). This experiment strongly indicates that JH is
crucial for successful larval-pupal development and completion of the
normal pupal molt.

Further evidence also shows that the disappearance of prepupal
JH is as vital as was its appearance. Application of JH, a JHA or
a selective JHE inhibitor (EPPAT), blocks shedding of the cuticle,
results in pupal-pupal molts or causes deformed pupal-adults.

Critical timing is thus involved is regulation of JH via degradation. Since the relationship between JH and JHE in regulation of
metamorphosis is so intertwined, it would seen that the most efficient mechanism to accomplish the delicate balance in timing of changes
in the JH titer is by direct induction of its own degradative enzyme.
Induction of JHE by JH via the brain (Jones et al. 1981) would involve
a lag time. The regulation of the prepupal peak of JHE was therefore
investigated. It was found that JHE was induced directly by JH.

Progressive allatectomy further and further past the initial
point of JH release allowed recovery of increasing levels of JHE activity. Reimplantation of the CA also restored the prepupal JHE peak.
Application of JH and JHA induced JHE activity in allatectomized larvae, further confirming the hypothesis.

In summary the rapid appearance and folllowing disappearance of
JH is crucial for the larval-pupal molt. Modulation of the delicate
balance is via direct induction of the degradative enzyme JH esterase.

Table 1.   The effect of time of allatectomy on
           successful larval ecdysis in T. ni

Time of allatectomy	% successful ecdysis
0500	11
0700	14
1000	24
1300	32
1500	74
1800	97

Day begins at lights on. Percentages are
corrected for failure to molt in sham larvae.

This work was supported by grant R0l ES02710-03 and B. D. Hammock Research Career Development Award ES 00107-04 both from the National institutes of Health Sciences, U.S.A.

Jones, G., Wing, K. D., Jones, D., and Hammock, B. D., 1981, Source and action of head factors regulating juvenile hormone esterase in larvae of the cabbage looper, Trichoplusia ni, J. Insect Physiol. 27:85-91.

Sparks, T. C. and Hammock, B. D., 1980, Comparative inhibition of the juvenile hormone esterases from Trichoplusia ni, Tenebrio molitor and Musca domestica, Pestic. Biochem. Physiol., 14: 290-32.

BIOASSAY AND PRELIMINARY PROPERTIES OF THE

HYPERTREHALOSEMIC HORMONE OF BLABERUS DISCOIDALIS COCKROACHES

Larry L. Keeley and Timothy K. Hayes

Department of Entomology, Texas A&M University
Texas Agricultural Experiment Station
College Station, TX 77843 USA

The hypertrehalosemic hormone (HTH) increases hemolymph trehalose levels in many insect species. The nature of HTH has remained in dispute, however, since it has not been isolated and characterized. Locust adipokinetic hormone (AKH) and crustacean red-pigment-concentrating hormone (RPCH) both increase trehalose levels in cockroaches (Mordue and Stone, 1977) and have similar structures. HTH and AKH appear chemically identical in locusts but different in cockroaches (Holwerda, 1977a,b). Recently, however, a second AKH (II) was suggested as the HTH of locusts (Loughton and Orchard, 1981). Obviously, the chemical nature of HTH is unclear. For this reason, we initiated studies to develop a bioassay for HTH that would be suitable for routine use in conjunction with isolation of the hormone.

An in vitro bioassay was developed based on placing fragments from the same fat body in separate flasks and comparing the quantities of trehalose produced by test samples exposed to corpora cardiaca-allata (CC+CA) extracts relative to control samples that lack gland extracts. Few previous studies have examined the HTH effect on the fat body in vitro (Friedman, 1967; Wiens and Gilbert, 1967; Downer, 1979; McClure and Steele, 1981), and the purpose of this report is to describe the properties of an in vitro system suitable as a HTH bioassay.

Methods and Materials: Adult male Blaberus discoidalis cockroaches >10 days old were used as the experimental animals. CC+CA extracts were used even though we confirmed that the CC is the major source of the hormone. Inclusion of the CA simiplified the surgical procedures with no adverse effects on the assay. The assay medium consisted of 153 mM NaCl, 2.7 mM KCl, 1.8 mM $CaCl_2$ and 5 mM N-2-hydroxyethylpiperazine-N'-2-ethanesulfonic acid

397

(HEPES),pH 7.2.   Fat body fragments were incubated for 2 h in
the medium either in the presence (test samples) or the absence
(control samples) of CC+CA extracts. A 95:5 O_2:CO_2 atmosphere was
used during the incubation. Aliquots were removed at the start and
end of the incubation period for carbohydrate analysis using the
anthrone procedure. The fat body fragments were dried to constant
weight and the specific activity was calculated for the fragment
as: ug trehalose produced/mg fat body dry wt per hr.

Results: In vivo studies confirmed the HTH response in adult male
B. discoidalis.   Injections of CC+CA extract (=1 gland
pair/animal) into intact cockroaches increased hemolymph trehalose
levels with no effect on glucose levels.   Injections of CC+CA
extract into decapitated animals increased the hypertrehalosemic
response by approx. 5-fold above the amount of response observed
for intact animals. Again, glucose levels were unaffected.   These
results confirm the specificity of the HTH effect by the CC-CA
complex in B. discoidalis.
     The nature of the carbohydrate produced in response to CC+CA
extracts was determined.   TLC analysis of the incubation medium
indicated that 97% of the carbohydrate produced was trehalose.
Glucose, although present, did not change in response to the
CC+CA extracts.   Test samples produced trehalose at a more rapid
rate than controls for 3 h, at which time the rate of the test
samples declined to the control rate. A two hour period was chosen
as the maximum bioassay time since it was well within the time for
the linear HTH response.   Comparisons of the specific activities
for fragments from the same fat body indicated a consistent
degree of increase relative to the control sample; however, it was
found that between different fat bodies specific activities for
high control samples overlapped the specific activities of low
test samples.   In order to reduce the degree of variability
inherent between fat bodies and their degree of responsiveness, it
was decided to use the ratio of specific activities between the
test and control samples.   This activation ratio standardized the
degree of responsiveness between different fat bodies and reduced
data variability.
     Studies on the relationship between adult age and HTH
response by the fat body indicated that tissues from 0-day-old
animals were 50% less responsive than tissues from animals that
were >5 days old. A dose-response was determined with CC+CA
extracts.   The threshold for the HTH response occurred at 0.01
CC+CA pair/ml of medium and was linear through 0.08 CC+CA pair/ml;
thereafter,   further additions of extract resulted in only an
asymptotic response.
     The specificity of the HTH response was assessed for the in
vitro fat body. No biogenic amine, including octopamine, produced
a hypertrehalosemic response nor did the vertebrate hormones
glucagon or insulin. Strong positive responses were observed with
AKH and RPCH. Comparison by reversed-phase TLC of the migration of

synthetic AKH and HTH activity of a CC+CA extract demonstrated different R_f values for the two hormonal agents and indicated that HTH of B. discoidalis and locust AKH are distinct hormones.

These studies provide a simple, direct and quick assay for the presence of HTH. Although AKH gives a false positive response, the data indicate that AKH is distinct from HTH in B. discoidalis. Biogenic amines did not give a response in this system. Studies are in progress to isolate and characterize the HTH of B. discoidalis.

Acknowledgments: This research was supported by NIH Biomedical Research Support Grant SO7RR07090 and by the Texas Agricultural Experiment Station.

REFERENCES

Downer R.G.H. (1979) Trehalose production in isolated fat body of the american cockroach, Periplaneta americana. Comp. Biochem. Physiol. 62C, 31–34.

Friedman S. (1967) The control of trehalose synthesis in the blowfly, Phormia regina Meig. J. Insect Physiol. 13, 397–405.

Holwerda D. A., van Doorn J. and Beenakkers A.M.T. (1977a) Characterization of the adipokinetic and hyperglycaemic substances from the locust corpus cardiacum. Insect Biochem. 7, 151–157.

Holwerda D.A., Weeda E. and van Doorn J.M. (1977b) Separation of the hyperglycemic and adipokinetic factors from the cockroach corpus cardiacum. Insect Biochem. 7, 477–481.

Loughton B. G. and Orchard I. (1981) The nature of the hyperglycaemic factor from the glandular lobe of the corpus cardiacum of Locusta migratoria. J. Insect Physiol. 27, 383–385.

McClure J.B. and Steele J.E. (1981) The role of extracellular calcium in hormonal activation of glycogen phosphorylase in cockroach fat body. Insect Biochem. 11, 605–613.

Mordue W. and Stone J.V. (1977) Relative potencies of locust adipokinetic hormone and prawn red-pigment-concentrating hormone in insect and crustacean systems. Gen. comp. Endocr. 33, 103–108.

Wiens A.W. and Gilbert L.I. (1967) Regulation of carbohydrate mobilization and utilization in Leucophaea maderae. J. Insect Physiol. 13, 779–794.

OOSTATIC HORMONE AND BIOGENIC AMINES AS INHIBITORS OF OVARIAN

MATURATION IN HOUSE FLIES AND MOSQUITOES*

Thomas J. Kelly, Charles W. Woods, Mark J. Birnbaum
and Alexej B. Bořkovec

Insect Reproduction Laboratory, ARS, USDA
Beltsville, MD USA

Oostatic mechanisms have been studied extensively in only a few species of insects: Rhodnius prolixus (Davey, 1978), Diploptera punctata (Tobe , 1980), and Musca domestica (Adams, 1980). In R. prolixus the inhibitory substance, antigonadotropin, acts directly on the ovary to inhibit juvenile hormone (JH) induced patency of the follicles. In D. punctata, the inhibitory substance suppresses JH synthesis by the corpus allatum. In M. domestica, Adams suggests that oostatic hormone prevents the release of egg development neurosecretory hormone. We have tested the effects of oostatic hormone from house flies in the autogenous mosquito, Aedes atropalpus, where the hormonal events controlling vitellogenesis are well-defined (Fuchs et al., 1980; Kelly and Fuchs, 1980; Masler et al., 1980; Kelly et al., 1981), and the necessary surgical manipulations are easy to perform.

With an in vivo bioassay (Fig. 1, legend) inhibitory activity was demonstrated in extracts of heads, thoraces and abdomens of mature M. domestica females with the major portion present in the abdomen, apparently associated with the ovaries (Kelly et al., 1983; Kelly, Birnbaum and Borkovec, in prep.). Fractionation by high pressure liquid chromatography (HPLC) produced a fraction eluting at 0-2.5 min, which was the only fraction to consistently show in vivo and in vitro inhibitory activity in extracts from ovaries (Fig. 1) and whole bodies of mature M. domestica females. Unfortunately, toxic factors associated with the Sep-pak purification step have complicated furter analysis.

---

*Mention of a commercial or proprietary product in this paper does not constitute an endorsement of this product by the USDA.

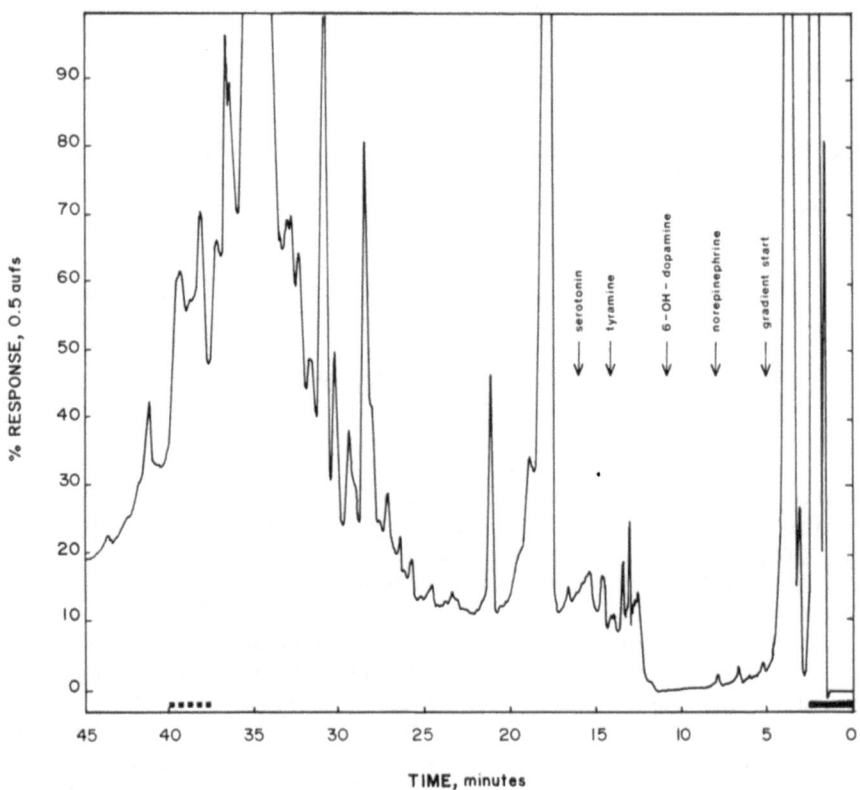

Fig. 1. HPLC fractionation of M. domestica mature ovary extract on a Shandon-C18 column (3.9 mm i.d. x 30 cm, 5 μm particle size) by gradient elution with 0-50% CH3CN in 0.1% TFA at 1.5 ml/min. The supernatant from 45 mature ovaries was fractionated following clean up by high-speed centrifugation, elution from a Sep-pak (Waters Associates, Milford, MA) with 80% CH3CN in 0.1% TFA and flash evaporation. Fractions were dried and resolubilized in Aedes saline (Hayes, 1953) to a concentration of 0.2 ovary equivalent per μl and 1 μl was injected into newly emerged A. atropalpus females or assayed in an in vitro ovarian ecdysteroid production assay (Birnbaum, Kelly and Imberski, in prep.). In the in vivo assay females were scored at 24 hr after emergence for the prescence of white, opaque yolk in the ovaries by dissection on a black background at 15X. UV absorbance at 214 mm (——); consistently high degree (>80%) of inhibitory activity (━); lower, less consistent degree of inhibitory activity (▪▪▪).

The report by Huybrechts and De Loof (1981) that 6-OH-dopamine inhibited yolk deposition in Sarcophaga bullata prompted us to investigate the possibility that M. domestica oostatic hormone might be a biogenic amine. Although a number of amines (e.g., 6-OH-dopamine, norepinephrine and tyramine) were inhibitory in vivo at 10 μg/insect and in vitro at 1 μg/μl, none of the biogenic amines tested cochromatographed on the Shandon-C18 column with the inhibitory material (Fig. 1).

Further analysis of this complex system is necessary to understand the oostatic mechanism in house flies and mosquitoes. The cross-reactivity of oostatic extracts from house flies and other species has not been examined but might be informative in determining the general nature of the oostatic mechanism.

References

ADAMS T. S. (1980) The role of ovarian hormones in maintaining cyclical egg production in insects. In Advances in Invertebrate Reproduction (Ed. by CLARK W. H. JR and ADAMS T. S.), p. 109-125, Elsevier/North Holland Biomedical Press, Amsterdam.

DAVEY K. G. (1978) Hormonal stimulation and inhibition in the ovary of an insect, Rhodnius prolixus. In Comparative Endocrinology (Ed. by GAILLARD P. J. and BOER H. H.), p. 13-16, Elsevier/North-Holland Biomedical Press, Amsterdam.

FUCHS M. S., SUNDLAND B. R. and KANG S-H. (1980) In vivo induction of ovarian development in Aedes atropalpus by a head extract from Aedes aegypti. Int. J. Invert. Reprod. 2, 121-129.

HAYES R. O. (1953) Determination of a physiological saline solution for Aedes aegypti (L.). J. Econ. Entomol. 46, 624-627.

HUYBRECHTS R. and DE LOOF A. (1981) Effect of ecdysterone on vitellogenin concentration in haemolymph of male and female Sarcophaga bullata. Int. J. Invert. Reprod. 3, 157-168.

KELLY T. J. and FUCHS M. S. (1980) In vivo induction of ovarian development in decapitated Aedes atropalpus by physiological levels of 20-hydroxyecdysone. J. Exp. Zool. 213, 25-32.

KELLY T. J., FUCHS M. S. and KANG S-H. (1981) Induction of ovarian development in autogenous Aedes atropalpus by juvenile hormone and 20-hydroxyecdysone. Int. J. Invert. Reprod. 3, 101-112.

KELLY T. J., WOODS C. W. and BIRNBAUM M. J. (1983) Physiological function and partial purification of an ovarian maturation inhibitor from the house fly, Musca domestica. Southwest. Entomol. Suppl. 5, 36-37.

MASLER E. P., FUCHS M. S., SAGE B. and O'CONNOR J. D. (1980) Endocrine regulation of ovarian development in the autogenous mosquito, Aedes atropalpus. Gen. Comp. Endocrinol. 41, 250-259.

TOBE S. S. (1980) Regulation of the corpora allata in adult female insects. In Insect Biology in the Future (Ed. by LOCKE M. and SMITH D. S.), p. 345-367, Academic Press, New York.

DEVELOPMENT OF GABA LEVELS IN THE CNS OF <u>MANDUCA</u> <u>SEXTA</u>

Timothy G. Kingan

Department of Biological Sciences
Columbia University
New York, NY   10027

The development of a neuronal population consists in part of its accumulation of neurotransmitter. Accumulation is controlled by substrate transport systems, activity of transmitter synthesizing and degradative enzymes, and a storage mechanism. This machinery in turn depends on an exogenous source of an appropriate precursor for production of transmitter. We have begun to study the processes by which the inhibitory neurotransmitter, γ-amino-butyric acid (GABA), accumulates in the central nervous system (CNS) of the moth <u>Manduca</u> <u>sexta</u>.

GABA LEVELS IN DEVELOPING AND ADULT CNS REGIONS

Amino acid determinations of CNS extracts were done by reverse phase liquid chromatography of the dansyl-derivatives. GABA was found throughout the CNS in a range of quantities paralleling earlier findings on the rates of synthesis in intact ganglia (Maxwell et al., 1978). Thus, in animals one day before eclosion, the lamina region of the optic lobe contains 7.8 pmole GABA per μg protein and the ptherothoracic ganglion contains 110 pmole per μg protein; 34.5 pmole/μg protein was found in the antennal lobe (AL). Because of our interest in biochemical correlates of insect olfactory development, we quantified the accumulation of GABA in the AL during adult development and after emergence (Kingan and Hildebrand, 1982). The AL of Stage 6 animals (emergence occurs at Stage 18) contains 22 pmole GABA; thereafter levels rise to a plateau of 207 pmole/AL at Stage 18, changing little in the final three days of development. Within 18 hours after emergence there is a 50% increase in GABA levels to 320 pmole/AL. Glutamate (+44%) and alanine (+30%), also increase while others, e.g., glycine, do not significantly change. GABA levels increase in the thoracic ganglion (+36%) as

well, but not in the abdominal ganglia. Thus some, but not all
CNS structures experience a rapid change in free amino acid levels
soon after eclosion. This suggests that at least one factor limiting accumulation in developing moths has been removed.

UPTAKE OF GLUTAMATE INTO GANGLIA IN VITRO

If the availability of substrate in glutamate decarboxylase
containing neurons is limiting in GABA accumulation, a change in
the capacity of CNS structures to concentrate glutamate might be
reflected in GABA levels. Therefore, we have begun to study glutamate uptake processes with in vitro preparations of isolated
ganglia. Initial experiments were carried out with abdominal ganglia from Stage 18 animals, providing the convenience of three
ganglia per animal. Uptake both in the presence and absence of
100 mM Na^+ (replaced with 100 mM Tris) is saturable. In the presence of Na^+ (plus Na^+) the apparent Km is 0.70 mM; without Na^+
(minus Na^+) the Km is 1.28 mM. To determine that the increase in
the plus Na^+ rate is due to a component distinct from that operating in the absence on Na^+ will require the testing of agents
that differentially affect the two components. Additional features
of ganglionic glutamate uptake characteristic of mediated transport
include 1) specificity: uptake was inhibited only by L-aspartic
acid of several amino acids tested, including D-glutamic acid and
2) energy dependence: uptake was poisoned by azide and dinitro-
phenol, but not by ouabain. When plus Na^+ rates in abdominal and
thoracic ganglia were compared in Stage 18 and Stage A1 (one day
after emergence) animals no statistically significant differences
could be detected. In contrast, the minus Na^+ rate in abdominal
ganglia increases 40% in Stage A1 animals at high glutamate concentration (3.5 mM) to a rate comparable to the plus Na^+ rate. The
significance of the apparent decreased Na^+ dependence in Stage A1
animals at high glutamate concentrations may be revealed by determining rates at intermediate Na^+ concentrations. Because hemolymph
glutamate is less than 0.2 mM, we believe that changes in glutamate
transport alone could not account for the observed CNS amino acid
increases.

GLUTAMATE DECARBOXYLASE ACTIVITY

A second possible limiting factor in GABA accumulation is the
enzyme responsible for its synthesis, glutamate decarboxylase (GAD).
We have partially characterized the enzyme by means of a filtration
assay utilizing the anion exchange resin AG1-X8 (Bio-Rad) to separate unreacted glutamate from GABA. Kinetic constants for the enzyme from Stage 18 CNS extracts have been determined; the apparent
Km is 11 mM. Comparing the specific activities of the enzyme from
Stage 18 and Stage A1 animals indicates that maximal rates and the
apparent affinity of the enzyme for substrate do not change.

LEVELS OF AMINO ACIDS IN HEMOLYMPH

The hemolymph concentrations of several amino acids were deter-mined in stage 18 and A1 animals. RPLC of the dansyl derivatives was employed, both alone and after preparative ion-exchange liquid chromatography (DC-5A, Dionex Corp.) of hemolymph extracts. Gluta-mate concentrations in the hemolymph are apparently unchanged after emergence of the moth (0.17 mmole/1 hemolymph). Other amino acids, however, do change. Serine increases three to six fold (to as much as 29 mM in one of two strains tested); proline increases over two fold (to 6-10 mM), and glutamine decreases 13% to 46% (to 11-19 mM). Aspartate, previously thought to be present by RPLC, could not be detected by the combined ion-exchange/RPLC procedure (0.04 mM, limit of sensitivity). Because of the high concentrations of glutamine and proline and their simple pathways to glutamate, we decided to compare the efficiency of conversions of these amino acids to gluta-mate by intact ganglia. In addition, we wanted to know if hemolymph proline and glutamine could then serve as a carbon source for GABA synthesis.

POSSIBLE METABOLIC ROUTES TO GABA

Intact abdominal ganglia were incubated with 0.5 mM ^{3}H-L-glutamic acid or a combination of 1.0 mM each ^{3}H-L-proline and 14C-L-glutamine. The extracts were fractionated by RPLC and peaks cor-responding to identified amino acids were collected for liquid scin-tillation counting. When ganglia are incubated in ^{3}H-L-glutamic cid a significant fraction of the acetone:1N HCO_2H soluble radioac-tivity in the ganglia is recovered in the glutamine (22%), proline (14%), and GABA (2.1%) peaks, in addition to the glutamate peak (23%). ^{3}H-L-proline is converted to glutamate (6.6%) and GABA (2.8%); 14C-L-glutamine can also be recovered as glutamate (4.7%) and GABA (1.5%). If recovered GABA and glutamate are expressed as a ratio (GABA/glutamate) in each of the three examples, the results, ^{3}H-L-glutamate (0.09), ^{3}H-L-proline (0.42), and ^{3}H-L-glutamine (0.31), suggest that proline and glutamine can more efficiently label a glutamate pool for GABA synthesis.

(Supported by a research contract from the U.S. Army Research Office to J.G. Hildebrand)

REFERENCES

Kingan T.G. and Hildebrand J.G. (1982) GABA in the antennal lobes
    of metamorphosing and mature Manduca sexta. Soc. Neurosci.
    Abstr. 8: 988.
Maxwell G.D., Tait J.F., and Hildebrand J.G. (1978) Regional
    synthesis of neurotransmitter candidates in the CNS of the
    moth Manduca sexta. Comp. Biochem. Physiol. 61C: 109-119.

HORMONAL CONTROL OF HAEMOLYMPH LIPID DURING FLIGHT

IN THE LOCUST, LOCUSTA MIGRATORIA

Angela B. Lange and Ian Orchard [*]

York University, Dept. of Biology, Downsview
Ontario, Canada, M3J 1P3: [*] University of Toronto, Dept.
of Zoology, Toronto, Ontario, Canada, M5S 1A1

The elevated haemolymph lipid levels characteristic of locust flight are hormonally controlled (see CHESSEMAN and GOLDSWORTHY, 1979), and a peptidergic adipokinetic hormone (AKH I) derived from the glandular lobe has been characterised, sequenced and synthesised (see BROOMFIELD and HARDY, 1977). CHEESEMAN and GOLDSWORTHY (1979) demonstrated the presence of an AKH in the haemolymph of flown locusts, and also measured the kinetics of hormone release during flight . However, the interpretation of their data has now been complicated by the recent finding of a second adipokinetic hormone (AKH II) in the glandular lobe (CARLSEN et al., 1979) and the demonstration that both AKH I and II are released during a 30 min flight (ORCHARD and LANGE, 1983a,b). A further complication arises because of the report that octopamine, which has been shown to have a direct adipokinetic action on the fat body (ORCHARD et al., 1982), is also released during the initial stages of flight (GOOSEY and CANDY, 1980).Thus there are at least three hormones released during flight which possess adipokinetic activity, namely AKH I, AKH II and octopamine.

It became apparent, that a re-examination of the time courses of release of AKH I and II during flight in Locusta was pertinent, and that a detailed comparison between the published release of octopamine, and the elevation in haemolymph lipid was required. Thus, day 15 adult virgin female locusts were flown for varying periods of time, and the haemolymph lipid measured. Methanolic extracts of haemolymph were also taken and subjected to TLC (ORCHARD and LANGE, 1983a) for separation of AKH I and II. Adipokinetic activity of these two TLC fractions was assayed in day 10 adult male locusts as previously described (ORCHARD and LANGE, 1983b).

A model for the hormonal control of haemolymph lipid during flight, incorporating the findings of this study, is presented in Figure 1. It is

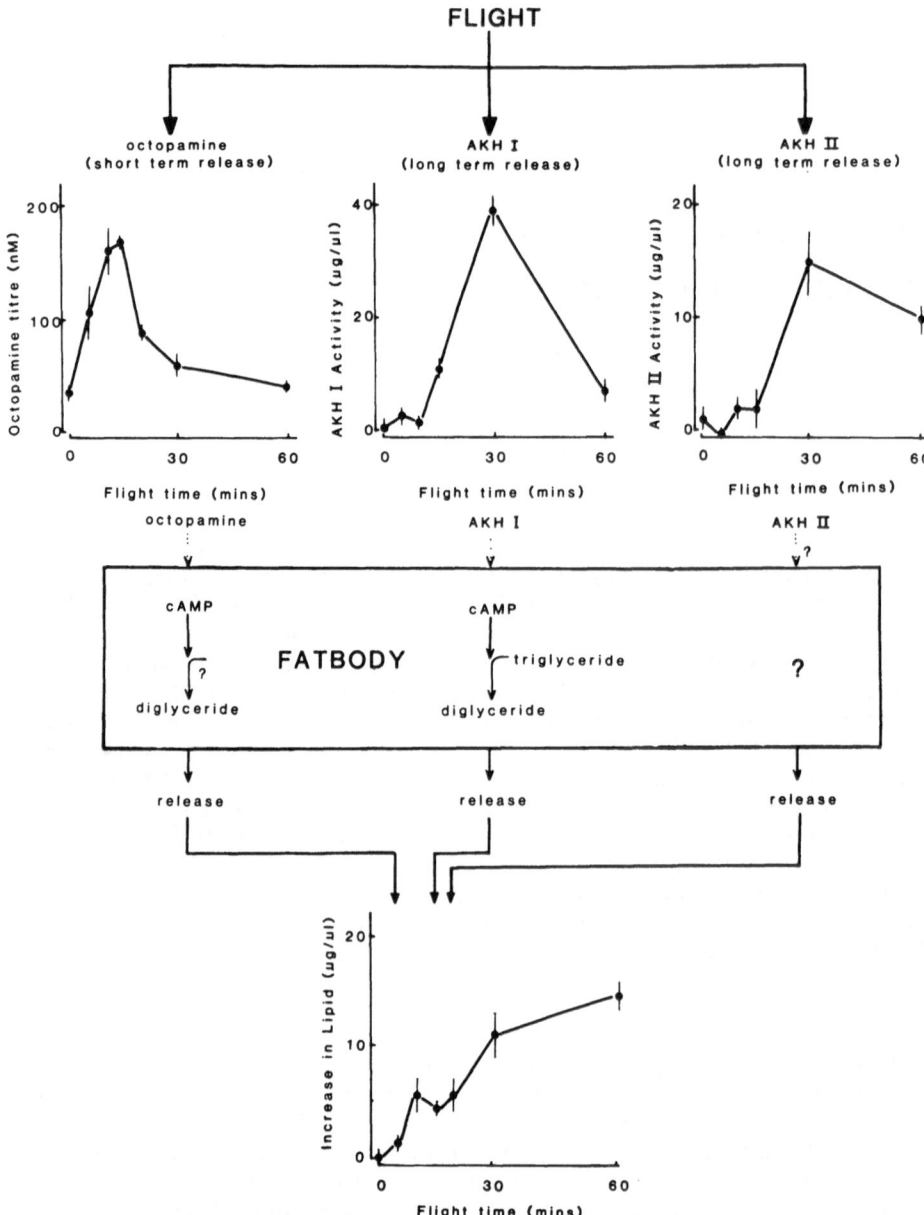

FIGURE 1. Events associated with flight in Locusta. Explanation in text. Octopamine titre redrawn from GOOSEY and CANDY (1980). AKH I and AKH II titre redrawn from ORCHARD and LANGE (1983b); AKH activity shown is increase in haemolymph lipid. Increase in haemolymph lipid during flight redrawn from ORCHARD and LANGE (1983b). Fat body model derived from ORCHARD et al. (1982); SPENCER and CANDY (1976).

apparent that haemolymph lipid increases biphasically during flight. Similar results were obtained by CHEESEMAN and GOLDSWORTHY (1979) although they chose to interpret the rise as a smooth one. Thus there is a rapid elevation in lipid evident between 5 and 10 mins after the start of flight which plateaus before a second rise which occurs after 20 - 30 mins. Of interest to us was the fact that this initial elevation in lipid occurred prior to the appearance of any detectable levels of haemolymph AKH I and II. In fact the release of AKH I is only evident after 10 - 15 mins of flight . There is then a rapid elevation between 10 - 30 mins, with the titre declining precipitously by 60 min. AKH II on the other hand is released after 15 - 30 min and remains relatively constant for a further 30 min. The appearance of these two AKH's coincided with the second phase of lipid elevation which commences after about 20 mins of flight. Thus it appears that neither of these hormones is responsible for the initial elevation in lipid level. A possible explanation for this initial rise became apparent when GOOSEY and CANDY (1980) demonstrated a short - lived elevation in haemolymph octopamine during the first 10 mins of flight in Schistocerca (see Fig. 1). Since ORCHARD et al. (1982) had shown octopamine to have a direct action upon the fat body via cAMP to release lipid, a correlation between the timing of octopamine elevation and the initial rise in haemolymph lipid became apparent. The plateau level of haemolymph lipid is maintained until AKH I and then AKH II are released. AKH I, like octopamine, acts on the fat body via cAMP (SPENCER and CANDY, 1976) resulting in a further elevation in haemolymph lipid. Presumably AKH II acts in a similar manner. Therefore during flight the increased lipid level, resulting in availability of fuels for flight, are a product of the integrated activities of octopamine, AKH I and AKH II.

REFERENCES

BROOMFIELD C. E. and HARDY P. M. (1977) The synthesis of locust adipokinetic hormone. Tetrahedron Lett. 25: 2201-2204.

CARLSEN J., HERMAN W. S., CHRISTENSEN M. and JOSEFSSON L. (1979) Characterisation of a second peptide with adipokinetic and red pigment-concentrating activity from the locust corpora cardiaca. Insect Biochem. 9: 497-501.

CHEESEMAN P. and GOLDSWORTHY G. J. (1979) The release of adipokinetic hormone during flight and starvation in Locusta. Gen. Comp. Endocrinol. 37: 35-43.

GOOSEY M. W. and CANDY D. J. (1980) The D-octopamine content of the haemolymph of the locust Schistocerca gregaria and its elevation during flight. Insect Biochem. 10 393-397.

ORCHARD I. and LANGE A. B. (1983a) Release of identified adipolinetic hormones during flight and following neural stimulation in Locusta migratoria. J. Insect Physiol. 29: 425-429.

ORCHARD I. and LANGE A. B. (1983b) The hormonal control of haemolymph lipid during flight in Locusta migratoria. J. Insect Physiol. (IN PRESS)

ORCHARD I., CARLISLE J.C., LOUGHTON B. G., GOLE J. W. D. and
    DOWNER R.   G. H.(1982)   In vitro studies on the effects of
    octopamine on locust fat body. Gen. Comp. Endocrinol. 48: 7-13.
SPENCER I. M. and CANDY D. J. (1976) Hormonal control of diacyl glycerol
    mobilisation from fat body of the desert locust, Schistocerca gregaria.
    Insect Biochem. 6: 289-296.

ENDOCRINE-LIKE CELLS IN THE MIDGUT OF AEDES AEGYPTI

Arden O. Lea and Mark R. Brown

Department of Entomology
University of Georgia
Athens, Georgia 30602

In the mosquito, Aedes aegypti, the trophocytes of the fat
body, which do not synthesize vitellogenin as long as the female
is fed sugar, begin vitellogenin synthesis within one hour after
taking blood (Raikhel & Lea, 1983). This synthesis is dependent
on the release of a neurosecretory hormone from the brain (Lea,
1967), and we have been attempting to identify the signal that
initiates its release. Because release must be very rapid, one
possibility is that the neurosecretory system is responding to
a chemical signal directly from the midgut.

Cells, with the ultrastructural characteristics of peptide
hormonogenic cells found in the vertebrate gut (Solcia et al.,
1981), are dispersed in the midgut of several insects (Nishiitsutsuji-Uwo and Endo, 1981), including A. aegypti (Hecker et al.,
1971). Also, investigators have shown, by immunocytochemistry
(Iwanaga et al., 1981; Duve and Thorpe, 1982) and radioimmune
assay (Tager and Kramer, 1980; Kramer et al., 1982), that such
vertebrate-like peptides as pancreatic polypeptide, glucagon/
glicentin, and insulin are present in the midgut of various
insects. Consequently, we have begun a systematic study of the
endocrine-like cells in the midgut of the female mosquito to
determine their distribution, ultrastructural characteristics,
and immunoreactivity to vertebrate peptide hormones.

The midgut epithelium consists of digestive cells, and
scattered regenerative and endocrine-like cells; all are separated
from the hemolymph by a thin basal lamina. The endocrine-like
cells are smaller (2-8 µm in width) than the digestive cells,
positioned basally in the epithelium, and often flasked-shaped.
Sometimes, the cells have a thin elongate neck, extending to the

midgut lumen, capped with a small tuft of microvilli. Frequently, lateral extensions of the basal part of a cell contact several digestive cells. The basal plasma membrane of the endocrine-like cells lies smoothly along the basal lamina, without forming a basal labyrinth that is characteristic of digestive cells. The nucleus fills most of the cell, in contrast to the digestive cells. Usually the cells are solitary, but occasionally they are associated with another endocrine-like cell or a regenerative cell. We estimate that there are at least 200 endocrine-like cells in the anterior (thoracic) and posterior (abdominal) regions of the midgut.

The most notable characteristic of an endocrine-like cell is the presence of secretory granules (approx. 100 nm in dia.) concentrated in the basal area of the cell. These granules are formed by the cell in a manner typical of cells that export peptides or proteins. Along the maturing face of Golgi complexes, the bulbous ends of the cisternae become filled with a dense matrix and separate from the cisternae as membrane-bound secretion granules. The granules release their contents into the intercellular space by exocytosis, as indicated by our observations of the membrane of granules fused to the plasma membrane and the outline of granules in the plasma membrane. Thus, the secretory product may contact neighboring digestive cells or cross through the basal lamina into the circulating hemolymph.

With light microscopy, we have observed cells that are immunoreactive to antisera against bovine pancreatic polypeptide [Bouin's fixative; Araldite embedment; primary antibody, 1:1000 (a gift of Dr. Chance, Eli Lilly Co.); secondary antibody conjugated to biotin, 1:500; avidin conjugated to peroxidase, 1:2500]. These cells resemble the endocrine-like cells just described, in that they are smaller than and positioned basally among the digestive cells of the posterior midgut. Currently, we are continuing to test antisera to other vertebrate peptides and are combining immunocytochemistry with electron microscopy to determine the specific location of the vertebrate-like peptides within the immunoreactive cells.

REFERENCES

DUVE H. and THORPE A.   (1982) The distribution of pancreatic polypeptide in the nervous system and gut of the blowfly, Calliphora vomitoria (Diptera). Cell Tissue Res. 227, 67-77.

HECKER H., FREYVOGEL T. A., BRIEGEL H., and STEIGER R. (1971) Ultrastructural differentiation of the midgut epithelium in female Aedes aegypti L. (Insecta, Diptera) imagines. Acta Trop. 28, 80-104.

IWANAGA T., FUJITA T., NISHIITSUTSUJI-UWO J., and ENDO Y. (1981) Immunohistochemical demonstration of PP-, somatostatin-, and enteroglucagon-immunoreactivities in the cockroach midgut. Biomed. Res. 2, 202-207.

KRAMER K. J., CHILDS C. N., SPIERS R. D., and JACOBS R. M. (1982) Purification of insulin-like peptides from insect hemolymph and royal jelly. Insect Biochem. 12, 91-98.

LEA A. O. (1967) The medical neurosecretory cells and egg maturation in aedine mosquitoes. J. Insect Physiol. 13, 419-429.

NISHIITSUTSUJI-UWO J. and ENDO Y. (1981) Gut endocrine cells in insects: the ultrastructure of the endocrine cells in the cockroach midgut. Biomed. Res. 2, 30-44.

RAIKHEL A. S. and LEA A. O. (1983) Previtellogenic development and vitellogenin synthesis in the fat body of a mosquito: an ultrastructural and immunocytochemical study. Tissue and Cell 15, 281-300.

SOLCIA E., CAPELLA C., BUFFA R., USELLINI L., FIOCCA R., and SESSA F. (1981) Endocrine cells of the digestive system. In Physiology of the Gastrointestinal Tract (Edited by JOHNSON L. R.) Vol. 1, pp. 39-58. Raven Press, New York.

TAGER H. S. and KRAMER K. J. (1980) Insect glucagon-like peptides: evidence for a high-molecular weight form in midgut from Mannduca sexta (L.). Insect Biochem. 10, 617-619.

CHOLINERGIC TITERS IN THE BRAIN OF <u>MANDUCA</u> <u>SEXTA</u> DURING LARVAL-

PUPAL DEVELOPMENT

D. S. Lester and L. I. Gilbert

Department of Biology
University of North Carolina
Chapel Hill, NC 27514

The larval brain of <u>Manduca</u> <u>sexta</u> possesses high cholinergic
activity which was characterized and titered for the last two larval
instars to determine if a correlation existed between the various
components of the cholinergic system and prothoracicotropic hormone
(PTTH) release. Initial screening revealed that cholinergic activity
was 50 to 100 times greater than that of any other neurotransmitter.
Isolated brains were incubated in medium containing $[^3H]$-choline and
choline uptake and acetylcholine (ACh) accumulation were measured
using high voltage paper electrophoresis. The level of choline accumulation remained relatively constant except for the last 3 days
before pupation when the brain grew rapidly in size. Numerous fluctuations in ACh accumulation occurred that did not correlate with
brain growth. The levels, intensity and frequency of ACh accumulation
increased during the fourth and fifth instars at times of PTTH release. During the head critical periods (HCP) signalling molting
in the fourth and fifth instars, there were 30 to 40% increases in
brain ACh accumulation lasting 6 to 8 hr. Analyses of the first HCP
of the fifth instar revealed 3 short increases, 4 to 6 hr, of lesser
magnitude that correlated with the 3 pulsatile releases of PTTH at
that time (Bollenbacher and Gilbert, personal communication). Such
abbreviated, temporal changes in ACh levels have not been reported
previously.

A sensitive cell-free radioenzymatic assay was employed to
measure the level of choline acetyltransferase (ChAT) activity during
the last two larval instars. No specific changes in activity were
found at times of PTTH release; its activity generally reflected the
growth of the brain. Acetylcholinesterase (AChE) activity was 5-8
times greater than ChAT activity in cell-free brain extracts as determined radioenzymatically. The level of activity remains relatively

417

constant until the last 3 days of the fifth instar when activity increases along with brain protein content. Specific fluctuations in
AChE were observed at the times of increased ACh accumulation; e.g.,
at the time of the fourth instar HCP and the second HCP of the fifth
instar. The nadir and peaks in AChE activity at the first HCP of
the fifth instar correlate respectively with the increases and decreases in ACh accumulation in the brain, i.e., when AChE activity
decreased, ACh accumulation increased.

The relative number of nicotinic binding sites was estimated
by determining the concentration of α-bungarotoxin binding sites in
the developing brain. During the HCP of the fourth instar and
second HCP of the fifth instar a decrease in binding sites was
noted, while large increases correlated with the release patterns
of PTTH and the increases in ACh accumulation evident during the
first HCP of the fifth instar.

Although the above does not prove that the cholinergic system
is involved in PTTH release, this possibility does exist and is
under investigation.

PROGRESSIVE CHANGES IN STAINABLE NEUROSECRETION IN BRAINS OF HELIOTHIS VIRESCENS THROUGHOUT THE LAST LARVAL INSTAR[1]

Marcia J. Loeb and James Dodson

Insect Reproduction Laboratory, ARS, USDA
Beltsville, MD USA

Changes in neurosecretory patterns in insect brain cell have been correlated with events such as molting, diapause, and egg development (reviewed by Raabe, 1982). This study provides a view of changing patterns of paraldehyde fuchsin (PAF) stainability in brain cells, which we interpret as changes in neurosecretory activity, and of the transport of PAF-stainable material in axons during the last instar of larval life of the tobacco budworm, Heliothis virescens.

Larvae were raised on artificial diet at 30°C and staged daily during the last instar as early (N, newly molted, S, slender, and P, puffy), mid instar (dig holes in medium and grow rapidly) and late instar (burrow into medium prior to pupation). Ten brains of each sex at each stage were fixed in Bouin's fluid, embedded in paraffin, cut at 5μm sections, and stained in Ewen's modification of PAF; a cell was considered neurosecretory (NS) if it stained dark purple (PAF-positive) (Ewen, 1962). Although this method may not reveal all NS cells (Raabe, 1982), it provides considerable indication of NS activity.

Brains of newly molted fifth instar larvae of H. virescens contained few faintly PAF-positive dorsolateral and dorsomedial cell bodies and axonal tracts indicating some NS activity (Fig. 1). It is probable that the ecdysteroid synthesis-inducing peptide, prothoracicotropic hormone (PTTH), was included in exported NS, since low titers of hemolymph ecdysteroids occurred then (Loeb, 1982). One day later (stage S), dorsomedial and medial cells in brains of both sexes (Fig. 2), and ventromedial cells of female brains were

---

[1] Mention of a commercial or proprietary product in this paper does not constitute an endorsement of this product by the USDA.

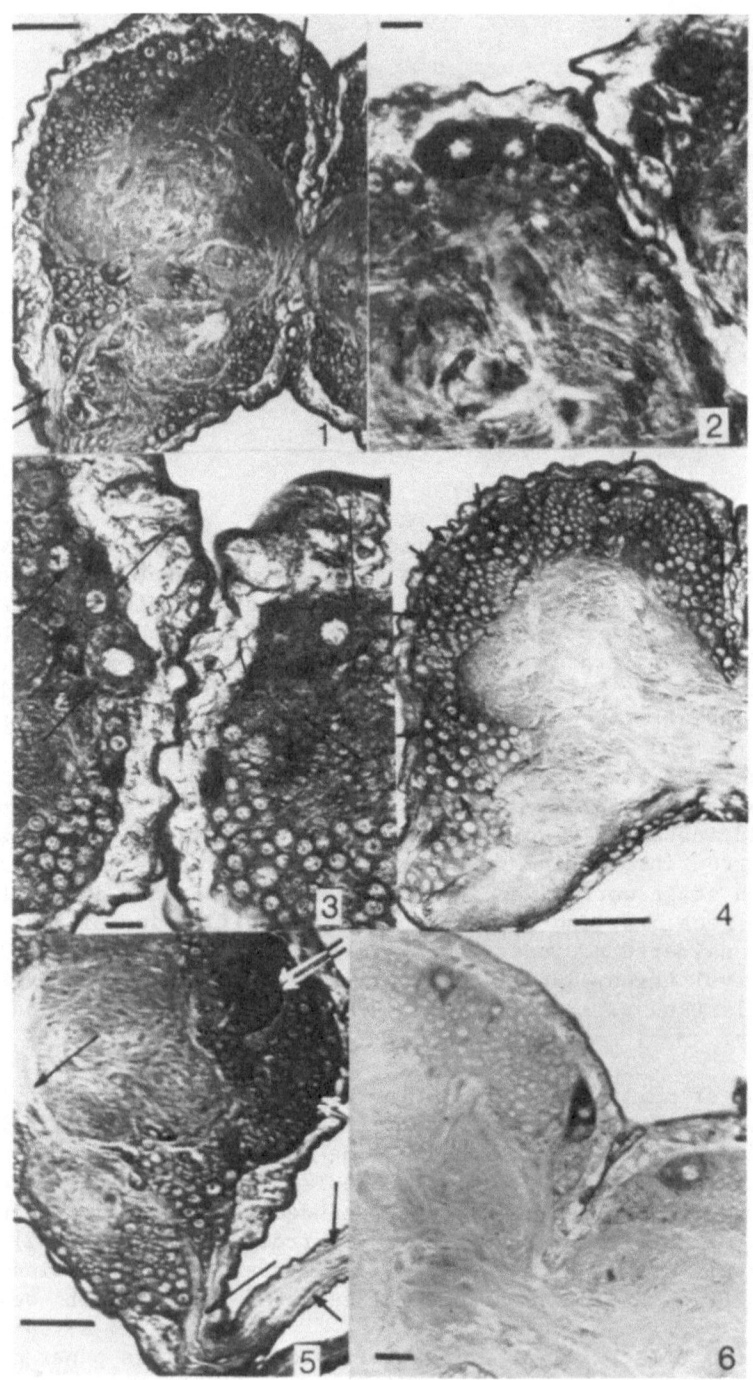

Figures 1-6.

enlarged and stained intensely with PAF; axons stained lightly. Synthesis, with predominant sequestration of NS within brain cells was indicated. Little, if any, PTTH was released, since ecdysteroid was not detectable in hemolymph at this time (Loeb, 1982). Next day (P) the large cells were visible but essentially empty of PAF-stainable material (arrows, Fig. 3); other cells near them stained, and axons were dark purple. Therefore, a massive release of NS from the "empty" cells was indicated at this stage. It corresponded to renewed appearance of ecdysteroid in the hemolymph as well as the subsequent change to the active, rapidly growing animal in mid-instar. In mid-instar, dorsomedial and dorsolateral cells corresponding to those seen earlier, as well as numerous groups of large (Fig. 4) and very small (double arrows, Fig. 5) newly PAF-responsive cells were stained (arrows, Fig. 4). PAF-positive medial cells were seen in male brains. Thus, much brain NS activity and transport occurred in mid-instar. Insect NS substances regulate a wide range of metabolic, enzymatic and muscular activities (review by Raabe, 1982) and must prevail in mid-instar. In contrast, most cortical cells in brains of first day late instar larvae stained weakly if at all, although axons remained dark, and by second day only a few enlarged dorsomedial and dorsolateral cells of male brains (Fig. 6) and medial cells of female brains were PAF-positive. Nerve tracts no longer stained with PAF. These events in late last instar may reflect slowing metabolism prior to pupation. Surprisingly, hemolymph ecdysteroid titers were 10 times previous levels despite the apparent lack of outward NS transport. Therefore, PTTH affecting late instar ecdysteroid titers may have already been released from the brain. Remaining PAF-positive cells may indicate accumulation of secretions for later release.

References

EWEN A. B. (1962) An improved aldehyde fuchsin staining technique for neurosecretory products in insects. Trans. Am. Micr. Soc. 81, 94-96.

LOEB M. J. (1982) Diapause and development in the tobacco budworm, Heliothis virescens: a comparison of haemolymph ecdysteroid titres. J. Insect Physiol. 28, 667-673.

RAABE M. (1982) Insect Neurohormones. Plenum Press, New York.

For clarity, only sections of male brains are shown. Scale bars: 50 μm. Fig. 1: Early stage N; Single arrow: PAF positive dorsomedial and dorsolateral cells; double arrow: PAF-positive axons. Fig. 2: Early stage S, PAF positive dorsomedial cells. Fig. 3: Early Stage P; arrows: 'empty' (PAF-neg.) dorsomedial cells. Fig. 4: Mid-stage; arrows: PAF-positive cells. Fig. 5: Mid stage; white arrows: small PAF-positive cells; black arrows: PAF-positive axons. Fig. 6: Late stage; PAF positive dorsomedial and dorsolateral cells.

NEUROPHYSIN-LIKE MOLECULES IN THE LOCUST

Barry G. Loughton, Nelly Singer and James A. Carlisle

Department of Biology, York University
Downsview, Ontario
Canada   M3J 1P3

Neurophysins are proteins synthesized by the cells of the vertebrate hypothalmus which are alcohol soluble, rich in cysteine, weigh appoximateley 10,000 daltons, bind reversibly to their appropriate neurohormone and are released from the neurosecretory granule with that hormone upon depolarization of the neurosecretory terminal. Though vertebrate and invertebrate neurosecretory cells stain in similar histochemical fashion it is only recently that evidence of neurophysin-like molecules in insects has been forthcoming.

Friedel et al. (1980) isolated a protein from the A cells of the pars intercerebralis of the locust which met several of the criteria set forth above for neurophysins.   Conformation with these criteria were extended by Orchard et al.   (1981).   Nevertheless the suposed prime function of the neurophysin, that of binding reversibly to the neurohormone and thereby stabilizing it within the neurosecretory granule could not be demonstrated, mainly because no insect hormone from the pars intercerebralis had been completely characterized. In this paper we will adduce indirect evidence of the association of neurophysin from the pars intercerebralis and a hormone and direct evidence of the binding of adipokinetic hormone (AKH) to a neurophysin-like molecule. Electrophoresis of alcohol extracts of storage lobe of the corpus cardiacum (CC) of locust yields two major fractions (fig. 1). Elution of these fractions (MW 10,000) and subsequent testing of the eluates for hypolipaemic activity demonstrated that they contain hypolipaemic activity. Hypolipaemic hormone can be shown to have a molecular weight of 1,000 by gel filtration on biogel P2. If the glandular lobes of the CC were treated in like fashion (fig.1) hyperlipaemic activity could be shown to be associated with the major protein fractions.   Adipokinetic hormone itself does not migrate in the electrophoretic field.   Thus hormones of low molecular weight are found associated with neuroproteins of 10,000 MW (table 1).

The two major alcohol soluble proteins of the glandular lobe are released from the cells by depolarization with 40 mm K^+ or octopamine. The release is Ca^{++} dependent (fig.2). Hyperlipaemic activity is also released from the glandular lobe of the CC by depolaization in high K^+ (ORCHARD AND LOUGHTON, 1981). Thus the cells of the glandular lobe of the CC contain an alcohol soluble molecule, 10,000 MW, which is released together with hyperlipaemic activity by 40 mm K^+ or octopamine. Thus these molecules also appear to behave like neurophysins.

Direct evidence of the association between synthetic AKH and the glandular lobe neuroprotein was obtained by immobilizing the neuroprotein on Sepharose-CN-Br. Known concentrations of AKH were incubated with the Sepharose + neuroprotein and the supernatant then tested for hypolipaemic activity. Controls were Sepharose-CN-Br to which storage lobe neuroprotein or bovine serum albumin had been coupled. Table 2 shows that hypolipaemic activity was reduced only after incubation with glandular lobe neuroproteins. The Sepharose was washed several times with saline and then incubated with 0.1 N HCl. Hypolipaemic activity was recovered only from the Sepharose coupled to glandular lobe neurophysin, thus demonstrating the reversible nature of the association between hormone and neuroprotein.

We believe that our evidence further strengthens the case for the presence of neurophysins in insects. The recognition of such molecules may provide the means for analysis of the synthesis of peptidergic neurohormones in insects.

Table 1
The effect of injection of extracts of fractions of storage and glandular lobe extracts separated by polyacrylamide gel electrophoresis.

fraction	% Decrease in haemolymph lipid (N=10)	% Increase in haemolymph lipid (N=4)
B_1	60	0
B_2	49	0
B_3	—	80

B_1 and B_2 are storage lobe neuroproteins eluted from polyacrylamide gels.
B_3 is glandular lobe neuroprotein eluted from polyacrylamide gels (see fig.1).

Table 2

The adipokinetic activity of the supernatant of Sepharose CN-Br co coupled to (1) glandular lobe neuroprotein, (2) storage lobe neuroprotein or (3) bovine serum albumin. 100 ul of synthetic AKH solution (0.12 ng/ul) was added to each preparation. Four hours later the supernatant was assayed for AKH activity. The supernatant was decanted and the Sepharose washed in saline then incubated for 4 hours in 100 ul 0.1 N HCl. The HCl supernatant was tested for adipokinetic activity.

	% Increase in Haemolymph Lipid of Assay Locusts	
	Incubation with AKH Solution	Incubation with 0.1N HCl
Glandular lobe neuroprotein	13	82
Storage lobe neuroprotein	82	0
Bovine Serum	102	0

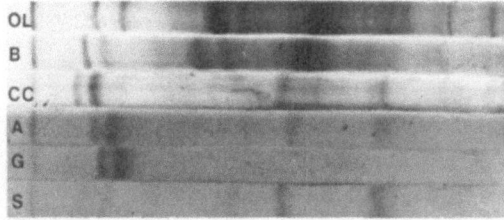

Fig. 1 Electropherogram of: extracts of OL- optic lobe (saline), B- Brain (saline), CC- corpus cardiacum (saline), A- corpus cardiacum (ethyl alcohol) G- glandular lobe (ethyl alcohol) and S- storage lobe (ethyl alcohol).

Fig. 2 Electropherogram of : E- extract of corpus cardiacum (CC) in 40 uM K^+ saline, W- normal saline perfusate of CC, CK- 40 mM K^+ saline minus Ca^{++} perfuate of CC, K- 40 mM K^+ saline perfusate of CC, O- normal saline + 10^{-6}mM octopamine perfusate of CC.

REFERENCES

FRIEDEL T., LOUGHTON B. G. AND ANDREW R. D. (1980) A neurosecretary protein from **Locusta migratoria. Gen. Comp. Endocrinol. 41,** 487-498

ORCHARD, I. AND LOUGHTON, B.G. (1981) The neural control of the release of hyperlipaemic hormone from the corpus cardiacum of **Locusta migratoria. Comp. Biochem. Physiol. 68A** 25-30

ORCHARD I., FRIEDEL T., AND LOUGHTON B.G. (1981) Release of neurosecretory protein from the corpora cardiaca of **Locusta migratoria** induced by high potassium saline and compound action potentials. **J. Insect Physiol. 27,** 297-304

PREPARATION OF EGG DEVELOPMENT NEUROSECRETORY HORMONE USING

REVERSE-PHASE LIQUID CHROMOTOGRAPHIC TECHNIQUES

Edward P. Masler[1], Henry H. Hagedorn[2], David H. Petzel[2]
and Alexej B. Borkovec[1]

[1]Insect Reproduction Laboratory, ARS, USDA, Beltsville, MD*
[2]Dept. of Entomology, Cornell University, Ithaca
NY 14853

Schneiderman and Gilbert (1964) illustrated the relationship between a brain hormone and ecdysteroid production by the prothoracic gland. This ecdysiotropic factor, or prothoracicotropic hormone (PTTH), has since been under intense study (Ishizaki and Suzuki, 1980; Gilbert et al., 1981). For the past 10 years, the steroidotropic factor from the heads of mosquitos, egg development neurosecretory hormone (EDNH, Lea, 1972) has also received much attention (Hagedorn et al., 1979; Hanaoka and Hagedorn, 1980; Masler and Hagedorn, 1981; Masler et al., 1983). Most significant to the isolation and analysis of these ecdysiotropic neuropeptides has been the development of reliable and sensitive in vitro bioassays for the detection of EDNH (Hagedorn et al., 1979) and PTTH (Bollenbacher et al., 1979). Of potentially equal significance will be the application of micro methods of protein analysis and separations, such as reverse-phase high-performance liquid chromatography (RP-HPLC), which will expedite the isolation of minute quantities of neuropeptides (Moore, 1982). We have combined the in vitro bioassay with RP-HPLC in our efforts to isolate EDNH. Aedes aegypti heads were collected and extracted in Aedes saline essentially as described by Hagedorn et al. (1979) except that heat-precipitation of crude head extract was replaced by processing of the extract on a low-pressure, C_{18}-silica cartridge (Waters Assoc.). This was essentially a solvent exchange step which removed large proteins, lipids and salts from the active sample (Bennett et al., 1981). Fractions were eluted from the cartridges using a step gradient of acetonitrile (CH_3CH) in water. The eluants were lyophilized, reconstituted in Aedes saline and EDNH activity was detected using the in vitro bioassay. EDNH activity eluted between 20% and 60% CH_3CN (Fig. 1). In subsequent

---

*Mention of a commercial or proprietary product in this paper does not constitute an endorsement of this product by the USDA.

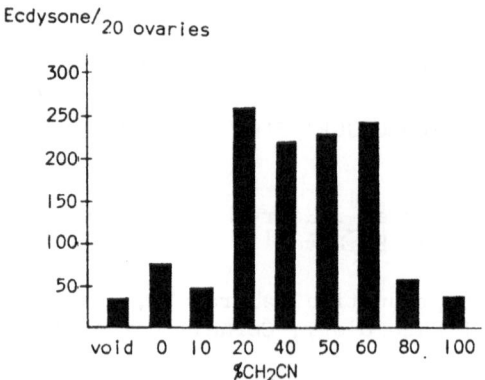

Fig. 1 Step-wise, low-pressure elution of EDNH activity from a
C_{18} reverse phase cartridge. EDNH activity expressed as
PG ecdysone produced by 20 A Aegypti ovaries per 6 hrs.

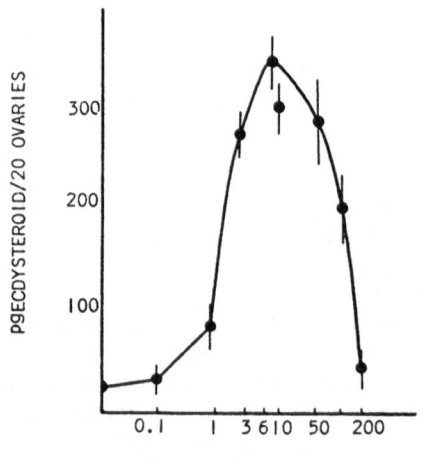

Fig. 2 Dose response of ecdysteroid production (±SE) by ovaries
from sugar-fed females exposed to head extract prepared by
low-pressure reverse phase chromatography. Dose expressed
as head equivalents in 50μl incubation.

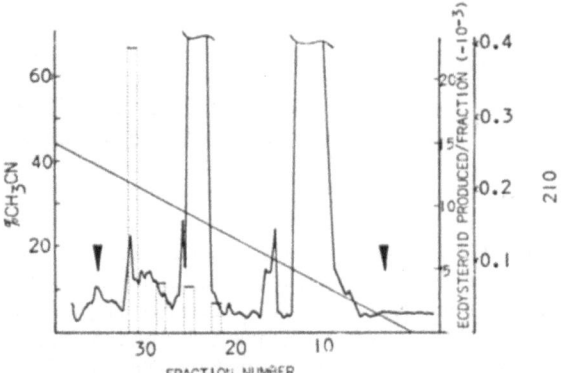

Fig. 3 Elution profile of EDNH activity on a μBondapak C_{18} column
        (3.9 x 300 mm) with a linear CH_3CN gradient. UV profile,
        solid line; EDNH activity, broken line; Arrows, inhibitor
        activity.

preparations, EDNH activity was routinely eluted with 60% CH_3CN.
When the active 60% fraction was tested in vitro, ovaries responded
in a dose-dependent manner (Fig. 2). High concentrations of the
eluted extract inhibited ecdysteroid production (Masler et al.,
1983) reminiscent of the inhibition observed by Hagedorn et al.
(1979) with crude head extract. For further purification, the 60%
CH_3CN eluant was lyophilized and dissolved in RP-HPLC starting solvent (0.075% trifluoroacetic acid, TFA) and fractionated on an analytical μBondapak C_{18} column (3.9 x 300 mm, Waters Assoc.) using a
gradient (0.075% TFA to 60% CH_3CN in 0.075% TFA) at a rate of increase of 0.56% CH_3CN/min. EDNH activity eluted near 35% CH_3CH
(Fig. 3). These methods were successfully applied to a preparative
scale μBondapak column (7.8 x 300 mm) which allowed processing of
50,000 heads/extraction. EDNH activity eluted in two fractions near
35% CH_3CN. Aliquots of each fraction were chromatographed using a
linear CH_3CN gradient on an analytical Aquapore RP-300 C_{18} column
(4.6 x 250 mm; Brownlee Labs). The EDNH activities from both fractions co-eluted, suggesting the presence of a single EDNH. When
samples of the preparative fractions were chromatographed isocrati-
cally at 30% CH_3CN on the RP-300, all EDNH activity eluted primarily
in a single peak, again suggesting a single EDNH. This activity was
sensitive to proteolysis by subtilisin. We are now attempting to increase yield to facilitate biochemical characterization of the peptide. In addition, we have found that RP-HPLC efficiently separates
(Fig. 3) the inhibitory activity observed in low-pressure fractions
(Fig. 2), from EDNH activity. The inhibitor elutes near 5% CH_3CN
and, to a lesser extent, near 40% CH_3CN (Fig. 3). We do not yet
know the chemical nature of the inhibitor but suspect it might be
an amine (Masler et al., 1983). There is precedent for the inhibition of ecdysteroid production in insects (Carlisle and Ellis, 1968)
and crustaceans (Kleinholz, 1976) by neural factors and we are exploring this observation further. It appears that RP-HPLC will be of
value not only to the isolation of insect steroidogenic neurohormones
but also to the isolation of their modulators.

REFERENCES

BENNETT H., BROWN C., BRUBAKER P. and SOLOMON S. (1981) A com-
    prehensive approach to the isolation and purification of peptide
    hormone using only reverse-phase liquid chromatography. In
    Biological/Biomedical Applications of Liquid Chromatography.
    (Ed. by HAWK, G. L.), pp. 197-210, Marcel Dekker, NY.
BOLLENBACHER W., AGUI N., GRANGER N. and GILBERT L. (1979) In
    vitro activities of insect prothoracic glands by the prothoracico-
    tropic hormone. Proc. Nat. Acad. Sci. USA 76, 5148-5142.
CARLISLE D. and ELLIS P. (1968) Hormonal inhibition of the pro-
    thoracic gland by the brain in locusts. Nature 220, 706-707.
GILBERT L., BOLLENBACHER W., AGUI N., GRANGER N., SEDLAK B., GIBBS
    D. and BUYS C. (1981) The prothoracicotrpies: Source of the
    prothoracicotropic hormone. Amer. Zool. 21, 641-653.
HAGEDORN H., SHAPIRO J., and HANAOKA K. (1979) Ovarian ecdysone
    secretion is controlled by a brain hormone in an adult mosquito,
    Nature 282, 92-94
HANAOKA K. and HAGEDORN H. (1980) Brain hormone control of ecdy-
    sone secretion by the ovary in a mosquito. In Progress in Ecdy-
    sone Research (Ed. by HOFFMANN, J.), pp. 467-480, Elsevier/North
    Holland, New York.
ISHIZAKI H. and SUZUKI A. (1980) Prothoracicotropic hormone. In
    Neurohormonal Techniques in Insects (ed. by MILLER, T.), pp.
    244-276, Springer-Verlag, New York.
KLEINHOLZ L. (1976) Crustacean neurosecretory hormones and physio-
    logical specificity. Amer. Zoo. 16, 151-166.
LEA A. (1972) Regulation of egg maturation in the mosquito by the
    neurosecretory system: the role of the corpus cardiacum. Gen.
    Comp. Endocrinol. Suppl. 3, 602-608.
MASLER E. and HAGEDORN H. (1981) Reverse-phase chromatography of
    a steroidogenic neuropeptide from mosquitos. First Int. Symp.
    HPLC Prot. Pep. Varian Assoc. Abs. 511.
MASLER E., HAGEDORN H., PETZEL D. and BORKOVEC A. (1983) Partial
    purification of egg development neurosecretory hormone with re-
    verse-phase liquid chromatographic techniques. Life Sci. (in
    press).
MOORE A. G. (1982) Reversed phase high pressure liquid chromato-
    graphy for the identification and purification of neuropeptides.
    Life Sci. 30, 995-1002.
SCHNEIDERMAN H. and GILBERT L. (1964) Control of growth and de-
    velopment in insects. Science 143, 325-333.

USE OF HPLC TO PURIFY PHYSIOLOGICALLY ACTIVE PEPTIDES IN HEAD EXTRACTS OF THE MOSQUITO AEDES AEGYPTI

E. P. Masler, D. H. Petzel, J. C. Williams, G. Wheelock, and H. H. Hagedorn

Department of Entomology, Cornell University, Ithaca NY 14853

Although several neurosecretory hormones are thought to exist in insects, few have been characterized chemically because of the large amounts required and the absence of rapid and reliable bioassays. The development of HPLC for peptide purification and the use of in vitro bioassays has had a dramatic effect on this field. We here describe our work on the purification of two neurosecretory hormones from the heads of mosquitoes: the egg development neurosecretory hormone (EDNH) (Lea, 1972; Hanaoka and Hagedorn, 1980) and the diuretic factor (DF) (Williams et al., 1983).

Assays: EDNH stimulates ecdysone production by ovaries and we established an in vitro assay for this hormone using incubated ovaries followed by analysis of the incubation medium using the ecdysone RIA (Hagedorn et al., 1979). We established an assay for DF using the change in transepithelial potential difference across the malpighian tubule as shown by Williams (1983).

Purification of EDNH and DF: Extracts of heads were prepared in saline and after centrifugation were applied to a low pressure C_{18} column. Activity was eluted with 60% CH_3CN, lyophilized, applied to a C_{18}-HPLC column and eluted with a 0-60% CH_3CN gradient containing 0.01% CF_3COOH. The elution profile at 210nm is shown in Fig. 1. DF eluted at 25% CH_3CN while EDNH eluted at 34% CH_3CN. A single pass through the HPLC column resulted in a 128-fold purification of EDNH and a yield of about 10%. We found that yields were increased about two-fold by coating glassware with silicon and including 0.1 mg/ml Bacitracin and 0.001% thiodiglycol during the assay. Both factors were stable to boiling and sensitive to proteases.

Peptides in <u>Aedes aegypti</u>

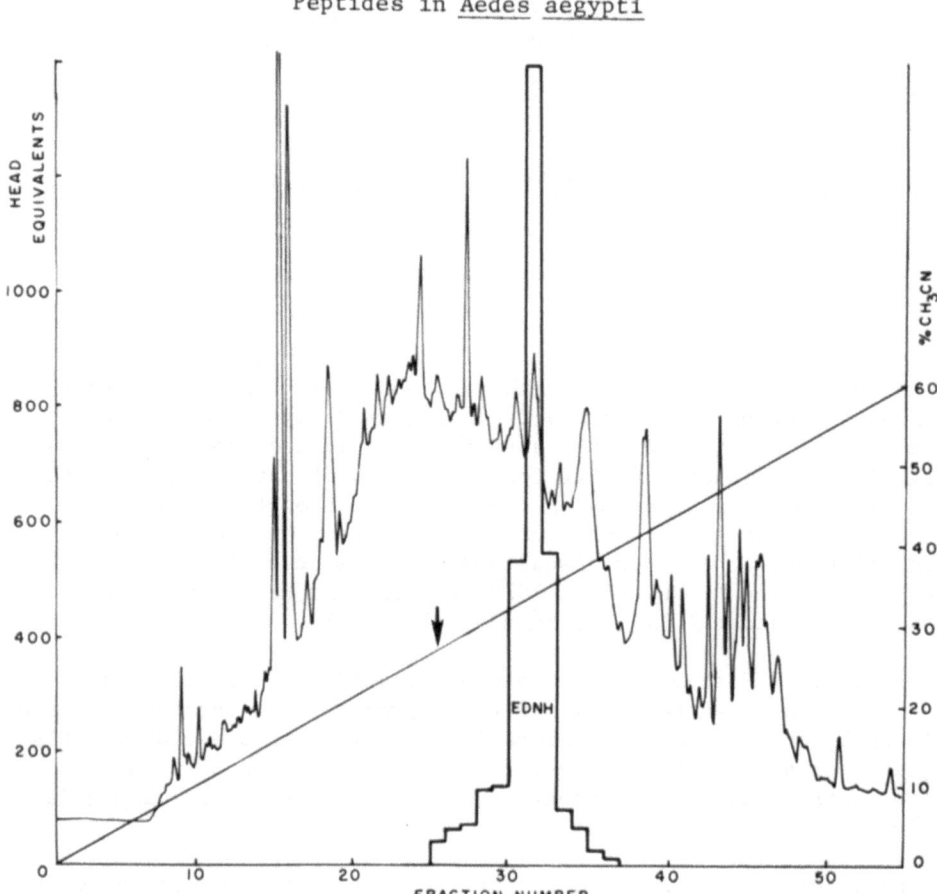

Fig. 1. Scan from the separation of head extract on a C_{18} HPLC
column using a 0-60% CH_3CN gradient. The eluent was scanned at
210 nm. 3 ml fractions were assayed for EDNH activity (histogram)
or DF activity (arrow).

Activity of EDNH and DF in vivo

The activity of both EDNH and DF was assayed by injection into the live animal. Injection of HPLC-purified DF into a non-blood-fed female resulted in a 4-fold increase in the rate of fluid excretion. By contrast, injection of HPLC-purified EDNH into the non-blood-fed female had no effect. However, females that were blood-fed and then decapitated two hours later (before the release of endogenous EDNH) did respond to injected EDNH as demonstrated by increased ecdysteroid levels in whole-body extracts. The need for a blood meal before EDNH is effective is presumably due to events occurring shortly after a blood meal as discussed in Hagedorn (1984).

Discussion: DF and EDNH are apparently released shortly after a blood meal, causing elimination of 40% of the excess water in the meal within one hour (Williams et al., 1983) and production of ecdysone by the ovary, resulting in vitellogenesis (Hagedorn et al., 1979). We have shown that reverse-phase HPLC can be used efficiently to separate physiologically active peptides that apparently regulate these events from crude extracts of whole heads. Biological activity is retained, and the use of volatile buffers allows one to measure activity in several in vivo and in vitro assays.

References

HAGEDORN H. H. (1984) Simplicity vs. complexity, a study of the problems involved in relating in vitro results to the live animal. In Invertebrate Systems in Vitro (Ed. KURSTAK E.), in press.

HANAOKA K. and HAGEDORN H. H. (1980) Brain hormone control of ecdysone secretion by the ovary in a mosquito. In Progress in Ecdysone Research (Ed. HOFFMANN J. A.), Elsevier/North Holland Biomedical Press, Amsterdam.

HAGEDORN H. H., SHAPIRO J. P., and HANAOKA K. (1979) Ovarian ecdysone secretion is controlled by a brain hormone in an adult mosquito. Nature 282, 92-94.

LEA A. O. (1972) Regulation of egg maturation in the mosquito by the neurosecretory system: the role of the corpus cardiacum. Gen. Comp. Endocrinol. Suppl. 3, 602-608.

WILLIAMS J. C. (1983) The malpighian tubule of the yellow fever mosquito: its function in vitro and in vivo. Ph.D. Thesis, Cornell University, Ithaca, NY.

WILLIAMS J. C., HAGEDORN H. H., and BEYENBACH K. W. (1983) Dynamic changes in flow rate and composition of urine during the cost post-bloodmeal diuresis in Aedes aegypti. J. Comp. Physiol., in press.

MORPHOLOGY OF THE NEUROENDOCRINE ELEMENTS OF THE RETROCEREBRAL

SYSTEM OF THE ADULT STABLE FLY, <u>STOMOXYS</u> <u>CALCITRANS</u>

Shirlee Meola

Veterinary Toxicology and Entomology
  Research Laboratory
United States Department of Agriculture
Agricultural Research Service
College Station, TX 77840

The retrocerebral system of the adult stable fly consists of a corpus allatum, a corpus cardiacum and their associated nerves, and in newly emerged flies, the degenerating prothoracic or ecdysial gland. These elements are arranged around the aorta in the typical ring gland of higher Diptera. Thus the corpus allatum lies on the dorso-medial surface of the aorta, connected by a pair of neuro-secretory nerves, the nervi corporis allati (NCA) to the corpus cardiacum beneath the aorta. When present, the prothoracic cells, encircle the NCA. Prior to entering the corpus cardiacum, the neuro-secretory, nervi corporis cardiaci (NCC) fuse with the recurrent nerve from the frontal ganglion, forming the cardiac-recurrent nerve (CR). The name CR was assigned to this compound nerve by Thomsen (1968) in his description of the retrocerebral system of <u>Calliphora</u>. The NCC component of the CR nerve is readily detectable since its products stain with paraldehyde-fuchsin. Fibers of the NCC were traced in the NCA and their terminals in the corpus allatum (Fig. 1) as well as their terminals on the periphery of the corpus cardiacum. The CR nerve extends through the neuropile of the corpus cardiacum then bifercates to form a dorsal and ventral pair of nerves that extend to the hypocerebral and ingluvial ganglia in the thorax. Fibers of the NCC also enter the aorta and ramify between the muscle fibers of the aorta, terminating in the lumen of this organ thus forming the aortic neurohemal site (Fig. 2).
    The corpus cardiacum of the stable fly is a ganglion with a medial neuropile composed of axons of the intrinsic neurosecretory cells and of afferent axons of the CR and hypocerebral ganglion. The intrinsic cells of the corpus cardiacum encircle the periphery of this organ. Their perikarya are ensheathed by layers of glial processes

435

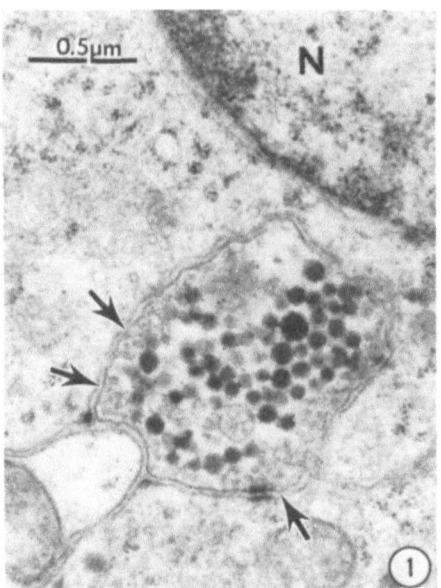 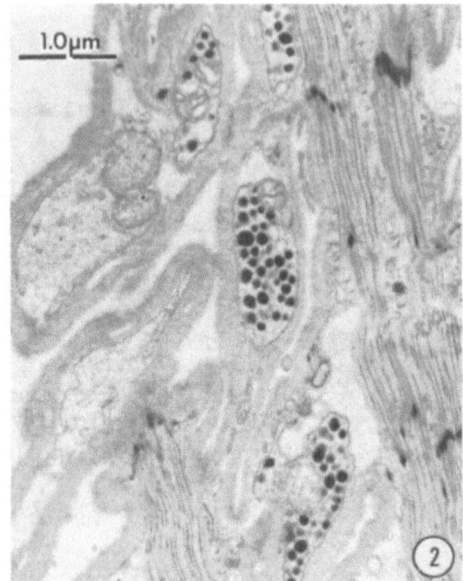

Fig. 1. A neurosecretory fiber of the NCA terminating on the
surface of an endocrine cell of the CA, in close proximity
to the nucleus (N). Synaptoid vesicles (arrows).

Fig. 2. Neurosecretory axons of the NCC extending between the muscle
fibers of the aorta, and terminating in the aortic release
site.

that is discontinuous. Thus portions of the perikarya are covered
only by the fiberous stroma. In these areas, short processes filled
with granules, project from the surface of the perikarya and terminate
in the hemocoel. Thus, unlike intrinsic cells of other species, these
appear to be multipolar. The secretion of these cells is not
fuchsinophilic. The intrinsic cells also form large, relatively short
axons that terminate in the sinuses of the lacunar neurohemal site of
the corpus cardiacum. The lacunae are extensive invaginations of the
stroma into the corpus cardiacum forming hemolymph channels confluent
with the hemocoel. The processes and terminals of the intrinsic
cells are readily distinguishable from other neuroendocrine axons,
since these cells synthesize both a typical spherical, electron dense
granule and an electron dense angular shaped granule that ranges
from square to rectangular profiles (Fig. 3). The proportion of
spherical to angular shaped granules varies with these neurons.
Occasionally a rod-shaped appendage extends from the angular type of
granule resulting in a club-like profile (Fig. 3). This type of
intrinsic neuroendocrine cell has not been previously reported in the
corpus cardiacum of insects, but its secretory granules are

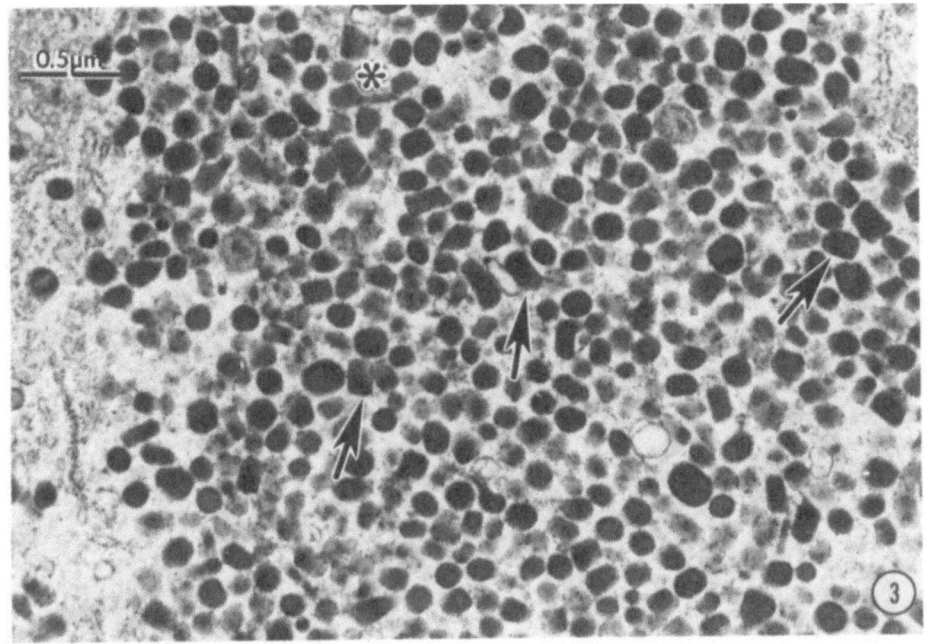

Fig. 3.   An intrinsic cell of the corpus cardiacum of the stable fly
           containing both spherical and angular (arrows) electron
           dense granules.   Club-shaped granule (asterisk).

structurally very similar to the Cr type of granule in the evolved
perisympathetic organs of Coleoptera and Heteroptera, that also
contain irregular shaped granules (see Figs. 16, 17, and 18 in
Baudry-Partiaoglou, 1983).

References

BAUDRY-PARTIAOGLOU N.   (1983)   Ultrastructure of perisympathetic
     organs in insects.   In Neurohemal Organs of Arthropods (Fd. by
     GUPTA A. P.), Charles C. Thomas Publisher, Springfield.
THOMSEN M.   (1968)   The neurosecretory system of the adult Calliphora
     erythrocephala.   IV. A histological study of the corpus cardiacum
     and its connections with the nervous system.   Z. Zellforsch.
     94, 205-219.

RECENT PROGRESS IN THE ISOLATION AND CHARACTERIZATION OF LOCUST

DIURETIC HORMONE

P. J. Morgan and W. Mordue

Department of Zoology
University of Aberdeen
Tillydrone Avenue, Aberdeen, U.K.

Diuretic hormone (DH) is a neurosecretory peptide with a central role to play in the regulation of insect water balance, and although its existence has been known for some time, the structure of this peptide still eludes us. This has in part been due to the lability of the hormone; but in Locusta is probably also due to the small amounts of available starting material, compared with a peptide such as adipokinetic hormone (AKH) (Morgan and Mordue 1983). These problems together with the rather slow, low resolution chromatographic techniques employed up till recently have hampered attempts to successfully isolate this hormone. In our recent work we have been investigating the use of High Performance Liquid Chromatography (HPLC) methods to isolate and purify locust DH. This paper reports on the progress made using these techniques.

In our initial attempts to isolate DH reversed-phase HPLC (using a Shandon ODS column) was employed to separate DH from methanol extracts of storage lobe material (Morgan and Mordue 1983). These experiments revealed that locust DH chromatographs as at least two separate peaks of activity, suggesting that locust DH might exist in at least two distinct molecular forms. This separation of DH into two peaks of activity has been confirmed using other makes of column, such as a Waters μ-Bondapak (see Mordue and Morgan, this proceedings) and a Phase-Sep. Spherisorb ODS 2 column, although the retention times of the DH peaks differ for each column. It was clear from these separations that methanol extracts of storage lobes are complex mixtures and that a chromatographic step prior to reversed-phase HPLC would be needed.

It was decided to investigate high performance size-exclusion chromatography (HPSEC) using a TSK 2000 SW column, with a predicted

439

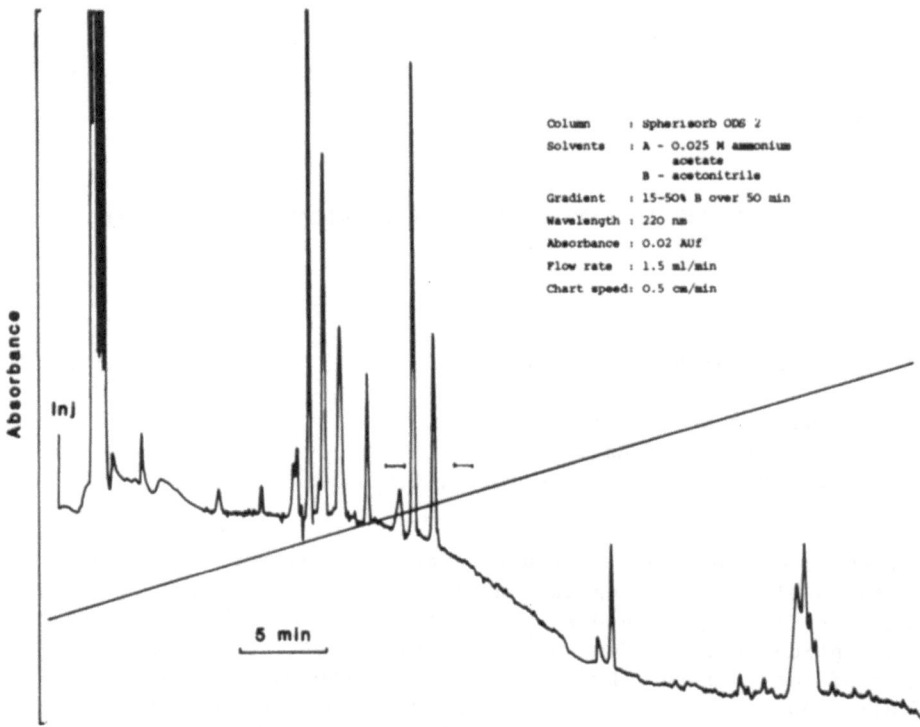

Fig. 1.   Chromatography of fraction 11 from TSK (2000 SW) separa-
           tion of 50 corpora cardiaca using reversed phase HPLC.
           Horizontal bars show regions of predicted DH I and DH II
           activity respectively.

fractionation range of 500 to 70,000 for proteins/peptides.   Using
a 30 cm column eluted in 0.1% trifluoroacetic acid (TFA) at a flow
rate of 1 ml/min, fractions of 1 ml were collected at 1 min inter-
vals and lyophilized prior to bioassay, (for bioassay method, see
Morgan and Mordue, 1983).   DH eluted in the fractions between 10-
12 min after injection (i.e. fractions 11 and 12).   Accurate molecular weight estimations are not generally possible using HPSEC
columns;   however by calibrating the column with a number of marker
molecules, it was found that diuretic hormone eluted close to AKH,
oxytocin and vasopressing, confirming previous estimates of a molecular weight of <2000, and suggesting a molecular weight close to
1000.

     From previous experiments using methanol extracted storage lobe
material chromatographed directly on a Spherisorb ODS 2 column without a preliminary separation step, it had been established that DH I
eluted in the fraction collected between 22-23 min, and DH II in the
fraction between 26-27 min and these fractions could be identified

with small peaks of absorbance.   Therefore when each of the fractions produced from the separation of corpora cardiaca extracts on
the TSK column were subsequently chromatographed on the Spherisorb
column using the linear gradient system described previously (Morgan
and Mordue 1983), it was possible to predict which peaks corresponded
to DH.   These results showed that the peaks corresponding to DH I
and II occurred in the TSK fractions which had eluted between 10-12
min, as predicted by the bioassay data.

However despite this two stage separation, it was apparent from
the separation profiles that the DH peaks were not resolved sufficiently to ensure the collection of uncontaminated hormone for amino
acid analysis.   Therefore modifications to the gradient system to
improve the resolution were investigated.   A slower linear gradient
was employed running from 15% to 50% acetonitrile over 50 min;   also
the buffer strength was reduced to 0.025 M ammonium acetate, allowing
the U.V. wavelength to be lowered from 230 to 220 nm.   The absorbance
profile and the elution conditions employed for the separation of a
TSK fraction 11 are shown in Fig. 1.   The regions of predicted DH I
and DH II activity are shown by solid horizontal bars.   A peak corresponding to DH I can be observed, but no peak for DH II.

Although the predicted DH I peak is sufficiently resolved from
neighbouring peaks it was not clear whether the peak was due only to
a single compound.   To test this a stepped gradient was employed
which further separated all the peaks.   However, the DH I peak was
not resolved into further peaks, suggesting that the absorbance peak
corresponding to DH I is due to a homogenous compound.

In this work bioassay has been used to identify the DH activity
from methanol extracts of corpora cardiaca separated directly either
on a reversed-phase HPLC (Spherisorb) or a HPSEC (TSK) column.
Using HPSEC as the preliminary separation step it has then been
possible to use chromatographic indices to identify the peaks corresponding to DH, and in this way to determine the optimum separation
methods to purify this peptide.   This approach has been adopted
mainly because of both the tedious and temperamental nature of the
locust DH bioassay.   It must now be established that the predicted
DH I peak possesses biological activity;   we will then be in a position to collect DH I for amino acid analysis.

Acknowledgement.   Both PJM and WM would like to thank the ARC
for their support of this work.

REFERENCE

MORGAN P. J. and MORDUE W. (1983) Separation and characteristics of
     diuretic hormone from the corpus cardiacum of Locusta. Comp.
     Biochem. Physiol. 75B(1), 75-80.

NEUROSECRETORY CONTROL OF FEEDING AND DIGESTION BY OVARIAN

ENDOCRINE FACTOR IN <u>DYSDERCUS CINGULATUS</u>

D. Muraleedharan

Department of Zoology
University of Kerala
Kariavattom-695 581, India

Ovaries of the adult insects are known to synthesize ecdysone under the influence of pars intercerebralis neurosecretory cells (PINSC) in some of the insects (Hagedorn et al., 1979; Richards, 1981). Hormonal factors from PINSC stimulate digestive enzyme activity in <u>Dysdercus cingulatus</u> through a secretogogue mechanism (Muraleedharan and Prabhu, 1979). Absence of the same hormonal factors prevent egg maturation in the same insect species (Jalaja, 1974). So it is of interest to elucidate whether this hormonal influence on vitellogenesis is a direct one or through its influence on food consumption and its subsequent digestion. The present paper reports the probable influence of ovarian endocrine factor on the activity of PINSC and also on food consumption and its subsequent digestion.

Ovariectomy was performed in adult females of <u>Dysdercus cingulatus</u> obtained from the stock colony maintained under controlled conditions in the laboratory. PINSC-activity was judged by taking into account 3 criteria, namely neurosecretory index, diameter of the neurosecretory cell and diameter of the nucleus of the neurosecretory cell; all taken from either stained paraffin sections or whole mount preparations of PINSC of experimental and corresponding sham-operated control insects 1-7 consecutive days after ovariectomy. Amount of food consumed and the midgut protease and invertase (two selected enzymes) activities were also estimated in the same operated and control groups of insects. Methanol extract of adult ovaries procured from 4-day old healthy adults were injected into the ovariectomized insects 48 hr after operation at a concentration of 3 μl (containing 0.6 part of a pair of ovaries) per individual. PINSC-activity, food consumption rate and midgut protease and invertase activities were estimated 24 hr after the injection in one group

443

and in the other group all these parameters were estimated after
giving a repeated dose 24 hr later and a subsequent elapse of 24
hr.   For each experiment mean value of 8 separate determinations
were taken.

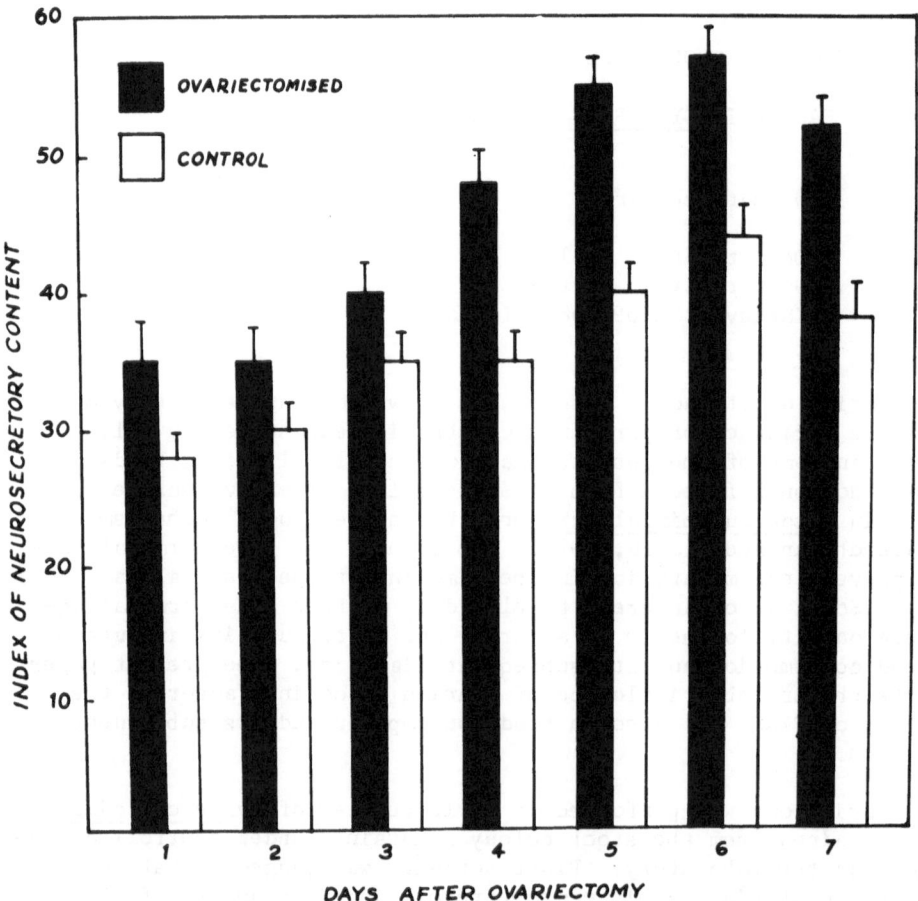

Fig. 1.   Effect of ovariectomy on neurosecretory index in D. cingulatus.   Each column represents mean value from 8 separate individuals
and the bars ± SEM.

     PINSC-activity, rate of food consumption and midgut protease
and invertase activity, all are significantly higher in the ovariectomised insects than that noticed in the respective control groups
1-7 consecutive days after ovariectomy (P < 0.01) (Figs. 1 to 3,
Table 1).   Injection of methanol extract of adult ovaries into ovariectomised insects brought down all the three parameters to a significant extent (P < 0.01) as opposed to that noticed in the respective
control groups (Fig. 4; Tables 2 and 3).

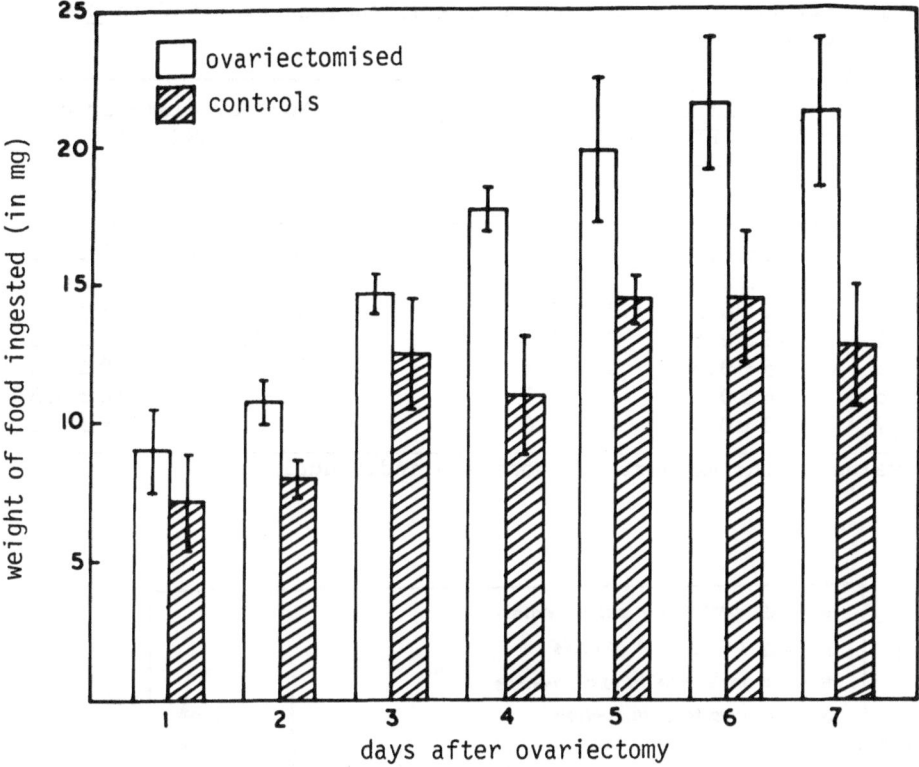

Fig. 2. Effect of ovariectomy on food consumption in D. cingulatus.
Each column denotes mean value from 8 separate individuals and the
bars represent ± SEM.

      Results of the present studies demonstrates that in D. cingulatus
PINSC become hyperactive as a result of ovariectomy and the injection
of methanol extract of adult ovaries brings down this activity to a
considerable extent. Many workers have reported ecdysone secretion
by the ovaries in different species of insects such as Bombyx mori
(Karlson and Stamm, 1956), Aedes aegypti (Schlaeger et al., 1974),
Leucophaea maderae (King and Marks, 1974), Locusta migratoria
(Hoffmann et al., 1975). Hormonal factors from PINSC have been reported to stimulate ecdysone secretion by insects such as Aedes
aegypti (Hagedorn et al., 1979), Gryllus bimaculatus (Hoffmann et
al., 1981). On the contrary, in Aedes atropalpus brain stimulates
CA to secrete JH and not to stimulate ovaries to secrete ecdysone
(Shapiro and Hagedorn, 1982). In the ovariectomised Locusta migratoria and in young ovaries of normal ones, very low levels of ecdysone
has been noted while substantial increase in ecdysone titre has been
noticed as egg maturation progressed. However, there is no report
hitherto on an inhibitory effect of ecdysone on PINSC in insects.
Since this inhibitory effect progresses as days pass by after

Table 1. Effect of ovariectomy on the dimensions of PINSC in D.
          cingulatus

Days after ovariectomy	Cell diamater of PINSC (μ)		Nuclear diameter of PINSC (μ)	
	EXP	CON	EXP	CON
1	20 ± 0.4	19 ± 0.5	7 ± 0.3	7 ± 0.2
2	20 ± 0.3	20 ± 0.4	7 ± 0.2	7 ± 0.2
3	22 ± 0.2	21 ± 0.5	8 ± 0.2	8 ± 0.2
4	23 ± 0.8	21 ± 0.6	9 ± 0.5	8 ± 0.3
5	25 ± 0.6	23 ± 0.8	11 ± 0.3	10 ± 0.2
6	26 ± 0.5	23 ± 0.5	11 ± 0.3	10 ± 0.3
7	25 ± 0.4	20 0.6	11 ± 0.2	9 ± 0.3

Each value is the mean from 5 separate individuals.

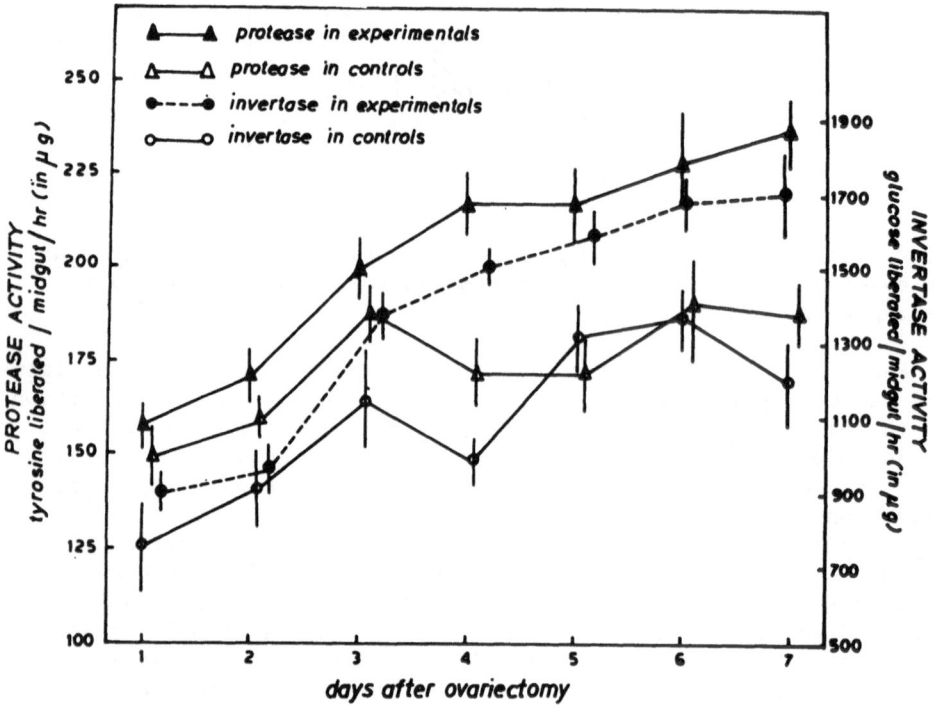

Fig. 3. Effect of ovariectomy on midgut protease and invertase
activity in D. cingulatus. Each point represents the mean value
from 8 separate determinations and the vertical lines are ± SEM.

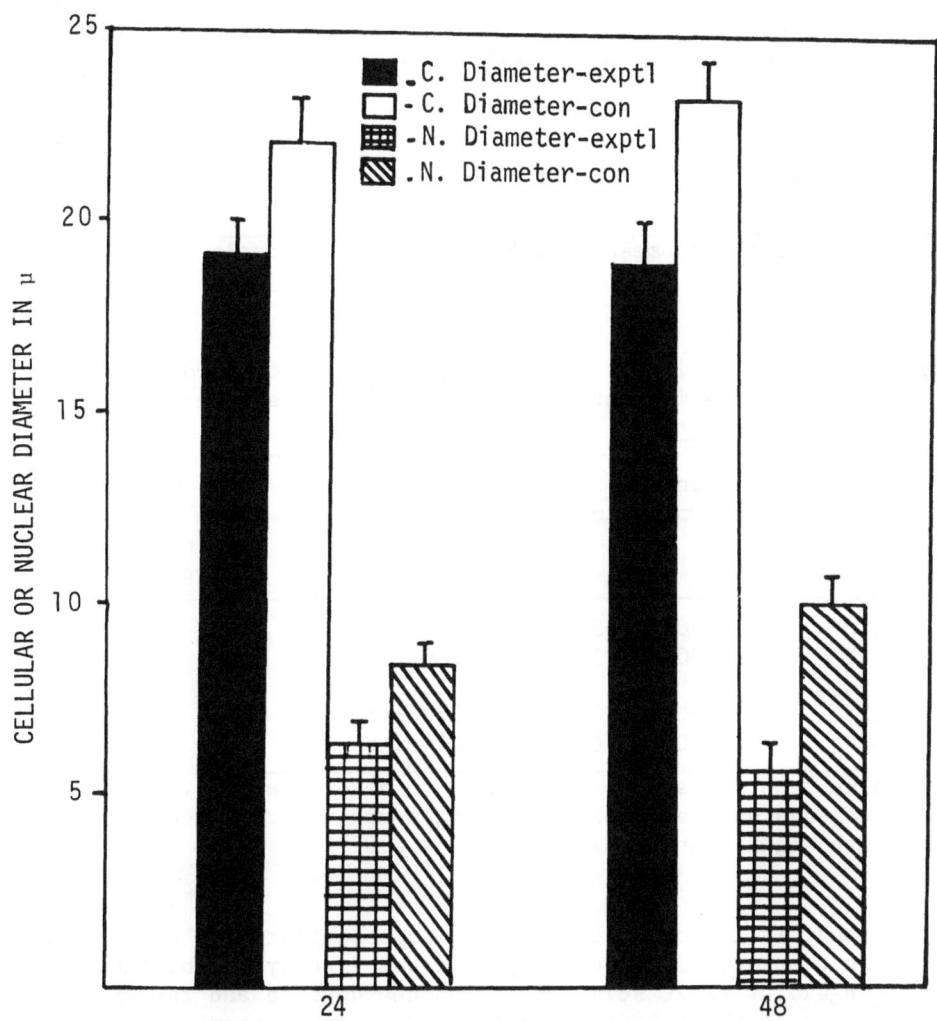

Fig. 4. Effect of injection of ovarian extract on the cell dimensions of pars intercerebralis neurosecretory cells in D. cingulatus. Each column represents mean value from 5 separate individuals and the bars represent ± SEM.

Table 2. Effect of injection of ovarian extract on the index of
          neurosecretory content in <u>D</u>. <u>cingulatus</u>

Time after first injection	Neurosecretory index of PINSC (arb. Units)
24 hr-EXP	29
24 hr-CON	42
48 hr-EXP	30
48 hr-CON	48

Each value represents mean from 5 separate individuals.

Table 3. Effect of injection of ovarian extract on food consumption
          and digestive enzyme activity in the castrated females of
          <u>D</u>. <u>cingulatus</u>

Time after 1st injection	Amount of food consumed (in mg)	Protease activity (μg tyrosine/ gut/hr)	Invertase activity (μg glucose/ gut/hr)
24 hr-EXP	12.00 ± 0.3	178 ± 9	1080 ± 60
24 hr-CON	15.00 ± 0.5	200 ± 7	1350 ± 43
48 hr-EXP	10.50 ± 0.8	170 ± 10	915 ± 50
48 hr-CON	18.50 ± 0.6	220 ± 8	1480 ± 40

Each value represents mean of 8 separate determinations. Mean
values of adjacent pairs of experimental and control of each
column are significantly different ($P < 0.01$).

ovariectomy, it is possible that the ecdysone titre may have a feed
back effect on the PINSC activity. It was already established that
in <u>D</u>. <u>cingulatus</u> PINSC stimulates food consumption and the present
finding that in the ovariectomised insects food consumption is very
much increased is in quite agreement with our earlier finding
(Muraleedharan and Prabhu, 1979). The increased amount of food
consumed naturally stimulates midgut protease and invertase activities and so the higher rate of enzyme activities noticed in the
castrated females of <u>D</u>. <u>cingulatus</u> also agrees with our earlier reports. So it may be concluded that in adult <u>D</u>. <u>cingulatus</u> females,
some ovarian humoral factor (probably ecdysone) inhibits PINSC activity resulting in a reduced rate of food consumption. Such a drop
on food consumption reduces midgut protease and invertase activities
through a secretogogue mechanism.

Acknowledgments

Research Grants provided by the University Grants Commission is gratefully acknowledged.

References

HAGEDORN H. H., SHAPIRO J. P., and HANAKOA K. (1979) Ovarian ecdysone secretion is controlled by a brain hormone in an adult mosquito. Nature Lond. 282, 92–94.

HOFFMANN J. A., KOOLMAN J., and BEYLER C. (1975) Role des glandes prothoraciques dans la production d'ecdysone au cours du dernier stade larvaire de locusta migratoria L. C. R. Acad. Sci. Paris (D) 280, 733–736.

HOFFMANN K. H., BEHRENS W., and RESSIN W. (1981) Effects of daily temperature cycle on ecdysteroid and cyclic nucleotide titres in adult female crickets, Gryllus bimaculatus. Physiol. Ent. 6, 375–385.

JALAJA M. (1974) Complete inhibition of vitellogenesis after extirpation of median neurosecretory cells in Dysdercus cingulatus. Curr. Sci. 43, 286–287.

KARLSON P. and STAMM D. (1956) Notiz uber den nachweis von Metamorphose hormon in den imagines von Bombyx mori. Hoppe-Seyler's Z. Physiol. Chem. 306, 109–111.

KING D. S. and MARKS E. P. (1974) The secretion and metabolism of alpha-ecdysone by the cockroach (Leucophaea maderae). Tissues in vitro Life Sci. 15, 147–154.

MURALEEDHARAN D. and PRABHU V. K. K. (1979) Role of the median neurosecretory cells in secretion of protease and invertase in the red cotton bug, Dysdercus cingulatus. J. Insect Physiol. 25, 237–240.

RICHARDS G. (1981) Insect hormones in development. Biol. Rev. 56, 501–549.

SCHLAEGER D. A., FUCHS M. S., and KANG S. H. (1974) Ecdysone-mediated stimulation of dopadecarboxylase activity and its relationship to ovarian development in Aedes aegypti. J. Cell Biol. 61, 454–465.

SHAPIRO J. P. and HAGEDORN H. H. (1982) Juvenile hormone and the development of ovarian responsiveness to a brain hormone in the mosquito, Aedes aegypti. Gen. Comp. Endocrinol. 46, 176–183.

IMMUNOHISTOCHEMICAL DEMONSTRATION OF BRAIN-MIDGUT ENDOCRINE SYSTEM IN THE COCKROACH

Junko Nishiitsutsuji-Uwo and Yasuhisa Endo

Shionogi Research Laboratories, Shionogi & Co., Ltd.
Fukushima-ku, Osaka, 553 Japan

In mammals, the brain and the gut, including pancreas, contain a large number of bioactive substances, the brain-gut peptides. According to conventional biology, they might be partly regards as neurotransmitters or neurosecretions and partly as gastro-enteropancreatic (GEP) hormones. Indeed, a number of bioactive peptides are shared both by neurons and certain endocrine cells.

In human GEP systems, there are many different endocrine cells believed to correspond to different types of peptide hormones. Recently, we have found similar types of cells are present in the insect midgut epithelium by electron microscope. Such endocrine types of cells or basal-granulated cells are extensively distributed in the epithelia of the midgut and enteric caeca of the adult cockroach, Periplaneta americana[1] and the larval, pupal[2] and adult midguts of all species of Lepidoptera so far observed.

Basal-granulated cells have electron-lucent cytoplasm and basally concentrated granules which are encircled by a distinct membrane derived from the Golgi complex. They are bowl-shaped, pyramidal or bottle-shaped cells standing on the basal lamina and extending a slender apical process to the gut lumen. Some of them are undoubtedly open-type cells having a tuft of microvilli at the luminal surface and are as tall as 70-100 μm.

In the case of the cockroach, for instance, at least six types of endocrine granules were identified in the basal-granulated cells and by reference to the granules the cells were temporarily classified into 4-6 cell types. By means of tannic acid fixation in combination with glutaraldehyde and osmium tetraoxide, exocytotic granule release was clearly demonstrated in all of these cell types

451

except for type II-a granules.[3] Thus, ultrastructurally the basal
granulated cells in the insect midguts seemed to be endocrine in
nature and correspond to the well established endocrine cells in
mammalian GEP system.

   In order to substantiate the concept of the endocrine nature
of these cells and the correspondency of insect brain-midgut system
to mammalian brain-GEP system, the peroxidase-antiperoxidase
immunocytochemical method, with the use of antisera to mammalian
bioactive substances has been applied to the brain-midgut system of
the cockroach.[4-7]

   In the midgut epithelium, three neuropeptide-like immunoreactivities in endocrine cells and three in nerve fibers were so
far identified. Pancreatic polypeptide (PP)-, somatostatin- and
enteroglucagon-like immunoreactivities were located in different
endocrine cells.[4] The PP-reactive cell was most numerous and
identified as the type II-b cell using the serial semithin-thin
section technique.[5]

   The nerve fibers innervating the outer longitudinal masculature
of the mdigut were found to contain vasoactive intestinal polypeptide (VIP)-, somatostatin- and PP-like immunoreactivities, each
occurring in different fibers.[4]

   As shown in Table, in the entire central and visceral nervous
systems, PP-, somatostatin- and VIP-like immunoreactivities were
also demonstrated.[6,7] Many other neuropeptides and bioactive monoamines were also found in the central nervous system and a few in
retrocerebral complex. Immunoreactive cells or cell-groups were
always bilaterally symmetrical and present in the defined region of
the ganglia. Different neuropeptide-like immunoreactivities may
locate, if not all, in different cells.

   Taking PP-like immunoreactivity as an example,[7] PP-positive
neuron somata were most numerous in the brain. No reactive cells
could be found in the retrocerebral complex and the second to the
fifth abdominal ganglia. The axons containing PP-like immunoreactivity were distributed issuing many branches in the entire brain-
retrocerebral complex, ventral cord and visceral nervous system.

   PP and somatostatin, for example, are known as common neural
and endocrine peptides in vertebrates. The present immunohistochemical studies were clearly demonstrated that the cockroach
endocrine and nervous element also possess peptides with as closely
related molecular structures as to be stainable with the antisera
produced by use of mammalian peptides.

   In conclusion, the data available thus far have demonstrated
that the peptidergic (and aminergic) brain and midgut system of

Table 1. Distribution of PP-, Somatostatin- and VIP-like
Immunoreactivities in the Nervous System of the Cockroach

	PP	Som	VIP
Brain			
Protocerebrum			
Median region	+++	++	-
Lateral region	++	++	+
Other region	++	+	(-)
Deutocerebrum	+	(-)	-
Tritocerebrum	+	+	+
Retrocerebral complex			
Corpora cardiaca (axons)	+++	+++	+
Corpora allata (axons)	+	+	±
Ventral cord			
Suboesophageal ganglion	++	++	+
Pro-, meso-, metathoracic ganglia	++	++	+
1st abdominal ganglion	+	a	a
2nd-5th abdominal ganglia	-	a	a
Terminal ganglion	++	++	+
Visceral nervous system			
Frontal ganglion	+	a	a
Recurrent nerve	++	a	a

a Not observed.   ( ) Not sure.

insects are analogous to the brain and GEP system of vertebrates.

Some parts of the present study have been done in collaboration
with Drs. T. Fujita and T. Iwanaga of Niigata University.

REFERENCES

1. J. Nishiitsutsuji-Uwo and Y. Endo, Biomed. Res. 2:30 (1981).
2. Y. Endo and J. Nishiitsutsuji-Uwo, Biomed. Res. 2:270 (1981).
3. Y. Endo and J. Nishiitsutsuji-Uwo, Cell Tissue Res. 222:515
    (1982).
4. T. Iwanaga, T. Fujita, J. Nishiitsutsuji-Uwo, and Y. Endo,
    Biomed. Res. 2:202 (1981).
5. Y. Endo, J. Nishiitsutsuji-Uwo, T. Iwanaga, and T. Fujita,
    Biomed. Res. 3:454 (1982).
6. T. Fujita, R. Yui, T. Iwanaga, J. Nishiitsutsuji-Uwo, Y. Endo,
    and N. Yanaihara, Peptides 2, suppl. 2:123 (1981).
7. Y. Endo, T. Iwanaga, T. Fujita, and J. Nishiitsutsuji-Uwo,
    Cell Tissue Res. 227:1 (1982).

PURIFICATION AND PROPERTIES OF PROTHORACICOTROPIC HORMONE (PTTH) FROM DEVELOPING ADULT BRAINS OF THE SILKWORM BOMBYX MORI

Masaji S. Nishimura and Junko Nishiitsutsuji-Uwo

Shionogi Research Laboratories, Shionogi & Co., Ltd.
Fukushima-ku, Osaka, 553 Japan

In insects, only two pairs of epithelial endocrine glands are well-known: the prothoracic (thoracic) glands and the corpora allata. On the other hand, numerous neurosecretory cells are grouped conspicuously within ganglia, especially in the brain, which produce various kinds of neuropeptide hormones controlling the insect physiology and moulting. Prothoracicotropic hormone (PTTH) is considered to be the most important cerebral neuropeptide. The extraction and purification of the PTTH has been initiated long time ago and either pupal brains or adult heads of silkworms have been used as the main source of the PTTH. However, considerable confusion currently exists concerning the chemical nature of the PTTH(s).[1-9]

During the past over ten years, we have also engaged in the purification of the PTTH both from the whole adult heads and the pupal brains of the silkworm, Bombyx mori. In 1972,[4] we reported the active principle extracted from adult heads appeared to be a peptide of 5,000 daltons or less and this conclusion is supported by more recent studies of Ishizaki and his co-workers,[5,6,8] who succeeded to isolate one of PTTHs in 1982.[9] However, insect head contains many important neuroendocrine organs such as corpora cardiaca-allata, suboesophageal and frontal ganglia besides the brain. Since the reports[1,2] on the molecular size of PTTH extracted from the silkworm brains is varying from 9,000 to 31,000, the PTTH(s) purified from the whole head might be a different PTTH(s) obtained from the brain.

The present paper reports briefly the purification of PTTH from dissected silkworm brain and some properties of the purified hormone.

As the source material, pupal (pharate adult) brains of <u>Bombyx</u> <u>mori</u> were used. The bioassay of PTTH (double or triple tests) was performed on brainless pupae of the Eri-silkworm, <u>Samia</u> <u>cynthia</u> <u>ricini</u> at any time between two to eight months after removal of the brain. The extraction and purification of the PTTH from the brain were repeated eight times on a scale of 27,000-180,000 brains. Each experiment was performed essentially the same as follows: step 1. acetone powder (AP) → step 2. extraction with 2% aqueous NaCl (CE) → step 3. heat treatment at 90°C for 3 min (HF) → step 4. phenol extract (PhE) → 5. chromatography on DEAE-Sephadex A-25 by stepwise elution with 0.01-0.5 M NaCl in a buffer (DS) → step 6. second chromatography on DEAE-Sephadex A-25 by linear gradient elution with 0.01-0.5 M NaCl in a buffer (2nd DS) → step 7. gel-filtration on Sephadex G-50 (G-50) → step 8. gel-filtration on Sephadex G-25 (G-25) → step 9. chromatography on DEAE-cellulose by gradient elution with 0.1-0.5 M NaCl in a buffer (DC). Desalting, if necessary, was done in any step through a Sephadex G-10 column with 0.02 M NH_4HCO_3.

A general profile of extraction and purification of the PTTH starting with average 90,000 silkworm pupal brains is summarized in Table 1 with reference to eight different experimental runs.

A pupal brain contained average 50 <u>Samia</u> units of PTTH and the first extract was considerably high in its purity (3 μg/<u>Samia</u>). Through 8-step procedure, the effective dose achieved to 28 ng/<u>Samia</u> and the yield of activity was 41%. Through a further one-step purification, the effective dose attained to several ng/<u>Samia</u>, which seemed to be comparable with the "highly purified PTTH" accomplished

Table 1.   Purification Profile of the PTTH Extracted from Silkworm Pupal Brain[a] (Mean of 8 Repeated Experiments)

Purifi-cation-step	Prepara-tions	Effective dose (ng/<u>Samia</u>)		Purifi-cation-fold	Recovery of PTTH (%)
2	CE	3010 ± 1476	(5986 -810)	1	100
		(50 ± 29 <u>Samia</u> units[b]/brain)			
3	HF	-	-	-	85 ± 10
4	PhE	344 ± 149	(600 - 88)	6 - 10	58 ± 6
5	DS	223 ± 130	(432 - 24)	8 - 34	47 ± 14
6	2nd DS	131 ± 70	(258 - 33)	13 - 41	39 ± 10
7	G-50	58 ± 35	(138 - 13.1)	35 - 209	38 ± 18
8	G-25	28 ± 23	(96 - 3.2)	63 - 815	41 ± 23

[a] Average 90,000 (27,000 - 180,000) pupal (pharate adult) brains were used for each experimental lot.
[b] One <u>Samia</u> unit of PTTH was defined as the minimum amount necessary to cause adult development in one debrained <u>Samia</u> <u>pupa</u>.

in several ten throusand-fold starting with the whole adult
head.[6,8,9]

Chemical properties of the PTTH was studied with Ampholine isoelectric focusing, dialysis, ultrafiltration and protease digestion
experiments. Our PTTH obtained from silkworm pharate adult brains
is an acidic peptide(s) and involves aromatic amino acids. The mol.
wt. of the real active principle or one of active molecular species
of the PTTH would, probably be less than 5,000 daltons.[4] This PTTH
seemed to be a similar or a close substance to our PTTH[4] and PTTH(s)
of Ishizaki and his co-workers purified from the whole head.[6,8,9]

In another lepidopteran species, <u>Manduca</u> <u>sexta</u>, the mol. wt of
the PTTH extracted from the larval head was reported to be 25,000,[11]
whereas PTTHs in the pupal brain, 7,000 and 22,000.[12] In the silkworm PTTH activity in the brain is very high both at the pharate
adult and adult stages,[13,14] when PTTH is apparently no longer
physiologically necessary as an ecdysiotropin. A possible function
of the PTTH obtained from the source of adult is discussed.

REFERECNES

1. H. Ishizaki and M. Ichikawa, <u>Biol</u>. <u>Bull</u>. 133:355 (1967).
2. M. Yamazaki and M. Kobayashi, <u>J</u>. <u>Insect</u> <u>Physiol</u>. 15:1981 (1969).
3. C.M. Williams, The present status of the brain hormone, <u>in</u>
    "Insects and Physiology," J.W.L. Beament and J.E. Treherne,
    ed., Oliver & Boyd, Edinburgh and London, pp. 133-139 (1967).
4. J. Nishiitsutsuji-Uwo, <u>Botyu-Kagaku</u> 37:93 (1972).
5. H. Ishizaki, A. Suzuki, A. Isogai, H. Nagasawa, and S. Tamura,
    <u>J</u>. <u>Insect</u> <u>Physiol</u>. 23:1219 (1977).
6. H. Nagasawa, A. Isogai, A. Suzuki, S. Tamura, and H. Ishizaki,
    <u>Develop</u>. <u>Growth</u> and <u>Differ</u>. 21:29 (1979).
7. M. Funatsu, Y. Aizono, N. Matsuo, G. Funatsu, and M. Kobayashi,
    XVI Int. Cong. Entomol. Abstracts p. 187. Kyoto, Japan, 3-9
    August, 1980.
8. H. Ishizaki, Y. Koide, A. Suzuki, Y. Hori, H. Nagasawa, A.
    Isogai, and S. Tamura, XVI Int. Cong. Entomol. Abstracts
    p. 157, Kyoto, Japan, 3-9 August, 1980.
9. S. Suzuki, H. Nagasawa, H. Kataoka, Y. Hori, A. Isogai, S.
    Tamura, F. Guo, X. Zhong, H. Ishizaki, M. Fujishita, and A.
    Mizoguchi, <u>Agric</u>. <u>Biol</u>. <u>Chem</u>. 46:1107 (1982).
10. J. Nishiitsutsuji-Uwo and M.S. Nishimura, <u>Insect</u> <u>Biochem</u>.
    (1983) in press.
11. T.G. Kingan, <u>Life</u> <u>Sci</u>. 28:2585 (1981).
12. L.I. Gilbert, W.E. Bollenbacher, N. Agui, N.A. Granger,
    B.J. Sedlak, D. Gibbs, and C.M. Buys, <u>Am</u>. <u>Zool</u>. 21:641 (1981).
13. H. Ishizaki, <u>Develop</u>. <u>Growth</u> and <u>Differ</u>. 11:1 (1969).
14. J. Nishiitsutsuji-Uwo and M.S. Nishimura, <u>Experientia</u> 31:1105
    (1975).

DOPAMINE, NORADRENALINE AND 5-HYDROXYTRYPTAMINE IN THE CEREBRAL

GANGLIA OF *PERIPLANETA AMERICANA*

Michael D. Owen and B. Duff Sloley

Department of Zoology, University of Western Ontario
London, Ontario, Canada   N6A 5B7

While dopamine (DA), noradrenaline (NA) and 5-hydroxytryptamine (5-HT) are well known as putative neurotransmitters in insect nervous systems (reviewed by Evans, 1980) little is known of the functional significance of the central aminergic system and the control of the levels of these amines in insect central nervous systems. The combination of high-performance liquid chromatography with electrochemical detection (reviewed by Kissinger *et al*, 1981) allows the separation and quantitation of the amounts of biogenic amines present in insect central nervous systems. We have applied this technique to studies of the natural levels of DA, NA and 5-HT in the cerebral ganglia (brain) of *Periplaneta americana*, followed by a quantitative examination of the effects of reserpine on these amine levels and an investigation of the effects of the organophosphorus insecticide "Dicrotophos" (Shell Chemical Co.).

In these experiments *P. americana* brains were dissected (without saline), homogenized (ultrasonically) in 0.05M perchloric acid containing 0.005M sodium metabisulphite, centrifuged (12,800g at 4°C) and the supernatant passed through a 0.22μm filter before being injected onto either Corasil CX or Zipax SCX (60cm x 2mm) ion exchange columns. (More sensitive measurements are possible after an initial isolation of the amine fraction followed by separation on C_{18} columns.) Eluting solutions were: (i) for DA and 5-HT on Zipax SCX columns - 0.05M perchloric acid, 0.01M potassium perchlorate, 0.003M sodium azide, pH 1.7, flow rate 0.8ml/min; (ii) for DA and 5-HT on Corasil CX columns - 0.17M citric acid, 0.033M sodium acetate, 0.04M sodium hydroxide, acetic acid to give pH 5.0, flow rate 1.25ml/min; (iii) for NA on Corasil CX columns - 0.85M citric acid, 0.016M sodium acetate, 0.02M sodium hydroxide, acetic acid to give pH 4.0, flow rate 0.8ml/min. Chromatographically separated

amines were quantified by oxidation (at a potential of -.07V relative to an Ag/AgCl reference electrode) in a thin layer detection
cell (BAS model TL-5 with an LC2A amplifier). DA, NA, 5-HT and
reserpine were from Sigma Chem. Co., reserpine solutions contained
the appropriate dose of alkaloid in 15µl (injection volume) and
were prepared by dissolving reserpine in 0.5ml glacial acetic acid
which was then diluted to 20ml with water. Dicrotophos, (Bidrin),
[α-3-(dimethoxyphosphinyloxy)-*N*,*N*-dimethylcrotonamide] (a gift
from Shell Chemical Co.) doses were applied topically in 10µl of
acetone. Brain acetylcholinesterase activity was assayed colori-
metrically (Ellman *et al*. 1961). The Falck-Hillarp paraformaldehyde-
amine condensation method was used for histochemical localisation
of amines.

Early experiments suggested that DA levels in female cockroach
brains were lower and more variable than those in males; suspecting
that variability in females might result from the involvement of
dopamine metabolism in the tanning of oothecae we measured DA and
5-HT levels in the brains of males (DA 5.3ng/brain, 5-HT 6.0ng/
brain), females extruding oothecae (DA 3.5ng/brain, 5-HT 7.4ng/
brain) and females 48hr after the release of oothecae (DA 3.1ng/
brain, 5-HT 6.5ng/brain). The DA levels in female cockroach brains
are significantly (p<0.05) lower than those in males, the differences
in DA levels at the two stages of oothecal development and 5-HT
levels in males and the two female stages are not significant.

Both DA and 5-HT levels in the brain of male cockroaches vary
in the first few weeks after adult ecdysis. When the adult first
emerges male brains contain about 5ng DA and 2ng 5-HT; during the
next six weeks these levels increase steadily to about 13ng DA and
10ng 5-HT, this peak level is followed by a rapid fall in amine
levels to levels which were then maintained through the duration
of our experiment (15 weeks) (5-7ng NA, 4-7ng 5-HT). We found no
diurnal variation in the levels of DA and 5-HT in the cerebral
ganglia of male *P. americana*.

While the *Rauwolfia* alkaloid reserpine is known to deplete the
amine content of insect nervous systems (references in Evans 1980)
there has been no previous study of the effective dose or of the
rate of insect recovery from reserpine treatment. We have measured
(Sloley and Owen 1982a) DA, NA and 5-HT in the brains of cockroaches
treated with varying doses of reserpine. We find that 20µg/animal
maximally depletes 5-HT while DA and NA are maximally reduced by
40µg/animal. We have followed the recovery of amine levels in the
brains of cockroaches injected with 40µg of reserpine. In these
experiments DA and 5-HT are depleted more rapidly (maximal depletion with 12hr) than NA (maximal depletion takes 24hr). DA and
5-HT levels also commence recovery earlier (3-4 days) than NA
levels (6 days) although complete recovery of the levels of all
three amines takes about three weeks.

Gardner and Brady (1977) suggest that the organophosphorus insecticide dicrotophos may reduce catecholamine levels in the *corpus centrale* of crickets. We have measured the acetylcholinesterase inhibition and DA and 5-HT levels in the brain of cockroaches treated with dicrotophos doses of 0 - 40µg/animal (Sloley and Owen 1982b). No significant change in DA and 5-HT levels in the cerebral ganglia followed such treatment; measurements of acetylcholinesterase inhibition show that the insecticide is reaching and acting upon the cerebral ganglia. Slight falls in amine levels (most marked after 30µg doses, less obvious after 40µg doses) were not statistically significant. Since the 30µg dose produced the greatest uncoordinated movement and tremor (the 40µg dose caused rapid immobilisation) we suggest that this amine reduction is a secondary consequence of stress and uncoordinated activity, rather than a direct effect of dicrotophos. Fluorescent histochemical tests did not show a reduction in the amine fluorescence in the *corpus centrale* of dicrotophos treated cockroaches.

References

Ellman G.L., Courney, K.D., Andrus, V. Jr., and Featherstone, R.M., 1961, A new and rapid colorimetric determination of acetylcholinesterase activity, Biochem. Pharmacol., 7:88-95.

Evans, P.D., 1980, Biogenetic amines in the insect nervous system, Adv. Insect Physiol., 15:317-473.

Gardner, F.E. and Brady, E.U., 1977, The effect of organophosphorus insecticides upon catecholamine fluorescence in the corpus centrale and frontal ganglion of the house cricket, *Acheta domesticus* (L.), Pestic. Biochem. Physiol., 7:466-473.

Kissinger, P.T., Brunlett, C.S. and Shoup, R.E., 1981, Neurochemical application of liquid chromatography with electrochemical detection, Life Sci., 28:455-465.

Sloley, B.D. and Owen, M.D., 1982a, The effects of reserpine on amine concentrations in the nervous system of the cockroach (*Periplaneta americana*), Insect Biochem., 12:469-476.

Sloley, B.D. and Owen, M.D., 1982b, The effects of organophosphorus poisoning on dopamine and 5-hydroxytryptamine concentrations in the brain of the cricket, *Acheta pennsylvanicus*, and the cockroach, *Periplaneta americana*, Pestic. Biochem. Physiol., 18:1-8.

SEPARATION AND ISOLATION OF A LOCUST DIURETIC HORMONE

Ada Rafaeli

The Hebrew University
Department of Entomology, Faculty of Agriculture
Rehovot 76-100, ISRAEL

Locusts possess a diuretic hormone which is produced by cerebral neurosecretory cells and stored in the corpora cardiaca. This hormone has been reported to be a peptide hormone as evidenced by its sensitivity to proteolysis (MORDUE and GOLDSWORTHY, 1969). Since extracellular cyclic-AMP has been shown to increase fluid secretion (RAFAELI and MORDUE, 1978; 1982), it has been assumed that diuretic hormone acts via cyclic-AMP as a second messenger. Direct evidence of increases in intracellular cyclic-AMP levels as a result of stimulation by diuretic hormone in locust Malpighian tubules has recently been reported (RAFAELI *et al.*, 1983). Preliminary investigations of the diuretic hormone of locusts (RAFAELI and MORDUE, 1978; 1982) showed it to be soluble and stable in methanol extracts of corpora cardiaca and soluble but unstable in aqueous solutions. The present report demonstrates the separation and isolation of diuretic hormone from methanolic extracts of corpora cardiaca using a reversed phase HPLC column.

Successful isolation of the locust diuretic hormone has failed in the past due to the lack of a sufficiently sensitive bioassay. The bioassay commonly used has been a semi-isolated preparation that includes the whole alimentary tract with attached tubules (MADDRELL and KLUNSUWAN, 1973). Using isolated Malpighian tubules from starved locusts, where large accumulations of diuretic hormone occur in the corpora cardiaca, tubule preparations were obtained with very low basal rates of fluid secretion. These tubules were found to be more sensitive to diuretic hormone, the lower the basal rate of fluid secretion the more sensitive the tubules became to diuretic hormone stimulation (RAFAELI, 1982). This bioassay was used in the present research to identify the active diuretic hormone peak.

463

A novel Malpighian tubule bioassay, recently developed (RAFAELI et al.,1983), was also exploited in the isolation of diuretic hormone. This bioassay is based on the knowledge that the increase in intracellular cyclic-AMP levels, as a result of diuretic hormone stimulation, is in such excess as to cause high levels of cyclic-AMP secretion into the lumen of the tubule and thus its excretion in the urine. Moreover, this bioassay was found to be highly sensitive when using tubules from starved locusts where basal levels of intracellular cyclic-AMP are very low and therefore a basal level of excretion of cyclic-AMP into the medium bathing the tubules is reduced to undetectable levels.

Diuretic hormone was separated, using a Spectra Physics 8000 high pressure liquid chromatogram and a 10 µm Lichrosorb RP-8 column (Alltech), from methanol extracts of corpora cardiaca of adult male locusts which have been starved for 48 hours. A typical elution profile of a methanol extract of 266 corpora cardiaca is shown in Fig. 1. As a reference point the position of the synthetic peptide hormone, adipokinetic hormone (AKH) is indicated. The adipokinetic hormone peak in extracts of corpora cardiaca, identified by its lipid mobilizing activity in vivo, corresponds in elution time to the synthetic counterpart (capacity factor k' synthetic = 7.345; k' corpora cardiaca = 7.346). The diuretic response by isolated Malpighian tubules to diuretic hormone (Fig. 1, lower insert) and the cyclic-AMP secretory response (Fig. 1, upper insert) is shown to correspond to an absorption peak at k' = 8.95.

Subsequent dilution tests showed that the purified diuretic hormone (herein termed diuresin) response is dose-dependent. Diuresin was found to be stable at room temperatures when kept overnight in sterile medium containing bovine serum albumen (BSA), and it can be subjected to boiling for 10 minutes without a significant change in activity. Its activity is, however, abolished after proteolysis by incubations in protease (0.02 U/µg, Sigma Type XI: Fungal Proteinase K) at 30°C for two hours and subsequently boiling for 5 minutes (Table 1).

Table 1. Diuresin activity under various conditions

CONDITION	% INCREASE IN FLUID RATE (mean + s.e.m.) n=5
In the presence of BSA	145 + 39
In the absence of BSA	71 + 20
At room temp. with BSA	100 + 42
At room temp. without BSA	43 + 29
After boiling	152 + 24
After proteolysis	0

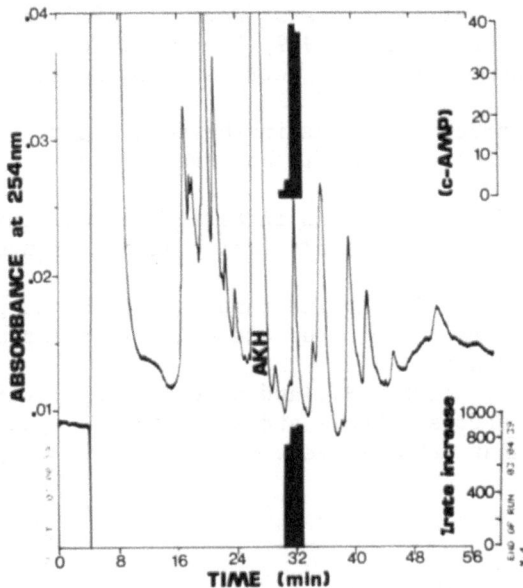

Fig. 1. Elution profile of a 266 corpora cardiaca methanol extract.
Column: 10 µm Lichrosorb RP-8; gradient: linear 0-20%
2-propanol in 0.1% trifluoroacetic acid, slope 2%/min
followed by 20-50% 2-propanol, slope 1%/min; flow: 1ml/min.

The active diuresin peak was further purified using a 10 µm
Lichrosorb RP-18 column and a linear gradient 10-50% 2-propanol at
0.3%/min. A small percentage of activity remained in the void volume
(30% increase in fluid secretion) but the majority of the activity
(177% increase in fluid secretion) was found to correspond to a
single absorption peak at k' = 30 (46% 2-propanol).
     The concentration of diuresin can be estimated by calculations
of peak areas compared to AKH. These calculations indicate that
100 corpora cardiaca yielded approximately 2 µg diuresin. However,
accurate determinations of hormone titres in storage tissues as well
as in haemolymph and target cells awaits full characterization of
this isolated hormone.

Acknowledgment - This work was supported within the framework of a
grant from the United States-Israel Binational Agricultural Research
and Development Fund (BARD) grant no. I-47-79 to Profs. S.W.
Applebaum and Y. Birk. I would like to thank Prof. Applebaum for his
support and helpful suggestions.

REFERENCES
MADDRELL S.H.P. and KLUNSUWAN S. (1973) Fluid secretion by *in vitro*
     preparations of the Malpighian tubules of the desert locust

Schistocerca gregaria. J. Insect Physiol. 19, 1369–1376.

MORDUE W. and GOLDSWORTHY G.J. (1969) The physiological effects of corpus cardiacum extracts in locusts. *Gen. Comp. Endocr.* 12, 306–369.

RAFAELI A. (1982) Induced increase in the sensitivity of the Malpighian tubules to diuretic hormone of locusts. *Gen. Comp. Endocr.* 46, 381.

RAFAELI A. and MORDUE W. (1978) An investigation of diuretic responses to hormones and other physiologically active agents in *Locusta Gen. Comp. Endocr.* 34, 110.

RAFAELI A. and MORDUE W. (1982) The response of the Malpighian tubules of *Locusta* to hormones and other stimulants.*Gen. Comp. Endocr.* 46, 130–135.

RAFAELI A., PINES M., STERN P.S., and APPLEBAUM S.W. (1983) Diuretic hormone stimulated synthesis and excretion of cyclic-AMP by locust Malpighian tubules. *Gen. Comp. Endocr. (in press).*

Note on proof: A similar separation, using different conditions was recently reported by Morgan P.J. and Mordue W. (1983) Separation and characteristics of diuretic hormone from the corpus cardiacum of *Locusta. Comp. Biochem. and Physiol.* 75B (1), 75–80.

NEURO-HORMONAL CONTROL OF SEX PHEROMONE PRODUCTION IN HELIOTHIS ZEA

Ashok K. Raina and Jerome A. Klun

Department of Entomology, University of Maryland, and
Organic Chemical Synthesis Laboratory, USDA, ARS
Bldg. 007, BARC-West, Beltsville, MD 20705

The sex pheromones of over 200 insect species have been
identified (Klassen et al. 1982), however, limited information is
available on the mechanisms that regulate sex pheromone production in
insects. Barth and his co-workers have contributed to the
understanding of the control of pheromone production in cockroaches
(Barth, 1961; Bell and Barth, 1970; Barth and Lester, 1973). In
these insects it was demonstrated that corpora allata (CA) are
responsible for the control of sex pheromone production and sexual
receptivity of the females. Riddiford and Williams (1971) working
with Hyalophora cecropia and Antheraea pernyi indicated that
"calling" in these moths was induced by environmental cues processed
by the brain followed by discharge of signals to the corpora cardiaca
(CC), which in turn stimulated release of a hormone from its
intrinsic cells. In almost all previous studies, pheromone
production was inferred by monitoring female calling behavior and/or
male sexual response to females rather than by direct chemical
analysis of pheromone titer.

We report the detection of a substance that regulates pheromone
titer in the bollworm, Heliothis zea. Female H. zea produce a
mixture of C_{16} aldehydes, of which (Z)-11-hexadecenal (Z-11-HDAL) is
a major constituent of the sex pheromone (ca. 92%) (Klun et al.,
1980). The pheromone is secreted from glandular cells situated in
the intersegmental membrane between the 8th and 9th abdominal
segments (Jefferson et al., 1968).

The Z-11-HDAL titer of the pheromone glands of individual
females was determined by open tubular capillary chromatography using
the internal standard method of quantitative analysis. Insects used
in this study were maintained under 65% r.h., 16:8 (L:D) with

467

temperatures at $26^{\circ}C$ and $20^{\circ}C$ in the respective phases. Maximal pheromone production under these conditions occurred during the 3rd night (54 hr post-emergence) and pheromone-titer analyses were performed on 54-hr old females. An average 54-hr old virgin female yielded about 120 ng of extractable Z-11-HDAL. Ligation of a female, between head and thorax with a fine cotton thread 15 min after emergence and to within 2 hr of the usual extraction time caused a reduction of over 99% in the extractable pheromone. Ligation 1 hr before extraction caused a 78% reduction in the pheromone-titer. This result indicated that the cephalic region exercises a continuous control over pheromone production in female H. zea.

In an effort to determine which portion of the neuroendocrine system was responsible for the control of pheromone production, H. zea females were ligated 1 hr after emergence. Next, brain and CC, brain alone, and CC alone were dissected from 50-hr virgin females. The tissues were homogenized in 30 µl of physiological saline and injected with a 50 µl syringe into the abdomens of ligated females. Three hours later the pheromone titer in these females was determined. Injection of brain + CC or brain homogenate alone increased the average Z-11-HDAL levels in ligated females to 139 (n=10) and 132 (n=8) ng/female respectively, compared to 0.6 ng/female in ligated control females that were not injected. In CC-injected females only a small increase (14 ng/female) was observed. These results indicated that the factor responsible for activating pheromone production was present in the brain, and that the CC probably act as a neurohemal organ. This was confirmed by injecting hemolymph from 50-hr unligated females into ligated females. This treatment resulted in the recovery of an average of 19 ng Z-11-HDAL/female (n=5).

We also conducted several experiments to determine if the CA were involved in control of pheromone production in H. zea. Injection of CA homogenates from 50-hr unligated females did not cause any significant increase in the pheromone-titer in ligated females. Topical application of 10 µg of the juvenile hormones JH II, and JH III, singly and in combination to ligated females did not induce pheromone production. Likewise, topical application of 100 µg/female of fluoromevalonate (an inhibitor of JH biosynthesis) (Quistad et al., 1981), to unligated females 8 hr after emergence did not suppress pheromone production (ave. 117 ng Z-11-HDAL/female, n=5). Thus we conclude that the CA in H. zea are not involved in the control of pheromone production.

We have found that pheromone glands of female H. zea do not contain extractable pheromone in the photophase. However, when cell-free brain-tissue homogenate from 46-hr females (2 hr before the onset of scotophase) was injected into ligated females it triggered a significant rise in pheromone titer (ave. 145 ng Z-11-HDAL/female, n=5). The results indicate that a factor(s) responsible for control of pheromone titer in the female is continuously present in the brain and it is probably secreted into the hemolymph at the onset of the scotophase to trigger pheromone production in the pheromone gland.

We speculate that the pheromone-gland activating hormone (PAH) produced in the brain is released into the hemolymph in response to exogenous factor(s) such as photoperiod. A continous activation of the process of pheromone production is required through this hormone. Histological studies to identify the site of this hormone production in the brain and elucidation of its chemistry are planned.

This research was supported, in part, by an USDA Cooperative Grant No. 58-32U4-1-299. Scientific article No. 3550, contribution No. 6625 of the Maryland Agricultural Experiment Station.

REFERENCES

BARTH R. H. (1961) Hormonal control of sex attractant production in the Cuban cockroach. Science, Wash. 133,1598-1599.

BARTH R. H. and LESTER R. J. (1973) Neuro-hormonal control of sexual behavior in insects. Ann. Rev. Ent. 18,445-472.

BELL W. J. and BARTH R. H. (1970) Quantitative effect of juvenile hormone on reproduction in the cockroach Byrsotria fumigata. J. Insect Physiol. 16,2303-2313.

JEFFERSON R. N., SHOREY H. H. and RUBIN R. E. (1968) Sex pheromones of noctuid moths. XVI. The morphology of the female sex pheromone glands of eight species. Ann. Ent. Soc. Am. 61,861-865.

KLASSEN W., RIDGWAY R. L. and INSCOE M. (1982) Chemical attractants in integrated pest management programs. In Insect Suppression with Controlled Release Pheromone Systems (Edited by KYDONIEUS A.F. and BEROZA M. pp.13-130. CRS Press, Boca Raton, Florida.

KLUN J. A., PLIMMER J. R., BIERL-LEONHARDT B. A., SPARKS A. N., PRIMIANI M., CHAPMAN O. L., LEE G. H. and LEPONE G. (1980) Sex pheromone chemistry of female corn earworm moth, Heliothis zea. J. Chem. Ecol. 6,165-175.

QUISTAD G. B., CERF D. C., SCHOOLEY D. A. and STAAL G. B. (1981) Fluoromevalonate acts as an inhibitor of insect juvenile hormone biosynthesis. Nature,Lond. 289,176-177.

RIDDIFORD L. M. and WILLIAMS C. M. (1971) Role of the corpora cardiaca in the behavior of saturnid moths. I. Release of sex pheromone. Biol. Bull. 140, 1-7.

THE FORMATION OF THE INSECT BLOOD-BRAIN BARRIER: EVIDENCE FROM

THE COCKROACH NERVE CORD AGAINST THE TIGHT JUNCTION HYPOTHESIS.

Stephen R. Shaw and Deborah B. Henken

Department of Psychology
Dalhousie University
Halifax, Nova Scotia, Canada

The blood-brain barrier of insects is thought to be formed, as in the vertebrates, by a single seal of tight junctions, connecting perineurial cells that encircle the CNS at its interface with the haemolymph (review: LANE, 1981a). This picture fails to explain several findings on the barrier in the eye, which is extensive rather than superficial, and which despite earlier claims (LANE, 1981b), appears not to involve tight junctions (SHAW, 1978, 1983). Since it has been claimed that the barrier in the eye may be different from that present elsewhere in the CNS, we have re-examined the system for which the perineurial tight junction hypothesis was first proposed, the insect nerve cord. Our findings on the structure of the cord in the cockroach Periplaneta americana both disagree with earlier anatomical interpretation, and fail to support the tight junction hypothesis.

Perineurial cells are the supposed sites of desmosomes, septate, gap and tight junctions, and are readily identifiable by their morphology and superficial location below the neural lamella. We found no obvious junctions of any type between these obliquely inclined columnar cells. Nor do they form a detectable barrier. Ionic lanthanum (La^{3+}) applied 1h before electronmicroscopic fixation, obviously penetrated between and behind them (Fig. 1); this is in disagreement with earlier claims (LANE, 1981a). We quantified this by logging the penetration of La^{3+} down through each of the 50 or so main entry points (clefts) to the perineurial layer on the entire circumference of a section through the cord, in the abdominal connectives. Tracer reached the deepest point we could follow in 83% of the clefts; even the 17% failure is attributed to the general lightness of the deposits and to tracer washout during fixation. Most significantly, La^{3+} extensively plated the surface of the next layer of cells under the

471

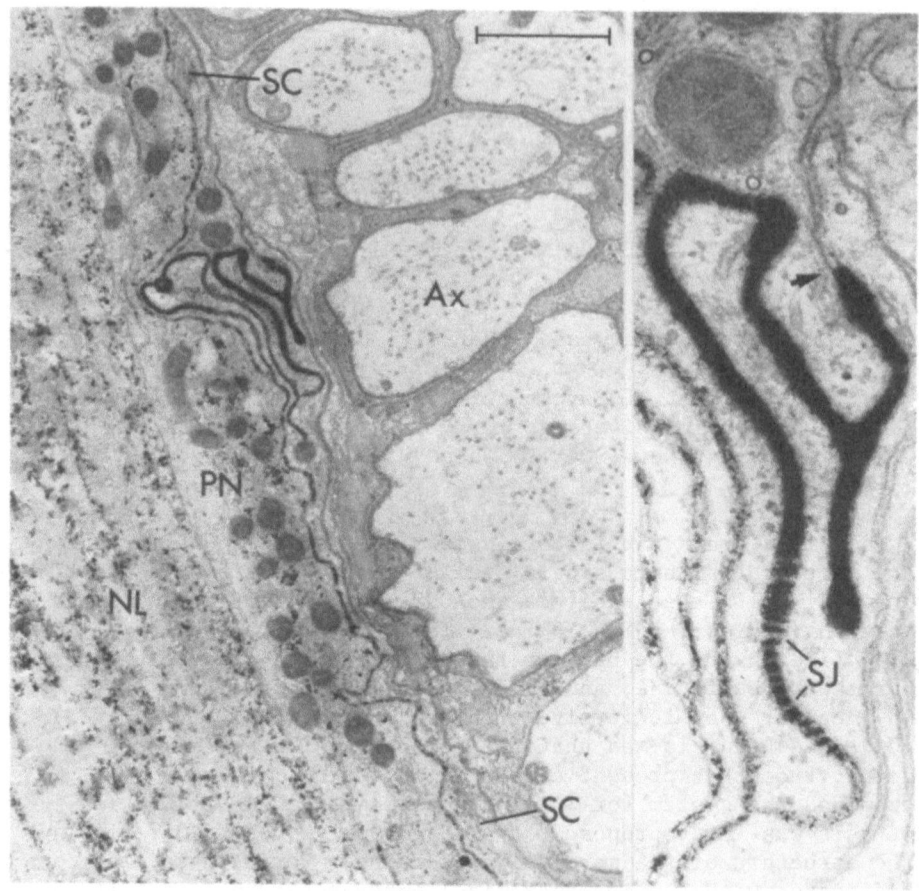

Fig. 1. A cockroach nerve cord after incubation in lanthanum, which
has surrounded the perineurial cells (PN). Between their infrequent
interlocking borders, the underlying sheath cells (SC) retain heavy
deposits of lanthanum, which fail near the innermost edge. Right:
one of a series of serial sections showing that the point of failure
(arrow) is not associated with a specialized intercellular junction.
NL, neural lamella; SJ, septate junction; Ax, axon. Scale bar, 1 μm.

perineurium.
     Below the perineurium lies a thin monolayer of distinctive
cells that have hitherto gone unrecognized (Fig. 1). Apart from
their deeper position, they are distinguishable from the cells
above and below by their breadth and thinness, by possessing
numerous microtubules, by making hemidesmosomes with local patches
of extracellular matrix material, and by sending processes down
between the neurones. These sheath cells halt the ingress of La^{3+}

and thus form the first line of defense of the barrier. They are most remarkable for their extent, since only 4-6 are responsible for completely encircling the cord. Where they meet, their borders interdigitate extensively making clefts up to 15μm long that retain particularly heavy deposits of La^{3+}, and are bridged by septate junctions. Tracer permeated extracellularly through much of the length of these junctions, but the deposits failed gradually near the inner borders of the sheath cells. In a survey of 10 sites, there was no exact correlation of the point of ultimate tracer stoppage with any particular junctional structure, although in several cases deposits ended inside a stretch of a narrow, septate-like junction. Serial sections were examined to ensure that tight junctions had not been overlooked in this region, and none at all were found.

The first stage of the barrier thus appears to be constructed through a straightforward dimensional restriction, occasioned by the extreme narrowing of the overall pathway available for extracellular diffusion, where this runs between the sheath cells. Additional diffusional retardation could be imposed by the extracellular matrix within these clefts, although the presence of La^{3+} deposits shows that if this occurs, its effect is not absolute, or else is confined to the innermost ends of the clefts.

In the eye, tracers introduced artificially 'behind' the barrier's edge still fail to penetrate significantly into the neural layers beneath, indicating that some additional process is restricting extracellular diffusion there (SHAW, 1983). We are currently trying to extend these experiments to the nerve cord. Thus far, the indications are that the barrier in the cord resembles that in the eye, in that neither the perineurial cells nor tight junctions are responsible for barrier formation. In light of recent clear evidence that tight junctions can be found in insects (CHI and CARLSON, 1980), their absence at the face of the blood-brain barrier is somewhat surprising.

REFERENCES

CHI, C. and CARLSON, S. D. (1980) Membrane specializations in the first optic neuropil of the housefly, Musca domestica L. J. Neurocytol. 9, 451-469.

LANE, N. J. (1981a) Tight junctions in arthropod tissues. Int. Rev. Cytol. 73, 243-318.

LANE, N. J. (1981b) Vertebrate-like tight junctions in the insect eye. Exp. Cell Res. 132, 482-488.

SHAW, S.R. (1978) The extracellular space and blood-eye barrier in an insect retina: an ultrastructural study. Cell Tiss. Res. 188, 35-61.

SHAW, S. R. (1983) Is the blood-brain barrier of insects just a single seal of tight junctions, as in vertebrates? Soc. Neurosci. Abstr. (in press).

HORMONE RELEASE IN <u>LOCUSTA</u> <u>MIGRATORIA</u> IN RELATION TO

INSECTICIDE POISONING SYNDROME

Gur Jai Pal Singh
Research Centre, Agriculture Canada
University Sub Post Office
London, Ontario, Canada   N6A 5B7

It is now well recognized that, in insects, neurotoxic insecticides provoke the release of various physiologically active substances including diuretic hormone and a cuticle plasticization
factor in <u>Rhodnius</u> <u>prolixus</u>, hyperglycaemic and adipokinetic hormones in <u>Schistocerca</u> <u>gregaria</u>, hyperglycaemic and hypoglycaemic
hormones in <u>Calliphora</u> <u>vicinia</u> (for references see Singh and
Orchard, 1982). In most of these studies the hormone release was
determined in insects poisoned with insecticides. Since most insecticides are neuroactive it has generally been assumed that hormone release at the onset of poisoning is due to the action of insecticides on the central nervous system (CNS).

The neurohaemal organs are exposed to the haemolymph and as
such are readily accessible to the toxic chemicals that enter the
haemolymph. Therefore it is possible that regardless of their
effect on the CNS insecticides present in the haemolymph may act
directly upon neurosecretory cells and provoke hormone release.

We have examined this possibility by studying the effects of
low concentrations of insecticides on the release of hyperlipaemic
hormone in <u>Locusta</u> <u>migratoria</u>. This hormone is synthesized in the
glandular lobe of locust corpus cardiacum (CC). Ultrastructural
evidence for bioresmethrin-induced release of neurosecretory material from the glandular lobe has been documented (Singh <u>et al</u>.,
1982). Incubation of isolated CC in low concentrations (1-10 μM)
or organochlorine and organophosphorus (Singh and Orchard, 1982)
and pyrethroid (Singh and Orchard 1983) insecticides increased the
spontaneous activity of the glandular lobe and provoked the release
of hyperlipaemic hormone as judged by the bioassay of the incubation
media. The concentrations of insecticides required to bring a not-

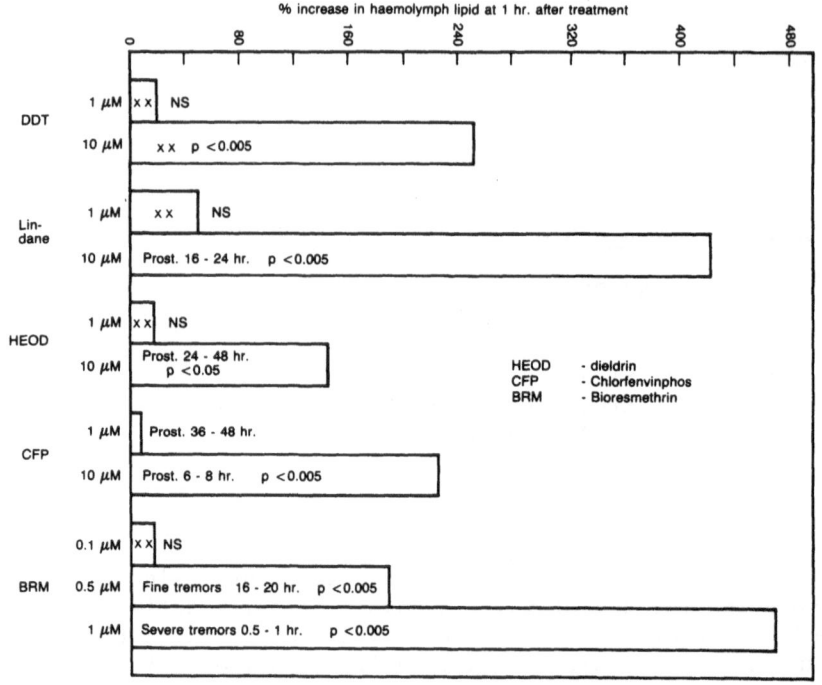

Fig. 1. Insecticide-induced release of hyperlipaemic hormone in
         L. migratoria in relation to symptoms of poisoning. The
         bars represent mean of four locusts, statistical analysis
         performed on raw data. xx - no poisoning symptoms obser-
         ved. Prost.-Prostrate.

able change in the spontaneous electrical activity of locust CC and
provoke hormone release were at least several fold less than that
required to modulate the ongoing electrical activity of the CNS.
These results clearly demonstrate that, compared with the CNS, the
neuroendocrine system is much more sensitive to the action of insecticides.

    Experiments were done to determine the relationship between
hormone release from the CC and insecticide poisoning syndrome in
intact locusts. Saline solutions of insecticides were injected into locusts (22 µl/locust) so that the final concentration of an insecticide in the haemolymph ranged from 0.1 µM to 10 µM. Treated
insects were left in jars up to 72 hr after treatment. One hr.
after the treatment the haemolymph lipid was determined as mentioned
elsewhere (Singh and Orchard, 1982). Elevation in the haemolymph
lipid indicates the release of hyperlipaemic hormone.

At a haemolymph conc. of 1 µM none of the organochlorine and organophosphorus compounds showed a significant effect (Fig. 1). However at an internal conc. of 10 µM all insecticides caused notable increase in the haemolymph lipid. Note that these insecticides provoked measureable release of hyperlipaemic hormone at a stage (1 hr. after treatment) when treated insects exhibited no poisoning symptoms. It would appear, therefore, that insecticide-induced hormone release precedes the onset of CNS poisoning.

The results obtained from treatment of locusts with bioresmethrin further clarify the role of neuroendocrine system in overall insecticide poisoning. At an internal concentration of 0.5 µM bioresmethrin induced significant hormone release without producing general symptoms of poisoning (Fig. 1). However an increase in the concentration of bioresmethrin to 1 µM caused severe tremors within 30-40 min. after treatment. In this case there was 470% increase in haemolymph lipid of treated locusts compared with that of 190% in locust with 0.5 µM haemolymph concentration of bioresmethrin. These results clearly show that following insecticide treatment of insects neurohormones are released before the poisoning symptoms become apparent and that the CNS hyperactivity induced by insecticide poisoning may augment the hormone release.

REFERENCES

Singh, G. J. P. and Orchard, I. (1982) Is insecticide induced release of insect neurohormones a secondary effect of hyperactivity of the central nervous system? Pestic. Biochem. Physiol. 17, 232-242.

Singh, G. J. P., Barker, J. F. and Kundu, S. C. (1982) Bioresmethrin-induced alterations in the ultrastructure of neurosecretory cells of insect corpora cardiaca. Pestic. Biochem. Physiol. 18, 158-168.

Singh, G. J. P. and Orchard, I. (1983) Action of bioresmethrin on the corpus cardiacum of Locusta migratoria. Pestic. Sci. (In press).

INSECTICIDE-INDUCED DEPLETION OF SYNAPTIC VESICLES IN INSECT CENTRAL NERVOUS SYSTEM

Gur Jai Pal Singh and Betty Singh

Agriculture Canada, Research Centre, University Sub Post Office, London, Ontario, Canada N6A 5B7

Though the field application of cyclodiene insecticides has been banned in western countries, problems related to their mode of action continue to intrigue insect toxicologists. Dieldrin and some of its metabolites have been extensively used to gain understanding of the mechanism of the neurotoxic action of cyclodiene compounds. In the insect central nervous system, the putative site of action of dieldrin is considered to be presynaptic terminals of cholinergic junctions (Shankland and Schroeder, 1973). Compared with dieldrin, some of its metabolites acted faster on the exposed nerve preparations. It has been considered, therefore, that to exert its action dieldrin may be metabolized to trans-aldrindiol in the nervous system. (For references see Singh and Singh, 1983).

We have examined the above hypothesis by studying dieldrin-induced transmitter release in sixth abdominal ganglion of Periplaneta americana at the ultrastructural level. The cockroaches were dissected as described elsewhere (Singh and Singh, 1983). The nerve cords (n=5) were treated with saline containing 2 µl/ml of ethanol (control). Separate groups (n=5) of exposed nerve cords were treated with 5 µM solutions of dieldrin and trans-aldrindiol. The incubation of both control and treated ganglia was carried out for periods of 10 min. and 2 hr., then the cockroach cavity was flooded with 1.5% glutaraldehyde in cacodylate buffer and the sixth abdominal ganglia were processed for electron microscopy as described by Singh and Singh (1983). Ultrathin sections were cut from the neuropile, stained with uranyl acetate and lead citrate and examined in a Jeol 100S electron microscope. Quantitative estimates of synaptic vesicle densities were determined from micrographs depicting axon terminals in different areas in the neuropile.

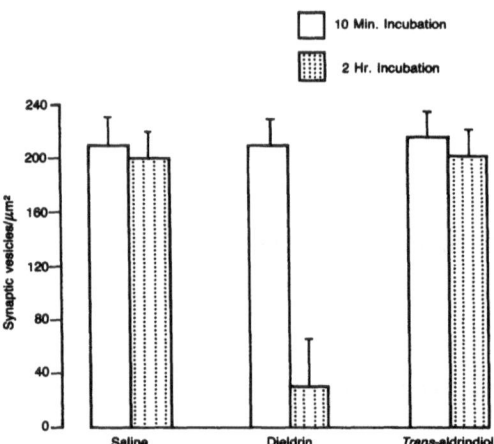

Fig. 1. Effect of dieldrin and <u>trans</u>-aldrindiol on synaptic vesicle population in the sixth abdomonal ganglia of <u>P.</u>
<u>americana</u>. Bars indicate mean $\pm$ SD (n=40).

In the neuropile of untreated ganglia the presynaptic axon
terminals are filled with either small lucent vesicles or dense
core vesicles or a mixture of both (see Singh and Singh, 1983).
Only small lucent vesicles were considered for quantitative analysis since these are known to contain acetylcholine. The 2 hr. incubation of ganglia in the saline had no effect on vesicle population (Fig. 1). Similarly a 10 min. incubation of ganglia in solutions of dieldrin and <u>trans</u>-aldrindiol had no effect on the number
of synaptic vesicles. There was, however, considerable depletion
of vesicles from axon terminals in the ganglia treated with dieldrin for 2 hr. (Fig. 1). Treatment of ganglia with <u>trans</u>-aldrin-
diol for 2 hr. had no significant effect on vesicle population.
The ultrastructural alterations in the neuropile following dieldrin
treatment were prevented by pretreatment of ganglia with 10mM Mg^{2+},
indicating, thereby, that dieldrin-induced transmitter release from
axon terminals is caused by an increase in the intracellular levels
of calcium. The mechanism of dieldrin-induced calcium influx is
obscure. There is no evidence to suggest membrane depolarization.

Depletion of synaptic vesicles in the neuropile of dieldrin-
treated ganglia was accompanied by mitochondrial swelling and the
formation of membraneous residual bodies (see Singh and Singh, 1983).
The mitochondria at nerve terminals are implicated in various biochemical processes. They provide energy for ATP-utilizing systems
involved in cation exchange and protein phosphorylation. At cholinergic terminals they are the main source of acetyl-COA which is
utilized in the synthesis of acetylcholine. Thus dieldrin by dam-
aging mitochondria at nerve terminals may block processes involved

in the synthesis and transport of acetylcholine. This, when combined with enhanced transmitter release, may have far reaching consequences leading to the failure of synaptic transmission.

Trans-aldrindiol did not cause the type of ultrastructural
alteration in the ganglia as observed following dieldrin treatment.
The differences in the action of these compounds upon the fine
structure of the neuropile were closely related to the intensity of
their effects on the electrical activity of the ganglia. The results of our study seem to indicate that observed electrophysiological effects of trans-aldrindiol upon the ganglia may not be due
to its direct action on cellular organelles critically involved in
synaptic transmission. On the other hand, neuronal lesions involving degeneration of cytoplasmic organelles occur in the neuropile
of dieldrin treated ganglia. The cellular organelles undergo des-
truction and digestion and undigested material accumulates as membraneous residual bodies. It may be possible, therefore, that the
mechanisms of action, on the nervous system, of dieldrin and trans-
aldrindiol may be different. To confirm this it may be necessary
to study the action of these compounds on preparations which are
less heterogenous and include a component of the nervous system
which is believed to be the target site.

REFERENCES

Shankland, D. L., and Shroeder, M. E. (1973) Pharmacological evi-
    dence for a discrete neurotoxic action of dieldrin (HEOD) in
    the American cockroach, Periplaneta americana (L.). Pestic.
    Biochem. Physiol. 3, 77-86.

Singh, G. J. P., and Singh, B. (1983) Action of dieldrin and
    trans-aldrindiol upon the ultrastructure of sixth abdominal
    ganglion of Periplaneta americana in relation to their electro-
    physiological effects. Pestic. Biochem. Physiol. (in press).

REGULATION OF JUVENILE HORMONE SYNTHESIS BY THE CORPORA ALLATA DURING

PREGNANCY IN THE VIVIPAROUS COCKROACH, DIPLOPTERA PUNCTATA

Barbara Stay and Susan M. Rankin

Department of Zoology
University of Iowa
Iowa City, IA 52242

Pregnancy in the viviparous cockroach, Diploptera punctata, refers to the 60 day period when embryos are nourished in the brood sac. During this time the corpora allata (CA) synthesize juvenile hormone (JH) at low rates (Tobe and Stay, 1977). The CA are held at low rates of synthesis in part by inhibitory sensory input from the brood sac (Roth and Stay, 1961). Removing embryos or releasing CA from neural inhibition by denervation in early pregnancy is followed by a long delay before CA are activated (Roth and Stay, 1961, Stay and Tobe, 1977). The ovary is important for regulating CA activity. Even when CA are denervated they do not show high rates of JH synthesis in the absence of the ovary (Stay et al., 1983). Furthermore, vitellogenic ovaries have increasing ability to stimulate the CA as they grow until almost the end of vitellogenesis (Rankin and Stay, 1983).

We now ask how the humoral environment of the early pregnant female can be changed to elicit increased rates of JH synthesis. Rates of JH synthesis were assayed by the in vitro radiochemical method of Feyereisen and Tobe (1981) following in vivo treatments of the CA in pregnant females at 16-20% of total gestation time, at 27° C. The rate of synthesis of CA from these 18-25 day old pregnant females is about 5-10 pmol h^{-1} per pair, and the basal oocytes, 0.38 to 0.42 mm in length, are not yet previtellogenic.

When the CA were completely denervated on Day 18 and a vitellogenic ovary (1.1 mm basal oocyte length) was implanted on Day 19, the mean rate of JH synthesis after 3 days was about twice that of saline injected controls; the implanted basal oocytes grew to 1.3 mm in length (Fig. 1A). One might expect much higher rates of synthesis by the stimulated glands in view of the ability of glands from 18-day

483

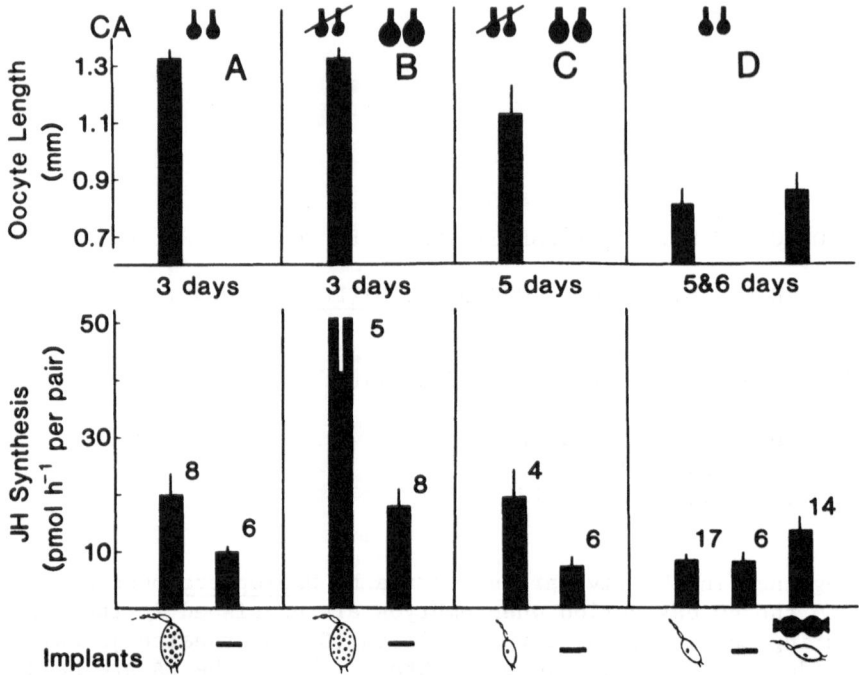

Fig. 1. Mean rates of JH synthesis and lengths of basal oocytes in
early pregnant females which were 18 or 19 days old when CA were
denervated (A,D) or removed and replaced with CA from 2 day mated
females (B,C) whose basal oocytes had begun vitellogenesis, (0.8 mm
long). In a second operation on Day 18 or 19 the females in A and
B were implanted with one ovary from a 4-day mated female with
vitellogenic oocytes 1.1 mm long (☜) or injected with saline (-);
those in C and D were implanted with an ovary from a 0-day mated
female with previtellogenic basal oocytes 0.6 mm long (☞), or with
saline (-), or with such an ovary and a brain (☞ ♠♠). The brains
were from females in the second gonadotropic cycle, with oocytes
1.2 mm long. Groups were assayed 3 (A,B), 5(C) or 5 and 6(D) days
after the second operation. The number of individual measurements
is shown beside each mean JH synthesis rate; the bars represent
standard errors of the means.

females to synthesize JH at 70 pmol h^{-1} per pair after 4 days in a
female during the first gonadotrophic cycle (Stay and Tobe, 1977).
To investigate whether higher rates of JH synthesis could be attained
in these pregnant females we removed the CA on Day 18, and on Day 19
implanted an ovary with 1.1 mm long basal oocytes and CA (from 2-day
mated females) with a rate of synthesis at implant of 25-30 pmol h^{-1}
per pair. After 3 days the rate of JH synthesis was 2.8x that of
the controls which received 2-day CA but no ovary (Fig. 1B). Thus,
a higher rate was attained in the presence of a stimulatory ovary but

the rate declined in the absence of one. A similar removal of host CA and substitution of CA from a 2-day mated female was made (on Day 19), but with an ovary containing previtellogenic basal oocytes 0.6 mm long (on Day 18). Oocytes become vitellogenic and stimulatory at 0.8 mm. After 5 days of interaction between the ovary and "pre-activated" CA, the rate of JH synthesis was no higher (Fig. 1C) than that shown in Fig. 1A. In the saline-injected controls the rate of JH synthesis declined further in 5 days (Fig. 1C) than in 3 days (Fig. 1B). The ability of the inactive CA of the pregnant female to respond to the 0.6 mm oocytes was tested by implanting such an ovary on Day 19 after denervating the CA on Day 18. The rate of JH synthesis after 5 and 6 days was not significantly different from that in saline-injected controls, although some of the implanted ovaries grew (Fig. 1D). In search of a factor which, along with the ovary, might activate these CA, a brain from a female in mid-second gonadotrophic cycle was implanted with the 0.6 mm ovary. The mean rate of JH synthesis was higher than that in animals with an ovary alone or with saline (Fig. 1D), but was not stimulated to the extent that was achieved by a larger ovary (Fig. 1A).

These experiments show that the rate of JH synthesis in early pregnant females can be increased above normal by denervating the CA and introducing a strongly stimulatory ovary. A previtellogenic ovary is able to maintain rates of JH synthesis in more active CA substituted for the host CA. Such an ovary can stimulate the denervated host CA when a brain is implanted along with the ovary. Further study is required to establish whether the brain releases an allatotropic factor under these conditions.

REFERENCES

Feyereisen, R. and Tobe, S.S. (1981) A rapid partition assay for routine analysis of juvenile hormone release by insect corpora allata. Anal. Biochem. 11, 372-374.
Rankin, S.M. and Stay, B. (1983) The changing effect of the ovary on rates of juvenile hormone synthesis in Diploptera punctata. Gen. comp. Endocr. (in press).
Roth, L.M. And Stay, B. (1961) Oocyte development in Diploptera punctata (Eschscholtz) (Blattaria). J. Insect Physiol. 25, 449-453.
Stay, B. and Tobe, S.S. (1977) Control of juvenile hormone biosynthesis during the reproductive cycle of a viviparous cockroach. I. Activation and inhibition of corpora allata. Gen. comp. Endocr. 33, 531-540.
Stay, B., Tobe, S.S., Mundall, E.C. and Rankin, S. (1983) Ovarian stimulation of juvenile hormone biosynthesis in the viviparous cockroach, Diploptera punctata. Gen. comp. Endocr. (in press).
Tobe, S.S. and Stay, B. (1977) Corpus allatum activity in vitro during the reproductive cycle of the viviparous cockroach, Diploptera punctata (Eschscholtz). Gen. comp. Endocr. 31, 138-147.

Thanks to J. Buschor, Berne, Switzerland for a helpful idea.

RAPID ISOLATION OF PROCTOLIN AND OTHER PHARMACOLOGICALLY ACTIVE CONSTITUENTS FROM COCKROACH TISSUES

Robert W. Steele and Alvin N. Starratt

Research Centre, Agriculture Canada
University Sub Post Office
London, Ontario N6A 5B7, Canada

Early investigations on the neuropeptide proctolin employed lengthy isolation procedures (Brown and Starratt, 1975) that deterred attempts at quantifying other myotropic peptides known to be present in cockroach tissues (Brown, 1965; Holman and Cook, 1972). The following procedures were developed specifically to provide a convenient first stage purification of such peptides. Only data on <u>Periplaneta americana</u> head extracts are presented, although other cockroach tissues have been similarly examined and several of the myotropic constituents have been found to be widely distributed.

Heads, in batches of 100, were homogenized in 50 ml ice-cold extraction media spiked with $[^3H\text{-}Tyr^2]$proctolin $(6.5 \times 10^5$ dpm; 5.8 ng), centrifuged at 30,000g for 30 min $(4°C)$, and the supernatants separated and dried. Residues were dissolved in water (5 ml) and chromatographed on Sep-Pak C_{18} cartridges using the procedures of Bennett et al. (1981). Retained materials were eluted with 5 ml acetonitrile-water (1:1, v/v) containing 10 mM trifluoroacetic acid (TFA) and redried. The residues, redissolved in water (200 µl), were chromatographed on a µBondapak C_{18} column and analyzed for radioactivity as described in Fig. 1. After removal of solvent, residues were redissolved in water (500 µl) and bioassayed (Starratt and Steele, 1983) for both excitatory and inhibitory bioactivity.

Four media (20% (v/v) TFA, methanol-water (90:10, v/v), methanol-water-acetic acid (90:9:1, v/v), and acetonitrile-water-acetic acid (60:10:10, v/v)), were compared (n=2) for

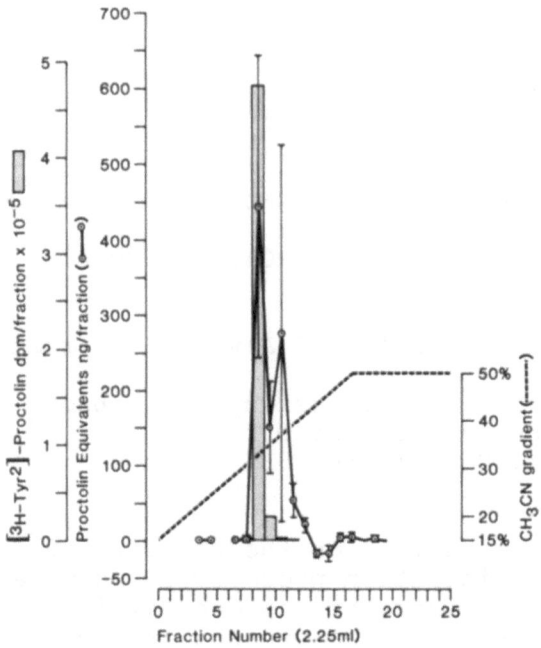

Fig. 1. HPLC fractionation of [^{3}H-Tyr^2]proctolin and the
pharmacologically active constituents in P. americana head extracts (see Text). Column: μBondapak
C$_{18}$ (30 x 0.39 cm). Elution: 20 min linear gradient
from 15% to 50% (v/v) acetonitrile containing 5 mM
heptafluorobutyric acid (1.5 ml/min). ^{3}H-Radio-
activity was determined by liquid scintillation
counting of 50 μl aliquots. Pharmacological
activity, ± SD (n=8), was determined by bioassay
after solvent removal.

extraction efficiency and the composite result is shown
in Fig. 1. As judged by [^{3}H-Tyr^2]proctolin measurements,
these extraction media were of similar effectiveness
giving recoveries after HPLC fractionation of 78.5 + 2.7%
SD (n=8). The TFA extracts differed from the others by
containing red-brown pigmented materials. TFA and acetonitrile-water-acetic acid extracts contained relatively
more of the proctolin-like bioactivity (PLB) that
comprised a secondary peak in fraction 11 (Fig. 1).

O'Shea et al. (1982) detected only 0.032 ng/head PLB
by radioimmunoassay. In contrast to this estimate, the
data in Fig. 1 show about 10 ng/head PLB. Our higher
estimate can be explained, at least in part, by PLB

contributed by constituents other than proctolin. Indeed, the constituents in fraction 11 (Fig. 1) were subsequently estimated to contribute about 40% of total PLB after complete separation from authentic proctolin was obtained by re-chromatography on the same μBondapak C_{18} column using isocratic elution with 15% (v/v) acetonitrile–water containing 5 mM TFA. Moreover, when fraction 9 (Fig. 1) was re-chromatographed under the same conditions, a major peak of PLB co-chromatographed with [^{3}H-Tyr^2]proctolin but a uniquely different constituent was separated and estimated to contribute about 25% to total PLB. Several additional minor peaks of PLB were also detected. The major peak was further characterized on the μBondapak C_{18} column using isocratic elution with 10% (v/v) acetonitrile in 20 mM ammonium acetate pH 4.5, and subsequently on a Micropak AX-10 column (30 x 0.4 cm) eluted with 75% (v/v) acetonitrile in 10 mM triethylammonium acetate pH 6.0 (Dizdaroglu et al., 1982). No additional myotropic constituents were fractionated by these procedures. This PLB, which was chromatographically indistinguishable from [^{3}H-Tyr^2]proctolin was estimated at 2-4 ng/head. Fig. 1 also shows the presence in fractions 14,15 of a previously undescribed inhibitory peptide which we have named neutrolin.

REFERENCES

BENNETT H.P.J., BROWNE C.A. and SOLOMON S. (1981) Purification of the two major forms of rat pituitary corticotropin using only reversed-phase liquid chromatography. Biochem. 20, 4530-4538.

BROWN B.E. (1965) Pharmacologically active constituents of the cockroach corpus cardiacum: resolution and some characteristics. Gen. comp. Endocr. 5, 387-401.

BROWN B.E. and STARRATT A.N. (1975) Isolation of proctolin, a myotropic peptide from Periplaneta americana. J. Insect Physiol. 21, 1879-1881.

DIZDAROGLU M., KRUTZSCH H. and SIMIC M.G. (1982) Separation of peptides by high-performance liquid chromatography on a weak anion-exchange bonded phase. J. Chromat. 237, 417-428.

HOLMAN G.M. and COOK B.J. (1972) Isolation, partial purification and characterization of a peptide which stimulates the hindgut of the cockroach, Leucophaea maderae (Fabr.). Biol. Bull. Woods Hole 142, 446-460.

O'SHEA M., ADAMS M.E. and BISHOP C.A. (1982) Identification of proctolin-containing neurons. Federation Proc. 41, 2940-2947.

STARRATT A.N. and STEELE R.W. (1983) In vivo inactivation of the insect neuropeptide proctolin in Periplaneta americana. Insect Biochem. In press.

CARDIOACCELERATORY PEPTIDES(CAPs) AND THE INSECT HEART

Nathan Tublitz and James W. Truman

Department of Zoology NJ-15
University of Washington
Seattle, WA   98195

The heart of the tobacco hawkmoth, Manduca sexta, is very
responsive to a number of putative neurotransmitter substances. We
have taken advantage of this fact to investigate the physiology and
biochemistry of two novel cardioregulatory neuropeptides isolated
from this insect.

To detect the activity of cardioactive factors, a highly sen-
sitive(threshold=0.1 picomoles serotonin) in vitro Manduca heart
bioassay was developed. A survey of the central nervous system(CNS)
of pharate adult moths revealed two factors from the ventral nerve
cord(VNC), each of which caused a dose-dependent increase in heart
rate when bioassayed on the isolated Manduca heart. These cardioacceleratory factors were heat-stable, small(ca. 500 and 1000 daltons, respectively), and sensitive to several proteases, suggesting
that each of these factors is a peptide. The two cardioacceleratory
peptides(CAPs) were localized to the perivisceral organs(PVOs), the
segmental neurohaemal release sites for the ventral ganglia.

One physiological function for these peptides occurs at adult
emergence of Manduca. In vivo recordings of heart rate during
adult emergence and subsequent wing-spreading behaviors showed that
heart rate increased immediately following emergence and remained
elevated throughout wing-spreading. The heart rate returned to pre-
emergence levels only after the wings were fully inflated.

Results from a number of diverse experiments indicated that the
CAPs were responsible for the increase in heart rate seen during
wing-spreading behavior. Two cardioacceleratory factors were found
in the haemolymph only during wing-spreading and these blood-borne
factors co-eluted with the two CAPs on gel filtration columns. In
addition, a significant depletion in the amount of CAP stored in
the VNC was seen during the time of wing-spreading. Furthermore,
measurements of CAP activity in the haemolymph showed that the CAP

491

titer peaked at the same time as the initiation of this behavior.
Finally, experimental manipulations which delayed wing-spreading
behavior also delayed CAP release.

We have recently identified the neurosecretory cells which
synthesize and release the two CAPs. Initial experiments, utilizing
microsurgical dissection of individual abdominal ganglia, implicated
a small number of midline neurosecretory cells that project to the
PVOs. From ganglion extirpation experiments, the list of putative
CAP-containing neurons was pared to three pairs of midline cells,
which differentiate post-embryonically, arising during the period
of metamorphosis from larva to adult. Other data, including the
growth kinetics of these neurons and the rate of accumulation of
the CAPs in the VNC during adult development, provided supporting
evidence for this interpretation. Direct evidence was obtained by
electrophysiological experiments:intracellular stimulation of a
single, midline neuron evoked the release of CAP bioactivity from
the PVO as assayed on the isolated Manduca heart. These data
stringly support the notion that the midline neurons contain and
release the CAPs.

In summary, two cardioacceleratory peptides(CAPs) were isolated
from the PVO of Manduca sexta, the tobacco hawkmoth. A physiological
role was assigned to these peptides; they are responsible for the
increase in heart rate seen during wing-spreading behavior in the
adult moth. We have also identified the neurons which synthesize
and release the CAPs. Future research will focus on mode of action
of the CAPs upon the heart and the possible involvement of these
two peptides in other behaviors.

IN VITRO ANALYSIS OF BRAIN ECDYSIOTROPIN OF THE DIAPAUSING AND

NON-DIAPAUSING SOUTHWESTERN CORN BORER, DIATRAEA GRANDIOSELLA DYAR

Chih-Ming Yin

Department of Entomology
University of Massachusetts
Amherst, MA 01003, USA

Ecdysiotropin (ET) stimulates the ecdysial gland to produce and secrete ecdysone. Target tissues then hydroxylate it at the C-20 position to form the molting hormone ecdysterone. A relationship between
diapause and ET secretion has been recognized for decades. For
example, pupal diapause has been traced to ecdysone deficiency due to
the lack of ET secretion (Williams, 1952). In contrast, different
opinions have been offered concerning ET activity in diapausing
larvae (see review in Chippendale, 1977). In the southwestern corn
borer (SWCB), histological evidence from brain sections (Yin and
Chippendale, 1973) and whole mount brain preparations (Yin, unpublished) indicate that ET producing cells of diapausing and non-diapausing larvae exhibit similar staining properties. Thus, cell activities can't be clearly deliniated and point to needs for further work.
My study utilized a recently developed in vitro bioassay for the ET
of the tobacco hornworm, Manduca sexta (Bollenbacher et al., 1979;
Carrow et al., 1981) to determine ET activities in diapausing and
mature non-diapasuing female SWCB. Laboratory rearing procedure for
SWCB has been published (Yin and Peng, 1981). Female non-diapausing
larvae (FNDL) were collected on days 1 to 5 post last ecdysis under a
$30^{\circ}C$, 12L:12D regime. Female diapausing larvae (FDL) were staged on
days 45, 60, 110, 170 of age under a $23^{\circ}C$, 12L:12D regime. Diapause-
bound larvae usually enter diapause around 45 days of age and reach
50% pupation in 200 days (Yin and Chippendale, 1973). Brain and EG
dissections, culturing conditions, and ET extractions will be published elsewhere (Yin and Wang, in preparation). The ecdysteroid radioimmunoassay has been described in detail (Bollenbacher et al., 1975).
Feasibility study
To develop an in vitro ET bioassay the following criteria have to be
met: 1) isolated EG continues to secrete ecdysone in the medium, 2)
member glands of an EG pair should secrete at comparable rates to
allow for the culture of one with ET preparation and the other (alone)
as a background control, and 3) cultured gland should remain respon-

493

sive to ET. EG from the SWCB meet all these requirements. Randomly selected EG from days 1 to 5 last instar FNDL produced 8.3 ± 5.9^a(7), 9.3 ± 6.7^a(9). 39.0 ± 19.8^a(7), 79.4 ± 35.5^a(7), and 202.5 ± 121.9^b (7) pg of $\propto$-ecdysone equivalent of ecdysteroid/gland/6 hr respectively. Means followed by different letters are significantly different (p= 0.05, Duncan's analysis) and the numbers of EG donors (each donated one gland) are shown in parenthesis. The rather large standard error (S.E.) indicated that EG from donors of the same age are not highly synchronized in their ecdysone secretion. Nevertheless, it is clear that day 5 EG produced significantly more than glands of other ages. Data also show that secretory rates of left and right glands are comparable. For 7 pairs (from 18 days old last instar FNDL) studied, hourly differences ranged from 2.8 to 20.8 pg (mean= 10.5 ± 6.5 pg). Thus the differences are small enough for the intended bioassay. EG of different ages responded to ET differently. Evoked ecdysone productions of days 1 to 5 EG were 52 ± 23^a, 370 ± 158^{ab}, 223 ± 113^a, 814 ± 581^b, and $5,528 \pm 2,996^c$ pg/gland/6 hr more than their respective controls. The results indicate that the intended ET-EG bioassay is feasible for the SWCB.

Axonal transport and brain storage of ET

   Day 4 last instar FNDL provided target EG for brain, brain extract, brain-CC, and brain-CC-CA in pilot tests. Only brains and extracts showed consistent tropic effects. Brain-EG cultures revealed axonal transport of ET from its producing cells to the neurohemal organs, while extract-EG cultures revealed total ET in the brain. Brains of days 1 to 3 last instar FNDL evoked ecdysone secretion in test EG with the highest effect from day 3 brains (Table 1). In contrast, days 4 and 5 brains had negative effects. Large S.E. again reflect the innate variability in the activities of brains and EG of the same age. Axonal transport of ET showed no relationship with hemolymph ecdysteroid titers at the same age (parenthesis, Table 1). In diapausing larvae, days 45, 60, and 170 brains evoked ecdysone production, while day 110 brains either elicited or inhibited the gland. Again no defined relationship can be found between axonal transport and hemolymph ecdysteroid titers.

   Regardless of brain age, all extracts from FNDL showed high ET activities. Thus, total ET does not reflect axonal transport. Total ET of FDL brains varied greatly during diapause but low axonal transport was not always associated to high ET remained in the brain (Table 1). But, high axonal transport in late diapausing larvae (day 110) was related to a depletion of ET in the brain.

   Lack of correlation between axonal transport and total brain ET, and between axonal transport and hemolymph ecdysteroid titer strongly suggests that 1) brain implantations caused ecdyses may reflect only axonal transport of ET but not normal neurohemal organ release in a in vivo molting process, and 2) brain activities estimated by comparing stainable material of the neurosecretory cells may only evaluate synthesis and storage of ET in the cells but not axonal transport (and neurohemal organ release). In systems comparable to the SWCB, conclusions concerning brain activity estimated by the above mentioned methods may need to be reexamined. Future experiments on brain-neurohemal organ activity should be designed to consider the activity at four levels, i.e. synthesis and storage in the secretory cells, axonal transport, neurohemal organ storage, and neurohemal organ release. From previous and present data, it is clear that the onset and maintenance of diapause in the SWCB are related to specific juvenile hormone (Yin and Chippendale, 1973, 1979), ecdysteroid (Yin, in preparation), and ET activities. I thank NSF (PCM 80-10948) and Massachusetts Agricultural Experiment Station (Hatch 461) for support, and Drs. L.I. Gilbert and W.E. Bollenbacher for anti-ecdysone anti-serum.

Table 1. Ecdysteroid secretion evoked by brain or brain extract from female last instar non-diapausing and diapausing SWCB.

Donor age	Brain Ecdysteroid*(mean ± S.E.)		Brain extract Ecdysteroid*(mean ± S.E.)
Non-diap. larvae			
Day 1	41.7 ± 29.8[ab]	(4.5 ± 1.5)**	598.5 ± 428.9[ab]
Day 2	17.7 ± 10.3[ab]	(7.2 ± 3.6)	676.3 ± 367.8[ab]
Day 3	137.9 ± 94.4[c]	(18.4 ± 4.7)	670.7 ± 480.5[a]
Day 4	-32.4 ± 30.8[a]	(23.3 ± 9.2)	700.9 ± 588.7[a]
Day 5	-17.1 ± 16.8[a]	(56.7 ± 16.8)***	902.6 ± 691.0[a]
Diap. larvae			
Day 45	84.5 ± 25.4[a]	(35.0 ± 31.7)**	837.5 ± 162.3[a]
Day 60	37.8 ± 23.5[a]	(15.4 ± 10.8)	45.8 ± 91.4[b]
Day 110	20.5 ± 60.5[a]	(9.0 ± 2.2)	441.8 ± 147.6[c]
Day 170	458.8 ± 249.0[b]	(13.5 ± 10.1)	76.8 ± 42.9[b]

* pg of α-ecdysone equival./gland/6 hr. ** Corresponding hemolymph ecdysteroid titer in ng/ml. *** From day 4.7 larvae. Means followed by different letters are significantly different(p=0.05, Duncan's test).

REFERENCES

Bollenbacher W.E., Vedeckis W.V., Gilbert L.I. and O'Connor J.D. (1975) Ecdysone titres and prothoracic gland activity during the larval-pupal development of Manduca sexta. Devel. Biol. 44, 46-53.

Bollenbacher W.E., Agui N., Granger N.A. and Gilbert L.I. (1979) In vitro activation of insect prothoracic glands by the prothoracicotropic hormone. Proc. Natl. Acad. Sci. USA 76, 5148-5152.

Carrow G.M., Calabrese R.L. and Williams C.M. (1981) Spontaneous and evoked release of prothoracicotropin from multiple neurohemal organs of the tobacco hornworm. Proc. Natl. Acad. Sci. USA 78, 5866-5870

Chippendale G.M. (1977) Hormonal regulation of larval diapause. A. Rev. Ent. 22, 121-138.

Williams C.M. (1952) Physiology of insect diapause IV. The brain and prothoracic glands as an endocrine system in the cecropia silkworm. Biol. Bull. Woods Hole, 103, 120-138.

Yin C.-M. and Chippendale G.M. (1973) Juvenile hormone regulation of the larval diapause of the southwestern corn borer, Diatraea grandiosella. J. Insect Physiol. 19, 2403-2420.

Yin C.-M. and Chippendale G.M. (1979) Diapause of the southwestern corn borer, Diatraea grandiosella: further evidence showing juvenile hormone to be the regulator. J. Insect Physiol. 25, 513-523.

Yin C.-M. and Peng W.-K. (1981) Simplified soy pulp-wheat germ diets for rearing the southwestern corn borer, Diatraea grandiosella Dyar. Ann. ent. Soc. Am. 74, 425-427.

PARTICIPANTS

ORGANIZING COMMITTEE

A. B. Borkovec, Chairman
R. E. Menzer, Secretary-Treasurer
L. L. Keeley, Program Chairman
A. B. DeMilo, Registrar

P. A. Cruickshank	M. J. Loeb
K. G. Davey	M. Ma
J. L. Frazier	E. P. Marks
D. B. Gelman	E. P. Masler
L. I. Gilbert	J. J. Menn
J. V. Gramlich	G. C. Pliszka
W. H. Grimes	H. A. Schneiderman
R. G. Hanson	J. de Wilde
T. J. Kelly	C. M. Williams
W. Klassen	C. W. Woods

REGISTRANTS

MICHAEL E. ADAMS
Zoecon Corp.
975 California Avenue
Palo Alto, CA 94304   USA

WILLIAM E. ALLISON
Dow Chemical Co.
Drawer H
Walnut Creek, CA 94596   USA

SHALOM APPLEBAUM
Hormone Research Laboratory
Univ. of California
San Francisco, CA 94143   USA

D. A. AVÉ
Dept. of Breeding &
   Biometry
Cornell Univ.
Ithaca, NY 14853   USA

HAFEZ M. AYAD
Union Carbide
   Agricultural Prod.
   Co., Inc.
P.O. Box 12014
Research Triangle
   Park, NC 27709 USA

JOHN F. BAKER
USDA Basic Biology Lab
1700 S.W. 23rd Drive
Gainesville, FL 32604 USA

DAVID J. BEADLE
School of Biological
    Sciences
Thames Polytechnic
London SE 18 6PF
United Kingdom

NANCY BECKAGE
Issaquah Health Research
    Institute
Issaquah, WA 98027 USA

ROBERT A. BELL
USDA/ARS
Otis Methods Development
    Center
Otis ANGB, MA 02542 USA

JACK A. BENSON
Ciba Geigy Ag. R1060.4.30
CH-4002 Basel
Switzerland

KEVIN L. BLAIR
Pesticide Research Center
Michigan State University
E. Lansing, MI 48824 USA

WALTER E. BOLLENBACHER
Dept. of Biology
Univ. of North Carolina
Chapel Hill, NC 27514 USA

ALEXEJ B. BOŘKOVEC
IRL/USDA
BARC-East 306
Beltsville, MD 20705 USA

EUGENE BRADY
Dept. of Entomology
Univ. of Georgia
Athens, GA 30602 USA

HEINZ BREER
Dept. of Zoophysiology
Univ. of Osnabruck
4500 Osnabruck
W. Germany

GERALD T. BROOKS
ARC Insect Chemistry and
    Physiology Group
Univ. of Sussex
Brighton BN1 9RQ
United Kingdom

JOHN J. BROWN
Dept. of Entomology
Washington State Univ.
Pullman, WA 99164 USA

MARK BROWN
Dept. of Entomology
Univ. of Georgia
Athens, GA 30602 USA

PHILIP S. CALLAHAN
USDA Basic Biology Lab
P.O. Box 14565
Gainesville, FL 32604
USA

JAMES A. CARLISLE
Dept. of Biology
York University
Downsview, Ontario
M3J 1P3
Canada

GRANT M. CARROW
The Biological
    Laboratories
Harvard University
Cambridge, MA 02138 USA

ANDREW C. CHEN
USDA/VTERL
P.O. Drawer GE
College Station, TX 77840
USA

DAVID J. CHITWOOD
IPL/USDA
BARC-East 467
Beltsville, MD 20705 USA

WENDELL L. COMBEST
Dept. of Biology
Wilson Hall 046A
Univ. of North Carolina
Chapel Hill, NC 27514   USA

BENJAMIN J. COOK
USDA/VTERL
P.O. Drawer GE
College Station, TX   77840
USA

DIANA LYNN COX
Dept. of Entomology
Univ. of Illinois
Urbana, IL   61801   USA

PHILIP A. CRUICKSHANK
FMC Corporation
Box 8
Princeton, NJ   08540   USA

DOUGLAS L. DAHLMAN
Dept. of Entomology
Univ. of Kentucky
Lexington, KY   40546
USA

KENNETH G. DAVEY
Dept. of Biology
York University
Downsview, Ontario
M3J 1PS
Canada

NORMAN T. DAVIS
Univ. of Connecticut
Storrs, CT   06268   USA

ALBERT B. DeMILO
USDA/IRL
BARC-East 306
Beltsville, MD   20705   USA

JAN de WILDE
Dept. of Entomology
Agricultural Univ.
6709 PD Wageningen
Netherlands

ELVIRA DOMAN
National Science
    Foundation
1800 G Street, NW
Washington, DC 20550   USA

ROGER G. H. DOWNER
Dept. of Biology
Univ. of Waterloo
Waterloo, Ontario
N2L 3G1
Canada

HANNE DUVE
School of Biological
    Sciences
Queen Mary College
Univ. of London
Mile End Road
London E1 4NS
United Kingdom

JOHN L. EATON
Dept. of Entomology
Virginia Polytechnic
    Institute St. Univ.
Blacksburg, VA   24061   USA

LEE EIDEN
National Institutes of
    Mental Health
Bldg. 10, Rm. 4N-312
Bethesda, MD   20205   USA

STANLEY R. FARKAS
Stauffer Chemical Co.
P.O. Box 760
Mt. View, CA   94042 USA

LAWRENCE H. FINLAYSON
Dept. of Zoology and
    Comparative Physiology
Univ. of Birmingham
P.O. Box 363
Birmingham B15 2TT
United Kingdom

JAMES L. FRAZIER
Du Pont Experimental Station
Biochemicals Dept.
Wilmington, DE 19098   USA

GERD GÄDE
Institut fur Zoophysiologie
Universitat Bonn, AVZI,
    Endeniche
Allee 11-13, Bonn
W. Germany

DEREK W. GAMMON
FMC Corporation
Box 8
Princeton, NJ   08540   USA

SEMAHAT GELDIAY
Ege Universitesi, Fen
    Fakultesi
Biyoloji Bolumu
Bornova, Izmir
Turkey

DALE B. GELMAN
IRL/USDA
BARC-East 306
Beltsville, MD   20705   USA

EDWIN G. GEMRICH
Dept. 9602-50-1
The Upjohn Company
Kalamazoo, MI   49001   USA

DANIEL GIBBS
Dept. of Biological Sciences
DePaul Univ.
Chicago, IL   60614   USA

LAWRENCE I. GILBERT
Dept. of Biology
Univ. of North Carolina
Chapel Hill, NC 27514
USA

J. V. GRAMLICH
American Cyanamid
P.O. Box 400
Princeton, NJ 08540 USA

NOELLE A. GRANGER
Dept. of Anatomy
Univ. of North Carolina
Chapel Hill, NC   27514
USA

WALTER H. GRIMES
Mobay Chemical Corp.
P.O. Box 4913
Kansas City, MO   64120
USA

H. H. HAGEDORN
Dept. of Entomology
Cornell University
Ithaca, NY   14853   USA

ABNER M. HAMMOND, JR.
Dept. of Entomology
Louisiana State Univ.
Baton Rouge, LA   70803
USA

FRANK E. HANSON
Dept. of Biology
UMBC
Catonsville, MD   21228
USA

RICHARD J. HART
The Wellcome Foundation
    Ltd.
Berkhamsted Hill
Berkhamsted Herts, HP4
    2QE
United Kingdom

NORBERT HAUNERLAUD
Dept. of Entomology
NYS Agricultural
    Experiment Station
Geneva, NY   14456   USA

DORA K. HAYES
LIL/USDA
BARC-East 307
Beltsville, MD   20705   USA

TIMOTHY K. HAYES
Dept. of Entomology
Texas A&M University
College Station, TX   77843
USA

ARTHUR HESS
Dept. of Anatomy
UMDNJ-Rutgers Medical
    School
Piscataway, NJ   08854   USA

JOHN G. HILDEBRAND
Dept. of Biological
    Sciences
Columbia University
New York, NY   10027   USA

G. MARK HOLMAN
VTERL/USDA
P.O. Drawer GE
College Station, TX 77841
USA

CALEB W. HOLYOKE
E.I. Du Pont de Nemours &
    Co., Inc.
Experimental Station
Biochemicals Dept.
B-324
Wilmington, DE   19898   USA

STEPHEN IRVING
ICI Plant Protection Div.
Jealott's Hill Research
    Station
Bracknell, Berkshire
    RG12 6EY
United Kingdom

HARUHIKO ITAGAKI
Dept. of Zoology
Duke University
Durham, NC   27706   USA

F. ROB JACKSON
Albert Einstein College
    of Medicine
Bronx, NY   10461   USA

HOWARD JAFFE
USDA/ARS
BARC-East 307
Beltsville, MD   20705   USA

ERIC B. JANG
USDA
P.O. Box 917
Stain Back Highway
Hilo, HI   96720   USA

KENT R. JENNINGS
American Cyanamid Co.
Insecticide Discovery
    Dept.
P.O. Box 400
Princeton, NJ   08540   USA

DAVY JONES
Dept. of Entomology
Univ. of Kentucky
Lexington, KY   40506   USA

GRACE JONES
Dept. of Entomology
Univ. of Kentucky
Lexington, KY   40506   USA

LARRY L. KEELEY
Dept. of Entomology
Texas A&M Univ.
College Station, TX   77843
USA

SCOTT T. KELLOG
The Salk Institute Bio-
    technology/Industrial
    Associates, Inc.
P.O. Box 85200
San Diego, CA   92138   USA

THOMAS J. KELLY
IRL/USDA
BARC-East 306
Beltsville, MD 20705 USA

TIMOTHY G. KINGAN
Dept. of Biological
    Sciences
Columbia University
New York, NY 10027 USA

WALDEMAR KLASSEN
USDA/NPS
BARC-West 005
Beltsville, MD 20705 USA

DAVID G. KUHN
American Cyanamid Co.
P.O. Box 400
Princeton, NJ 08540 USA

THOMAS A. LAJINESS
S.C. Johnson & Son, Inc.
Mail Station 401
Racine, WI 53403 USA

ANGELA B. LANGE
Dept. of Biology
York Univ.
4700 Keele Street
Downsview, Ontario
M3J 1P3
Canada

ARDEN O. LEA
Dept. of Biology
Univ. of Georgia
Athens, GA 30602 USA

WOLFGANG LEICHT
Bayer AG
Biologische Forschung
Pflanzenshutz
509 Leverkusen
W. Germany

DAVID LESTER
Dept. of Biology
Univ. of North Carolina
Chapel Hill, NC 27514 USA

ROBERT J. LITTLE, JR.
American Cyanamid Co.
P.O. Box 400
Princeton, NJ 08540 USA

MARCIA LOEB
IRL/USDA
BARC-East 306
Beltsville, MD 20705 USA

BARRY G. LOUGHTON
Dept. of Biology
York Univ.
Downsview, Ontario
M3J 1P3
Canada

ALBERT E. LUND
E.I. Du Pont de Nemours
    & Co., Inc.
Biochemicals Dept.
B-335
Experimental Station
Wilmington, DE 19898 USA

MICHAEL MA
Dept. of Entomology
Univ. of Maryland
College Park, MD 20742
USA

NABIL A. MANSOUR
Dept. of Plant
    Protection
Div. of Pesticide Chem.
Univ. of Alexandria
El-Chtby, Alexandria
Egypt

EDWIN P. MARKS
USDA/MRRL
P.O. Box 5674
State University Station
Fargo, ND 58103 USA

PAMELA G. MARRONE
Monsanto Agricultural
    Products Co.
800 N. Lindbergh Blvd.
St. Louis, MO 63167 USA

EDWARD P. MASLER
USDA/IRL
BARC-East 306
Beltsville, MD 20705 USA

FUMIO MATSUMURA
Michigan State University
Pesticide Research Center
East Lansing, MI 48824
USA

RICHARD T. MAYER
USDA/VTERL
P.O. Drawer GE
College Station, TX 77841
USA

JULIUS J. MENN
Zoecon Corp.
975 California Avenue
Palo Alto, CA 94304 USA

ROBERT E. MENZER
Dept. of Entomology
Univ. of Maryland
College Park, MD 20742 USA

SHIRLEE MEOLA
USDA/VTERL
P.O. Drawer GE
College Station, TX 77841
USA

NANCY S. MILBURN
Dept. of Biology
Tufts Univ.
Medford, MA 02155 USA

WILLIAM MORDUE
Dept. of Zoology
Univ. of Aberdeen
Tillydrone Ave.
Aberdeen AB9 2TN
United Kingdom

MICHAEL A. MORELLI
Rohm and Haas Co.
727 Norristown Road
Spring House, PA 19477 USA

PETER J. MORGAN
Dept. of Zoology
Univ. of Aberdeen
Tillydrone Ave.
Aberdeen AB9 2TN
United Kingdom

D. MURALEEDHARAN
Dept. of Zoology
Univ. of Kerala
Kariavattom 695581
Trivandrum
India

LARRY L. MURDOCK
Dept. of Entomology
Purdue Univ.
West Lafayette, IN 47907
USA

JAMES A. NATHANSON
Harvard University
Massachusetts General
  Hospital
Boston, MA 02114 USA

JUDD O. NELSON
Dept. of Entomology
Univ. of Maryland
College Park, MD 20742 USA

JUNKO NISHIITSUTSUJI-UWO
Shionogi Research
  Laboratories
Fukushima-ku, Osaka 553
Japan

MARTHA A. O'BRIEN
Dept. of Biology
Univ. of North Carolina
Chapel Hill, NC   27514   USA

IAN ORCHARD
Dept. of Zoology
Univ. of Toronto
Toronto, Ontario M5S 1A1
Canada

MICHAEL D. OWEN
Dept. of Zoology
Univ. of Western Ontario
London, Ontario N6A 5B7
Canada

LARS-ERIK KRUSE PEDERSEN
Research Dept.
Cheminova A/S
P.O. Box 9
DK-7620
Lemvig, Denmark

YVES PICHON
Centre National de la
    Recherche Scientifique
Laboratoire de
    Neurobiologie Cellulaire
Dept. of Biophysique
F-91 190 Gif sur Yvette
France

ADA RAFAELI
Dept. of Entomology
The Hebrew Univ.
Rehovot 76-100
Israel

ASHOK K. RAINA
USDA/OCSL
BARC-West 007
Beltsville, MD   20705   USA

DAVID M. ROUSH
FMC Corp.
Box 8
Princeton, NJ   08540   USA

DAVID B. SATTELLE
ARC Unit of Insect
    Neurophysiology
    and Pharmacology
Dept. of Zoology
Downing Street
Cambridge CB2 3EJ
United Kingdom

ROBERT M. SCARBOROUGH
Zoecon Corp.
975 California Avenue
Palo Alto, CA   94304   USA

STEPHEN P. SCHMIDT
Union Carbide
    Agricultural
    Products Co., Inc.
Box 12014
T. W. Alexander Dr.
Research Triangle
    Park, NC   27709   USA

HOWARD A. SCHNEIDERMAN
Monsanto Co.
800 North Lindbergh Blvd.
St. Louis, MO   63167   USA

MARK E. SCHROEDER
Shell Development Co.
P.O. Box 4248
Modesto, CA   95352   USA

STEPHEN R. SHAW
Dept. of Life Sciences/
    Psychology
Dalhousie Univ.
Halifax, Nova Scotia
B3H 4J1
Canada

RAVINDER S. SIDHU
Monsanto Co.
800 N. Lindbergh Blvd.
St. Louis, MO   63167
USA

DON L. SILHACEK
USDA/IAL
P.O. Box 14565
Gainesville, FL   32604   USA

G. J. P. SINGH
Agriculture Canada
Univ. Sub Post Office
London, Ontario N6A 5B7
Canada

WENDY A. SMITH
Dept. of Biology
Univ. of North Carolina
Chapel Hill, NC   27514 USA

THOMAS SMYTH, JR.
Pennsylvania State Univ.
2 Patterson Bldg.
University Park, PA 16802 USA

DAVID M. SODERLUND
Dept. of Entomology
Cornell Univ.
Geneva, NY   14456   USA

ALVIN N. STARRATT
Agriculture Canada
Univ. Sub Post Office
London, Ontario N6A 5B7
Canada

BARBARA STAY
Dept. of Zoology
Univ. of Iowa
Iowa City, Iowa   52242   USA

JOHN E. STEELE
Dept. of Zoology
University of Western Ontario
London, Ontario N6A 5B7
Canada

R. W. STEELE
Agriculture Canada
Univ. Sub Post Office
London, Ontario N6A 5B7
Canada

R. R. STEWART
FMC Corp.
Agricultural Chemical
    Group
Box 8
Princeton, NJ   08540 USA

ALAN THORPE
School of Biological
    Sciences
Queen Mary College
Univ. of London
Mile End Road
London E1 4NS
United Kingdom

AVERETT S. TOMBES
George Mason University
Graduate School
Fairfax, VA   22032   USA

NATHAN TUBLITZ
Dept. of Zoology
Univ. of Washington
Seattle, WA   98195   USA

BRUCE L. UMMINGER
National Science
    Foundation
Regulatory Biology
    Program
1800 G Street, NW
Washington, DC   20550
USA

SARALEE NEUMANN VISSCHER
Dept. of Biology
Montana State Univ.
Bozeman, MT   59717   USA

RENÉE M. WAGNER
USDA/VTERL
P.O. Drawer GE
College Station, TX   77841
USA

BRUCE A. WEBB
Dept. of Entomology
Univ. of Kentucky
Lexington, KY   40506   USA

LAVERN R. WHISENTON
Dept. of Biology
Univ. of North Carolina
Chapel Hill, NC   27514   USA

W. KEITH WHITNEY
American Cyanamid Co.
P.O. Box 400
Princeton, NJ   08540   USA

CARROLL M. WILLIAMS
The Biological Laboratories
Harvard Univ.
Cambridge, MA   02138   USA

JUDITH H. WILLIS
National Science Foundation
1800 G Street, NW
Washington, DC   20550   USA

ROBERT A. WIRTZ
Dept. of Entomology
Walter Reed Army Institute
   of Research
Washington, DC   20307   USA

CHARLES W. WOODS
USDA/IRL
BARC-East 306
Beltsville, MD   20705   USA

CHIH-MING YIN
Dept. of Entomology
Univ. of Massachusetts
Amherst, MA   01003   USA

BARBARA ZAIN
National Science
   Foundation
1800 G Street, NW
Washington, DC   20550
USA

WALTER M. ZECK
Mobay Chemical Corp.
P.O. Box 1508
Vero Beach, FL   32960
USA

AUTHOR INDEX